Denis Diderot

The System of Nature

Or, Laws of the Moral and Physical World

Denis Diderot

The System of Nature
Or, Laws of the Moral and Physical World

ISBN/EAN: 9783744726382

Printed in Europe, USA, Canada, Australia, Japan

Cover: Foto ©berggeist007 / pixelio.de

More available books at **www.hansebooks.com**

THE

SYSTEM OF NATURE:

OR,

LAWS OF THE MORAL AND PHYSICAL WORLD.

BY BARON D'HOLBACH,
AUTHOR OF "GOOD SENSE," ETC.

A NEW AND IMPROVED EDITION, WITH NOTES BY DIDEROT.

TRANSLATED, FOR THE FIRST TIME,
BY H. D. ROBINSON.

TWO VOLUMES IN ONE.

VOL. I.

STEREOTYPE EDITION.

BOSTON:
PUBLISHED BY J. P. MENDUM,
84 Washington Street.
1868.

ADVERTISEMENT.

TO THE PUBLIC.

To expose superstition, the ignorance and credulity on which it is based, and to ameliorate the condition of the human race, is the ardent desire of every philanthropic mind.

Mankind are unhappy, in proportion as they are deluded by imaginary systems of theology. Taught to attach much importance to belief in religious doctrines, and to mere forms and ceremonies of religious worship, the slightest disagreement among theological dogmatists is oftentimes sufficient to inflame their minds, already excited by bigotry, and to lead them to anathematize and destroy each other without pity, mercy, or remorse.

The various theological systems in which mankind have been misled to have *faith*, are but fables and falsehoods imposed by visionaries and fanatics on the ignorant, the weak, and the credulous, as historical truths ; and for unbelief of which, millions have perished at the stake, or pined in gloomy dungeons : and such will ever be the case, until the mists of superstition, and the influence of priestcraft, are exposed by the light of knowledge and the power of truth.

Many honest and talented philanthropists have directed their powerful intellects against the religious dogmas which have caused so much misery and persecution among mankind. Owing, however, to the combined power and influence of kings and priests, many of those learned and liberal works have been either destroyed or buried in oblivion, and the characters of the writers assailed by the unsparing and relentless rancour of *pious* abuse.

To counteract and destroy, if possible, these sources of mischief and misery, is the intention of the publisher in issuing the SYSTEM OF NATURE; and this truly able work of a celebrated author, whose writings, owing to religious intolerance, have been kept in comparative obscurity, is now offered to the public in a form which unites the various advantages of neatness of typography and cheapness of price.

The publisher commends to all Liberals this translation of BARON d'HOLBACH's SYSTEM OF NATURE, because it is estimated as one of

the most able expositions of theological absurdities which has ever
been written. It is in *reality* a *System of Nature*. Man is here considered in all his relations both to his own species and those spiritual
beings which are supposed to exist in the imaginary Utopia of religious
devotees. This great work strikes at the root of all the errours and
evil consequences of religious superstition and intolerance. It inculcates the purest morality; instructing us to be kind one to another, in
order to live happily in each other's society—to be tolerant and for-
bearing, because belief is involuntary, and mankind are so organized
that all *cannot* think alike—to be indulgent and benevolent, because
kindness begets kindness, and hence each individual becomes interested for the happiness of every other, and thus all contribute to human
felicity.

Let those who declare the immorality of sceptical writings, read the
System of Nature, and they will be undeceived. They will then learn
that the calumniated sceptics are incited by no other motives than the
most praiseworthy benevolence; that far from endeavouring to increase
that misery which is incidental to human life, they only wish to heal
the animosities caused by religious dissensions, and to show men that
their true polar star is to be happy, and endeavour to render others
so. But above all, let those read this work who seek to come at a
" knowledge of the truth ;"—let those read it whose minds are har-
assed by the fear of death, or troubled by the horrible tales of a
sanguinary and vengeful God. Let them read this work, and their
doubts will vanish if there is any potency in the spear of Ithuriel.

If the most profound logic, the acutest discrimination, the keenest
and most caustic sarcasm, can reflect credit on an author, then we may
justly hail Baron d'Holbach as the greatest among philosophers, and
an honour to infidels. He is the author of many celebrated works besides the SYSTEM OF NATURE,* among which we may number, GOOD
SENSE, THE NATURAL HISTORY OF SUPERSTITION, LETTERS TO
EUGENIA, and other famous publications. He is described by bio-
graphers as " a man of great and varied talents, generous and kind-
hearted."† And the Reverend Laurence Sterne, informs us in his
Letters, that he was rich, generous, and learned, keeping an open
house several days in the week for indigent scholars. Davenport,
ubi sup., page 324, says, " His works are numerous, and were all published anonymously." It is, no doubt, on this account that the *Système de la Nature was first attributed to Helvétius, and then to Mira-*

---

* A person by the name of Robinet, wrote a work of a similar tendency,
called De la Nature, which should not be confounded with that of Baron
d Holbach.

† ·Vide R. A. Davenport's Dictionary of Biography, Boston edition, page
324, Article, HOLBACH. Perhaps it may be well to add that he was born in
1723, in Heidesheim, Germany, though he was educated at Paris, where he
spent the greatest part of his life. He was a distinguished member of many
European academies, and peculiarly conversant with mineralogy. He died in
1789.

beau. But this important question has been set to rest by Baron Grimm, from whose celebrated correspondence we make the following extracts, under the date of August 10th, 1789:—

"I became acquainted with the Baron d'Holbach only a few years before his death ; but, to know him, and to feel that esteem and veneration with which his noble character inspired his friends, a long acquaintance was not necessary. I therefore shall endeavour to portray him as he appeared to me ; and I fain would persuade myself, that if his manes could hear me, they would be pleased with the frankness and simplicity of my homage.

"I have never met with a man more learned—I may add, more *universally* learned, than the Baron d'Holbach ; and I have never seen any one who cared so little to pass for learned in the eyes of the world. Had it not been for the sincere interest he took in the progress of science, and a longing to impart to others what he thought might be useful to them, the world would always have remained ignorant of his vast erudition. His learning, like his fortune, he gave away, but never crouched to public opinion.

"'The French nation is indebted to Baron d'Holbach for its rapid progress in natural history and chymistry. It was he who, 30 years ago, translated the best works published by the Germans on both these sciences, till then, scarcely known, or at least, very much neglected in France. His translations are enriched with valuable notes, but those who availed themselves of his labour ignored to whom they were indebted for it ; and even now it is scarcely known.

'There is no longer any indiscretion in stating that Baron d'Holbach is the author of the work which, eighteen years ago, made so much noise in Europe, of the far-famed SYSTEM OF NATURE. His self-love was never seduced by the lofty reputation his work obtained. If he was so fortunate as to escape suspicion, he was more indebted for it to his own modesty, than to the prudence and discretion of his friends. As to myself, I do not like the doctrines taught in that work, but those who have known the author, will, in justice, admit, that no private consideration induced him to advocate that system : he became its apostle with a purity of intention, and an abnegation of self, which in the eyes of faith, would have done honour to the apostles of the holiest religion.

"His *Système Social*, and his *Morale Universelle*, did not create the same sensation as the *Système de la Nature* ; but those two works show that, after having pulled down what human weakness had erected as a barrier to vice, the author felt the necessity of rebuilding another founded on the progress of reason, a good education, and wholesome laws.

"It was natural for the Baron d'Holbach to believe in the empire of reason, for his passions (and we always judge others by ourselves), were such, as in all cases to give the ascendency to virtue and correct principles. It was impossible for him to hate any one ; yet he could not, without an effort, dissimulate his profound horrour for priests, the

panders of despotism, and the promoters of superstition. Whenever
he spoke of these, his naturally good temper forsook him.
" Among his friends, the Baron d'Holbach numbered the celebrated
Helvétius, Diderot, d'Alembert, Naigeon, Condillac, Turgot, Buffon,
J. J. Rousseau, Voltaire, &c.; and in other countries, such men as
Hume, Garrick, the Abbate Galiani, &c. If so distinguished and
learned a society was calculated to give more strength and expansion
to his mind, it has also been justly remarked, that those illustrious
men could not but learn many curious and useful things from him ;
for he possessed an extensive library, and the tenacity of his memory
was such as to enable him to remember without effort every thing he
had once read."

However, the most praiseworthy feature in d'Holbach's character,
was his benevolence ; and we now conclude this sketch with the following pithy anecdote related by Mr. Naigeon, in the Journal of
Paris :—

" Among those who frequented d'Holbach's house, was a literary
gentleman, who, for some time past, appeared musing and in deep
melancholy. Pained to see his friend in that state, d'Holbach called
on him. ' I do not wish,' said d'Holbach, ' to pry into a secret you
did not wish to confide to me, but I see you are sorrowful, and your
situation makes me both uneasy and unhappy. I know you are not
rich, and you may have wants which you have hid from me. I bring
you ten thousand francs which are of no use to me. You will certainly not refuse them if you feel any friendship for me ; and by-and-
by, when you find yourself in better circumstances, you will return
them.' This friend, moved to tears by the generosity of the action,
assured him that he did not want money, that his chagrin had another
cause, and therefore could not accept his offer ; but he never forgot
the kindness which prompted it, and to him I am indebted for the facts
I have just related."

We have no apologies to make for republishing the System of
Nature at this time ; the work will support itself, and needs no advocate ; it has never been answered, because, in truth, it is, indeed, un-
answerable. It demonstrates the fallacy as well of the religion of
the Pagan as the Jew—the Christian as the Mahometan. It is a
guide alike to the philosopher emancipated from religious thraldom,
and the poor votary misled by the follies of superstition.

All Christian writers on Natural Theology have studiously avoided
even the mention of this masterly production : knowing their utter inability to cope with its powerful reasoning, they have wisely passed
it by in silence. Henry Lord Brougham, it is true, in his recent
Discourse of Natural Theology, has mentioned this extraordinary
treatise, but with what care does he evade entering the lists with this
distinguished writer ! He passes over the work with a haste and
sophistry that indicates how fully conscious he was of his own weakness and his opponent's strength. " There is no book of an Atheistical

description," says his lordship, "which has ever made a greater impression than the famous *Système de la Nature*."

* * * * *

" It is impossible to deny the merits of the Système de la Nature. The work of a great writer it unquestionably is ; but its merit lies in the extraordinary eloquence of the composition, and the skill with which words are substituted for ideas ; and assumptions for proofs, are made to pass current," &c. It is with a few pages of *such* empty declamation that his lordship attacks and condemns this eloquent and ·logical work.*

We do not wish to detain the reader longer from its perusal by lengthening out our preface, and have only to remark, in conclusion, that when Baron d'Holbach finished this work, he might have said with more truth, and far less vanity than Horace :—

> " Exegi monumentum ære perennius,
> Regalique situ pyramidum altius ;
> Quod non imber edax, non Aquilo impotens
> Possit diruere, aut innumerabilis
> Annorum series, et fuga temporum."—et seq.
> Q. Hor. Flac. Car. Lib. III. 30, v. 1–5.

---

* Vide A Discourse of Natural Theology, by Henry Lord Brougham, F.R.S., &c. Philadelphia: Carey, Lea, and Blanchard. 1835. Pages 146 and 147.

THE source of man's unhappiness is his ignorance of Nature. The pertinacity with which he clings to blind opinions imbibed in his infancy, which interweave themselves with his existence, the consequent prejudice that warps his mind, that prevents its expansion, that renders him the slave òf fiction, appears to doom him to continual errour. , He resembles a child destitute of experience, full of idle notions : a dangerous leaven mixes itself with all his knowledge : it is of necessity obscure, it is vacillating and false :—He takes the tone of his ideas on the authority of others, who are themselves in errour, or else have an interest in deceiving him. To remove this Cimmerian darkness, these barriers to the improvement of his condition ; to disentangle him from the clouds of errour that envelop him, that obscure the path he ought to tread ; to guide him out of this Cretan labyrinth, requires the clue of Ariadne, with all the love she could bestow on Theseus. It exacts more than common exertion ; it needs a most determined, a most undaunted courage—it is never effected but by a persevering resolution to act, to think for himself; to examine with rigour and impartiality the opinions he has adopted. He will find that the most noxious weeds have sprung up beside beautiful flowers ; entwined themselves around their stems, overshadowed them with an exuberance of foliage, choked the ground, enfeebled their growth, diminished their petals, dimmed the brilliancy of their colours ; that deceived by the apparent freshness of their verdure, by the rapidity of their exfoliation, he has given them cultivation, watered them, nurtured them, when he ought to have plucked out their very roots.

Man seeks to range out of his sphere : notwithstanding the reiterated checks his ambitious folly experiences, he still attempts the impossible ; strives to carry his researches beyond the visible world ; and hunts out misery in imaginary regions. He would be a metaphysician before he has become a practical philosopher. He quits the contemplation of realities to meditate on chimeras. He neglects experience to feed on conjecture, to indulge in hypothesis. He dares not cultivate his reason, because from his earliest days he has been taught to consider it criminal. He pretends to know his fate in the indistinct abodes of another life, before he has considered of the means by which he is to render himself happy in the world he inhabits : in short, man dis-

dams the study of Nature, except it be partially : he pursues phantoms
that resemble an *ignis-fatuus*, which at once dazzle, bewilder, and
affright : like the benighted traveller led astray by these deceptive exhalations of a swampy soil, he frequently quits the plain, the simple
road of truth, by pursuing of which, he can alone ever reasonably hope
to reach the goal of happiness.

The most important of our duties, then, is to seek means by which
we may destroy delusions that can never do more than mislead us.
The remedies for these evils must be sought for in Nature herself; it
is only in the abundance of her resources, that we can rationally expect to find antidotes to the mischiefs brought upon us by an ill-di-
rected, by an overpowering enthusiasm. It is time these remedies
were sought ; it is time to look the evil boldly in the face, to examine
its foundations, to scrutinize its superstructure : reason, with its faithful
guide experience, must attack'in their entrenchments those prejudices
to which the human race has but too long been the victim. For this
purpose reason must be restored to its proper rank,—it must be rescued
from the evil company with which it is associated. It has been too
long degraded—too long neglected—cowardice has rendered it subservient to delirium, the slave to falsehood. It must no longer be held
down by the massive chains of ignorant prejudice.

Truth is invariable—it is requisite to man—it can never harm him—
his very necessities, sooner or later, make him sensible of this ; oblige
him to acknowledge it. Let us then discover it to mortals—let us exhibit its charms—let us shed its effulgence over the darkened road ; it
is the only mode by which man can become disgusted with that disgraceful superstition which leads him into errour, and which but too
often usurps his homage by treacherously covering itself with the mask
of truth—its lustre can wound none but those enemies to the human
race whose power is bottomed solely on the ignorance, on the darkness in which they have in almost every climate contrived to involve
the mind of man.

Truth speaks not to these perverse beings :—her voice can only be
heard by generous minds accustomed to reflection, whose sensibilities
make them lament the numberless calamities showered on the earth
by political and religious tyranny—whose enlightened minds contemplate with horrour the immensity, the ponderosity of that series of misfortunes with which errour has in all ages overwhelmed mankind.

To errour must be attributed those insupportable chains which tyrants,
which priests have forged for all nations. To errour must be equally
attributed that abject slavery into which the people of almost every
country have fallen. Nature designed they should pursue their happiness by the most perfect freedom. To errour must be attributed
those religious terrours which, in almost every climate, have either petrified man with fear, or caused him to destroy himself for coarse or
fanciful beings. . To errour must be attributed those inveterate hatreds,
those barbarous persecutions, those numerous massacres, those dread-

2

ful tragedies, of which, under pretext of serving the interests of heaven, the earth has been but too frequently made the theatre. It is errour consecrated by religious enthusiasm, which produces that ignorance, that uncertainty in which man ever finds himself with regard to his most evident duties, his clearest rights, the most demonstrable truths. In short, man is almost every where a poor degraded captive, devoid either of greatness of soul, of reason, or of virtue, whom his inhuman gaolers have never permitted to see the light of day.

Let us then endeavour to disperse those clouds of ignorance, those mists of darkness which impede man on his journey, which obscure his progress, which prevent his marching through life with a firm, with a steady step. Let us try to inspire him with courage—with respect for his reason—with an inextinguishable love for truth—to the end that he may learn to know himself—to know his legitimate rights—that he may learn to consult his experience, and no longer be the dupe of an imagination led astray by authority—that he may renounce the prejudices of his childhood—that he may learn to found his morals on his nature, on his wants, on the real advantage of society—that he may dare to love himself—that he may learn to pursue his true happiness by promoting that of others—in short, that he may no longer occupy himself with reveries either useless or dangerous—that he may become a virtuous, a rational being, in which case he cannot fail to become happy.

If he must have his chimeras, let him at least learn to permit others to form theirs after their own fashion; since nothing can be more immaterial than the manner of men's thinking on subjects not accessible to reason, provided those thoughts be not suffered to imbody themselves into actions injurious to others: above all, let him be fully persuaded that it is of the utmost importance to the inhabitants of this world to be JUST, KIND, and PEACEABLE.

Far from injuring the cause of virtue, an impartial examination of the principles of this work will show that its object is to restore truth to its proper temple, to build up an altar whose foundations shall be consolidated by morality, reason, and justice: from this sacred fane, virtue guarded by truth, clothed with experience, shall shed forth her radiance on delighted mortals; whose homage flowing consecutively shall open to the world a new era, by rendering general the belief that happiness, the true end of man's existence, can never be attained but BY PROMOTING THAT OF HIS FELLOW CREATURE.

In conclusion:—Warned by old age and weak limbs that death is fast approaching, the author protests most solemnly that, in his labours, his sole object has been to promote the happiness of his fellow creatures; and his only ambition, to merit the approbation of the few partizans of Truth who honestly and sincerely seek her. He writes not for those who are deaf to the voice of reason, who judge of things only by their vile interest or fatal prejudices: his cold remains will fear neither their clamours nor their resentments, so terrible to those who, whilst living, dare proclaim the TRUTH.

THE SYSTEM OF NATURE.

OF NATURE AND HER LAWS—OF MAN—OF THE SOUL AND ITS FACULTIES—OF THE DOCTRINE OF IMMORTALITY—ON HAPPINESS.

CHAPTER I.

Of Nature.

MEN will always deceive themselves by abandoning experience to follow imaginary systems. Man is the work of Nature: he exists in Nature: he is submitted to her laws: he cannot deliver himself from them; nor can he step beyond them even in thought. It is in vain his mind would spring forward beyond the visible world, an imperious necessity always compels his return. For a being formed by Nature, and circumscribed by her laws, there exists nothing beyond the great whole of which he forms a part, of which he experiences the influence. The beings which he pictures to himself as above nature, or distinguished from her, are always chimeras formed after that which he has already seen, but of which it is impossible he should ever form any correct idea, either as to the place they occupy, or of their manner of acting. There is not, there can be nothing out of that Nature which includes all beings.

Instead, therefore, of seeking out of the World he inhabits for beings who can procure him a happiness denied to him by Nature, let man study this Nature, let him learn her laws, contemplate her energies, observe the immutable rules by which she acts :—let him apply these discoveries to his own felicity and submit in silence to her mandates, which nothing can alter :—let him cheerfully consent to ignore causes hid from him by an impenetrable veil:—let him without murmuring yield to the decrees of a 'universal necessity, which can never be brought within his comprehension, nor ever emancipate him from those laws imposed on him by his essence.

The distinction which has been so often made between the *physical* and the *moral* man is evidently an abuse of terms. Man is a being purely physical: the moral man is nothing more than this physical being considered under a certain point of view, that is to say, with relation to some of his modes of action, arising out of his particular organization. But is not this organization itself the work of Nature ? The motion or impulse to action of which he is susceptible, is that not physical ? His visible actions, as well as the invisible motion interiorly excited by his will or his thoughts, are equally the natural effects, the necessary consequences, of his peculiar mechanism, and the impulse he receives from those beings by whom he is surrounded. All that the human mind has successively invented with a view to change or perfect his being, and to render himself more happy, was only a necessary consequence of man's peculiar essence, and that of the beings who act upon him. The object of all his institutions, of all his reflections, of all his knowledge, is only to procure that happiness towards which he is incessantly impelled by the peculiarity of his nature. All that he does, all that he thinks, all that he is, all that he will be, is nothing more than what Universal Nature has made him. His ideas, his will, his actions, are the necessary effects of those qualities infused into him by Nature, and of those circumstances in which she has placed him. In short, *art* is nothing but Nature acting with the tools she has made.

Nature sends man naked and destitute into this world which is to be his abode: he quickly learns to cover his nakedness, to shelter himself from the inclemency of the weather, first with rude huts and the skins of the beasts of the forest; by degrees he mends their appearance, renders them more convenient: he establishes manufactories of cloth, of cotton, of silk; he digs clay, gold, and other fossils from the bowels of the earth, converts them into bricks for his house, into vessels for his use, gradually improves their shape, augments their beauty. To a being elevated above our terrestrial globe, who should contemplate the human species through all the changes he undergoes in his progress towards civilization, man would not appear less subjected to the laws of Nature when naked in the forest painfully seeking his sustenance, than when living in civilized society surrounded with comforts; that is to say, enriched with greater experience, plunged in luxury, where he every day invents a thousand new wants and discovers a thousand new modes of satisfying them. All the steps taken by man to regulate his existence, ought only to be considered as a long succession of causes and effects, which are nothing more than the development of the first impulse given him by nature.

The same animal by virtue of his organization passes successively from the most simple to the most complicated wants; it is nevertheless the consequence of his nature. The butterfly whose beauty we admire, whose colours are so rich, whose appearance is so brilliant, commences as an inanimate unattractive egg; from this, heat produces a worm, this becomes a chrysalis, then changes into that winged insect decorated with the most vivid tints: arrived at this stage he reproduces, he propagates: at last despoiled of his ornaments he is obliged to disappear, having fulfilled the task imposed on him by Nature, having described the circle of mutation marked out for beings of his order.

The same progress, the same change takes place in vegetables. It is by a succession of combinations originally interwoven with the energies of the aloe, that this plant is insensibly regulated, gradually expanded, and at the end of a great number of years produces those flowers which announce its dissolution.

It is equally so with man, who in all his motion, all the changes he un-

dergoes, never acts but according to laws peculiar to his organization, and to the matter of which he is composed.

The *physical man*, is he who acts by causes our senses make us understand.

The *moral man*, is he who acts by physical causes, with which our prejudices preclude us from becoming acquainted.

The *wild man*, is a child destitute of experience, who is incapable of pursuing his happiness, because he has not learnt how to oppose resistance to the impulses he receives from those beings by whom he is surrounded.

The *civilized man*, is he whom experience and social life have enabled to draw from nature the means of his own happiness; because he has learned to oppose resistance to those impulses he receives from exterior beings, when experience has taught him they would be injurious to his welfare.

The *enlightened man*, is man in his maturity, in his perfection; who is capable of pursuing his own happiness; because he has learned to examine, to think for himself, and not to take that for truth upon the authority of others, which experience has taught him examination will frequently prove erroneous.

The *happy man*, is he who knows how to enjoy the benefits of nature; in other words, he who thinks for himself; who is thankful for the good he possesses; who does not envy the welfare of others; who does not sigh after imaginary benefits always beyond his grasp.

The *unhappy man*, is he who is incapacitated to enjoy the benefits of nature; that is, he who suffers others to think for him; who neglects the absolute good he possesses, in a fruitless search after imaginary benefits; who vainly sighs after that which ever eludes his pursuit.

It necessarily results, that man in his researches ought always to fall back on experience, and natural philosophy: These are what he should consult in his religion—in his morals—in his legislation—in his political government —in the arts—in the sciences—in his pleasures—in his misfortunes. Experience teaches that Nature acts by simple, uniform, and invariable laws.

It is by his senses man is bound to this universal Nature; it is by his senses he must penetrate her secrets; it is from his senses he must draw experience of her laws. Whenever, therefore, he either fails to acquire experience or quits its path, he stumbles into an abyss, his imagination leads him astray.

All the errours of man are physical errours: he never deceives himself but when he neglects to return back to nature, to consult her laws, to call experience to his aid. It is for want of experience he forms such imperfect ideas of matter, of its properties, of its combinations, of its power, of its mode of action, or of the energies which spring from its essence. Wanting this experience, the whole universe to him is but one vast scene of illusion. The most ordinary results appear to him the most astonishing phenomena; he wonders at every thing, understands nothing, and yields the guidance of his actions to those interested in betraying his interests. He is ignorant of Nature, he has mistaken her laws; he has not contemplated the necessary routine which she has marked out for every thing she contains. Mistaken the laws of Nature, did I say? He has mistaken himself: the consequence is, that all his systems, all his conjectures, all his reasonings, from which he has banished experience, are nothing more than a tissue of errours, a long chain of absurdities.

All errour is prejudicial: it is by deceiving himself that man is plunged in misery. He neglected Nature; he understood not her laws; he formed gods of the most preposterous kinds: these became the sole objects of his hope, the creatures of his fear, and he trembled under these visionary deities; under the supposed influence of imaginary beings created by himself; under the terrour inspired by blocks of stone; by logs of wood; by flying fish; or else under the frowns of men, mortal as himself, whom his distempered fancy had elevated above that Nature of which alone he is capable of forming any idea. His very posterity laughs to scorn his folly, because experience has convinced them of the absurdity of his groundless fears, of his misplaced worship. Thus has passed away the ancient mythology, with all the trumpery attributes attached to it by ignorance.*

Man did not understand that Nature, equal in her distributions, entirely destitute of goodness or malice, follows only necessary and immutable laws, when she either produces beings or destroys them, when she causes those to suffer, whose organization creates sensibility; when she scatters among them good and evil; when she subjects them to incessant change —he did not perceive it was in the bosom of Nature herself, that it was in her abundance he ought to seek to satisfy his wants; for remedies against his pains; for the means of rendering himself happy: he expected to derive these benefits from imaginary beings, whom he erroneously imagined to be the authors of his pleasures, the cause of his misfortunes. From hence it is clear that to his ignorance of Nature, man owes the creation of those illusive powers under which he has so long trembled with fear; that superstitious worship, which has been the source of all his misery.

For want of clearly understanding his own peculiar nature, his proper tendency, his wants, and his rights, man has fallen in society, from FREEDOM into SLAVERY. He had forgotten the design of his existence, or else he believed himself obliged to smother his natural desires of his heart, and to sacrifice his welfare to the caprice of chiefs, either elected by himself, or submitted to without examination. He was ignorant of the true policy of association —of the true object of government; he disdained to listen to the voice of Nature, which loudly proclaimed that the price of all submission is protection and

* It is impossible to peruse the ancient and modern theological works without feeling disgusted at the contemptible invention of those gods *which* have been made objects of terrour or love to mankind. To begin with the inhabitants of India and Egypt, of Greece and Rome, what littleness and foolery in their worship—what rascality and infamy in their priests! Are our own any better? No! Cicero said, that two Augurs could not look at each other without laughing; but he little thought that a time would come when a set of *mean wretches*,† assuming the title of *Reverend*, would endeavour to persuade their fellow men that they represented the Divinity on earth!   † *Des misérables.*

happiness: the end of all government the benefit of the governed, not the exclusive advantage of the governours. He gave himself up without reserve to men like himself, whom his prejudices induced him to contemplate as beings of a superior order, as gods upon earth: these profited by his ignorance, took advantage of his prejudices, corrupted him, rendered him vicious, enslaved him, made him miserable. Thus man, intended by Nature for the full enjoyment of freedom, to patiently investigate her laws, to search into her secrets, to always cling to his experience, has, from a neglect of her salutary admonitions, from an inexcusable ignorance of his own peculiar essence, fallen into servitude, and has been wickedly governed.

Having mistaken himself, he has remained ignorant of the necessary affinity that subsists between him and the beings of his own species: having mistaken his duty to himself, it followed, as a consequence, he has mistaken his duty to others. He made an erroneous calculation of what his felicity required; he did not perceive, what he owed to himself, the excesses he ought to avoid, the passions he ought to resist, the impulses he ought to follow, in order to consolidate his happiness, to promote his comfort, to further his advantage. In short, he was ignorant of his true interests; hence his irregularities, his intemperance, his shameful voluptuousness, with that long train of vices to which he has abandoned himself, at the expense of his preservation, at the risk of his permanent felicity.

It is, therefore, ignorance of himself, that has prevented man from enlightening his morals. The depraved govern ments to which he had submitted, felt an interest in preventing the practice of his duties, even when he knew them.

Man's ignorance has endured so long, he has taken such slow, such irresolute steps to ameliorate his condition, only because he has neglected to study Nature, to scrutinize her laws, to search out her resources, to discover her properties. His sluggishness finds its account in permitting himself to be guided by precedent, rather than to follow experience which demands

activity; to be led by routine, rather than by his reason which exacts reflection. From hence may be traced the aversion man betrays for every thing that swerves from those rules to which he has been accustomed: hence his stupid, his scrupulous respect for antiquity, for the most silly, the most absurd institutions of his fathers: hence those fears that seize him, when the most advantageous changes are proposed to him, or the most probable attempts are made to better his condition. He dreads to examine, because he has been taught to hold it a profanation of something immediately connected with his welfare; he credulously believes the interested advice, and spurns at those who wish to show him the danger of the road he is travelling.

This is the reason why nations linger on in the most scandalous lethargy, groaning under abuses transmitted from century to century, trembling at the very idea of that which alone can remedy their misfortunes.

It is for want of energy, for want of consulting experience, that medicine, natural philosophy, agriculture, painting, in short, all the useful sciences have so long remained under the shackles of authority, have progressed so little: those who profess these sciences, for the most part prefer treading the beaten paths, however inadequate to their end, rather than strike out new ones: they prefer the ravings of their imagination, their gratuitous conjectures, to that laborious experience which alone can extract her secrets from Nature.

In short, man, whether from sloth or from terrour, having renounced the evidence of his senses, has been guided in all his actions, in all his enterprises, by imagination, by enthusiasm, by habit, by prejudice, and above all, by authority, which knew well how to deceive him. Thus, imaginary systems have supplied the place of experience—of reflection—of reason. Man, petrified with his fears, inebriated with the marvellous, or benumbed with sloth, surrendered his experience: guided by his credulity, he was unable to fall back upon it, he became consequently inexperienced: from thence he gave birth to the most ridiculous opinions, or else adopted without examination, all those chimeras, all those idle notions offered to him by men whose interest it was to fool him to the top of his bent. Thus, because man has forgotten Nature, has neglected her ways—because he has disdained experience—because he has thrown by his reason—because he has been enraptured with the marvellous, with the supernatural—because he has unnecessarily *trembled*, man has continued so long in a state of infancy ; and these are the reasons there is so much trouble in conducting him from this state of childhood to that of manhood. He has had nothing but the most jejune hypotheses, of which he has never dared to examine either the principles or the proofs, because he has been accustomed to hold them sacred, to consider them as the most perfect truths, of which it is not permitted to doubt, even for an instant. His ignorance rendered him credulous : his curiosity made him swallow large draughts of the marvellous: time confirmed him in his opinions, and he passed his conjectures from race to race for realities ; a tyrannical power maintained him in his notions, because by those alone could society be enslaved. At length the whole science of man became a confused mass of darkness, falsehood and contradictions, with here and there a feeble ray of truth, furnished by that Nature of which he can never entirely divest himself, because, without his knowledge, his necessities are continually bringing him back to her resources.

Let us then, raise ourselves above these clouds of prejudice, contemplate the opinions of men, and observe their various systems ; let us learn to distrust a disordered imagination ; let us take experience, that faithful monitor, for our guide ; let us consult Nature, explore her laws, dive into her stores; let us draw from herself our ideas of the beings she contains; let us fall back on our senses, which errour, interested errour has taught us to suspect; let us consult that reason, which, for the vilest purposes, has been so shamefully calumniated, so cruelly disgraced ; let us attentively examine the visible world, and let us try if it will not enable us to form a tolerable judgment of the invisible territory of the intellectual world : perhaps it may be found that there has been no sufficient reason for

distinguishing them, and that it is not without motives that two empires have been separated, which are equally the inheritance of nature.

The universe, that vast assemblage of every thing that exists, presents only matter and motion : the whole offers to our contemplation nothing but an immense, an uninterrupted succession of causes and effects; some of these causes are known to us, because they strike immediately on our senses ; others are unknown to us, because they act upon us by effects, frequently very remote from their original cause.

An immense variety of matter, combined under an infinity of forms, incessantly communicates, unceasingly receives a diversity of impulses. The different properties of this matter, its innumerable combinations, its various methods of action, which are the necessary consequence of these combinations, constitute for man, what he calls the *essence* of beings: it is from these diversified essences that spring the orders, the classes, or the systems, which these beings respectively occupy, of which the sum total makes up that which is called NATURE.

Nature, therefore, in its most extended signification, is the great whole that results from the assemblage of matter under its various combinations, with that diversity of motions which the universe offers to our view. Nature, in a less extended sense, or considered in each individual, is the whole that results from its essence; that is to say, the properties, the combination, the impulse, and the peculiar modes of action, by which it is discriminated from other beings. It is thus that MAN is, as a whole, the result of a certain combination of matter, endowed with peculiar properties, competent to give, capable of receiving, certain impulses, the arrangement of which is called *organization*, of which the essence is, to feel, to think, to act, to move, after a manner distinguished from other beings with which he can be compared. Man, therefore, ranks in an order, in a system, in a class by himself, which differs from that of other animals, in whom we do not perceive those properties of which he is possessed. The different systems of beings, or if they will, their *particular natures*, depend on the general system of the great whole, or that universal nature, of which they form a part; to which every thing that exists is necessarily submitted, and attached.

Having described the proper definition that should be applied to the word NATURE, I must advise the reader, once for all, that whenever, in the course of this work, the expression occurs, that "Nature produces such or such an effect," there is no intention of personifying that nature, which is purely an abstract being; it merely indicates, that the effect spoken of, necessarily springs from the peculiar properties of those beings which compose the mighty macrocosm. When, therefore, it is said, *Nature demands that man should pursue his own happiness,* it is to prevent circumlocution, to avoid tautology; it is to be understood that it is the property of a being that feels, that thinks, that wills, that acts, to labour to its own happiness; in short, *that* is called *natural* which is conformable to the essence of things, or to the laws which Nature prescribes to the beings she contains, in the different orders they occupy, under tHe various circumstances through which they are obliged to pass. Thus health is *natural* to man in a certain state; disease is *natural* to him under other circumstances; dissolution, or if they will, death, is a *natural* state for a body, deprived of some of those things, necessary to maintain the existence of the animal, &c. By ESSENCE is to be understood, that which constitutes a being such as it is; the whole of the properties, or qualities, by which it acts as it does. Thus, when it is said, it is the *essence* of a stone to fall, it is the same as saying, that its descent, is the necessary effect of its gravity, of its density, of the cohesion of its parts, of the elements of which it is composed. In short, the *essence* of a being, is its particular, its individual nature.

CHAPTER II.

Of Motion, and its Origin.

MOTION is an effect by which a body either changes, or has a tendency to change its position: that is to say, by which it successively corresponds with different parts of space, or changes its relative distance to other bodies. It is motion alone that establishes the relation between our senses and exterior or interior beings: it is only by motion, that these beings are impressed upon us—that we know their existence— that we judge of their properties—that we distinguish the one from the other— that we distribute them into classes.

The beings, the substances, or the various bodies, of which nature is the assemblage, are themselves effects of certain combinations—effects which become causes in their turn. A *cause* is a being which puts another in motion, or which produces some change in it. The *effect* is the change produced in one body by the motion or presence of another.

Each being, by its essence, by its peculiar nature, has the faculty of producing, is capable of receiving, has the power of communicating a variety of motion. Thus some beings are proper to strike our organs: these organs are competent to receiving the impression, are adequate to undergoing changes by their presence. Those which cannot act on any of our organs, either immediately and by themselves, or mediately, by the intervention of other bodies, exist not for us; since they can neither move us, nor consequently furnish us with ideas: they can neither be known to us, nor of course be judged of by us. To know an object, is to have felt it; to feel it, it is requisite to have been moved by it. To see, is to have been moved by something acting on the visual organs; to hear, is to have been struck by something on our auditory nerves. In short, in whatever mode a body may act upon us, whatever impulse we may receive from it, we can have no other knowledge of it than by the change it produces in us.

Nature, as we have already said, is the assemblage of all the beings, and consequently, of all the motion of which we have a knowledge, as well as of many others of which we know nothing, because they have not yet become accessible to our senses. From the continual action and re-action of these beings, result a series of causes and effects; or a chain of motion guided by the constant and invariable laws peculiar to each being: which are necessary or inherent to its particular nature, which make it always act or move after a determinate manner. The different principles of this motion, are unknown to us, because we are in many instances, if not in all, ignorant of what constitutes the essence of beings. The elements of bodies escape our senses; we know them only in the mass: we are neither acquainted with their intimate combination, nor the proportion of these combinations; from whence must necessarily result their mode of action, their impulse, or their different effects.

Our senses, bring us generally acquainted with two sorts of motion in the beings that surround us. The one is the motion of the mass, by which an entire body, is transferred from one place to another. Of the motion of this genus we are perfectly sensible.—Thus, we see a stone fall, a ball roll, an arm move or change its position. The other, is an internal or concealed motion, which always depends on the peculiar energies of a body: that is to say, on its *essence*, or the combination, the action, and re-action of the minute, of the insensible particles of matter, of which that body is composed. This motion we do not see; we know it only by the alteration, or change, which, after some time, we discover in these bodies or mixtures. Of this genus is that concealed motion which fermentation produces in the particles that compose flour, which, however scattered, however separated, unite, and form that mass which we call *bread*. Such, also, is the imperceptible motion, by which we see a plant or animal enlarge, strengthen, undergo changes, and acquire new qualities, without our eyes being competent to follow its progression, or to perceive the causes which have produced these effects. Such, also, is the internal motion that takes place in man, which is called his *intellectual faculties*, his *thoughts*, his *passions*, his *will*. Of these we have no other mode of judging than by their action; that is, by those sensible effects which either accompany or follow them. Thus, when we see a man run away, we judge him to be interiorly actuated by the passion of fear.

Motion, whether visible or concealed, is styled *acquired* when it is impressed on one body by another; either by a cause to which we are a stranger, or by an exterior agent which our senses

No. I.—3

enable us to discover. Thus we call that *acquired motion*, which the wind gives to the sails of a ship. That motion, which is excited in a body containing within itself the causes of those changes we see it undergo, is called *spontaneous.*—Then it is said, this body acts or moves by its own peculiar energies. Of this kind is the motion of the man who walks, who talks, who thinks. Nevertheless, if we examine the matter a little closer, we shall be convinced, that, strictly speaking, there is no such thing as spontaneous motion in any of the various bodies of Nature; seeing they are perpetually acting one upon the other; that all their changes are to be attributed to the causes, either visible or concealed, by which they are moved. The will of man, is secretly moved or determined by some exterior cause producing a change in him: we believe he moves of himself, because we neither see the cause that determined him, the mode in which it acted, nor the organ that it put in motion.

That is called *simple motion*, which is excited in a body by a single cause. *Compound motion*, that, which is produced by two or more different causes; whether these causes are equal or unequal, conspiring differently, acting together or in succession, known or unknown.

Let the motion of beings be of whatsoever nature it may, it is always the necessary consequence of their essence, or of the properties which compose them, and of those causes of which they experience the action. Each being can only move, and act, after a particular manner; that is to say, conformably to those laws which result from its peculiar essence, its particular combination, its individual nature: in short, from its specific energies, and those of the bodies from which it receives an impulse. It is this that constitutes the invariable laws of motion: I say *invariable*, because they can never change without producing confusion in the essence of things. It is thus that a heavy body must necessarily fall, if it meets with no obstacle sufficient to arrest its descent; that a sensible body must naturally seek pleasure, and avoid pain; that fire must necessarily burn, and diffuse light.

Each being, then, has laws of motion that are adapted to itself, and constantly acts, or moves according to these laws; at least when no superior cause interrupts its action. Thus, fire ceases to turn combustible matter, as soon as sufficient water is thrown into it to arrest its progress. Thus, a sensible being ceases to seek pleasure, as soon as he fears that pain will be the result.

The communication of motion, or the medium of action, from one body to another, also follows certain and necessary laws: one being can only communicate motion to another by the affinity, by the resemblance, by the conformity, by the analogy, or by the point of contact which it has with that other being. Fire can only propagate when it finds matter analogous to itself: it extinguishes when it encounters bodies which it cannot embrace; that is to say, that do not bear towards it a certain degree of relation or affinity.

Every thing in the universe is in motion; the essence of matter is to act: if we consider its parts attentively, we shall discover that not a particle enjoys absolute repose. Those which appear to us to be without motion, are, in fact, only in relative or apparent rest; they experience such an imperceptible motion, and expose it so little on their surfaces, that we cannot perceive the changes they undergo.* All that appears to us to be at rest, does not, however, remain one instant in the same state. All beings are continually breeding, increasing, decreasing, or dispersing, with more or less tardiness or rapidity. The insect called *ephemeron*, is produced, and perishes in the same day; consequently, it experiences the great changes of its being very rapidly. Those combinations which form the most solid bodies, and which, to our eyes, appear to enjoy the most perfect repose, are nevertheless decomposed and dissolved in the course of time. The hardest stones, by degrees, give way to the contact of air. A mass of

iron, which time, and the action of the atmosphere, has gnawed into rust, must have been in motion from the moment of its formation in the bowels of the earth, until the instant we behold it in this state of dissolution.

Natural philosophers, for the most part, seem not to have sufficiently reflected on what they call the *nisus;* that is to say, the incessant efforts one body is making on another, but which, notwithstanding, appear, to our superficial observation, to enjoy the most perfect repose. A stone of five hundred weight seems at rest on the earth, nevertheless, it never ceases for an instant to press with force upon the earth, which resists or repulses it in its turn. Will the assertion be ventured, that the stone and the earth do not act? Do they wish to be undeceived? They have nothing to do, but interpose their hand betwixt the earth and the stone; it will then be discovered, that, notwithstanding its seeming repose, the stone has power adequate to bruise it. Action cannot exist in bodies without re-action. A body that experiences an impulse, an attraction, or a pressure of any kind, if it resists, clearly demonstrates by such resistance, that it re-acts; from whence it follows, there is a concealed force, called by philosophers *vis inertia,* that displays itself against another force; and this clearly demonstrates, that this inert force is capable of both acting and re-acting. In short, it will be found, on close investigation, that those powers which are called *dead,* and those which are termed *live* or *moving,* are powers of the same species, which only display themselves after a different manner.†

---

* This truth, which is still denied by many metaphysicians, has been conclusively established by the celebrated Toland, in a work which appeared in the beginning of the eighteenth century, entitled *Letters to Screna.* Those who can procure this scarce work will do well to refer to it, and their doubts on the subject, if they have any, will be removed.

---

† *Actioni æqualis et contraria est reactio.* V. BILFINGER, DE DEO, ANIMA ET MUNDO. § ccxviii. page 241. Upon which the Commentator adds,—Reactio dicitur actio patientis in agens, seu corporis in quod agitur actio in illud quod in ipsum agit. Nulla autem datur in corporibus actio sine reactione, dum enim corpus ad motum sollicitatur, resistit motui, atque hâc ipsâ resistentiâ reagit in agens. Nisus se exerens adversus nisum agentis, seu vis illa corporis, quatenus resistit, internum resistentiæ principium, vocatur vis inertiæ, seu passiva. Ergo corpus reagit vi inertiæ. Vis igitur inertiæ et vis motrix in corporibus una eademque est vis, diverso tamen modo se exercns. Vis autem inertiæ consistit in nisi adversus nisum agentis se exerente, &c. IBIDEM.

May we not go farther yet, may we not say, that in those bodies, or masses, of which the whole appears to us to be at rest, there is, notwithstanding, a continual action and reaction, constant efforts, uninterrupted impulse, and continued resistance? In short, a *nisus*, by which the component particles of these bodies press one upon another, reciprocally resisting each other, acting, and reacting incessantly? that this reciprocity of action, this simultaneous reaction, keeps them united, causes their particles to form a mass, a body, a combination, which, viewed in its whole. has the semblance of complete rest, although no one of its particles ever *really* ceases to be in motion for a single instant? These bodies appear to be at rest, simply by the equality of the motion of the powers acting in them.

Thus bodies that have the appearance of enjoying the most perfect repose, really receive, whether upon their surface, or in their interior, continual impulsion from those bodies by which they are either surrounded or penetrated, dilated or contracted, rarefied or condensed; in short, from those which compose them: whereby their particles are constantly acting, and reacting, or in continual motion, the effects of which are ulteriorly displayed by very remarkable changes. Thus heat rarefies and dilates metals, which clearly demonstrates, that a bar of iron, from the variation of the atmosphere alone, must be in unceasing motion; and that not a single particle in it can be said to enjoy rest, even for a single moment. Indeed, in those hard bodies, the particles of which are contiguous, which are closely united, how is it possible to conceive, that air, cold or heat, can act upon one of these particles, even exteriorly, without the motion being successively communicated to those which are most intimate and minute in their union? How, without motion, should we be able to conceive the manner in which our sense of smelling is affected by emanations escaping from the most compact bodies, of which all the particles appear to be at perfect rest? How could we, even by the aid of a telescope, see the most distant stars, if there was not a progressive motion of light from these stars to the retina of our eye?

Observation and reflection ought to convince us, that every thing in Nature is in continual motion: that not one of its parts enjoys true repose: that Nature acts in all; that she would cease to be Nature if she did not act; and that, without unceasing motion, nothing could be preserved, nothing could be produced, nothing could act. Thus, the idea of Nature necessarily includes that of motion. But, it will be asked, from whence did she receive her motion? Our reply is, from herself, since she is the great whole, out of which, consequently, nothing can exist. We say this motion is a manner of existence, that flows, necessarily, out of the essence of matter; that matter moves by its own peculiar energies; that its motion is to be attributed to the force which is inherent in itself; that the variety of motion, and the phenomena which result, proceed from the diversity of the properties, of the qualities, and of the combinations, which are originally found in the primitive matter, of which Nature is the assemblage.

Natural philosophers, for the most part, have regarded as inanimate, or as deprived of the faculty of motion, those bodies which are only moved by the interposition of some agent, or exterior cause; they have considered themselves justified in concluding, that the matter which constitutes these bodies, is perfectly inert in its nature. They have not relinquished this errour, although they must have observed, that whenever a body is left to itself, or disengaged from those obstacles which oppose themselves to its descent, it has a tendency to fall, or to approach the centre of the earth, by a motion uniformly accelerated; they have rather chosen to suppose an imaginary exterior cause, of which they themselves had no correct idea, than admit that these bodies held their motion from their own peculiar nature. •

In like manner, although these philosophers saw above them an infinite number of immense globes, moving with great rapidity round a common centre, still they clung fast to their opinions; and never ceased to suppose chimerical causes for these movements, until the immortal NEWTON demonstrated that it was the effect of the gravitation of these celestial bodies towards each other.* A very simple observation would have sufficed to make the philosophers anterior to Newton feel the insufficiency of the causes they admitted to operate with such powerful effect: they had enough to convince themselves in the clashing of one body against another which they could contemplate, and in the known laws of that motion, which these always communicate by reason of their greater or less density: from whence they ought to have inferred, that the density of *subtile* or *ethereal* matter being infinitely less than that of the planets, it could only communicate to them a very feeble motion.

If they had viewed Nature uninfluenced by prejudice, they must have been long since convinced, that matter acts by its own peculiar energy, and needs not any exterior impulse to set it in motion. They would have perceived, that whenever mixed bodies were placed in a capacity to act on each other, motion was instantly engendered, and that these mixtures acted with a force capable of producing the most surprising effects. If filings of iron, sulphur and water be mixed together, these bodies thus capacitated to act on each other, are heated by degrees, and ultimately produce a violent combustion. If flour be wetted with water, and the mixture closed up, it will be found, after some little lapse of time, by the aid of a microscope, to have produced organized beings that enjoy life, of which the water and the

* Natural philosophers, and Newton himself, have considered the cause of *gravitation* to be inexplicable; yet it appears that it may be deduced from the motion of matter by which bodies are diversely determined. Gravitation is only a mode of moving—a tendency towards a centre. But, to speak correctly, all motion is relative *gravitation*: that which falls relatively to us, ascends with relation to other bodies. Hence it follows, that every motion in the universe is the effect of *gravitation*; for, in the universe, there is neither up nor *down*, nor positive centre. It appears that the weight of bodies depend on the configuration, both exterior and interior, which gives them that motion called *gravitation*. A ball of lead being spherical, falls quickly; but this ball being reduced into very thin plates, will be sustained for a longer time in the air; and the action of fire will cause this lead to rise in the atmosphere. Here the same lead, variously modified, will act after modes entirely different.

flour were believed incapable :† it is thus that inanimate matter can pass into life, or animate matter, which is in itself only an assemblage of motion Reasoning from analogy, the production of a man, independent of the ordinary means, would not be more marvellous than that of an insect with flour and water. Fermentation and putrefaction evidently produce living animals. We have here the principle; and with proper materials, principles can always be brought into action. That generation which is styled *equivocal*, is only so for those who do not reflect, or who do not permit themselves attentively to observe the operations of Nature.

The generation of motion, and its development, as well as the energy of matter, may be seen more especially in those combinations in which fire, air, and water, find themselves in union. These elements, or rather these mixed bodies, are the most volatile, the most fugitive of beings; nevertheless, in the hands of Nature they are the principal agents employed to produce the most striking phenomena. To these are to be ascribed the effects of thunder, the eruption of volcanoes, earthquakes, &c. Art offers an agent of astonishing force in gunpowder, the instant it comes in contact with fire. In fact, the most terrible effects result from the combination of matter which is generally believed to be dead and inert.

These facts incontestably prove, that motion is produced, is augmented, is accelerated in matter, without the concurrence of any exterior agent: it is, therefore, reasonable to conclude, that motion is the necessary consequence of immutable laws, resulting from the essence, from the properties inherent in the different elements, and the various combinations of these elements. Are we not justified, then, in concluding from these examples, that there may be an infinity of other combinations, with which we are unacquainted, competent to produce a great variety of motion in matter, without being under the necessity of recurring for the explanation to agents who are more difficult

† See the Microscopical Observations of Mr. Needham, which fully confirm the above statement of the author.

to comprenend than even the effects which are attributed to them?

If man had paid proper attention to what passed under his view, he would not have sought out of Nature a power distinguished from herself, to set her in action, and without which he believes she cannot move. If, indeed, by Nature is meant a heap of dead matter, destitute of properties, purely passive, we must unquestionably seek out of this Nature the principle of her motion: but, if by Nature, be understood what it really is, a whole, of which the numerous parts are endowed with diverse, and various properties; which oblige them to act according to these properties; which are in a perpetual reciprocity of action and reaction; which press, which gravitate towards a common centre, whilst others diverge and fly off towards the periphery, or circumference; which attract, and repel, which unite, and separate; which by continual approximation, and constant collision, produce and decompose all the bodies we behold; then I say, there is no necessity to have recourse to supernatural powers to account for the formation of things, and those phenomena which are the result of motion.

Those who admit a cause exterior to matter, are obliged to suppose, that this cause produced all the motion by which matter is agitated in giving it existence. This supposition rests on another, namely, that matter could begin to exist; a hypothesis that, until this moment, has never been demonstrated by any thing like solid proof. To produce from nothing, or the *Creation*, is a term that cannot give us the most slender idea of the formation of the universe; it presents no sense, upon which the mind can fasten itself.*

Motion becomes still more obscure, when creation, or the formation of matter. is attributed to a *spiritual* being, tnat is to say, to a being which has no analogy, no point of contact, with it; to a being which has neither extent, nor parts, and cannot, therefore, be susceptible of motion, as we understand the term; this being only the change of one body relatively to another body, in which the body moved, presents successively different parts to different points of space. Moreover, as all the world are nearly agreed that matter can never be totally annihilated, or cease to exist, how can we understand, that that which cannot cease to be, could ever have had a beginning?

If, therefore, it be asked, whence came matter? it is a very reasonable reply to say, it has always existed. If it be inquired, whence proceeds the motion that agitates matter? the same reasoning furnishes the answer; namely, that, as motion is coeval with matter, it must have existed from all eternity, seeing that motion is the necessary consequence of its existence, of its essence, of its primitive properties, such as its extent, its gravity, its impenetrability, its figure, &c. By virtue of these essential, constituent properties, inherent in all matter, and without which it is impossible to form an idea of it, the various matter of which the universe is composed must, from all eternity, have pressed against each other; have gravitated towards a centre; have clashed; have come in contact; have been attracted; have been repelled; have been combined; have been separated; in short, must have acted and moved according to the essence and energy peculiar to each genus, and to each of its combin-

* In fact, the human mind is not adequate to conceive a moment when all was nothing, or when all shall have passed away; even admitting this to be a truth, it is no truth for us, because by the very nature of our organization we cannot admit positions as facts, of which no evidence can be adduced that has relation to our senses: we may, indeed, consent to believe it, because others say it; but will any rational being be satisfied with such an admission? Can any moral good spring from such blind confidence? Is it consistent with sound doctrine, with philosophy, with reason? Do we, in fact, pay any respect to the understanding of another when we say to him, I will believe this, because in all the at-

tempts you have ventured for the purpose of proving what you say, you have entirely failed; and have been at last obliged to acknowledge, *you know nothing about the matter?* What moral reliance ought we to have on such people? Hypothesis may succeed hypothesis; sys'em may destroy system; a new set of ideas may overturn the ideas of a former day. Other Galileos may be condemned to deatt—other Newtons may arise—we may reason; we may argue; we may dispute; we may quarrel; we may punish; we may destroy; we may even exterminate those who differ from us in opinion; but when we have done all this, we shall be obliged to fall back on our original darkness; to confess, ations. Existence supposes properties in the thing that exists: whenever it has properties, its mode of action must necessarily flow from those properties which constitute its mode of being. Thus, when a body is ponderous, it must fall; when it falls, it must come in collision with the bodies it meets in its descent; when it is dense, when it is solid, it must, by reason of this density, communicate motion to the bodies with which it clashes; when it has analogy or affinity with these bodies, it must unite with them; when it has no point of analogy with them, it must be repulsed.

From which it may be fairly inferred, that, in supposing, as we are under the necessity of doing, the existence of matter, we must suppose it to have some kind of properties, from which its motion, or modes of action, must necessarily flow. To form the universe *Descartes* asked but matter and motion: a diversity of matter sufficed for him; variety of motion was the consequence of its existence, of its essence, of its properties: its different modes of action would be the necessary consequence of its different modes of being. Matter without properties, would be a

that that which has no relation with our senses, which cannot manifest itself to us by some of the ordinary modes by which other things are manifested, has no existence for us; is not comprehensible by us; can never entirely remove our doubts; can never seize on our steadfast belief; seeing it is that of which we cannot form even an idea; in short, that it is *that*, which as long as we remain what we are, must be hidden from us by a veil which no power, no faculty, no energy we possess, is able to remove. All who are not enslaved by prejudice, agree to the truth of the position: that *nothing can be made of nothing.* Many theologians have acknowledged nature to be an active whole. Almost all the ancient philosophers were agreed to regard the world as eternal. OCELLUS LUCANUS, speaking of the universe, says: "*it has always been, and it always will be.*" VATABLE and GROTIUS assure us, that, to render correctly the Hebrew phrase in the first chapter of *Genesis*, we must say: "*When God made heaven and earth, matter was without form:*" if this be true, and every Hebraist can judge for himself, then the word which has been rendered *created,* means only to fashion, form, arrange. We know that the Greek words *create* and *form,* have always indicated the same thing. According to St. JEROME, *creare* has the same meaning as *condere,* to found, to build. The bible does not any where say in a clear

mere nothing: therefore, as soon as matter exists, it must act; as soon as it is various, it must act variously; if it cannot commence to exist, it must have existed from all eternity; if it has always existed, it can never cease to be: if it can never cease to be, it can never cease to act by its own energy. Motion is a manner of being, which matter derives from its peculiar existence.

The existence then of matter is a fact; the existence of motion is another fact. Our visual organs point out to us matter with different essences, forming a variety of combinations, endowed with various properties that discriminate them. Indeed, it is an errour to believe that matter is a homogeneous body, of which the parts differ from each other only by their various modifications. Among the individuals of the same species that come under our notice, no two are exactly alike, and it is therefore evident that the difference of situation alone, will necessarily carry a diversity more or less sensible, not only in the modifications, but also in the essence, in the properties, in the entire system of beings.*

If this principle be properly weighed, manner, that the world was made of nothing. TERTULLIAN, and the father PETAU, both admit that, "*this is a truth established more by reasoning, than by authority.*" ST. JUSTIN seems to have contemplated matter as eternal, since he commends PLATO for having said that "*God in the creation of the world only gave impulse to matter, and fashioned it.*" BURNET and PYTHAGORAS were entirely of this opinion, and even the church service may be adduced in support; for although it admits by implication a beginning, it expressly denies an end: "*As it was in the beginning, is now, and ever shall be, world without end.*" It is easy to perceive, that that which cannot cease to exist, must have always been.

* Those who have observed nature closely, know that two grains of sand are not strictly alike. As soon as the circumstances or the modifications are not the same for the beings of the same species, there cannot be an exact resemblance between them. SEE CHAP. VI. This truth was well understood by the profound and subtle LEIBNITZ. This is the manner in which one of his disciples explained himself: Ex principio indiscernibilium patet elementa rerum materialium singula singulis esse dissimilia, adeò que unum ab altero distingui, convenienter omnia extra se invicem existere, in quo differunt a punctis mathematicis, cum illa uti hæc nunquam coincidere possint. BILFINGER, DE DEO, ANIMA ET MUNDO, page 276.

and experience seems always to pro-duce evidence of its truth, we must be convinced, that the matter, or primi-tive elements which enter the compo-sition of bodies, are not of the same na-ture, and, consequently, can neither have the same properties, nor the same modifications; and if so, they cannot have the same mode of moving, and acting. Their activity or motion, al-ready different, can be diversified to in-finity, augmented or diminished, acce-lerated or retarded, according to the combinations, the proportions, the pres-sure, the density, the volume of the matter that enters their composition. The element of fire, is visibly more active and more inconstant than that of earth. This is more solid and pon-derous than fire, air, or water. Ac-cording to the quality of the elements which enter the composition of bodies, these must act diversely, and their mo-tion must in some measure partake the motion peculiar to each of their constitu-ent parts. Elementary fire appears to be in nature the principle of activity; it may be compared to a fruitful leaven, that puts the mass into fermentation and gives it life. Earth appears to be the principle of solidity in bodies, from its impenetrability, and by the firm co-herence of its parts. Water is a me-dium, to facilitate the combination of bodies, into which it enters itself as a constituent part. Air is a fluid, whose business it seems to be, to furnish the other elements with the space requisite to exercise their motion, and which is, moreover, found proper to combine with them. These elements, which our senses never discover in a pure state; which are continually and reci-procally set in motion by each other; which are always acting and re-acting; combining and separating; attracting and repelling; are sufficient to explain to us the formation of all the beings we behold. Their motion is uninterrupt-edly, and reciprocally, produced from each other; they are alternately causes and effects. Thus, they form a vast circle of generation and destruction, of combination and decomposition, which could never have had a beginning, and which can never have an end. In short, nature is but an immense chain of causes and effects, which unceasing-ly flow from each other. The motion

of particular beings depends on the general motion, which is itself main-tained by individual motion. This is strengthened or weakened—accelera-ted or retarded—simplified or compli-cated—procreated or destroyed, by a variety of combinations and circum-stances, which every moment change the directions, the tendency, the modes of existing and of acting, of the differ-ent beings that receive its impulse.*

If we desire to go beyond this, to find the principle of action in matter and to trace the origin of things, it is for ever to fall back upon difficulties; it is absolutely to abridge the evidence of our senses, by which alone we can judge of and understand the causes acting upon them, or the impulse by which they are set in action.

Let us, therefore, content ourselves with saying *that* which is supported by our experience, and by all the evi-dence we are capable of understanding; against the truth of which, not a shadow of proof such as our reason can admit, has ever been adduced; which has been maintained by philosophers in every age; which theologians themselves have not denied, but which many of them have upheld; namely, that *matter always existed; that it moves by virtue of its essence; that all the phenomena of Nature is ascribable to the diversi-fied motion of the variety of matter she contains; and which, like the phenix, is continually regenerating out of her own ashes.†*

* If it were true that every thing has a tendency to form one unique or single mass, and in that unique mass the instant should arrive when all was in *nisus*, all would eter-nally remain in this state—to all eternity there would be but one effort, and this would be eternal and universal death. Natural philo-sophers understand by *nisus* the effort of one body against another body, without local translation. This granted, there could be no cause of dissolution, for, according to chy-mists, bodies act only when dissolved. *Cor-pora non agunt nisi sint soluta.*

† Omnium quæ in sempiterno isto mundo semper fuerunt futuraque sunt, aiunt principium fuisse nullum, sed orbem esse quemdam gene-rantium nascentiumque, in quo uniuscujusque geniti initium simul et finis esse videtur.— *V. Censorin. De Die Natali.*

The poet Manilius expresses himself in the same manner in these beautiful lines:—

Omnia mutantur mortali legi creata,
Nec se cognoscunt terræ vertentibus annis,
Exutas variam faciem per sæcula gentes

CHAPTER III.

Of Matter:—Of its various Combinations; Of its diversified Motion; or, of the Course of Nature.

WE know nothing of the elements of bodies, but we know some of their properties or qualities; and we distinguish their various matter by the effect or change produced on our senses; that is to say, by the variety of motion their presence excites in us. In consequence, we discover in them extent, mobility, divisibility, solidity, gravity, and inert force. From these general and primitive properties, flow a number of others, such as density, figure, colour, ponderosity, &c. Thus, relatively to us, matter is all that affects our senses, in any manner whatever; the various properties we attribute to matter, are founded on the different impressions we receive, on the changes they produce in us.

A satisfactory definition of matter has not yet been given. Man, deceived and led astray by his prejudices, formed but vague, superficial, and imperfect notions concerning it. He looked upon it as a unique being, gross and passive, incapable of either moving by itself, of forming combinations, or of producing any thing by its own energies; whilst he ought to have contemplated it as a *genus* of beings, of which the individuals, although they might possess some common properties, such as extent, divisibility, figure, &c., should not, however, be all ranked in the same class, nor comprised under the same general denomination.

An example will serve more fully to explain what we have just asserted, throw its correctness into light, and facilitate the application. The properties common to all matter, are, extent, divisibility, impenetrability, figure, mobility, or the property of being moved in mass. Fire, beside these general

At manet incolumis mundus suaque omnia servat,
Quæ nec longa dies auget, m) luitque senectus,
Nec motus puncto currit, cur susque fatigat:
Idem semper erit, quoniam semper fuit idem.
                        Manilii Astronom. Lib. I.

This also was the opinion of PYTHAGORAS, such as it is set forth by Ovid, in the fifteenth Book of his Metamorphoses, verse 165, and the following:—
Omnia mutantur, nihil interit; errat et illinc.
Huc venit, hinc illuc, &c.

properties common to all matter, enjoys also the peculiar property of being put into activity by a motion producing on our organs of feeling the sensation of heat, and by another, which communicates to our visual organs the sensation of light. Iron, in common with matter in general, has extent and figure; is divisible, and moveable in mass: if fire be combined with it in a certain proportion, the iron acquires two new properties, namely, those of exciting in us similar sensations of heat and light, which the iron had not before its combination with the igneous matter. These distinguishing properties are inseparable from matter, and the phenomena that result, may, in the strictest sense of the word, be said to result necessarily.

If we only contemplate the paths of nature; if we trace the beings in this nature under the different states through which, by reason of their properties, they are compelled to pass, we shall discover that it is to motion, and motion alone, that is to be ascribed all the changes, all the combinations, all the forms, in short, all the various modifications of matter. That it is by motion every thing that exists is produced, experiences change, expands, and is destroyed. It is motion that alters the aspect of beings, that adds to, or takes away from their properties; which obliges each of them, by a consequence of its nature, after having occupied a certain rank or order, to quit it to occupy another, and to contribute to the generation, maintenance, and decomposition of other beings, totally different in their bulk, rank, and essence.

In what experimental philosophers have styled the *three orders of nature*, that is to say, the *mineral*, the *vegetable*, and the *animal* worlds, they have established, by the aid of motion, a transmigration, an exchange, a continual circulation in the particles of matter. Nature has occasion in one place for those particles which, for a time, she has placed in another. These particles, after having, by particular combinations, constituted beings endued with peculiar essences, with specific properties, with determinate modes of action, dissolve and separate with more or less facility; and combining in a new manner, they form new beings. The attentive observer sees this law execute itself in a manner more or less prominent through all the beings by which he is surrounded. He sees nature full of *erratic germs*, some of which expand themselves, whilst others wait until motion has placed them in their proper situation, in suitable wombs or matrices, in the necessary circumstances to unfold, to increase, to render them more perceptible by the addition of other substances of matter analogous to their primitive being. In all this we see nothing but the effect of motion, necessarily guided, modified, accelerated or slackened, strengthened or weakened, by reason of the various properties that beings successively acquire and lose; which, every moment, infallibly produces alterations in bodies, more or less marked. Indeed these bodies cannot be, strictly speaking, the same in any two successive moments of their existence ; they must, every instant, either acquire or lose : in short, they are obliged to undergo continual variations in their essences, in their properties, in their energies, in their masses, in their qualities, in their mode of existence.

Animals, after they have been expanded in, and brought out of the wombs that are suitable to the elements of their machine, enlarge, strengthen, acquire new properties, new energies, new faculties ; either by deriving nourishment from plants analogous to their being, or by devouring other animals whose substance is suitable to their preservation ; that is to say, to repair the continual deperdition, or loss, of some portion of their own substance that is disengaging itself every instant. These same animals are nourished, preserved, strengthened, and enlarged by the aid of air, water, earth, and fire. Deprived of air, or of the fluid that surrounds them, that presses on them, that penetrates them, that gives them their elasticity, they presently cease to live. Water combined with this air, enters into their whole mechanism, of which it facilitates the motion. Earth serves them for a basis, by giving solidity to their texture : it is conveyed by air and water, which carry it to those parts of the body with which it can combine. Fire itself, disguised and enveloped under an infinity of forms, continually received into the animal, procures him

heat, continues him in life, renders him capable of exercising his functions. The aliments, charged with these various principles, entering into the stomach, re-establish the nervous system, and restore, by their activity, and the elements which compose them, the machine which begins to languish, to be depressed, by the loss it has sustained. Forthwith the animal experiences a change in his whole system ; he has more energy, more activity; he feels more courage ; displays more gaiety ; he acts, he moves, he thinks, after a different manner; all his faculties are exercised with more ease.* From this it is clear, that what are called the elements, or primitive parts of matter, when variously combined, are, by the agency of motion, continually united to, and assimilated with the substance of animals : that they visibly modify their being, have an evident influence over their actions, that is to say, upon the motion they undergo, whether visible or concealed.

The same elements, which under certain circumstances serve to nourish, to strengthen, to maintain the animal, become, under others, the principles of his weakness, the instruments of his dissolution, of his death : they work his destruction, whenever they are not in that just proportion, which renders them proper to maintain his existence: thus, when water becomes too abundant in the body of the animal, it enervates him, it relaxes the fibres, and impedes the necessary action of the other elements: thus, fire admitted in excess, excites in him disorderly motion, destructive of his machine : thus, air, charged with principles not analogous to his mechanism, brings upon him dangerous diseases and contagion. In fine, the aliments modified after certain

---

* We may here remark, that all spirituous substances (that is to say, those containing a great proportion of inflammable and igneous matter, such as wine, brandy, liquors, &c.) are those that accelerate most the organic motion of animals, by communicating to them heat. Thus, wine generates courage, and even wit. In spring and summer myriads of insects are hatched, and a luxuriant vegetation springs into life, because the matter of fire is then more abundant than in winter. This *igneous matter* is evidently the cause of • fermentation, of generation, and of life—the Jupiter of the ancients.

No. I.—4

modes, instead of nourishing destroy the animal, and condu ce to his ruin: the animal is preserved no longer than these substances are analogous to his system. They ruin him when they want that just equilibrium that renders them suitable to maintain his existence.

Plants, that serve to nourish and restore animals, are themselves nourished by earth; they expand on its bosom, enlarge and strengthen at its expense, continually receiving into their texture, by their roots and their pores, water, air, and igneous matter: water visibly reanimates them whenever their vegetation, or genus of life, languishes; it conveys to them those analogous principles by which they are enabled to reach perfection; air is requisite to their expansion, and furnishes them with water, earth, and igneous matter with which it is charged. By these means they receive more or less of the inflammable matter; and the different proportions of these principles, from whence numerous combinations, from whence result an infinity of properties, a variety of forms, constitute the various families and classes into which botanists have distributed plants: it is thus, we see the cedar, and the hyssop, develop their growth; the one, rises to the clouds; the other, creeps humbly on the earth. . Thus, by degrees, from an acorn springs the majestic oak, accumulating with time its numerous branches, and overshadowing us with its foliage. Thus, a grain of corn, after having drawn its own nourishment from the juices of the earth, serves, in its turn, for the nourishment of man, into whose system it conveys the elements or principles by which it has been itself expanded—combined and modified in such a manner, as to render this vegetable proper to assimilate and unite with the human frame: that is to say, with the fluids and solids of which it is composed.

. The same elements, the same principles, are found in the formation of minerals, and also in their decomposition, whether natural or artificial. We find that earth diversely modified, wrought and combined, serves to increase their bulk, and give them more or less density and gravity. Air and water contribute to make their parti-

cles cohere: the igneous matter, or inflammable principle, tinges them with colour, and sometimes, plainly indicates its presence by the brilliant scintillation, which motion elicits from them. These stones and metals, these bodies so compact and solid, are disunited, are destroyed, by the agency of air, water, and fire which the most ordinary analysis is sufficient to prove, as well as a multitude of experience to which our eyes are the daily evidence.

Animals, plants, and minerals, after a lapse of time, give back to nature—that is to say, to the general mass of things, to the universal magazine—the elements or principles which they have borrowed. The earth retakes that portion of the body of which it formed the basis and the solidity; the air charges itself with those parts that are analogous to it, and with those particles which are light and subtile; water carries off that which is suitable to liquescency; fire bursting its chains, disengages itself, and rushes into new combinations with other bodies. The elementary particles of the animal being thus dissolved, disunited, and dispersed, assume new activity, and form new combinations: thus, they serve to nourish, to preserve, or destroy new beings—among others, plants, which, arrived at their maturity, nourish and preserve new animals; these, in their turn, yielding to the same fate as the first.

Such is the invariable course of Nature: such is the eternal circle of mutation, which all that exists is obliged to describe. It is thus that motion generates, preserves for a time, and successively destroys one part of the universe by the other; whilst the sum of existence remains eternally the same. Nature, by its combinations, produces suns, which place themselves in the centre of so many systems: she forms planets, which, by their peculiar essence, gravitate and describe their revolutions round these suns: by degrees the motion is changed altogether, and becomes eccentric: perhaps the day may arrive when these wondrous masses will disperse, of which man, in the short space of his existence, can only have a faint and transient glimpse.

It is clear, then, that the continual

notion inherent in matter, changes and destroys all beings ; every instant depriving them of some of their properties to substitute others : it is motion which, in thus changing their actual essence, changes also their order, their direction, their tendency, and the laws which regulate their mode of acting and being : from the stone formed in the bowels of the earth by the intimate combination and close coherence of similar and analogous particles, to the sun, that vast reservoir of igneous particles, which sheds torrents of light over the firmament ; from the benumbed oyster, to the thoughtful and active man, we see an uninterrupted progression, a perpetual chain of motion and combination, from which is produced beings, that only differ from each other by the variety of their elementary matter : and by the numerous combinations of these elements spring modes of action and existence, diversified to infinity. In generation, in nutrition, in preservation, we see nothing more than matter variously combined, of which each has its peculiar motion, regulated by fixed and determinate laws, which oblige them to submit to necessary changes. We shall find in the formation, in the growth, in the instantaneous life of animals, vegetables and minerals, nothing but matter, which, combining, accumulating, aggregating, and expanding by degrees, forms beings, who are either feeling, living, vegetating, or else destitute of these faculties; and having existed some time under one particular form, they are obliged to contribute by their ruin to the production of other forms.*

CHAPTER IV.

Of the Laws of Motion common to all the Beings of Nature—Of Attraction and Repulsion—Of inert Force—Of Necessity.

MAN is never surprised at those effects of which he thinks he knows the cause; he believes he does know the cause as

soon as he sees them act in a uniform and determinate manner, or when the motion excited is simple : the descent of a stone, that falls by its own peculiar weight, is an object of meditation only to the philosopher, to whom the mode by which the most immediate causes act, and the most simple motion, are no less impenetrable mysteries than the most complex motion, and the manner by which the most complicated causes give impulse. The uninformed are seldom tempted either to examine the effects which are familiar to them, or to recur to first principles. They think they see nothing in the descent of a stone which ought to elicit their surprise, or become the object of their research : it requires a Newton to feel that the descent of heavy bodies is a phenomenon worthy his whole, his most serious attention : it requires the sagacity of a profound experimental philosopher, to discover the laws by which heavy bodies fall, by which they communicate to others their peculiar motion. In short, the mind that is most practised in philosophical observation, has frequently the chagrin to find, that the most simple and most common effects escape all his researches, and remain inexplicable to him.

When any extraordinary, any unusual effect is produced, to which our eyes have not been accustomed ; or when we are ignorant of the energies of the cause, the action of which so forcibly strikes our senses, we are tempted to meditate upon it, and take it into our consideration. The European, accustomed to the use of *gunpowder*, passes it by, without thinking much of its extraordinary energies ; the workman, who labours to manufacture it, finds nothing marvellous in its properties, because he daily handles the

---

* Destructio unius, generatio alterius. Thus, to speak strictly, nothing in nature is either born, or dies, according to the common acceptation of those terms. This truth was felt by many of the ancient philosophers. PLATO tells us, that according to an old tradition, "the living were born of the dead, the

same as the dead did come of the living; and that this is the constant routine of nature." He adds from himself, "Who knows if to live, be not to die; and if to die, be not to live?" This was the doctrine of PYTHAGORAS, a man of great talent and no less note. EMPEDOCLES says, "There is neither birth nor death for any mortal, but only a combination and a separation of that which was combined, and this is what amongst men they call birth and death." Again he remarks, "Those are infants, or short-sighted persons with very contracted understandings, who imagine any thing is born which did not exist before, or that any thing can die or perish totally."

matter that enters its composition. The American, who had never beheld its operation, looked upon it as a divine power, and its energies as supernatural. The uninformed, who are ignorant of the true cause of *thunder*, contemplate it as the instrument of celestial vengeance. The experimental philosopher considers it as the effect of the electric matter, which, nevertheless, is itself a cause which he is very far from perfectly understanding.*

Be this as it may, whenever we see a cause act, we look upon its effect as natural: when this cause becomes familiar to the sight, when we are accustomed to it, we think we understand it, and its effects surprise us no longer. Whenever any unusual effect is perceived without our discovering the cause, the mind sets to work, becomes uneasy; this uneasiness increases in proportion to its extent: as soon as it is believed to threaten our preservation, we become completely agitated: we seek after the cause with an earnestness proportioned to our alarm; our perplexity augments in a ratio equivalent to the persuasion we are under how essentially requisite it is we should become acquainted with the cause that has affected us in so lively a manner. As it frequently happens that our senses can teach us nothing respecting this cause which so deeply interests us, which we seek with so much ardour; we have recourse to our imagination; this, disturbed with alarm, enervated by fear, becomes a suspicious, a fallacious guide: we create chimeras, fictitious causes, to whom we give the credit, to whom we ascribe the honour of those phenomena by which we have been so much alarmed. It is to this disposition of the human mind that must be attributed, as will be seen in the sequel, the religious errours of man, who, despairing of the capability to trace the natural causes of those

perplexing phenomena to which he was the witness, and sometimes the victim, created in his brain, heated with terrour, imaginary causes, which have become to him a source of the most extravagant folly.

In nature, however, there can be only natural causes and effects; all the motion excited in this nature follows constant and necessary laws: the natural operations to the knowledge of which we are competent, of which we are in a capacity to judge, are of themselves sufficient to enable us to discover those which elude our sight; we can at least judge of them by analogy. If we study nature with attention, the modes of action which she displays to our senses will teach us not to be disconcerted by those which she refuses to discover. Those causes which are the most remote from their effects, unquestionably act by intermediate causes; by the aid of these, we can frequently trace out the first. If in the chain of these causes we sometimes meet with obstacles that oppose themselves to our research, we ought to endeavour by patience and diligence to overcome them; when it so happens we cannot surmount the difficulties that occur, we still are never justified in concluding the chain to be broken, or that the cause which acts is *supernatural.* Let us, then, be content with an honest avowal, that Nature contains resources of which we are ignorant; but never let us substitute phantoms, fictions, or imaginary causes, senseless terms, for those causes which escape our research; because, by such means, we only confirm ourselves in ignorance, impede our inquiries, and obstinately remain in errour.

In spite of our ignorance with respect to the meanderings of Nature, of the essence of beings, of their properties, their elements, their combinations, their proportions, we yet know the simple and general laws according to which bodies move, and we see clearly, that some of these laws, common to all beings, never contradict themselves: although, on some occasions, they appear to vary, we are frequently competent to discover that the cause becoming complex, from combination with other causes, either impedes, or prevents its mode of action, being such as in its primitive state we had a right to expect.

---

* It required the keen, the penetrating mind of a FRANKLIN, to throw light on the nature of this subtile fluid; to develop the means by which its effects might be rendered harmless; to turn to useful purposes a phenomenon that made the ignorant tremble, that filled their minds with terrour, their hearts with dismay, as indicating the anger of the gods: impressed with this idea, they prostrated themselves, they sacrificed to Jupiter or Jehovah, to deprecate their wrath.

We know that active, igneous matter, applied to gunpowder, must necessarily cause it to explode: whenever this effect does not follow the combination of the igneous matter with the gunpowder, whenever our senses do not give us evidence of the fact, we are justified in concluding, either that the powder is damp, or that it is united with some other substance that counteracts its explosion. We know that all the actions of man have a tendency to render him happy: whenever, therefore, we see him labouring to injure or destroy himself, it is just to infer that he is moved by some cause opposed to his natural tendency; that he is deceived by some prejudice; that, for want of experience, he is blind to consequences: that he does not see whither his actions will lead him.

If the motion excited in beings was always simple; if their actions did not blend and combine with each other, it would be easy to know the effect a cause would produce. I know that a stone, when descending, ought to describe a perpendicular: I also know, that if it encounters any other body which changes its course, it is obliged to take an oblique direction; but if its fall be interrupted by several contrary powers which act upon it alternately, I am no longer competent to determine what line it will describe. It may be a parabola, an ellipsis, spiral, circular, &c.; this will depend on the impulse it receives, and the powers by which it is impelled.

The most complex motion, however, is never more than the result of simple motion combined: therefore, as soon as we know the general laws of beings, and their action, we have only to decompose and to analyze them, in order to discover those of which they are combined: experience teaches us the effects we are to expect. Thus it is clear, the simplest motion causes that necessary junction of different matter of which all bodies are composed: that matter varied in its essence, in its properties, in its combinations, has each its several modes of action, or motion, peculiar to itself: the whole motion of a body is consequently the sum total of each particular motion that is combined.

Amongst the matter we behold, some is constantly disposed to unite, whilst

other is incapable of union; that which is suitable to unite, forms combinations more or less intimate, possessing more or less durability: that is to say, with more or less capacity to preserve their union and to resist dissolution. Those bodies which are called *solids*, receive into their composition a great number of homogeneous, similar, and analogous particles, disposed to unite themselves; with energies conspiring or tending to the same point. The primitive beings, or elements of bodies, have need of support, of props, that is to say, of the presence of each other, for the purpose of preserving themselves; of acquiring consistence, or solidity; a truth which applies with equal uniformity to what is called *physical*, as to what is termed *moral*.

It is upon this disposition in matter and bodies with relation to each other, that is founded those modes of action which natural philosophers designate by the terms *attraction, repulsion, sympathy, antipathy, affinities, relations.*[*] Moralists describe this disposition under the names of *love, hatred, friendship, aversion.* Man, like all the beings in nature, experiences the impulse of attraction and repulsion; the motion excited in him differing from that of other beings, only because it is more concealed, and frequently so hidden, that neither the causes which excite it, nor their mode of action are known.

Be this as it may, it is sufficient for us to know, that by an invariable law certain bodies are disposed to unite with more or less facility, whilst others

---

[*] This system of attraction and repulsion is very ancient, although it required a Newton to develop it. That love, to which the ancients attributed the unfolding or disentanglement of chaos, appears to have been nothing more than a personification of the principle of attraction. All their allegories and fables upon chaos, evidently indicate nothing more than the accord or union that exists between analogous and homogeneous substances, from whence resulted the existence of the universe: while discord or repulsion, which they called *sore*, was the cause of dissolution, confusion, and disorder. There can scarcely remain a doubt but this was the origin of the doctrine of the two principles. According to Diogenes Laërtius, the philosopher, Empedocles asserted. *"that there is a kind of affection, by which the elements unite themselves; and a sort of discord, by which they separate or remove themselves."*

cannot combine. Water combines itself readily with salt, but will not blend with oil. Some combinations are very strong, cohering with great force, as metals; others are extremely feeble, their cohesion slight, and easily decomposed, as in fugitive colours. Some bodies, incapable of uniting by themselves, become susceptible of union by the agency of other bodies, which serve for common bonds or *mediums*. Thus, oil and water, naturally heterogeneous, combine and make soap, by the intervention of alkaline salt. From matter diversely combined, in proportions varied almost to infinity, result all physical and moral bodies; the properties and qualities of which are essentially different, with modes of action more or less complex: which are either understood with facility, or difficult of comprehension, according to the matter that has entered into their composition, and the various modifications this matter has undergone.

It is thus, from the reciprocity of their attraction, that the primitive, imperceptible particles of matter which constitute bodies, become perceptible, and form compound substances, aggregate masses, by the union of similar and analogous matter, whose essences fit them to cohere. The same bodies are dissolved, or their union broken, whenever they undergo the action of matter inimical to their junction. Thus by degrees are formed plants, metals, animals, men; each grows, expands, and increases, in its own system, or order; sustaining itself in its respective existence by the continual attraction of analogous matter, to which it becomes united, and by which it is preserved and strengthened. Thus, certain aliments become fit for the sustenance of man; whilst others destroy his existence: some are pleasant to him, strengthen his habit; others are repugnant to him, weaken his system: in short, never to separate physical from moral laws—it is thus that men, mutually attracted to each other by their reciprocal wants, form those unions which we designate by the terms *marriage, families, societies, friendships, connexions :* it is thus that virtue strengthens and consolidates *them ;* that vice relaxes, or totally dissolves them.

Of whatever nature may be the combination of beings, their motion has always one direction or tendency: without direction we could not have any idea of motion: this direction is regulated by the properties of each being: as soon as they have any given properties, they necessarily act in obedience to them; that is to say, they follow the law invariably determined by these same properties, which, of themselves, constitute the being such as he is found, and settle his mode of action, which is always the consequence of his manner of existence. But what is the general direction, or common tendency, we see in all beings? What is the visible and known end of all their motion? It is to preserve their actual existence—to strengthen their several bodies—to attract that which is favourable to them—to repel that which is injurious to them—to avoid that which can harm them, to resist impulsions contrary to their manner of existence and to their natural tendency.

To exist, is to experience the motion peculiar to a determinate essence : to preserve this existence, is to give and receive that motion from which results the maintenance of its existence:—it is to attract matter suitable to corroborate its being,—to avoid that by which it may be either endangered, or enfeebled. Thus, all beings of which we have any knowledge, have a tendency to preserve themselves each after its own peculiar manner: the stone, by the firm adhesion of its particles, opposes resistance to its destruction. Organized beings preserve themselves by more complicated means, but which are, nevertheless, calculated to maintain their existence against that by which it may be injured. Man, both in his physical and in his moral capacity, is a living, feeling, thinking, active being, who every instant of his duration strives equally to avoid that which may be injurious, and to procure that which is pleasing to him, or that which is suitable to his mode of existence.*

Conservation, then, is the common point to which all the energies, all the powers, all the faculties of being, seem

* St. Augustine admits this tendency for self-preservation in all beings, whether organized or not.—See his tractate *De Civitate Dei,* lib. xi. cap. 28.

continually directed. Natural philosophers call this direction, or tendency, *self-gravitation*. *Newton* calls it *inert force*. Moralists denominate it, in man, *self-love;* which is nothing more than the tendency he has to preserve himself—a desire of happiness—a love of his own welfare—a wish for pleasure—a promptitude in seizing on every thing that appears favourable to his conservation—a marked aversion to all that either disturbs his happiness, or menaces his existence—primitive sentiments common to all beings of the human species, which all their faculties are continually striving to satisfy; which all their passions, their wills, their actions, have eternally for their object and their end. This self-gravitation, then, is clearly a necessary disposition in man and in all other beings, which, by a variety of means, contributes to the preservation of the existence they have received as long as nothing deranges the order of their machine or its primitive tendency.

Cause always produces effect; there can be no effect without cause. Impulse is always followed by some motion more or less sensible, by some change more or less remarkable in the body which receives it. But motion, and its various modes of displaying itself, is, as has been already shown, determined by the nature, the essence, the properties, the combinations of the beings acting. It must then be concluded, that motion, or the modes by which beings act, arises from some cause; and as this cause is not able to move or act but in conformity with the manner of its being, or its essential properties, it must equally be concluded, that all the phenomena we perceive are necessary; that every being in nature, under the circumstances in which it is placed and with the given properties it possesses, cannot act otherwise than it does.

Necessity is the constant and infallible connexion of causes with their effects. Fire, of necessity, consumes combustible matter placed within its sphere of action: man, of necessity, desires, either that which really is, or appears to be useful to his welfare. Nature, in all the phenomena she exhibits, necessarily acts after her own peculiar essence; all the beings she

contains necessarily act each after its individual essence: it is by motion that the whole has relation with its parts, and these with the whole: it is thus that in the universe every thing is connected; it is itself but an immense chain of causes and effects, which flow without ceasing one from the other. If we reflect a little, we shall be obliged to acknowledge, that every thing we see is necessary; that it cannot be otherwise than it is; that all the beings we behold, as well as those which escape our sight, act by certain and invariable laws. According to these laws heavy bodies fall, light bodies rise; analogous substances attract each other; beings tend to conserve themselves; man cherishes himself; loves that which he thinks advantageous, detests that which he has an idea may prove unfavourable to him. In fine, we are obliged to admit that there can be no independ' '.nt energy—no isolated cause—no detached action, in a nature where all the beings are in a reciprocity of action—who without interruption mutually impel and resist each other—who is herself nothing more than an eternal circle of motion given and received according to necessary laws.

Two examples will serve to throw the principle here laid down, into light—one shall be taken from physics, the other from morals.

In a whirlwind of dust, raised by the impetuous elements, confused as it appears to our eyes; in the most frightful tempest, excited by contrary winds, when the waves roll high as mountains; there is not a single particle of dust, or drop of water, that has been placed by *chance;* that has not a sufficient cause for occupying the place where it is found; that does not, in the most rigorous sense of the word, act after the manner in which it ought to act; that is, according to its own peculiar essence, and that of the beings from whom it receives impulse. A geometrician, who exactly knew the different energies acting in each case, with the properties of the particles moved, could demonstrate, that, after the causes given, each particle acted precisely as it ought to act, and that it could not have acted otherwise than it did.

In those terrible convulsions that sometimes agitate political societies, shake their foundations, and frequently produce the overthrow of an empire —there is not a single action, a single word, a single thought, a single will, a single passion in the agents, whether they act as destroyers or as victims, that is not the necessary result of the causes operating; that does not act as of necessity it must act from the peculiar situation these agents occupy in the moral whirlwind. This could be evidently proved by an understanding capacitated to seize and to rate all the actions and reactions of the minds and bodies of those who contributed to the revolution.

In fact, if all be connected in nature; if all motion be produced the one from the other, notwithstanding their secret communications frequently elude our —. we ought to feel convinced that there .: no ᴄause, however minute, however remote, that does not sometimes produce the greatest and the most immediate effects on man. It may perhaps be in the arid plains of Lybia, that are amassed the first elements of a storm or tempest, which, borne by the winds, approximate our climate, render our atmosphere dense, which operating on the temperament, may influence the passions of a man whose circumstances shall have capacitated him to influence many others, and who shall decide after his will the fate of many nations.

Man, in fact, finds himself in nature, and makes a part of it : he acts according to laws which are peculiar to him ; he receives, in a manner more or less distinct, the action, the impulse of the beings who surround him ; who themselves act after laws that are peculiar to their essence. It is thus that he is variously modified ; but his actions are always the result of his own peculiar energy, and that of the beings who act upon him, and by whom he is modified. This is what gives such variety to his determinations; what frequently produces such contradiction in his thoughts, his opinions, his will, his actions; in short, that motion, whether concealed or visible, by which he is agitated. We shall have occasion, in the sequel, to place this truth, at present so much contested, in a

broader light: it will be sufficient for our present purpose to prove, generally, that every thing in nature is necessary, that nothing to be found in it can act otherwise than it does.

It is motion alternately communicated and received, that establishes the connexion and the relation between the different orders of beings : when they are in the sphere of reciprocal action, attraction approximates them ; repulsion dissolves and separates them ; the one conserves and strengthens them ; the other enfeebles and destroys them. Once combined, they have a tendency to preserve themselves in that mode of existence, by virtue of their *inert force:* in this they cannot succeed, because they are exposed to the continual influence of all other ·beings who act upon them perpetually and in succession: their change of form, their dissolution is requisite to the preservation of nature herself: this is the sole end we are able to assign her; to which we see her tend incessantly ; which she follows without interruption by the destruction and reproduction of all subordinate beings, who are obliged to submit to her laws, and to concur, by their mode of action, to the maintenance of her active existence, so essentially requisite to the GREAT WHOLE.

Thus, each being is an individual, who, in the great family, executes the necessary task assigned to him. All bodies act according to laws inherent in their peculiar essence, without the capability to swerve, even for a single instant, from those according to which Nature herself acts. This is the central power, to which all other powers, all other essences, all other energies, are submitted; she regulates the motion of beings; by the necessity of her own peculiar essence, she makes them concur by various modes to the general plan : this plan appears to be nothing more than the life, action, and maintenance of the whole, by the continual change of its parts. This object she obtains in removing them one by the other : by that which establishes, and by that which destroys the relation subsisting between them ; by that which gives them, and by that which deprives them of their forms, combinations, proportions, qualities, according to which they act for a time, and after a given mode; these are afterwards taken from them, to make them act after a different manner. It is thus that nature makes them expand and change, grow and decline, augment and diminish, approximate and remove, forms them and destroys them, according as she finds it requisite to maintain the whole, towards the conservation of which this nature is herself essentially necessitated to have a tendency.

This irresistible power, this universal necessity, this general energy, is, then, only a consequence of the nature of things, by virtue of which every thing acts without intermission, after constant and immutable laws; these laws not varying more for the whole, than for the beings of which it is composed. Nature is an active, living whole, whose parts necessarily concur, and that without their own knowledge, to maintain activity, life, and existence. Nature acts and exists necessarily: all that she contains necessarily conspires to perpetuate her active existence.*

---

* This was the decided opinion of Plato, who says, "*Matter and necessity are the same thing; this necessity is the mother of the world.*" In point of fact we cannot go beyond this aphorism, *Matter acts because it exists, and exists to act.* If it be inquired how, or why, matter exists? We answer, we know not: but reasoning by analogy of what we do not know by that which we do, we are of opinion it exists necessarily, or because it contains within itself a sufficient reason for its existence. In supposing it to be created or produced by a being distinguished from it, or less known than itself, we must still admit that this being is necessary, and includes a sufficient reason for his own existence. We have not then removed any of the difficulty, we have not thrown a clearer light on the subject, we have not advanced a single step; we have simply laid aside an agent of which we know some of the properties, to have recourse to a power of which it is utterly impossible we can form any distinct idea, and whose existence cannot be demonstrated. As therefore these must be at best but speculative points of belief, which each individual, by reason of its obscurity, may contemplate with different optics and under various aspects; they surely ought to be left free for each to judge after his own fashion: the Deist can have no just cause of enmity against the Atheist for his want of faith; and the numerous sects of each of the various persuasions spread over the face of the earth ought to make it a creed, to look with an eye of complacency on the deviation of the other; and rest upon that great moral axiom, which

No. II.—5

We shall see in the sequel, how much man's imagination has laboured to form an idea of the energies of that nature he has personified and distinguished from herself: in short, we shall examine some of the ridiculous and pernicious inventions which for want of understanding nature, have been imagined to impede her course, to suspend her eternal laws, to place obstacles to the necessity of things.

CHAPTER V.

Of Order and Confusion—Of Intelligence—Of Chance.

THE observation of the necessary, regular, and periodical motion in the universe, generated in the mind of man the idea of *order*. This term, in its primitive signification, represents to him nothing more than a mode of considering, a facility of perceiving, together and separately, the different relations of a whole, in which is discovered by its manner of existing and acting, a certain affinity or conformity with his own. Man, in extending this idea to the universe, carried with him those methods of considering things which are peculiar to himself: he has consequently supposed there really existed in nature affinities and relations, which he classed under the name of *order*; and others, which appeared to him not to conform to those which he has ranked under the term *confusion.*

It is easy to comprehend that this idea of order and confusion can have no absolute existence in nature, where every thing is necessary; where the whole follows constant and invariable laws; and which oblige each being, in every moment of its duration, to submit to other laws which themselves flow from its own peculiar mode of existence. It is, therefore, in his imagination alone man finds the model of that which he terms order, or confusion, which, like all his abstract, metaphysical ideas, supposes nothing be-

is strictly conformable to nature, which contains the nucleus of man's happiness—"*Do not unto another, that which you do not wish another should do unto you;*" for it is evident, according to their own doctrines, that out of all their multifarious systems, one only can be right.

yond his reach. Order, however, is never more than the faculty of conforming himself with the beings by whom he is environed, or with the whole of which he forms a part.

Nevertheless, if the idea of order be applied to nature, it will be found to be nothing but a series of action, or motion, which man judges to conspire to one common end. Thus, in a body that moves, order is the chain of action, the series of motion proper to constitute it what it is, and to maintain it in its actual state. Order, relatively to the whole of nature, is the concatenation of causes and effects necessary to her active existence, and to the maintaining her eternally together; but, as it has been proved in the preceding chapter, every individual being is obliged to concur to this end in the different ranks they occupy; from whence it is a necessary deduction, that what is called the *order of nature*, can never be more than a certain manner of considering the necessity of things, to which all, of which man has any knowledge, is submitted. That which is styled *confusion*, is only a relative term used to designate that series of necessary action, that chain of requisite motion, by which an individual being is necessarily changed or disturbed in its mode of existence, and by which it is instantaneously obliged to alter its manner of action: but no one of these actions, no part of this motion, is capable, even- for a single instant, of contradicting or deranging the general order of nature, from which all beings derive their existence, their properties, the motion peculiar to each.

What is termed confusion in a being, is nothing more than its passage into a new class, a new mode of existence, which necessarily carries with it a new series of action, a new chain of motion, different from that of which this being found itself susceptible in the preceding rank it occupied. That which is called order in nature, is a mode of existence, or a disposition of its particles strictly *necessary*. In every other assemblage of causes and effects, or of worlds, as well as in that which we inhabit, some sort of arrangement, some kind of order, would necessarily be established. Suppose the most discordant and the most heterogeneous substances

were put into activity; by a concatenation of necessary phenomena they would form amongst themselves a complete order, a perfect arrangement of some sort. This is the true notion of a property which may be defined an aptitude to constitute a being such as it is actually found, such as it is, with respect to the whole of which it makes a part.

Thus, I repeat, order is nothing but necessity, considered relatively to the series of actions, or the connected chain of causes and effects that it produces in the universe. What is, in fact, the motion in our planetary system, the only one of which man has any distinct idea, but order; but a series of phenomena, operated according to necessary laws, regulating the bodies of which it is composed? In conformity to these laws, the sun occupies the centre; the planets gravitate towards it, and describe round it, in regulated periods, continual revolutions: the satellites of these planets gravitate towards those which are in the centre of their sphere of action, and describe round them their periodical route. One of these planets, the earth, which man inhabits, turns on its own axis, and by the various aspects which its annual revolution obliges it to present to the sun, experiences those regular variations which are called *seasons*. By a necessary series of the sun's action upon different parts of this globe, all its productions undergo vicissitudes: plants, animals, men, are in a sort of lethargy during *Winter*: in *Spring*, these beings appear to reanimate, to come, as it were, out of a long drowsiness. In short, the mode in which the earth receives the sun's beams, has an influence on all its productions; these rays, when darted obliquely, do not act in the same manner as when they fall perpendicularly; their periodical absence, caused by the revolution of this sphere on itself, produces *night* and *day*. In all this, however, man never witnesses more than necessary effects, flowing from the essence of things, which, whilst that shall remain the same, can never be contradicted. These effects are owing to gravitation, attraction, centrifugal power, &c.*

* Centrifugal force is a philosophical term, used to describe that force by which all bodies

OF ORDER AND CONFUSION.

On the other hand, this *order*, which man admires as a supernatural effect, is sometimes disturbed or changed into what he calls *confusion:* this confusion itself is, however, always a necessary consequence of the laws of nature, in which it is requisite for the maintenance of the whole that some of her parts should be deranged, and thrown out of the ordinary course. It is thus *comets* present themselves so unexpectedly to man's wondering eyes; their eccentric motion disturbs the tranquillity of his planetary system; they excite the terrour of the uninformed, to whom every thing unusual is marvellous. The natural philosopher himself conjectures that, in former ages, these comets have overthrown the surface of this mundane ball, and caused great revolutions on the earth. Independent of this extraordinary *confusion*, he is exposed to others more familiar to him: sometimes the seasons appear to have usurped each other's place—to have quitted their regular order; sometimes the discordant elements seem to dispute among themselves the dominion of the world; the sea bursts its limits; the solid earth is shaken, is rent asunder; mountains are iu a state of conflagration; pestilential diseases destroy men, sweep off animals; sterility desolates a country; then affrighted man utters piercing cries, offers up his prayers to recall order, and tremblingly raises his hands towards the Being he supposes to be the author of all these calamities: and yet, the whole of this afflicting confusion are necessary effects, produced by natural causes, which act according to fixed, to permanent laws, determined by their own peculiar essence, and the universal essence of nature, in which every thing must necessarily be changed, be moved, be dissolved; where that which is called *order* must sometimes be disturbed, and be altered into a new mode of existence, which, to his mind, appears *confusion*.

What is called the *confusion of nature*, has no existence: man finds order in every thing that is conformable to his own mode of being; confusion in every thing by which it is opposed:

which move round any other body in a circle or an ellipsis, do endeavour to fly off from the axis of their motion in a tangent to the periphery or circumference of it.

nevertheless, in nature all is in order, because none of her parts are ever able to emancipate themselves from those invariable and necessary rules, which flow from their respective essences: there is not, there cannot be, confusion in a whole, to the maintenance of which what is called confusion is absolutely requisite; of which the general course can never be deranged where all the effects produced are the consequence of natural causes, that, under the circumstances in which they are placed, act only as they infallibly are obliged to act.

It thus follows that there can be neither monsters nor prodigies, wonders nor miracles in nature: those which are designated as *monsters*, are certain combinations with which the eyes of man are not familiarized, but which are not less the necessary effects of natural causes. Those which he terms *prodigies, wonders*, or *supernatural* effects, are phenomena of nature with whose mode of action he is unacquainted—of which his ignorance does not permit him to ascertain the principles—whose causes he cannot trace, but which his heated imagination makes him foolishly attribute to fictitious causes, which, like the idea of order, have no existence but in himself; for, out of nature, none of these things can have existence.

As for those effects, which are called *miracles*, that is to say, contrary to the immutable laws of nature, such things are impossible; because nothing can for an instant suspend the necessary course of beings, without arresting the entire of nature, and disturbing her in her tendency. There have neither been wonders nor miracles in nature, except for those who have not sufficiently studied this nature, and who consequently do not feel that her laws can never be contradicted, even in the minutest of her parts, without the whole being annihilated, or at least, without changing her essence, or her mode of action.*

* A miracle, according to some metaphysicians, is an effect produced by a power not to be found in nature.—Miraculum vocamus effectum qui nullas sui vires sufficientes in naturâ agnoscit.—*See Bilfinger, De Deo, Animo et Mundo.* From this it has been concluded that the cause must be looked for beyond or out of nature; but reason bids us not to recur to *supernatural causes*, to explain

Order and confusion, then, are only relative terms, by which man designates tne state in which particular beings find themselves. He says, a being is in order when all the motion it undergoes conspires to favour its tendency to self-preservation, and is conducive to the maintenance of its actual existence; that it is in confusion, when the causes which move it disturb the harmony of its existence, or have a tendency to destroy the equilibrium necessary to the conservation of its actual state. Nevertheless, confusion, as we have shown, is nothing but the passage of a being into a new order; the more rapid the progress, the greater the confusion for the being that is submitted to it: that which conducts man to what is called death, is, for him, the greatest of all possible confusion. Yet this death is nothing more than a passage into a new mode of existence: it is in the order of nature.

The human body is said to be in order, when its various component parts act in that mode from which results the conservation of the whole, which is the end of his actual existence.* He is said to be in health, when the fluids and solids of his body concur towards this end. ·He is said to be in confusion, or in ill health, whenever this tendency is disturbed; when any of the constituent parts of his body cease to concur to his preservation, or to fulfil his peculiar functions. This it is that happens in a state of sickness, in which, however, the motion excited in the human machine is as necessary, is regulated by laws as certain, as natural, as invariable, as that which concurs to·produce health. Sickness merely produces in him a new order of motion, a new series of action, a new chain of things. Man dies: to us this appears the greatest confusion he can experience; his body is no longer what it was—its parts cease to concur to the same end—his blood has lost its circulation—he is deprived of feeling—his ideas have

the phenomena we behold, before we have become fully acquainted with *natural causes*—in other words, with the powers and capabilities which nature herself contains.

* In other words, when all the impulse he receives, all the motion he communicates, tends to preserve his health and to render him happy, by promoting the happiness of his fellow men.

vanished—he thinks no more—his desires have fled—death is the epoch, is the cessation of his human existence.—His frame becomes an inanimate mass by the substraction of those principles by which it was animated; its tendency has received a new direction, and the motion excited in its ruins conspires to a new end. To that motion, the harmony of which produced life, sentiment, thought, passions, and health, succeeds a series of motion of another species, which, nevertheless, follows laws as necessary as the first: all the parts of the dead man conspire to produce what is called dissolution, fermentation, putrefaction; and these new modes of being, of acting, are just as natural to man, reduced to this state, as sensibility, thought, the periodical motion of the blood, &c. were to the living man: his essence having changed, his mode of action can no longer be the same. To that regulated motion, to that necessary action, which conspired to the production of life, succeeds that determinate motion, that series of action, which concur to produce the dissolution of the dead carcass, the dispersion of its parts, and the formation of new combinations, from which result new beings: and this, as we have before seen, is the immutable order of ever-active nature.†

It cannot, then, be too often repeated, that, relatively to the great whole, all the motion of beings, all their modes of action, can never be but in order, that is to say, are always conformable to nature: that in all the stages through which beings are obliged to pass, they invariably act after a mode necessarily subordinate to the universal whole. Nay, each individual being always acts in order; all its actions, the whole system of its motion, are the necessary consequence of its peculiar mode of existence, whether that be momentary

† "We have accustomed ourselves to think," says an anonymous author, "that life is the contrary of death; and this appearing to us under the idea of absolute destruction, we have been eager at least to exempt the soul from it, as if the soul, or mind, were essentially any thing else but the result of life, whose opposites are *animate* and *inanimate*. Death is so little opposed to life, that it is the principle of it. From the body of a single animal that ceases to live, a thousand other living beings are formed." See *Miscellaneous Dissertations*: Amsterdam. 1740 pp. 252, 253.

or durable. Order, in political society, is the effect of a necessary series of ideas, of wills, of actions, in those who compose it, whose movements are regulated in a manner either calculated to maintain its indivisibility, or to hasten its dissolution. Man constituted or modified in the manner we term virtuous, acts necessarily in that mode from whence results the welfare of his associates: the man we style wicked, acts necessarily in that mode from whence springs the misery of his fellows: his nature and his modification being essentially different, he must necessarily act after a different mode: his individual order is at variance, but his relative order is complete: it is equally the essence of the one to promote happiness, as it is of the other to induce misery.

Thus order and confusion in individual beings, are nothing more than the manner of man's considering the natural and necessary effects which they produce relatively to himself. He fears the wicked man; he says that he will carry confusion into society, because he disturbs its tendency; because he places obstacles to its happiness. He avoids a falling stone, because it will derange in him the order necessary to his conservation. Nevertheless, order and confusion are always, as we have shown, consequences equally necessary to either the transient or durable state of beings. It is in order that fire burns, because it is of its essence to burn; for the wicked to do mischief, because it is of his essence to do mischief: on the other hand, it is in order that an intelligent being should remove himself from whatever can disturb his mode of existence. A being, whose organization renders him sensible, must, in virtue of his essence, fly from every thing that can injure his organs, that can place his existence in danger.

Man calls those beings *intelligent* who are organized after his own manner, in whom he sees faculties proper for their preservation, suitable to maintain their existence in the order that is convenient to them, enabling them to take the necessary measures towards this end with a consciousness of the motion they undergo. From hence it will be perceived, that the faculty called intelligence, consists in a capability to

act conformably to a known end in the being to which it is attributed. He looks upon those beings as deprived of intelligence in whom he finds no conformity with himself; in whom he discovers neither the same organization, nor the same faculties: of which he knows neither the essence, the end to which they tend, the energies by which they act, nor the order that is convenient to them. The whole cannot have a distinct end, because there is nothing out of itself to which it can have a tendency. If it be in himself that he arranges the idea of *order*, it is also in himself that he draws up that of *intelligence*. He refuses to ascribe it to those beings who do not act after his own manner: he accords it to all those whom he supposes to act like himself: the latter he calls intelligent agents; the former blind causes; that is to say, intelligent agents who act by CHANCE—a word void of sense, but which is always opposed to that of intelligence, without attaching to it any determinate or certain idea.*

In fact, he attributes to *chance* all those effects of which the connexion they have with their causes is not seen. Thus man uses the word *chance* to cover his ignorance of those natural causes which produce visible effects, by means of which he cannot form an idea; or that act by a mode of which he does not perceive the order; or whose system is not followed by actions conformable to his own. As soon as he sees, or believes he sees the order of action, he attributes this order to an intelligence; which is nothing more than a quality borrowed from himself,

* We always compare the intelligence of other beings with our own, and if it be not the same, we deny its existence, which is a very gross errour; for, although a being may appear deprived of our own intelligence, he nevertheless has one peculiar to his organization, which leads him, with the greatest impulse possible, towards an end we do not see; and all beings, with regard to the end Nature proposes to herself, are provided with that degree of intelligence necessary to obtain it. To assume that a being is deprived of intelligence, is merely to say that his intelligence is not like ours, and that we do not understand it:—to say that a being acts by *chance*, is merely to confess that we do not see its end, and the place it occupie. in the universal chain of existences. It is quite certain that all beings are possessed of intelligence, albeit we may not understand it; anc it is no less certain that all beings tend to ar end, albeit we may not perceive it.

from his own peculiar mode of action, and from the manner in which he is himself affected.

Thus an *intelligent being* is one who thinks, who wills, who acts, to compass an end. If so, he must have organs and an aim conformable to those of man: therefore, to say that nature is governed by an intelligence, is to affirm that she is governed by a being furnished with organs; seeing that without this organic construction he can neither have sensations, perceptions, ideas, thoughts, will, plan, nor self-understood action.

Man always makes himself the centre of the universe: it is to himself that he relates all he beholds. As soon as he believes he discovers a mode of action that has a conformity with his own, or some phenomenon that interests his feelings, he attributes it to a cause that resembles himself, that acts after his manner, that has similar faculties with those he himself possesses, whose interests are like his own, whose projects are in unison with, and have the same tendency as those he himself indulges: in short, it is from himself, from the properties which act ate him, that he forms the model of this cause. It is thus that man beholds out of his own species nothing but beings who act differently from himself; yet, believes that he remarks in nature an order analogous to his own peculiar ideas: views, conformable to those, which he himself has. He imagines that nature is governed by a cause, whose intelligence is conformable to his own; to whom he ascribes the honour of the order which he believes he witnesses: of those views that fall in with those that are peculiar to himself; of an aim which quadrates with that which is the great end of all his own actions. It is true that man, feeling his incapability to produce the vast, the multiplied effects, of which he witnesses the operation when contemplating the universe, was under the necessity of making a distinction between himself and the cause which he supposed to be the author of such stupendous effects; he believed he removed every difficulty by exaggerating in this cause all those faculties of which he was himself in possession. It was thus, and by degrees, he arrived at

forming an idea of that intelligent cause which he has placed above nature to preside over her action, and to give her that motion of which he has chosen to believe she was in herself incapable. He obstinately persists in always regarding this nature as a heap of dead, inert, formless matter, which has not within itself the power of producing any of those great effects, of those regular phenomena, from which emanates what he styles the *order of the universe.** From whence it may be deduced, that it 'is for want of being acquainted with the powers of nature, with the properties of matter, that man has multiplied beings without necessity : that he has supposed the universe, under the empire of an intelligent cause, of which he is, and perhaps always will be, himself the model : and he only rendered this cause more inconceivable, when he extended in it his own faculties too much. He either annihilates, or renders it altogether impossible, when he would attach to it incompatible qualities, which he is obliged to do to enable him to account for the contradictory and disorderly effects he beholds in the world. In fact, he sees confusion in the world ; yet, notwithstanding this confusion contradicts the plan, the power, the wisdom, the bounty of this intelligence, and the miraculous order which he ascribes to it, he says the extreme beautiful arrangement of the whole obliges him to suppose it to be the work of a sovereign intelligence.†

It will, no doubt, be argued, that as nature contains and produces intelligent beings, either she must be herself

---

* Anaxagoras is said to have been the first who supposed the universe created and governed by an Intelligence. Aristotle reproaches him with having made an automaton of this intelligence; that is, with ascribing to it the production of things only when he was at a loss, for good reasons, to account for their appearance.—See Bayle's Dictionary, *Art. Anaxagoras, Note E.*

† Unable to reconcile this seeming confusion with the benevolence he attaches to this cause, he had recourse to another effort of his imagination ; he made a new cause, to whom he ascribed all the evil, all the misery, resulting from this confusion : still, his own person served for the model, to which he added those deformities which he had learned to hold in disesteem : in multiplying these counter or destroying causes, he peopled Pandemonium

intelligent, or else she must be governed by an intelligent cause. We reply, intelligence is a faculty peculiar to organized beings, that is to say, to beings constituted and combined after a determinate manner, from whence results certain modes of action, which are designated under various names, according to the different effects which these beings produce: wine has not the properties called *wit* and *courage;* nevertheless, it is sometimes seen that it communicates those qualities to men who are supposed to be in themselves entirely devoid of them. It cannot be said that nature is intelligent after the manner of any one of the beings she contains; but she can produce intelligent beings, by assembling matter suitable to form the particular organization, from whose peculiar modes of action will result the faculty called intelligence, who shall be capable of producing those effects which are the necessary consequence of this property. I therefore repeat, that to have intelligence, designs, and views, it is requisite to have ideas: to the production of ideas, organs or senses are necessary: this is what is neither said of nature, nor of the causes he has supposed to preside over her actions. In short, experience proves beyond a doubt that matter, which is regarded as inert and dead, assumes sensible action, intelligence, and life, when it is combined after particular modes.

From what has been said, it must be concluded, that *order* is never more than the necessary, the uniform connexion of causes with their effects; or that series of action which flows from the peculiar properties of beings so long as they remain in a given state—that *confusion* is nothing more than the change of this state—that, in the universe, all is necessarily in order; because every thing acts and moves according to the properties of the beings it contains—that, in nature, there cannot be either confusion, or real evil, since every thing follows the laws of its natural existence—that there is neither *chance*, nor any thing fortuitous in this nature, where no effect is produced without a sufficient cause; where all causes act necessarily according to fixed, to certain laws, which are themselves dependant on the esse -

tial properties of these causes, as well as on the combination or modification which constitutes either their transitory or permanent state—that intelligence is a mode of acting, a method of existence, natural to some particular beings—that, if this intelligence should be attributed to nature, it would then be nothing more than the faculty of conserving herself in active existence by necessary means. In refusing to nature the intelligence he himself enjoys—in rejecting the intelligent cause which is supposed to be the contriver of this nature, or the principle of that *order* he discovers in her course, nothing is given to *chance*, nothing to a blind cause; but every thing he beholds is attributed to real, to known causes, or to such as are easy of comprehension. All that exists is acknowledged to be a consequence of the inherent properties of eternal matter, which, by contact, by blending, by combination, by change of form, produces order and confusion, and all those varieties which assail his sight—it is himself who is blind, when he imagines blind causes—man only manifested his ignorance of the powers and laws of nature, when he attributed any of its effects to *chance*. He did not show a more enlightened mind when he ascribed them to an intelligence, the idea of which is always borrowed from himself, but which is never in conformity with the effects which he attributes to its intervention—he only imagined words to supply the place of things, and believed he understood them by thus obscuring ideas which he never dared either define or analyze.

CHAPTER VI.

Of Man—Of his Distinction into Moral and Physical—Of his Origin.

LET us now apply the general laws we have scrutinized, to those beings of nature who interest us the most. Let us see in what man differs from the other beings by which he is surrounded. Let us examine if he has not certain points in conformity with them that oblige him, notwithstanding the different properties they respectively possess, to act in certain respects according to the universal laws to which every thing is submitted. Finally, let us inquire if the ideas he has formed of himself in meditating on his own peculiar mode of 'existence, be chimerical, or founded in reason.

Man occupies a place amidst that crowd, that multitude of beings, of 'which nature is the assemblage. His essence, that is to say, the peculiar manner of existence by which he is distinguished from other beings, renders him susceptible of various modes of action, of a variety of motion, some of which are simple and visible, others concealed and complicated. His life itself is nothing more than a long series, a succession of necessary and connected motion, which operates perpetual and continual changes in his machine; which has for its principle either causes contained within himself, such as blood, nerves, fibres, flesh, bones, in short, the matter, as well solid as fluid, of which his body is composed—or those exterior causes, which, by acting upon him, modify him diversely; such as the air with which he is encompassed, the aliments by which he is nourished, and all those objects from which he receives any impulse whatever by the impression they make on his senses.

Man, like all other beings in nature, tends to his own preservation—he experiences inert force—he gravitates upon himself—he is attracted by objects that are analogous, and repelled by those that are contrary to him—he seeks after some—he flies or endeavours to remove himself from others. It is this variety of action, this diversity of modification of which the human being is susceptible, that has been designated under such different names, by such varied nomenclature. It will be necessary, presently, to examine these closely and in detail.

However marvellous, however hidden, however complicated, may be the modes of action which the human frame undergoes, whether interiorly or exteriorly; whatever may be, or appear to be the impulse he either receives or communicates, examined closely, it will be found that all his motion, all his operations, all his changes, all his various states, all his revolutions, are constantly regulated by the same laws, which nature has prescribed to all the beings she brings

forth—which she develops—which she enriches with faculties—of which she increases the bulk—which she conserves for a season—which she ends by decomposing or destroying—thus obliging them to change their form. /

Man, in his origin, is an imperceptible point, a speck, of which the parts are without form; of which the mobility, the life, escapes his senses; in short, in which he does not perceive any sign of those qualities called *sentiment, feeling, thought, intelligence, force, reason,* &c. Placed in the womb suitable to his expansion, this point unfolds, extends, increases by the continual addition of matter he attracts that is analogous to his being, which consequently assimilates itself with him. Having quitted this womb, so appropriate to conserve his existence, to unfold his qualities, to strengthen his habit; so competent to give, for a season, consistence to the weak rudiments of his frame; he becomes adult: his body has then acquired a considerable extension of bulk, his motion is marked, his action is visible, he is sensible in all his parts; he is a living, an active mass; that is to say, he feels, thinks, and fulfils the functions peculiar to beings of his species. But how has he become sensible? Because he has been by degrees nourished, enlarged, repaired by the continual attraction that takes place within himself of that kind of matter which is pronounced inert, insensible, inanimate; although continually combining itself with his machine, of which it forms an active whole, that is living, that feels, judges, reasons, wills, deliberates, chooses, elects; with a capability of labouring, more or less efficaciously, to his own individual preservation; that is to say, to the maintenance of the harmony of his natural existence.

All the motion and changes that man experiences in the course of his life, whether it be from exterior objects, or from those substances contained within himself, are either favourable or prejudicial to his existence; either maintain its order, or throw it into confusion; are either in conformity with, or repugnant to the essential tendency of his peculiar mode of being. He is compelled by nature to approve of some, to disapprove of others; some of necessity render him happy, others contribute to his misery; some become the objects of his most ardent desire, others of his determined aversion: some elicit his confidence, others make him tremble with fear.

In all the phenomena man presents, from the moment he quits the womb of his mother, to that wherein he becomes the inhabitant of the silent tomb, he perceives nothing but a succession of necessary causes and effects, which are strictly conformable to those laws common to all the beings in nature. All his modes of action—all his sensations—all his ideas—all his passions— every act of his will—every impulse he either gives or receives, are the necessary consequences of his own peculiar properties, and those which he finds in the various beings by whom he is moved. Every thing he does—every thing that passes within himself, are the effects of inert force—of self-gravitation—of the attractive or repulsive powers contained in his machine—of the tendency he has, in common with other beings, to his own individual preservation; in short, of that energy which is the common property of every being he beholds. Nature, in man, does nothing more than show, in a decided manner, what belongs to the peculiar nature by which he is distinguished from the beings of a different system or order.

The source of those errours into which man has fallen when he has contemplated himself, has its rise, as will presently be shown, in the opinion he has entertained, that he moved by himself—that he always acts by his own natural energy—that in his actions, in the will that gave him impulse, he was independent of the general laws of nature, and of those objects which, frequently without his knowledge, and always in spite of him, are, in obedience to these laws, continually acting upon him. If he had examined himself attentively, he must have acknowledged, that none of the motion he underwent was spontaneous—he must have discovered, that even his birth depended on causes wholly out of the reach of his own powers—that it was without his own consent he entered into the system in which he occupies a place— that, from the moment in which he is

No. II.—6

born, until that in which he dies, he is continually impelled by causes which, in spite of himself, influence his frame, modify his existence, dispose of his conduct. Would not the slightest reflection have sufficed to prove to him, that the fluids and the solids of which his body is composed, as well as that concealed mechanism, which he believes to be independent of exterior causes, are, in fact, perpetually under the influence of these causes; that without them he would find himself in a total incapacity to act? Would he not have seen, that his temperament, his constitution, did in nowise depend on himself—that his passions are the necessary consequence of this temperament—that his will is influenced—his actions determined by these passions; and consequently by opinions which he has not given to himself? His blood more or less heated or abundant, his nerves more or less braced, his fibres more or less relaxed, give him dispositions either transitory or durable, which are at every moment decisive of his ideas, of his desires, of his fears, of his motion, whether visible or concealed. And the state in which he finds himself, does it not necessarily depend on the air which surrounds him diversely modified; on the various properties of the aliments which nourish him; on the secret combinations that form themselves in his machine, which either preserve its order, or throw it into confusion? In short, had man fairly studied himself, every thing must have convinced him, that in every moment of his duration, he was nothing more than a passive instrument in the hands of necessity.

Thus it must appear, that where all the causes are linked one to the other, where the whole forms but one immense chain, there cannot be any independent, any isolated energy; any detached power. It follows, then, that nature, always in action, marks out to man each point of the line he is bound to describe. It is nature that elaborates, that combines the elements of which he must be composed.—It is nature that gives him his being, his tendency, his peculiar mode of action.—It is nature that develops him, expands him, strengthens him, and preserves him for a season, during which he is obliged to

fulfil the task imposed on him.—It is
nature, that in his journey through life,
strews on the road those objects, those
events, those adventures, that modify
him in a variety of ways, and give him
impulses which are sometimes agreeable and beneficial, at others prejudicial
and disagreeable.—It is nature, that in
giving him feeling, has endowed him
with capacity to choose the means, and
to take those methods that are most
conducive to his conservation.—It is
nature, who, when he has finished his
career, conducts him to his destruction,
and thus obliges him to undergo the
constant, the universal law, from the
operation of which nothing is exempted.
It is thus, also, motion brings man forth
out of the womb, sustains him for a
season, and at length destroys him, or
obliges him to return into the bosom of
nature, who speedily reproduces him,
scattered under an infinity of forms, in
which each of his particles will, in the
same manner, run over again the different stages, as necessarily as the whole
had before run over those of his preceding existence.

The beings of the human species, as
well as all other beings, are susceptible
of two sorts of motion: the one, that of
the mass, by which an entire body, or
some of its parts, are visibly transferred
from one place to another; the other,
internal and concealed, of some of
which man is sensible, while some
takes place without his knowledge, and
is not even to be guessed at but by
the effect it outwardly produces. In a
machine so extremely complex as man,
formed by the combination of such a
multiplicity of matter, so diversified in
its properties, so different in its proportions, so varied in its modes of action,
the motion necessarily becomes of the
most complicated kind, its dullness, as
well as its rapidity, frequently escapes
the observation of those themselves in
whom it takes place.

Let us not, then, be surprised, if when
man would account to himself for his
existence, for his manner of acting,
finding so many obstacles to encounter,
he invented such strange hypotheses
to explain the concealed spring of his
machine—if when this motion appeared
to him to be different from that of other
bodies, he conceived an idea that he
moved and acted in a manner altogether

distinct from the other beings in nature.
He clearly perceived that his body, as
well as different parts of it, did act;
but, frequently, he was unable to discover what brought them into action:
he then conjectured he contained within
himself a moving principle distinguished from his machine, which secretly
gave an impulse to the springs which
set this machine in motion; that moved
him by its own natural energy; and
that consequently he acted according to
laws totally distinct from those which
regulated the motion of other beings.
He was conscious of certain internal
motion which he could not help feeling;
but how could he conceive that this
invisible motion was so frequently competent to produce such striking effects?
How could he comprehend that a
fugitive idea, an imperceptible act of
thought, could frequently bring his
whole being into trouble and confusion? He fell into the belief, that he
perceived within himself a substance
distinguished from that self, endowed
with a secret force, in which he supposed existed qualities distinctly differing from those of either the visible
causes that acted on his organs, or
those organs themselves. He did not
sufficiently understand, that the primitive cause which makes a stone fall, or
his arm move, are perhaps as difficult
of comprehension, as arduous to be
explained, as those internal impulses
of which his thought or his will are
the effects. Thus, for want of meditating nature—of considering her under
her true point of view—of remarking
the conformity and noticing the simul-
taneity of the motion of this fancied
motive-power with that of his body
and of his material organs—he conjectured he was not only a distinct
being, but that he was set apart, with
different energies, from all the other
beings in nature; that he was of a more
simple essence, having nothing in common with any thing that he beheld.*

* "We must," says an anonymous writer,
"define life, before we can reason upon the
soul: but this is what I esteem impossible,
because there are things in nature so simple
that imagination cannot divide them, nor reduce them to any thing more simple than
themselves: such is *life, whiteness,* and *light,*
which we have not been able to define but by
their effects."—*See Miscellaneous Dissertations, printed at Amsterdam,* 1740, page 232.—

It is from thence his notions of *spi-rituality, immateriality, immortality,* have successively sprung; in short, all those vague unmeaning words he has invented by degrees, in order to subtilize and designate the attributes of the unknown power which he believes he contains within himself, and which he conjectures to be the concealed principle of all his visible actions.* To crown the bold conjectures he ventured to make on this internal motive-power, he supposed that different from all other beings, even from the body that served to envelop it, it was not bound to undergo dissolution; that such was its perfect simplicity, that it could not be decomposed, nor even change its form; in short, that it was by its essence exempted from those revolutions to which he saw the body subjected, as well as all the compound beings with which nature is filled.

Thus man became double; he looked upon himself as a whole, composed by the inconceivable assemblage of two distinct natures, which had no point of analogy between themselves: he distinguished two substances in himself; one evidently submitted to the influence of gross beings, composed of coarse inert matter: this he called *body:*—the other, which he supposed to be simple, and of a purer essence, was contemplated as acting from itself, and giving motion to the body with which it found itself so miraculously united: this he called *soul or spirit:* the functions of the one he denominated *physical, corporeal, material;* the functions of the other he styled *spiritual, intellectual.* Man, considered relatively to the first, was termed the *physical man;* viewed with relation to the last, he was designated the *moral man.*

These distinctions, although adopted by the greater number of the philoso-

Life is the assemblage of motion natural to an organized being, and motion can only be a property of matter.

* When man once imbibes an idea he cannot comprehend, he meditates upon it until he has given it a complete personification. Thus he saw, or fancied he saw, the igneous matter pervade every thing; he conjectured that it was the only principle of life and activity; and proceeding to imbody it, he gave it his own form, called it JUPITER, and ended by worshipping this image of his own creation as the power from whom he derived every good he experienced, every evil he sustained.

phers of the present day, are only founded on gratuitous suppositions. Man has always believed he remedied his ignorance of things by inventing words to which he could never attach any true sense or meaning. He imagined he understood matter, its properties, its faculties, its resources, its different combinations, because he had a superficial glimpse of some of its qualities: he has, however, in reality done nothing more than obscure the faint ideas he has been capacitated to form of this matter, by associating it with a substance much less intelligible than itself. It is thus speculative man, in forming words, in multiplying beings, has only plunged himself into greater difficulties than those he endeavoured to avoid, and thereby placed obstacles to the progress of his knowledge: whenever he has been deficient of facts, he has had recourse to conjecture, which he quickly changed into fancied realities. Thus, his imagination no longer guided by experience, was lost, without hope of return, in the labyrinth of an ideal and intellectual world, to which he had himself given birth; it was next to impossible to withdraw him from this delusion, to place him in the right road of which nothing but experience can furnish him the clue. Nature points out, that in man himself, as well as in all those objects which act upon him, there is nothing more than matter endowed with various properties, diversely modified, and acting by reason of these properties: that man is an organized whole, composed of a variety of matter; that, like all the other productions of nature, he follows general and known laws, as well as those laws or modes of action which are peculiar to himself, and unknown.

Thus, when it shall be inquired, what is man?

We say, he is a material being, organized after a peculiar manner; conformed to a certain mode of thinking, of feeling, capable of modification in certain modes peculiar to himself, to his organization, to that particular combination of matter which is found assembled in him.

If, again, it be asked, what origin we give to beings of the human species?

We reply, that, like all other beings, man is a production of nature, who resembles the.n in some respects, and finds himself submitted to the same laws; who differs from them in other respects, and follows particular laws determined by the diversity of his conformation.

If, then, it be demanded, whence came man?*

We answer, our experience on this head does not capacitate us to resolve the question; but that it cannot interest us, as it suffices for us to know that man exists, and that he is so constituted as to be competent to the effects we witness.

But it will be urged, has man always existed? Has the human species existed from all eternity, or is it only an instantaneous production of nature? Have there been always men like ourselves? Will there always be such? Have there been, in all times, males and females? Was there a first man, from whom all others are descended? Was the animal anterior to the egg, or did the egg precede the animal? Is this species without beginning? Will it also be without end? The species itself, is it indestructible, or does it pass away like its individuals? Has man always been what he now is, or has he, before he arrived at the state in which we see him, been obliged to pass under an infinity of successive developments? Can man at last flatter himself with having arrived at a fixed being, or must the human species again change? If man is the production of nature, it will perhaps be asked, Is this nature competent to the production of new beings, and to make the old species disappear? Adopting this supposition, it may be inquired, why nature does not produce under our eyes new beings, new species?

It would appear on reviewing these questions, to be perfectly indifferent, as to the stability of the argument we have used, which side was taken: for want of experience, hypothesis must settle a curiosity that always endeavours to

spring forward beyond the boundaries prescribed to our mind. This granted, the contemplator of nature will say, that he sees no contradiction in supposing the human species, such as it is at the present day, was either produced in the course of time, or from all eternity: he will not perceive any advantage that can arise from supposing that it has arrived by different stages, or successive developments, to that state in which it is actually found. Matter is eternal, and necessary, but its forms are evanescent and contingent. It may be asked of man, is he any thing more than matter combined, of which the form varies every instant?

Notwithstanding, some reflections seem to favour the supposition, and to render more probable the hypothesis that man is a production formed in the course of time; who is peculiar to the globe he inhabits, and the result of the peculiar laws by which it is directed; who, consequently, can only date his formation as coeval with that of his planet. Existence is essential to the universe, or to the total assemblage of matter essentially varied that presents itself to our contemplation; but the combinations, the forms, are not essential. This granted, although the matter of which the earth is composed has always existed, this earth may not always have had its present form and its actual properties—perhaps, it may be a mass detached in the course of time from some other celestial body;—perhaps, it is the result of the spots or encrustations which astronomers discover in the sun's disk, which have had the faculty to diffuse themselves over our planetary system—perhaps, the sphere we inhabit may be an extinguished or a displaced comet, which heretofore occupied some other place in the regions of space, and which, consequently, was then competent to produce beings very different from those we now behold spread over its surface, seeing that its then position, its nature, must have rendered its productions different from those which, at this day, it offers to our view.

Whatever may be the supposition adopted, plants, animals, men, can only be regarded as productions inherent in and natural to our globe, in the position or in the circumstances in which it is actually found: these productions would

* Theologians will, without hesitation, answer this question in the most dogmatic and positive manner. Not only they will tell you *whence* man came, but also *how* and *who* brought him into existence; and what he said and what he did when he first walked the earth. However, true philosophy says—"*I do not know.*"

be changed, if this globe, by any revolution, should happen to shift its situation. What appears to strengthen this hypothesis, is, that on our ball itself, all the productions vary by reason of its different climates: men, animals, vegetables, minerals, are not the same on every part of it: they vary sometimes in a very sensible manner, at very inconsiderable distances. The elephant is indigenous to, or a native of the torrid zone: the reindeer is peculiar to the frozen climates of the north: Indostan is the womb that matures the diamond; we do not find it produced in our own country: the pineapple grows in the common atmosphere of America; in our climate it is never produced until art has furnished a sun analogous to that which it requires. Lastly, man, in different climates, varies in his colour, in his size, in his conformation, in his powers, in his industry, in his courage, in the faculties of his mind. But, what is it that constitutes climate? It is the different position of parts of the same globe relatively to the sun; positions that suffice to make a sensible variety in its productions.

There is, then, sufficient foundation to conjecture, that, if by any accident our globe should become displaced, all its productions would of necessity be changed; for, causes being no longer the same, or no longer acting after the same manner, the effects would necessarily no longer be what they now are: all productions, that they may be able to conserve themselves, or maintain their actual existence, have occasion to co-order themselves with the whole from which they have emanated: without this, they would no longer be in a capacity to subsist. It is this faculty of co-ordering themselves,—this relative adaptation, which is called the *order of the universe*, the want of it is called *confusion*. Those productions which are treated as *monstrous*, are such as are unable to co-order themselves with the general or particular laws of the beings who surround them, or with the whole in which they find themselves placed: they have had the faculty in their formation to accommodate themselves to these laws; but these very laws are opposed to their perfection: for this reason, they are unable to subsist. It is thus, that, by a certain

analogy of conformation which exists between animals of different species, mules are easily produced; but these mules cannot propagate their species. Man can live only in air, fish only in water. Put the man into the water, the fish into the air, not being able to co-order themselves with the fluids which surround them, these animals will quickly be destroyed. Transport, by imagination, a man from our planet into *Saturn*, his lungs will presently be rent by an atmosphere too rarefied for his mode of being, his members will be frozen with the intensity of the cold; he will perish for want of finding elements analogous to his actual existence: transport another into *Mercury*, the excess of heat will quickly destroy him.

Thus, every thing seems to authorize the conjecture that the human species is a production peculiar to our sphere, in the position in which it is found: that, when this position may happen to change, the human species will, of consequence, either be changed, or will be obliged to disappear; for then, there would not be that with which man could co-order himself with the whole, or connect himself with that which can enable him to subsist. It is this aptitude in man to co-order himself with the whole, that not only furnishes him with the idea of order, but also makes him exclaim, *Whatever is, is right*, whilst every thing is only that which it can be, and the whole is necessarily what it is, and whilst it is positively neither good nor bad. It is only requisite to displace a man to make him accuse the universe of confusion.

These reflections would appear to contradict the ideas of those who are willing to conjecture that the other planets, like our own, are inhabited by beings resembling ourselves. But if the Laplander differs in so marked a manner from the Hottentot, what difference ought we not rationally to suppose between an inhabitant of our planet and one of Saturn or of Venus?

However, if we are obliged to recur, by imagination, to the origin of things, to the infancy of the human species, we may say, that it is probable man was a necessary consequence of the disentangling of our globe, or one of the results of the qualities, of the properties, of the energies of which it is susceptible in its present position;—that he was born male and female;—that his existence is co-ordinate with that of the globe, under its present position;—that as long as this co-ordination shall subsist, the human species will conserve himself, will propagate himself, according to the impulse and the primitive laws which he has originally received—that, if this co-ordination should happen to cease; if the earth, displaced, should cease to receive the same impulse, the same influence, on the part of those causes which actually act upon it and give it energy; that then, the human species would change to make place for new beings suitable to co-order themselves with the state that should succeed to that which we now see subsist.

In thus supposing changes in the position of our globe, the primitive man did, perhaps, differ more from the actual man than the quadruped differs from the insect. Thus, man, the same as every thing else that exists on our planet, as well as in all the others, may be regarded as in a state of continual vicissitude: thus, the last term of the existence of man, is, to us, as unknown, as indistinct, as the first: there is, therefore, no contradiction in the belief, that the species vary incessantly; and it is as impossible to know what he will become, as to know what he has been.

With respect to those who may ask, why nature does not produce new beings? we inquire of them in turn, upon what foundation they suppose this fact? What is it that authorizes them to believe this sterility in nature? Know they, if, in the various combinations which she is every instant forming, nature be not occupied in producing new beings without the cognizance of these observers? Who has informed them that this nature is not actually assembling in her immense elaboratory the elements suitable to bring to light generations entirely new, that will have nothing in common with those of the species at present existing?* What

absurdity, then, or what want of just inference would there be to imagine, that man, the horse, the fish, the bird, will be no more! Are these animals so indispensably requisite to nature, that without them she cannot continue her eternal course? Does not all change around us? Do we not change ourselves? Is it not evident that the whole universe has not been, in its anterior eternal duration, rigorously the same that it now is; that it is impossible, in its posterior eternal duration, it can be rigidly in the same state that it now is for a single instant? How, then, pretend to divine the infinite succession of destruction, of reproduction, of combination, of dissolution, of metamorphosis, of change, of transposition, which may eventually take place? Suns encrust themselves, and are extinguished; planets perish and disperse themselves in the vast plains of air; other suns are kindled; new planets form themselves, either to make revolutions round these suns, or to describe new routes; and man, an infinitely small portion of the globe, which is itself but an imperceptible point in the immensity of space, vainly believes it is for himself this universe is made; foolishly imagines he ought to be the confidant of nature; confidently flatters himself he is eternal, and calls himself KING OF THE UNIVERSE!!

O man! wilt thou never conceive that thou art but an ephemeron? All changes in the universe: nature contains no one constant form, yet thou pretendest that thy species can never disappear; that thou shalt be exempted from the universal law, that wills all shall experience change! Alas! in thy actual being, art thou not submitted to continual alterations? Thou, who in thy folly arrogantly assumest to thyself the title of KING OF NATURE! Thou, who measurest the earth and the heavens! Thou, who in thy vanity imaginest that the whole was made because thou art intelligent! there requires but a very slight accident, a single atom to be displaced, to make thee perish; to degrade thee; to ravish from thee this intelligence of which thou appearest so proud.

---

* How do we know that the various beings and productions said to have been created at the same time with man, are not the posterior and spontaneous production of Nature? Four thousand years ago man became acquainted with the lion:—well! what are four thousand years? Who can prove that the lion, seen for

the first time by man four thousand years ago, had not then been in existence thousands of years? or again, that this lion was not produced thousands of years after the proud biped who arrogantly calls himself king of the universe?

If all the preceding conjectures be refused ; if it be pretended that nature acts by a certain quantum of immutable and general laws ; if it be believed that men, quadrupeds, fish, insects, plants, are from all eternity, and will remain eternally what they now are : if it be contended, that from all eternity the stars have shone in the immense regions of space ; if it be insisted that we must no more demand why man is such as he appears than ask why nature is such as we'behold her, or why the world exists ; we shall no longer oppose such arguments. Whatever may be the system adopted, it will perhaps reply equally well to the difficulties with which our opponents endeavour to embarrass the way : examined closely, it will be perceived they make nothing against those truths which we have gathered from experience. It is not given to man to know every thing : it is not given him to know his origin : it is not given him to penetrate into the essence of things, nor to recur to first principles ; but it is given him to have reason, to have honesty, to ingenuously allow he is ignorant of that which he cannot know, and not to substitute unintelligible words and absurd suppositions for' his uncertainty. Thus we say to those who, to solve difficulties, pretend that the human species descended from a first man and a first woman, created by a God, that we have some ideas of nature, but that we have none of the Divinity nor of creation, and that to use these words, is only in other terms to acknowledge our ignorance of the powers of nature, and our inability to fathom the means by which she has been capacitated to produce the phenomena we behold.*

Let us then conclude, that man has no reason to believe himself a privileged being in nature, for he is subject to the same vicissitudes as all her other productions. His pretended prerogatives have their foundation in errour. Let him but elevate himself, by his thoughts, above the globe he inhabits, and he will look upon his own species

with the same eyes he does all the other beings in nature. He will then clearly perceive that in the same manner each tree produces its fruit in consequence of its species, so each man acts by reason of his particular energy, and produces fruit, actions, works, equally necessary : he will feel, that the illusion which gives him such an exalted opinion of himself, arises from his being, at one and the same time a spectator and a part of the universe. He will acknowledge, that the idea of excellence which he attaches to his being, has no other foundation than his own peculiar interest, and the predilection he has in favour of himself.†

CHAPTER VII.

Of the Soul, and of the Spiritual System.

Man, after having gratuitously supposed himself composed of two distinct independent substances, having no common properties relatively with each other, has pretended, as we have seen, that that which actuated him interiorly, that motion which is invisible, that impulse which is placed within himself, is essentially different from those which act exteriorly. The first he designated, as we have already said, by the name of a *spirit*, or a *soul*. If, however, it be asked, what is a spirit ? the moderns will reply, that the whole fruit of their metaphysical researches is limited to learning that this motive-power, which they state to be the spring of man's action, is a substance of an unknown nature, so simple, so indivisible, so deprived of extent, so invisible, so impossible to be discovered by the senses, that its parts cannot be separated, even by abstraction or thought. But how can we conceive such a substance, which is only the negation of every thing of which we have a knowledge ? How form to ourselves an idea of a substance void of extent, yet acting on our senses ; that is to say, on material organs which

---

* Ut Tragici poetæ confugiunt ad Deum aliquem, cum aliter explicare argumenti exitum non possunt. *Cicero, de Divinatione* Lib. 2. He again says, magna stultitia est earum rerum Deos facere effectores, causas rerum non quærere.—*Ib.*

† In Nature nothing is mean or contemptible, and it is only pride, originating in a false idea of our superiority, which causes our contempt for some of her productions. In the eyes of Nature, however, the oyster that vegetates at the bottom of the sea is as dear and perfect as the proud biped who devours it.

nave extent? How can a being without extent be moveable and put matter in action? How can a substance, devoid of parts, correspond successively with different parts of space?

At any rate all men are agreed in this position, that motion is the successive change of the relations of one body with other bodies, or with the different parts of space. If that, which is called *spirit*, be susceptible of communicating or receiving motion; if it acts—if it gives play to the organs or body—to produce these effects it necessarily follows, that this being changes successively its relation, its tendency, its correspondence, the position of its parts, either relatively to the different points of space, or to the different organs of the body which it puts in action; but to change its relation with space and with the organs to which it gives impulse, this spirit must have extent, solidity, consequently these parts: whenever a substance possesses these qualities, it is what we call *matter*, and can no longer be regarded as a simple pure being in the sense attached to it by the moderns.*

* A very cogent question presents itself on this occasion: if this distinct substance, said to form one of the component parts of man, be really what it is reported, and if it be not, it is not what it is described; if it be unknown, if it be not pervious to the senses; if it be invisible, by what means did the metaphysicians themselves become acquainted with it? How did they form ideas of a substance, that, taking their own account of it, is not, under any of its circumstances, either directly or by analogy cognizable to the mind of man? If they could positively achieve this, there would no longer be any mystery in nature: it would oe as easy to conceive the time when all was nothing, when all shall have passed away, to account for the production of every thing we behold, as to dig in a garden, or read a lecture. Doubt would vanish from the human species; there could no longer be any difference of opinion, since all must necessarily be of one mind on a subject so accessible to every inquirer.

But it will be replied, the materialist himself admits, the natural philosophers of all ages have admitted, elements, atoms, beings simple and indivisible, of which bodies are composed:—granted; they have no more: they have also admitted that many of these atoms, many of these elements, if not all, are unknown to them: nevertheless, these simple beings, these atoms of the materialist, are not the same thing with the spirit, or the soul of the metaphysician. When the natural philosopher talks of atoms; when he describes

Thus it will be seen that those who have supposed in man an immaterial substance, distinguished from his body, have not thoroughly understood themselves; indeed they have done nothing more than imagined a negative quality of which they cannot have any correct idea: matter alone is capable of acting on our senses, and without this action nothing would be capable of making itself known to us. They have not seen that a being without extent, is neither in a capacity to move itself, nor has the capability of communicating motion to the body, since such a being, having no parts, has not the faculty of changing its relation, or its distance, relatively to other bodies, nor of exciting motion in the human body, which is itself material. That which is called our soul, moves itself with us; now motion is a property of matter—this soul gives impulse to the arm; the arm, moved by it, makes an impression, a blow, that follows the general law of motion: in this case, the force remaining the same, if the mass was twofold, the blow would be double. This soul again evinces its materiality in the invincible obstacles it encounters on the

them as simple beings, he indicates nothing more than that they are homogeneous, pure, without mixture: but then he allows that they have extent consequently parts are separable by thought, although no other natural agent with which he is acquainted is capable of dividing them—that the simple beings of this genus are susceptible of motion, can impart action, receive impulse, are material, are placed in nature, are indestructible; that consequently, if he cannot know them from themselves, he can form some idea of them by analogy; thus he has done that intelligibly which the metaphysician would do unintelligibly: the latter, with a view to render man immortal, finding difficulties to his wish, from seeing that the body decayed—that it submitted to the great, the universal law—has, to solve the difficulty, to remove the impediment, given him a soul, distinct from the body, which he says is exempted from the action of the general law: to account for this, he has called it a *spiritual being*, whose properties are the negation of all known properties, consequently inconceivable: had he, however, had recourse to the atoms of the former; had he made this substance the last possible term of the division of matter, it would at least have been intelligible; it would also have been immortal, since, according to the reasonings of all men, whether metaphysicians, theologians, or natural philosophers, an atom is an indestructible element, that must exist to all eternity.

part of the body. If the arm be moved by its impulse when nothing opposes it, yet this arm can no longer move when it is charged with a weight beyond its strength. Here then is a ·mass of matter that annihilates the impulse given by a spiritual cause, which spiritual cause having no analogy with matter, ought not to find more difficulty in moving the whole world than in moving a single atom, nor an atom than the universe. From this it is fair to conclude that such a substance is a chimera; a being of the imagination: nevertheless such is the being the metaphysicians have made the contriver and the author of nature! !*

As soon as I feel an impulse or experience motion, I am under the necessity to acknowledge extent, solidity, density, impenetrability in the substance I see move, or from which I receive impulse : thus, when action is attributed to any cause whatever, I am obliged to consider it *material.* I may be ignorant of its individual nature, of its mode of action, of its generic properties; but I cannot deceive myself in general properties which are common to all matter: besides this ignorance will only be increased, when I shall take that for granted of a being of which I am precluded from forming

any idea, which moreover deprives it completely of the faculty of moving and acting. Thus, a spiritual substance, that moves itself, that gives an impulse to matter, that acts, implies a contradiction, which necessarily infers a total impossibility.

The partizans of spirituality believe they answer the difficulties they have themselves accumulated, by saying, " *The soul is entire, is whole under each point of its extent.*" If an absurd answer will solve difficulties, they have done it ; for after all it will be found, that this point, which is called soul, however insensible, however minute, must yet remain something.† But if as much solidity had appeared in the answer as there is a want of it, it must be acknowledged, that in whatever manner the spirit or the soul finds itself in its extent, when the body moves forward, the soul does not remain behind ; if so, it has a quality in common with the body peculiar to matter, since it is transferred from place to place jointly with the body. Thus, if even the soul should be immaterial, what conclusion must be drawn ? Entirely submitted to the motion of the body without this body it would remain dead and inert. This soul would only be part of a twofold machine, necessarily

* As man, in all his speculations, takes himself for the model, he no sooner imagined a spirit within himself, than giving it extent, he made it universal, then ascribed to it all those causes with which his ignorance prevents him from becoming acquainted : thus he identified himself with the supposed author of nature ; then availed himself of the supposition to explain the connexion of the soul with the body. His self-complacency prevented his perceiving that he was only enlarging the circle of his errours, by pretending to understand that which it is more than probable he will never know: his self-love prevented him from feeling, that, whenever he punished another for not thinking as he did, he committed the greatest injustice, unless he was satisfactorily able to prove that other wrong—himself right: that if he himself was obliged to have recourse to hypothesis, to gratuitous suppositions, whereon to found his doctrine, that from the very fallibility of his nature these might be erroneous: thus GALILEO was persecuted, because the metaphysicians and the theologians of his day chose to make others believe what it was evident they did not themselves understand. As to our modern metaphysicians, they may dream of a *universal spirit* after the manner of the human soul—of an *infinite intelligence* after the

manner of a finite intelligence; but in so doing they do not perceive that this *spirit* or *intelligence,* whether they suppose it finite or infinite, will not be more convenient or fit to move matter.

† According to this answer an infinity of unextended substance, or the same unextended substance repeated an infinity of times, would constitute a substance that has extent, which is absurd; for, according to this principle, the human soul would then be as infinite as God, since it is assumed that God is a being without extent, who is an infinity of times whole in each part of the universe—and the same is stated of the human soul ; from whence we must necessarily conclude that God and the soul of man are equally infinite, unless we suppose unextended substances of *different* extents, or a God without extent more extended than the human soul. Such are, however, the rhapsodies which some of our theological metaphysicians would have thinking beings believe! With a view of making the human soul immortal, these theologians have spiritualized it, and thus rendered it an unintelligible being; had they said that the soul was the minutest division of matter, it would then have been intelligible—and immortal too, since it would have been an *atom,* an indissoluble element

No. II.—7

impelled forward by a concatenation or connexion with the whole. It would resemble a bird, which a child conducts at its pleasure by the string with which it is bound.

Thus, it is for want of consulting experience, and by not attending to reason, that man has obscured his ideas upon the concealed principle of his motion. If, disentangled from prejudice, he would contemplate his soul, or the moving principle that acts within him, he would be convinced that it forms part of his body; that it cannot be distinguished from it but by abstraction; and that it is only the body itself considered relatively with some of its functions, or with those faculties of which its nature and its peculiar organization renders it susceptible. He will also perceive that this soul is obliged to undergo the same changes as the body; that it is born and expands itself with it; that, like the body, it passes through a state of infancy, a period of weakness, a season of inexperience; that it enlarges and strengthens itself in the same progression; that, like the body, it arrives at an adult age, reaches maturity; that it is then it obtains the faculty of fulfilling certain functions, enjoys reason, and displays more or less wit, judgment, and manly activity; that like the body, it is subject to those vicissitudes which exterior causes oblige it to undergo by their influence; that, conjointly with the body, it suffers, enjoys, partakes of its pleasures, shares its pains, is sound when the body is healthy, diseased when the body is oppressed with sickness; that, like the body, it is continually modified by the different degrees of density in the atmosphere; by the variety of the seasons; by the various properties of the aliments received into the stomach: in short, he would be obliged to acknowledge that at some periods, it manifests visible signs of torpor, decrepitude, and death.

In despite of this analogy, or rather this continual identity of the soul with he body, man has been desirous of distinguishing their essence: he has therefore made the soul an inconceivable being; but in order that he might form to himself some idea of it, he was after all obliged to have recourse to material beings and to their manner of acting. In fact, the word *spirit* presents to the mind no other ideas than those of breathing, of respiration, of wind. Thus, when it is said, the *soul is a spirit*, it really means nothing more than that its mode of action is like that of breathing, which, though invisible in itself, or acting without being seen, produces, nevertheless, very visible effects. But breath is a material cause—it is air modified; it is not therefore a simple, a pure substance, such as the moderns designate under the name of *spirit*.*

Although the word *spirit* is so very ancient among men, the sense attached to it by the moderns is quite new; and the idea of spirituality, as admitted at this day, is a recent production of the imagination. Neither Pythagoras nor Plato, however heated their brain, and however decided their taste for the marvellous, appear to have understood by *spirit* an immaterial substance, or one without extent, such as that of which the moderns have formed the human soul, and the concealed author of motion. The ancients, by the word *spirit*, were desirous to define matter of an extreme subtilty, and of a purer quality than that which acted grossly on our senses. In consequence, some have regarded the soul as an ethereal substance; others as igneous matter: others again have compared it to light. Democritus made it consist in motion, consequently gave it a mode of existence. Aristoxenes, who was himself a musician, made it harmony. Aristotle regarded the soul as the moving faculty upon which depended the motion of living bodies.

The earliest doctors of Christianity had no other idea of the soul than that it was material.† Tertullian, Arnobius,

---

* The Hebrew word *Ruach*, signifies breath, respiration. The Greek word Πνευμα, means the same thing, and is derived from πνιωω, *spiro*. Lactantius states that the Latin word *anima* comes from the Greek word ἀνεμος which signifies wind. Some metaphysicians fearful of seeing too far into human nature, have compounded man of three substances, *body, soul*, and intellect—Σωμα, ψυχη, Νες.— See *Marc. Antonin.*, *Lib.* iii. § 16.

† According to Origen, ασωματος, *incorporeus*, an epithet given to God, signifies a substance more subtile than our of gross bodies. Tertullian says, Quis autem negabit deum esse corpus, etsi deus spiritus? The same Tertullian says, Nos autem animam

Clement of Alexandria, Origen, Saint Justin, Irenæus, have never spoken of it other than as a corporeal substance. It was reserved for their successors, at a great distance of time, to make the human soul, and the soul of the world, *pure spirits;* that is to say, immaterial substances, of which it is impossible to form any accurate idea: by degrees this incomprehensible doctrine of spirituality, conformable without doubt to the views of theologians who make it a principle to annihilate reason, prevailed over the others:* this doctrine was believed divine and supernatural, because it was inconceivable to man. Those who dared believe *that the soul was material,* were held as rash, inconsiderate madmen, or else treated as enemies to the welfare and happiness of the human race. When man had once renounced experience and abjured his reason, he did nothing more, day after day, than subtilize the ravings of his imagination: he pleased himself by continually sinking deeper into the most unfathomable depths of errour; and he felicitated himself on his discoveries, on his pretended knowledge, in an exact ratio as his understanding became enveloped with the clouds of ignorance. Thus, in consequence of man's reasoning upon false principles, the soul, or moving principle within him, as well as the concealed moving principle of Nature, have been made mere chimeras, mere beings of the imagination.†

corporalem et hic profitemur, et in suo volumine probamus, habentem proprium genus substantiæ, soliditatis, per quam quid et sentire et pati possit. V. *De Resurrectione Carnis.*

* The system of spirituality, such as it is admitted at this day, owes all its pretended proofs to Descartes. Although before him the soul had been considered spiritual, he was the first who established that *" that which thinks ought to be distinguished from matter ;"* from whence he concludes that the soul, or that which thinks in man, is a spirit—that is to say, a simple and indivisible substance. Would it not have been more consistent with logic and reason to have said that, since man, who is matter and who has no idea but of matter, enjoys the faculty of thought, matter can think—that is, it is susceptible of that particular modification called *thought.—See Bayle's Dictionary,* Art. *Pomponatius* and *Simonides.*

† Although there is so little reason and philosophy in the system of spirituality, yet we must confess that it required deep cunning on the part of the selfish theologians who invented it. To render man susceptible of

Therefore the doctrine of spirituality offers nothing but vague ideas—or rather is the absence of all ideas. What does it present to the mind, but a substance which possesses nothing of which our senses enable us to have a knowledge? Can it be truth, that man is able to figure to himself a being not material, having neither extent nor parts, which, nevertheless, acts upon matter without having any point of contact, any kind of analogy with it, and which itself receives the impulse of matter by means of material organs, which announce to it the presence of other beings? Is it possible to conceive the union of the soul with the body, and to comprehend how this material body can bind, enclose, constrain, determine a fugitive being which escapes all our senses? Is it honest to solve these difficulties by saying there is a mystery in them; that they are the effects of an omnipotent power more inconceivable than the human soul and its mode of acting? When, to resolve these problems, man is obliged to have recourse to miracles, and to make the Divinity interfere, does he not avow his own ignorance?

Let us not, then, be surprised at those subtle hypotheses, as ingenious as they are unsatisfactory, to which theological prejudice has obliged the most profound modern speculators to recur, when they have undertaken to reconcile the spirituality of the soul with the physical action of material beings on this incorporeal substance, its reaction upon these beings, and its union with the body. When the human mind permits itself to be guided by authority without proof—to be led forward by enthusiasm—when it renounces the evidence of its senses; what can it do more than sink into errour?‡

rewards and punishments after death, it was necessary to exempt some portion of him from corruption and dissolution—a doctrine extremely useful to priests, whose great aim is to intimidate, govern, and plunder the ignorant—a doctrine which enables them even to perplex many enlightened persons, who are equally incapable of comprehending the *"sublime truths"* about the soul and the Divinity! These honest priests tell us, that this *immaterial* soul shall be burnt, or, in other words, shall experience in hell the action of the *material* element of fire, and we believe them upon their word!!!

‡ Those who wish to form an idea of the shackles imposed by theology on the genius

If man wishes to form to himself clear ideas of his soul, let him throw himself back on his experience; let him renounce his prejudices; let him avoid theological conjecture; let him tear the sacred bandage with which he has been blindfolded only to confound his reason. Let the natural philosopher, let the anatomist, let the physician, unite their experience and compare their observations, in order to show what ought to be thought of a substance so disguised under a heap of absurdities: let their discoveries teach moralists the true motive-power that ought to influence the actions of man—legislators, the true motives that should excite him to labour to the welfare of society—sovereigns, the means of rendering truly happy the subjects committed to their charge. Physical souls have physical wants, and demand physical and real happiness, far preferable to that variety of fanciful chimeras with which the mind of man has been fed during so many ages. Let us labour to perfect the morality of man; let us make it agreeable to him; and we shall presently see his morals become better, himself become happier; his mind become calm and serene; his will determined to virtue by the natural and palpable motives held out to him. By the diligence and care which legislators shall bestow on natural philosophy, they will form citizens of sound understanding, robust and well constituted, who, finding themselves happy, will be themselves accessary to that useful impulse so necessary to general happiness. When the body is suffering, when nations are unhappy, the mind cannot be in a proper state. *Mens sana in corpore sano,* a sound mind in a sound body, this always makes a good citizen.

The more man reflects, the more he will be convinced that the soul, very far from being distinguished from the body, is only the body itself considered relatively to some of its functions, or to some of the modes of existing or acting of which it is susceptible whilst it en-

of philosophers born under the " *Christian dispensation,*" let them read the metaphysical romances of Leibnitz, Descartes, Malebranche, Cudworth, etc. and coolly examine the ingenious but rhapsodical systems entitled *the Pre-established harmony of occasional causes; Physical pre-motion,* etc.

joys life. Thus, the soul is man considered relatively to the faculty he has of feeling, of thinking, and of acting in a mode resulting from his peculiar nature; that is to say, from his properties, from his particular organization; from the modifications, whether durable or transitory, which the beings who act upon him cause his machine to undergo.*

Those who have distinguished the soul from the body, appear only to have distinguished their brain from themselves. Indeed, the brain is the common centre where all the nerves, distributed through every part of the body, meet and blend themselves: it is by the aid of this interior organ that all those operations are performed which are attributed to the soul: it is the impulse, the motion, communicated to the nerve, which modifies the, brain: in consequence, it reacts, and gives play to the bodily organs, or rather it acts upon itself, and becomes capable of producing within itself a great variety of motion, which has been designated *intellectual faculties.*

From this it may be seen, that some philosophers have been desirous to make a spiritual substance of the brain.; but it is evidently ignorance that has both given birth to, and accredited this system, which embraces so little of the natural. It is from not having studied himself that man has supposed he was compounded with an agent essentially different from his body: in examining this body he will find that it is quite useless to recur to hypothesis to explain the various phenomena it presents; for

* When a theologian, obstinately bent on admitting into man two substances essentially different, is asked why he multiplies beings without necessity, he will reply, " *Because thought cannot be a property of matter.*" If, then, it be inquired of him, " *Cannot God give to matter the faculty of thought?*" he will answer, " *No! seeing that God cannot do impossible things!*" But this is atheism, for, according to his principles, it is as impossible that spirit or thought can produce matter, as it is impossible that matter can produce spirit or thought: it must, therefore, be concluded against him, that the world was not made by a spirit, any more than a spirit was made by the world; that the world is eternal, and if an eternal spirit exists, then we have two eternal beings, which is absurd. If, therefore, there is only one eternal substance, it is the world, whose existence cannot be doubted or denied.

nypothesis can do nothing more than lead him out of the right road. What obscures this question, arises from this, that man cannot see himself: indeed, for this purpose it would be requisite that he could be at one and the same moment both within and without himself. Man may be compared to an Eolian harp, that issues sounds of it self, and should demand what it is that causes it to give them forth ? it does not perceive that the sensitive qualit) of its chords causes the air to brace them ; that being so braced, it is rendered sonorous by every gust of wind with which it comes in contact.

The more experience we collect, the more we shall be convinced that the word *spirit* conveys no one sense even to those that invented it; consequently, cannot be of the least use either in physics or morals. What modern metaphysicians believe and understand by the word, is in truth nothing more than an *occult* power, imagined to explain *occult* qualities and actions, but which, in fact, explains nothing. Savage nations admit of spirits to account to themselves for those effects which to them appear marvellous, and the cause of which they ignore. In attributing to *spirits* the phenomena of nature, as well as those of the human body, do we, in fact, do any thing more than reason like savages ? Man has filled nature with *spirits*, because he has almost always been ignorant of the true causes of those effects by which he was astonished. Not being acquainted with the powers of nature, he has supposed her to be animated by a *great spirit:* not understanding the energy of the human frame, he has, in like manner, conjectured it to be animated by a *spirit :* from this it would appear, that whenever he wished to indicate the unknown cause of .the phenomena he knew not how to explain in a natural manner, he had recourse to the word *spirit.* It was according to these principles, that when the Americans first beheld the terrible effects of gunpowder, they ascribed the cause to their Spirits or Divinities: it is by adopting these principles that we now believe in Angels and Demons, and that our ancestors believed in a plurality of Gods, in ghosts, in genii, &c., and pursuing the same track, we

ought to attribute to *spirits* gravitation, electricity, magnetism, &c., &c.*

CHAPTER VIII.

Of the Intellectual Faculties ; they are all de rived from the Faculty of Feeling.

To convince ourselves that the faculties called *intellectual*, are only certain modes of existence, or determinate manners of acting which result from the peculiar organization of the body, we have only to analyze them: we shall then see, that all the operations which are attributed to the soul, are nothing more than certain modifications of the body, of which a substance that is without extent, that has no parts, that is immaterial, is not susceptible.

The first faculty we behold in the living man, that from which all his others flow, is *feeling :* however inexplicable this faculty may appear on a first view, if it be examined closely, it will be found to be a consequence of the essence, a result of the properties of organized beings; the same as *gravity, magnetism, elasticity, electricity,* &c. result from the essence or nature of some others ; and we shall also find that these last phenomena are not less inexplicable than that of feeling. Nevertheless, if we wish to define to ourselves a precise idea of it, we shall find that feeling is a particular manner of being moved peculiar to certain organs of animated bodies, occasioned by the presence of a material object that acts upon these organs, and which transmits the impulse or shock to the brain.

* It is evident that the notion of *spirits,* imagined by savages and adopted by the ignorant, is calculated to retard the progress of knowledge, since it precludes our researches into the true cause of the effects which we see, by keeping the human mind in apathy and sloth. This state of ignorance may be very useful to crafty theologians, but very injurious to society. This is the reason, however, why in all ages priests have persecuted those who have been the first to give natural explanations of the phenomena of nature—as witness Anaxagoras, Aristotle, Galileo, Descartes—and, more recently, Richard Carlile, William Lawrence, Robert Taylor, and Abner Kneeland; to which we may add the name of the learned and venerable Thomas Cooper M. D., lately president of Columbia College, South Carolina.

Man only feels by the aid of nerves dispersed through his body, which is itself, to speak correctly, nothing more than a great nerve ; or may be said to resemble a large tree, of which the branches experience the action of the root communicated through the trunk. In man the nerves unite and loose themselves in the brain ; that intestine is the true seat of feeling: like the spider suspended in the centre of his web, it is quickly warned of all the changes that happen to the body, even at the extremities to which it sends its filaments and branches. Experience enables us to ascertain that man ceases to feel in those parts of his body of which the communication with the brain is intercepted ; he feels very little, or not at all, whenever this organ is itself deranged or affected in too lively a manner.*

However this may be, the sensibility of the brain, and of all its parts, is a fact. If it be asked, whence comes this property ? We shall reply, it is the result of an arrangement, of a combination, peculiar to the animal ; insomuch, that coarse and insensible matter ceases to be so by animalizing itself, that is to say, by combining and identifying itself with the animal. It is thus that milk, bread, wine, change themselves in the substance of man, who is a sensible being : this insensible matter becomes sensible in combining itself with a sensible whole. Some philosophers

think that sensibility is a universal quality of matter: in this case it would be useless to seek from whence this property is derived, as we know it by its effects. If this hypothes·s be admitted, in like manner as two kinds of motion are distinguished in nature, the one called live force, the other dead, or inert force, two sorts of sensibility will be distinguished—the one active or live, the other inert or dead. Then to animalize a substance, is only to destroy the obstacles that prevent its being active or sensible. In fact, sensibility is either a quality which communicates itself like motion, and which is acquired by combination ; or this sensibility is a property inherent in all matter: in both, or either case, an unextended being, without parts, such as the human soul is said to be, can neither be the cause of it, nor submitted to its operation.†

The conformation, the arrangement, the texture, the delicacy of the organs, as well exterior as interior, which compose men and animals, render their parts extremely mobile, and make their machine susceptible of being moved with great facility. In a body, which is only a heap of fibres, a mass of nerves, contiguous one to the other, and united in a common centre, always ready to act ; in a whole, composed of fluids and of solids, of which the parts are in equilibrium ; of which the smallest touch each other, are active, rapid in their motion, communicating reciprocally,

* A proof of this is afforded in the Transactions of the Royal Academy of Sciences at Paris: they inform us of a man, who had his scull taken off, in the room of which his brain was re-covered with skin ; and in proportion as a pressure was made by the hand on his brain, the man fell into a kind of insensibility which deprived him of all feeling. Bartolin says, the brain of a man is twice as big as that of an ox. This observation had been already made by Aristotle. In the dead body of an idiot dissected by Willis, the brain was found smaller than ordinary : he says, the greatest difference he found between the parts of the body of this idiot, and those of wiser men, was, that the plexus of the intercostal nerves, which is the mediator between the brain and the heart, was extremely small, accompanied by a less number of nerves than usual. According to Willis, the ape is of all animals that which has the largest brain, relatively to his size : he is also, after man, that which has the most intelligence ; and this is further confirmed by the name he bears in the soil to which he is indigenous, which is *orang outang,* or the man beast. There is, therefore,

every reason to believe that it is entirely in the brain that consists the difference that is found not only between man and beasts, but also between the man of wit and the fool ; between the thinking man and he who is ignorant; between the man of sound understanding and the madman. And again, a multitude of experience proves that those persons who are most accustomed to use their intellectual faculties, have their brain more extended than others: the same has been remarked of watermen or rowers, that they have arms much larger than other men.

† All the parts of nature enjoy the capability to arrive at animation; the obstacle is only in the state not in the quality. Life is the perfection of nature: she has no parts which do not tend to it, and which do not ·attain it by the same means. Life, in an insect, a dog, a man, has no other difference than that this act is more perfect, relatively to ourselves. in proportion to the structure of the organs: if, therefore, it be asked, what is requisite to animate a body ? we reply, it needs no foreign aid, it is sufficient that the power of nature be joined to its organization.

alternately and in succession, the impressions, the oscillations, the shocks they receive; in such a composition, I say, it is not at all surprising that the slightest impulse propagates itself with celerity; that the shocks excited in its remotest parts make themselves quickly felt in the brain, whose delicate texture renders it susceptible of being itself very easily modified. Air, fire, water, agents the most inconstant, possessing the most rapid motion, circulate continually in the fibres, incessantly penetrate the nerves, and without doubt contribute to that incredible celerity with which the brain is acquainted with what passes at the extremities of the body.

Notwithstanding the great mobility of which man's organization renders him susceptible; although exterior as well as interior causes are continually acting upon him, he does not always feel in a distinct, in a decided manner, the impulse given to his senses: indeed, he does not feel it until it has produced some change, or given some shock to his brain. Thus, although completely environed by air, he does not feel its action until it is so modified as to strike with a sufficient degree of force on his organs and his skin, through which his brain is warned of its presence. Thus, during a profound and tranquil sleep, undisturbed by any dream, man ceases to feel. In short, notwithstanding the continued motion that agitates his frame, man does not appear to feel when this motion acts in a convenient order; he does not perceive a state of health, but he discovers a state of grief or sickness; because, in the first, his brain does not receive too lively an impulse, whilst in the others his nerves are contracted, shocked, agitated, with violent and disorderly motion, thus giving notice that some cause acts strongly upon them, and impels them in a manner that bears no analogy with their natural habit: this constitutes in him that peculiar mode of existing which he calls *grief*.

On the other hand, it sometimes happens that exterior objects produce very considerable changes on his body, without his perceiving them at the moment. Often, in the heat of battle, the soldier perceives not that he is dangerously wounded; because at the time the rapidity, the multiplicity of impetuous motions that assail his brain, do not permit him to distinguish the particular change a part of his body has undergone by the wound. In short, when a great number of causes are simultaneously acting on him with too much vivacity, he sinks under their accumulated pressure,--he swoons—he loses his senses—he is deprived of feeling.

In general, feeling only obtains when the brain can distinguish distinctly the impressions made on the organs with which it has communication; it is the distinct shock, the decided modification, man undergoes, that constitutes *conscience*.* From whence it will appear, that *feeling* is a mode of being, or a marked change, produced on our brain by the impulse communicated to our organs, whether by interior or exterior agents, and by which it is modified either in a durable or transient manner. In fact, it is not always requisite that man's organs should be moved by an exterior object to enable him to be conscious of the changes effected in him: he can feel them within himself by means of an interior impulse; his brain is then modified, or rather, he renews within himself the anterior modifications. We should not be astonished that the brain should be necessarily warned of the shocks, of the impediments, of the changes that may happen to so complicated a machine as the human body, in which all the parts are contiguous to the brain—to a whole, in which all the sensible parts concentrate themselves in this brain, and are by their essence in a continual state of action and reaction.

When a man experiences the pains of the gout, he is conscious of them; in other words, he feels interiorly that it has produced very distinct changes in him, without his perceiving that he has received an impulse from any exterior cause; nevertheless, if he will recur to the true source of these changes, he will find that they have been wholly produced by exterior agents; they have been the consequence either of his temperament, of the organization received from his parents, or of the aliments with which his frame has been nourished,

---

* Doctor Clarke says, *Conscience is the act of reflecting, by means of which I know that I think; and that my thoughts, or my actions, belong to me, and not to another.—See his letter against Dodwell.*

besides a thousand trivial, inappreciable causes, which, congregating themselves by degrees, produce in him the gouty humour, the effect of which is to make him feel in a very acute manner. The pain of the gout engenders in his brain an idea or modification which it acquires the faculty of representing or reiterating to itself, even when he shall be no longer tormented with the gout: his brain, by a series of motion interiorly excited, is again placed in a state analogous to that in which it was when he really experienced this pain: but if he had never felt it, he would have had no idea of this excruciating disease.

The visible organs of man's body, by the intervention of which his brain is modified, take the name of *senses*. The various modifications which his brain receives by the aid of these senses, assume a variety of names. *Sensation*, *perception*, *idea*, are terms that designate nothing more than the changes produced in this interior organ, in consequence of impressions made on the exterior organs by bodies acting on them: these changes, considered by themselves, are called *sensations;* they adopt the term . *perception*, when the brain is warned of their presence; *ideas*, is that state of them in which the brain is able to ascribe them to the objects by which they have been produced.

Every *sensation*, then, is nothing more than the shock given to the organs; every *perception*, is this shock propagated to the brain: every *idea*, is the image of the object to which the sensation and the perception is to be ascribed. From whence it will be seen, that if the senses be not moved, there can neither be sensations, perceptions, nor ideas: and this will be proved to those who yet doubt so demonstrable and striking a truth.

It is the extreme mobility of which man is capable, owing to his peculiar organization, which distinguishes him from other beings that are called insensible or inanimate: and the different degrees of mobility of which the individuals of his species are susceptible, discriminate them from each other, making that incredible variety and that infinity of difference which is to be found, as well in their corporeal faculties as in those which are mental or intellectual. From this mobility, more

or less remarkable in each human being, results wit, sensibility, imagination, taste, &c. For the present, however, let us follow the operation of the senses: let us examine in what manner they are acted upon and are modified by exterior objects:—we will afterwards scrutinize the reaction of the interior organ or brain.

The eyes are very delicate, very moveable organs, by means of which the sensation of light, or colour, is experienced: these give to the brain a distinct perception, in consequence of which man forms an idea generated by the action of luminous or coloured bodies: as soon as the eyelids are opened, the retina is affected in a peculiar manner; the fluid, the fibres, the nerves, of which they are composed, are excited by shocks which they communicate to the brain, and to which they delineate the images of the bodies from which they have received the impulse; by this means an idea is acquired of the colour, the size, the form, the distance of these bodies: it is thus that may be explained the mechanism of *sight.*

The mobility and the elasticity of which the skin is rendered susceptible by the fibres and nerves which form its texture, account for the rapidity with which this envelope to the human body is affected when applied to any other body: by their agency the brain has notice of its presence, of its extent, of its roughness, of its smoothness, of its surface, of its pressure, of its ponderosity,. &c.—qualities from which the brain derives distinct perceptions, which breed in it a diversity of ideas; it is this that constitutes the *touch.*

The delicacy of the membrane by which the interior of the nostrils is covered, renders them easily susceptible of irritation, even by the invisible and impalpable corpuscles that emanate from odorous bodies: by this means sensations are excited, the brain has perceptions, and generates ideas: it is this that forms the sense of *smelling.*

The mouth, filled with nervous, sensible, moveable, and irritable glands, saturated with juices suitable to the dissolution of saline substances, is affected in a very lively manner by the aliments which pass through it; these glands transmit to the brain the impressions received : it is from this mechanism that results *taste.*

The ear, whose conformation fits it to receive the various impulses of air diversely modified, communicates to the brain the shocks or sensations; these breed the perception of sound, and generate the idea of sonorous bodies : it is this that constitutes *hearing.*

Such are the only means by which man receives sensations, perceptions, ideas. . These successive modifications of his brain are effects produced by objects that give impulse to his senses; they become themselves causes producing in his mind new modifications, which are denominated *thought, reflection, memory, imagination, judgment, will, action;* the basis, however, of all these is sensation.

To form a precise notion of *thought,* it will be requisite to examine step by step what passes in man during the presence of any object whatever. Suppose, for a moment, this object to be a peach : this fruit makes, at the first view, two different impressions on his eyes; that is to say, it produces two modifications, which are transmitted to the brain, which on this occasion experiences two new perceptions, has two new ideas or modes of existence, designated by the terms *colour* and *rotundity;* in consequence, he has an idea of a body possessing roundness and colour: if he places his hand on this fruit, the organ of feeling having been set in action, his hand experiences three new impressions, which are called *softness, coolness, weight,* from whence result three new perceptions in the brain, and consequently three new ideas : if he approximates this peach to his nose, the organ of *smelling* receives an impulse, which, communicated to the brain, a new perception arises, by which he acquires a new idea called *odour :* if he carries this fruit to his mouth, the organ of taste becomes affected in a very lively manner; this impulse communicated to the brain, is followed by a perception that generates in him the idea of *flavour.* In reuniting all these impressions, or these various modifications of his organs, which have been consequently transmitted to his brain, that is to say, in combining the different

no. II.—8

sensations, perceptions, and ideas, that result from the impulse he has received, he has the idea of a whole, which he designates by the name of a peach, with which he can then occupy his thoughts.*

What has been said is sufficient to show the generation of sensations, of perceptions, of ideas, with their associations, or connexion in the brain: it will be seen that these various modifications are nothing more than the consequence of successive impulsions, which the exterior organs transmit to the interior organ, which enjoys the faculty of thought, that is to say, to feel in itself the different modifications it has received, or to perceive the various ideas which it has generated—to combine them—to separate them—to extend them—to abridge them—to compare them—to renew them, &c. From whence it will be seen, that thought is nothing more than the perception of certain modifications which the brain either gives to itself, or has received from exterior objects.

Indeed, not only the interior organ perceives the modifications it receives from without, but again it has the faculty of modifying itself—of considering the changes which take place in it, the motion by which it is agitated in its peculiar operations, from which it imbibes new perceptions, new ideas. It is the exercise of this power to fall back upon itself, that is called *reflection.*

From this it will appear, that for man to think and to reflect, is to feel, or perceive within himself the impres-

---

* From this it is sufficiently proved that thought has a commencement, a duration, an end, or rather, a generation, a succession, a dissolution, like all the other modifications of matter; like them, thought is excited, is determined, is increased, is divided, is compounded, is simplified, &c. If, therefore, the soul, or the principle that thinks, be indivisible, how does it happen that the soul has the faculty of memory and of forgetfulness; is capacitated to think successively, to divide, to abstract, to combine, to extend its ideas, to retain them, to lose them? How can it cease to think? If forms appear divisible in matter, it is only in considering them by abstraction, after the method of geometricians; but this divisibility of form exists not in nature, in which there is neither a point, an atom, nor form perfectly regular; it must therefore be concluded, that the forms of matter are not less indivisible than thought.

sions, the sensations, the ideas, which have been furnished to his brain by those objects which give impulse to his senses in consequence of the various changes which his brain produced on itself.

Memory is the faculty which the brain has of renewing in itself the modifications it has received, or rather, to restore itself to a state similar to that in which it has been placed by the sensations, the perceptions, the ideas, produced by exterior objects, in the exact order it received them, without any new action on the part of these objects, or even when these objects are absent; the brain perceives that these modifications assimilate with those it formerly experienced in the presence of the objects to which it relates, or attributes them. Memory is faithful when these modifications are precisely the same; it is treacherous when they differ from those which the organs have exteriorly experienced.

Imagination in man is only the faculty which the brain has of modifying itself, or of forming to itself new perceptions upon the model of those which it has anteriorly received through the action of exterior objects on the senses. The brain, then, does nothing more than combine ideas which it has already formed, and which it recalls to itself to form a whole, or a collection of modifications, which it has not received, although the individual ideas, or the parts of which this ideal whole is composed, have been previously communicated to it. It is thus, man forms to himself the idea of *Centaurs,** of *Hyppogriffs,*† of Gods,‡ and Demons.¶ By memory, the brain renews in itself the sensations, the perceptions, the ideas, which it has received, and represents to itself the objects which have actually moved its organs. By imagination it combines them variously; forms objects or wholes in their place, which have not moved its organs, although it is perfectly acquainted with the elements or ideas of which it composes them. It is thus that man, by combining a great number of ideas bor-

rowed from himself, such as justice, wisdom, goodness, intelligence, &c., has, by the aid of imagination, formed an imaginary whole, which he has called God.

Judgment, is the faculty which the brain possesses of comparing with each other the modifications it receives, the ideas it engenders, or which it has the power of awakening within itself, to the end that it may discover their relations or their effects.

Will, is a modification of the brain, by which it is disposed to action, that is to say, to give such an impulse to the organs of the body as can induce it to act in a manner that will procure for itself what is requisite to modify it in a mode analogous to its own existence, or to enable it to avoid that by which it can be injured. To will is to be disposed to action. The exterior objects, or the interior ideas, which give birth to this disposition, are called *motives*, because they are the springs or movements which determine it to act, that is to say, which give play to the organs of the body. Thus *voluntary actions* are the motion of the body, determined by the modification of the brain. Fruit hanging on a tree, through the agency of the visual organs modifies the brain in such a manner as to dispose the arm to stretch itself forth to cull it; again it modifies it in another manner, by which it excites the hand to carry it to the mouth.

All the modifications which the interior organ or the brain receives; all the sensations—all the perceptions—all the ideas that are generated by the objects which give impulse to the senses, or which it renews within itself by its own peculiar faculties, are either favourable or prejudicial to man's mode of existence, whether that be transitory or habitual: they dispose the interior organ to action, which it exercises by reason of its own peculiar energy: this action is not, however, the same in all the individuals of the human species, depending much on their respective temperaments. From hence the *passions* have their birth: these are more or less violent: they are, however, nothing more than the motion of the will, determined by the objects which give it activity—consequently, composed of the analogy or

---

* A being composed of a man and a horse.
† A being composed of a horse with wings.
‡ A *nondescript!*
¶ A *gentleman* with two horns, a tail, and a cloven foot.

of the discordance which is found between these objects and man's peculiar mode of existence, or the force of his temperament. From this it results, that the passions are modes of existence or modifications of the brain, which either attract or repel those objects by which man is surrounded; that consequently they are submitted in their action to the physical laws of attraction and repulsion.

The faculty of perceiving, or of being modified, as well by itself as by exterior objects, which the brain enjoys, is sometimes designated by the term *understanding*. To the assemblage of the various faculties of which this interior organ is susceptible, is applied the name of *intelligence*. To a determined mode, in which the brain exercises the faculties peculiar to itself, is given the appellation of *reason*. The dispositions, or the modifications of the brain, some of them constant, others transitory, which give impulse to the beings of the human species, causing them to act, are styled *wit, wisdom, goodness, prudence, virtue, &c.*

In short, as there will be an opportunity presently to prove, all the intellectual faculties, that is to say, all the modes of action attributed to the soul, may be reduced to the modifications, to the qualities, to the modes of existence, to the changes produced by the motion of the brain, which is visibly in man the seat of feeling—the principle of all his actions. These modifications are to be attributed to the objects that strike on his senses; of which the impression is transmitted to the brain, or rather to the ideas which the perceptions caused by the action of these objects on his senses have there generated, and which it has the faculty to reproduce. This brain moves itself in its turn, reacts upon itself, gives play to the organs, which concentrate themselves in it, or which rather are nothing more than an extension of its own peculiar substance. It is thus the concealed motion of the interior organ renders itself sensible by outward and visible signs. The brain, affected by a modification which is called *fear*, diffuses a paleness over the countenance, excites a tremulous motion in the limbs, called trembling.

The brain, affected by a sensation of *grief*, causes tears to flow from the eyes, even without being moved by any exterior object; an idea which it retraces with great strength, suffices to give it very lively modifications, which visibly have an influence on the whole frame.

In all this nothing more is to be perceived than the same substance which acts diversely on the various parts of the body. If it be objected, that this mechanism does not sufficiently explain the principles of the motion, or the faculties of the soul; we reply, that it is in the same situation as all the other bodies of nature, in which the most simple motion, the most ordinary phenomena, the most common modes of action, are inexplicable mysteries, of which we shall never be able to fathom the first principles. Indeed, how can we flatter ourselves we shall ever be enabled to compass the true principle of that gravity by which a stone falls? Are we acquainted with the mechanism which produces attraction in some substances, repulsion in others? Are we in a condition to explain the communication of motion from one body to another? But it may be fairly asked; are the difficulties that occur, when attempting to explain the manner in which the soul acts, removed, by making it a *spiritual being*, a substance of which we have not, nor cannot form one idea, which consequently must bewilder all the notions we are capable of forming to ourselves of this being? Let us then be contented to know that the soul moves itself, modifies itself, in consequence of material causes, which act upon it, which give it activity; from whence the conclusion may be said to flow consecutively, that all its operations, all its faculties, prove that it is itself *material*.

●

———

CHAPTER IX.

Of the Diversity of the Intellectual Faculties; they depend on Physical Causes, as do their Moral Qualities. The Natural Principles of Society.—Of Morals.—Of Politics.

NATURE is under the necessity to diversify all her works. Elementary matter, different in its essence, must necessarily form different beings, various in their combinations, in their properties, in their modes of action, in their manner of existence. There is not, neither can there be, two beings, two combinations, which are mathematically and rigorously the same; because the place, the circumstances, the relations, the proportions, the modifications, never being exactly alike, the beings that result can never bear a perfect resemblance to each other: and their modes of action must of necessity vary in something, even when we believe we find between them the greatest conformity.

In consequence of this principle, which every thing we see conspires to prove to be a truth, there are not two individuals of the human species, who have precisely the same traits; who think exactly in the same manner; who view things under the same identical point of sight; who have decidedly the same ideas; consequently no two of them have uniformly the same system of conduct. The visible organs of man, as well as his concealed organs, have indeed some analogy, some common points of resemblance, some general conformity, which makes them appear, when viewed in the gross, to be affected in the same manner by certain causes; but the difference is infinite in the detail. The human soul may be compared to those instruments of which the chords, already diversified in themselves by the manner in which they have been spun, are also strung upon different notes: struck by the same impulse, each chord gives forth the sound that is peculiar to itself, that is to say, that which depends on its texture, its tension, its volume, on the momentary state in which it is placed by the circumambient air. It is this that produces the diversified spectacle, the varied scene, which the moral world offers to our view: it is from this that results the striking contrariety that is to be found in the minds, in the faculties, in the passions, in the energies, in the taste, in the imagination, in the ideas, in the opinions of man: this diversity is as great as that of his physical powers: like them it depends on his temperament, which is as much varied as his physiognomy. This variety gives birth to that continual series of action and reaction which constitutes

the life of the moral world: from this discordance results the harmony which at once maintains and preserves the human race.

The diversity found among the individuals of the human species, causes inequalities between man and man: this inequality constitutes the support of society. If all men were equal in their bodily powers, in their mental talents, they would not have any occasion for each other: it is the variation of his faculties, the inequality which this places him in with regard to his fellows, that renders man necessary to man: without these he would live by himself, he would remain an isolated being. From whence it may be perceived that this inequality, of which man so often complains without cause; this impossibility each man finds when in an isolated state, when left to himself, when unassociated with his fellow men, to labour efficaciously to his own welfare, to make his own security, to ensure his own conservation, places him in the happy situation of associating with his like, of depending on his fellow associates, of meriting their succour, of propitiating them to his views, of attracting their regard, of calling in their aid to chase away, by common and united efforts, that which would have the power to trouble or derange the order of his existence. In consequence of man's diversity and of the inequality that results, the weaker is obliged to seek the protection of the stronger: this, in his turn, recurs to the understanding, to the talents, to the industry of the weaker, whenever his judgment points out he can be useful to him: this natural inequality furnishes the reason why nations distinguish those citizens who have rendered their country eminent services; and it is in consequence of his exigencies that man honours, that he recompenses those whose understanding, whose good deeds, whose assistance, whose virtues, have procured for him real or supposed advantages, pleasures, or agreeable sensations of any sort: it is by this means that genius gains an ascendency over the mind of man, and obliges a whole people to acknowledge its power. Thus, the diversity, the inequality of the faculties, as well corporeal, as mental or intellectual, render man necessary to his fellow man, makes him a social being, and incontestably proves to him the necessity of morals.

According to this diversity of faculties, the individuals of the human species are divided into different classes, each in proportion to the effects produced, to the different qualities that may be remarked: all these varieties in man flow from the individual properties of his mind, or from the particular modification of his brain. It is thus that wit, imagination, sensibility, talents, &c. diversify to infinity the differences that are to be found in man. It is thus that some are called good, others wicked; some are denominated virtuous, others vicious; some are ranked as learned, others as ignorant; some are considered reasonable, others unreasonable, &c.

If all the various faculties attributed to the soul are examined, it will be found that like those of the body they are to be ascribed to physical causes, to which it will be very easy to recur. It will be found that the powers of the soul are the same as those of the body; that they always depend on the organization of this body, on its peculiar properties, on the permanent or transitory modifications that it undergoes; in a word, on its temperament.

Temperament, is, in each individual, the habitual state in which he finds the fluids and the solids of which his body is composed. This temperament varies by reason of the elements or matter that predominates in him; in consequence of the different combinations, of the various modifications, which this matter, diversified in itself, undergoes in his machine. Thus, in one the blood is superabundant; in another, the bile; in a third, phlegm, &c.

It is from nature—from his parents—from causes, which from the first moment of his existence have unceasingly modified him, that man derives his temperament. It is in his mother's womb that he has attracted the matter which, during his whole life, shall have an influence on his intellectual faculties—on his energies—on his passions—on his conduct. The very nourishment he takes, the quality of the air he respires, the climate he inhabits, the education he receives, the ideas that are presented to him, the opinions he imbibes, modify

this temperament. As these circumstances can never be rigorously the same in every point for any two men, it is by no means surprising that such an amazing variety, so great a contrariety, should be found in man, or that there should exist as many different temperaments as there are individuals in the human species. •

Thus, although man may bear a general resemblance, he differs essentially, as well by the texture of his fibres, the disposition of his nerves, as by the nature, the quality, the quantity of matter that gives them play, and sets his organs in motion. Man, already different from his fellow, by the elasticity of his fibres, the tension of his nerves, becomes still more distinguished by a variety of other circumstances: he is more active, more robust, when he receives nourishing aliments, when he drinks wine, when he takes exercise; whilst another, who drinks nothing but water, who takes less juicy nourishment, who languishes in idleness, shall be sluggish and feeble.

All these causes have necessarily an influence on the mind, on the passions, on the will, in a word, on what are called the intellectual faculties. Thus, it may be observed, that a man of a sanguine constitution is commonly lively, ingenious, full of imagination, passionate, voluptuous, enterprising; whilst the phlegmatic man is dull, of a heavy understanding, slow of conception, inactive, difficult to be moved, pusillanimous, without imagination, or possessing it in a less lively degree, incapable of taking any strong measures, or of willing resolutely.

If experience was consulted in the room of prejudice, the physician would collect from morals the key to the human heart: and in curing the body, he would sometimes be assured of curing the mind. Man, in making a spiritual substance of his soul, has contented himself with administering to it spiritual remedies, which either have no influence over his temperament, or do it an injury. The doctrine of the spirituality of the soul has rendered morals a conjectural science, that does not furnish a knowledge of the true motives which ought to be put in activity in order to influence man to his welfare. If, calling experience to his assistance, man sought out the elements which form the basis of his temperament, or of the greater number of the individuals composing a nation; he would then discover what would be most proper for him, that which could be most convenient to his mode of existence, which could most conduce to his true interest;—what laws would be necessary to his happiness—what institutions would be most useful for him—what regulations would be most beneficial. In short, morals and politics would be equally enabled to draw from *materialism* advantages which the dogma of spirituality can never supply, of which it even precludes the idea. Man will ever remain a mystery to those who shall obstinately persist in viewing him with eyes prepossessed by theology, or to those who shall pertinaciously attribute his actions to a principle of which it is impossible to form to themselves any distinct idea. When man shall be seriously inclined to understand himself, let him sedulously endeavour to discover the matter that enters into his combination, which constitutes his temperament; these discoveries will furnish him with the clue to the nature of his desires, to the quality of his passions, to the bent of his inclinations, and will enable him to foresee his conduct on given occasions; will indicate the remedies that may be successfully employed to correct the defects of a vicious organization and of a temperament as injurious to himself as to the society of which he is a member.

Indeed, it is not to be doubted that man's temperament is capable of being corrected, of being modified, of being changed, by causes as physical as the matter of which it is constituted. We are all in some measure capable of forming our own temperament: a man of a sanguine constitution, by taking less juicy nourishment, by abating its quantity, by abstaining from strong liquor, &c., may achieve the correction of the nature, the quality, the quantity, the tendency, the motion of the fluids, which predominate in his machine. A bilious man, or one who is melancholy, may, by the aid of certain remedies, diminish the mass of this bilious fluid; he may correct the blemish of his humours by the assistance of exer-

cise; he may dissipate his gloom by the gaiety which results from increased motion. A European transplanted into Hindostan will by degrees become quite a different man in his humours, in his ideas, in his temperament, and in his character.

Although but few experiments have been made with a view to learn what constitutes the temperament of man, there are still enough if he would but deign to make use of them, or if he would vouchsafe to apply to useful purposes the little experience he has gleaned. It would appear, speaking generally, that the igneous principle which chymists designate under the name of *phlogiston*, or inflammable matter, is that which in man yields him the most active life, furnishes him with the greatest energy, affords the greatest mobility to his frame, supplies the greatest spring to his organs, gives the greatest elasticity to his fibres, the greatest tension to his nerves, the greatest rapidity to his fluids. From these causes, which are entirely material, commonly result the dispositions or faculties, called sensibility, wit, imagination, genius, vivacity, &c., which give the tone to the passions, to the will, to the moral actions of man. In this sense, it is with great justice we apply the expressions, " warmth of soul," " ardency of imagination," " fire of genius," &c.*

It is this fiery element, diffused in different doses, distributed in various proportions, through the beings of the human species, that sets man in motion, gives him activity, supplies him with animal heat, and which, if we may be allowed the expression, renders him more or less alive. This igneous matter, so active, so subtile, dissipates itself with great facility, then requires to be reinstated in his system by means of aliments that contain it, which thereby become proper to restore his machine, to lend new warmth to the brain, to furnish it with the elasticity requisite

---

* It would not be unreasonable to suppose, that what physicians call the nervous fluid, which so promptly gives notice to the brain, of all that happens to the body, is nothing more than electric matter; that the various proportions of this matter, diffused through his system, is the cause of that great diversity to be discovered in the human being, and in the faculties he possesses.

..o tne performance of those functions which are called intellectual. It is this ardent matter, contained in wine, in strong liquor, that gives to the most torpid, to the dullest, to the most sluggish man, a vivacity, of which, without it, he would be incapable, and which urges even the coward on to battle. When this fiery element is too abundant in man, whilst he is labouring under certain diseases, it plunges him into delirium ; when it is in too weak, or in too small a quantity, he swoons, he sinks to the earth. This igneous matter diminishes in his old age, it totally dissipates at his death.*

If the intellectual faculties of man, or his moral qualities, be examined according to the principles here laid down, the conviction must be complete, that they are to be attributed to material causes, which have an influence more or less marked, either transitory or durable over his peculiar organization. But where does he derive this organization except it be from the parents from whom he receives the elements of a machine necessarily analogous to their own? From whence does he derive the greater or less quantity of igneous matter, or vivifying heat, which gives the tone to his mental qualities? It is from the mother, who bore him in her womb, who has communicated to him a portion of that fire with which she was herself animated, which circulated through her veins with her blood: it is from the aliments that have nourished him: it is from the climate he inhabits: it is from the atmosphere that surrounds him: for, all these causes have an influence over his fluids, over his solids, and decide on his natural dispositions. In examining these dispositions, from whence his faculties depend, it will ever be found that they are *corporeal* and *material.*

The most prominent of these dispositions in man, is that physical sensibility from which flows all his intel-

lectual or moral qualities. To feel, according to what has been said, is to receive an impulse, to be moved, and to have a consciousness of the changes operated on his system. To have sensibility, is nothing more than to be so constituted as to feel promptly, and in a very lively manner, the impressions of those objects which act upon him. A sensible soul, is only man's brain ' disposed in a mode to receive the motion communicated to it with facility and with promptness, by giving an instantaneous impulse to the organs. Thus, the man is called *sensible,* whom the sight of the distressed, the contemplation of the unhappy, the recital of a melancholy tale, the witnessing of an afflicting catastrophe, or the idea of a dreadful spectacle, touches in so lively a manner as to enable the brain to give play to his lachrymal organs, which cause him to shed tears ;' a sign by which we recognise the effect of extreme anguish ·in the human being. The man in whom musical sounds excite a degree of pleasure, or produce very remarkable effects, is said to have a *sensible* or a fine ear. In short, when it is perceived that eloquence,—the beauty of the arts,—the various objects that strike his senses, excite in him very lively emotions, he is said to possess a soul full of sensibility.†

Wit is a consequence of this physical sensibility; indeed, wit is nothing more than the facility which some beings of the human species possess of seizing with promptitude, of developing with quickness a whole, with its different relations to other objects. *Genius,* is the facility with which some men comprehend this whole, and its various relations, when they are difficult to be known, but useful to forward great and mighty projects. *Wit,* may be compared to a piercing eye, which perceived things quickly. *Genius,* is an eye that comprehends at one view all the points

---

* If we reflect a little we shall find that *heat* is the principle of life. It is by means of heat that beings pass from inaction into motion—from repose into fermentation—from a state of torpor into that of active life. This is proved by the egg, which heat hatches into a chicken ; and this example, among thousands which we might cite, must suffice to establish the fact, that without heat, there is no generation.

† Compassion depends on physical sensibility, which is never the same in all men. How absurd, then, to make compassion the source of all our moral ideas, and of those feelings which we experience for our fellow creatures. Not only all men are not alike sensible, but there are many in whom sensibility has not been developed—such as in kings, priests, statesmen,—

"And the hired bravoes who defend
The tyrant's throne—the bullies of his fear!"

of an extended horizon, or what the French term *coup d'œil*. *True wit*, is that which perceives objects with their relations, such as they really are. *False wit*, is that which catches at relations which do not apply to the object, or which arises from some blemish in the organization. *True wit* resembles the direction on a hand-post.

Imagination, is the faculty of combining with promptitude ideas or images; it consists in the power man possesses of reproducing with ease the modifications of his brain; of connecting them, and of attaching them to the objects to which they are suitable. When imagination does this, it gives pleasure; its fictions are approved, it embellishes nature, it is a proof of the soundness of the mind, it aids truth: when, on the contrary, it combines ideas not formed to associate themselves with each other; when it paints nothing but disagreeable phantoms, it disgusts. Thus poetry, calculated to render nature more pathetic, more touching, pleases when it adorns the object it portrays with all those beauties with which it can with propriety be associated. True, it only creates ideal beings, but as they move us agreeably, we forgive the illusions it has held forth on account of the pleasure we have reaped from them. The hideous chimeras of superstition displease, because they are nothing more than the productions of a distempered imagination, which can only awaken afflicting sensations.

Imagination, when it wanders, produces fanaticism—religious terrours—inconsiderate zeal—phrensy—the most enormous crimes. When imagination is well regulated, it gives birth to a strong predilection for useful objects—an energetic passion for virtue—an enthusiastic love of our country—the most ardent friendship: the man who is divested of imagination, is commonly one in whose torpid constitution phlegm predominates over that sacred fire, which is the great principle of his mobility, of his warmth of sentiment, and which vivifies all his intellectual faculties. There must be enthusiasm for transcendent virtues as well as for atrocious crimes. Enthusiasm places the soul, or brain, in a state similar to that of drunkenness; both the one and the other excite in man that rapidity of

motion which is approved when good results, but which is called folly, delirium, crime, fury, when it produces nothing but disorder.

The mind is out of order, it is incapable of judging sanely, and the imagination is badly regulated, whenever man's organization is not so modified as to perform its functions with precision. At each moment of his existence man gathers experience; every sensation he has, furnishes a fact that deposites in his brain an idea, which his memory recalls with more or less fidelity: these facts connect themselves, these ideas are associated, and their chain constitutes *experience* and *science*. Knowledge, is that consciousness which arises from reiterated experience, made with precision of the sensations, of the ideas, of the effects which an object is capable of producing, either in ourselves or in others. All science must be founded on truth. Truth itself rests on the constant and faithful relation of our senses. Thus *truth* is that conformity or perpetual affinity which man's senses, when well constituted, when aided by experience, discover to him, between the objects of which he has a knowledge, and the qualities with which he clothes them. In short, truth is nothing more than the just, the precise association of his ideas. But how can he without experience, assure himself of the accuracy of this association? How, if he do not reiterate this experience, can he compare it? If his senses are vitiated, how is it possible they can convey to him, with precision, the sensations, the facts, with which they store his brain? It is only by multiplied, by diversified, by repeated experience, that he is enabled to rectify the errours of his first conceptions.

Man is in errour every time his organs, either originally defective in their nature, or vitiated by the durable or transitory modifications which they undergo, render him incapable of judging soundly of objects. Errour consists in the false association of ideas, by which qualities are attributed to objects which they do not possess. Man is in errour, when he supposes those beings really to have existence which have no local habitation but in his own imagination: he is in errour, when he associates the idea of happiness with objects capable of injuring him, whether immediately or by remote consequences which he cannot foresee. But how can he foresee effects of which he has not yet any knowledge? It is by the aid of experience. By the assistance which this experience affords it is known, that analogous, or like causes, produce analogous or like effects: memory, by recalling these effects, enables him to form a judgment of those he may expect, whether it be from the same causes, or from causes that bear a relation to those of which he has already experienced the action. From this it will appear, that *prudence, foresight,* are faculties that grow out of experience. If he has felt that fire excited in his organs a painful sensation, this experience suffices him to foresee that fire so applied, will eventually excite the same sensations. If he has discovered that certain actions, on his part, stirred up the hatred, and elicited the contempt of others, this experience sufficiently enables him to foresee, that every time he shall act in a similar manner, he will be either hated or despised.

The faculty man has 'of gathering experience, of recalling it to himself, of foreseeing effects, by which he is enabled to avoid whatever may have the power to injure him, or procure that which may be useful to the conservation of his existence and his felicity, which is the sole end of all his actions, whether corporeal or mental, constitutes that which in one word is designated under the name of *reason.* Sentiment, imagination, temperament, may be capable of leading him astray; may have the power to deceive him; but experience and reflection will place him again in the right road, and teach him what can really conduct him to happiness. From this it will appear, .hat *reason* is man's nature modified by experience, moulded by judgment, regulated by reflection : it supposes a sober temperament, a sound mind, a well regulated imagination, a knowledge of truth grounded upon tried experience; in fact, prudence and foresight: and this proves, that, although nothing is more common than the assertion that *man is a reasonable being,* yet there are but a very small number of the individuals who compose the human

No. III.—9

species who really enjoy the faculty of reason, or who combine the dispositions and the experience by which it is constituted.

It ought not then to excite surprise that the individuals of the human race who are in a capacity to make true experience, are so few in number. Man, when he is born, brings with him organs susceptible of receiving impulse, and of collecting experience; but whether it be from the vice of his system, the imperfection of his organization, or from those causes by which it is modified, his experience is false, his ideas are confused, his images are badly associated, his judgment is erroneous, his brain is saturated with vicious systems, which necessarily have an influence over his conduct, and continually disturb his reason.

Man's senses, as it has been shown, are the only means by which he is enabled to ascertain whether his opinions are true or false, whether his conduct is useful to himself, and whether it is advantageous or disadvantageous. But that his senses may be competent to make a faithful relation, or be in a capacity to impress true ideas on his brain, it is requisite they should be sound ; that. is to say, in the state necessary to maintain his existence in that order which is suitable to his preservation and his permanent felicity. It is also indispensable that his brain itself should be healthy, or in the proper state to enable it to fulfil its functions with precision and to exercise its faculties with vigour. It is necessary that memory should faithfully retrace its anterior sensations and ideas, to the end, that he may be competent to judge or to foresee the effects he may have to hope or to fear from those actions to which he may be determined by his will. If his interior or exterior organs be defective, whether by their natural conformation, or from those causes by which they are regulated, he feels but imperfectly, and in a manner less distinct than is requisite ; his ideas are either false or suspicious; he judges badly ; he is in a delusion, or in a state of ebriety that prevents his grasping the true relation of things. In short, if his memory be faulty, if it be treacherous, his reflection is void; his imagination leads him astray ; his mind deceives him; whilst the sensibility of his organs, simultaneously assailed by a crowd of impressions, oppose him to prudence, to foresight, and to the exercise of his reason. On the other hand, if the confirmation of his organs, as it happens with those of a phlegmatic temperament, does not permit him to move, except with feebleness and in a sluggish manner, his experience is slow, and frequently unprofitable. The tortoise and the butterfly are alike incapable of preventing their destruction. The stupid man and he who is intoxicated, are in that state which renders it impossible for them to attain the end they have in view.

But what is the aim of man in the sphere he occupies? It is to preserve himself and to render his existence happy. It becomes, then, of the utmost importance that he should understand the true means which reason points out, which prudence teaches him to use, in order that he may always and with certainty arrive at the end which he proposes to himself. These are his natural faculties, his mind, his talents, his industry, his actions determined by those passions of which his nature renders him susceptible, and which give more or less activity to his will. Experience and reason show him again that the men with whom he is associated, are necessary to him—are capable of contributing to his happiness and to his pleasures, and are competent to assist him by those faculties which are peculiar to them: experience teaches him the mode he must adopt to induce them to concur in his designs—to determine them to will and to act in his favour. This points out to him the actions they approve—those which displease them—the conduct which attracts them—that which repels them—the judgment by which they are swayed—the advantages that occur, the prejudicial effects that result to him from their various modes of existence and manner of acting. This experience furnishes him with the ideas of virtue and of vice—of justice and of injustice—of goodness and of wickedness—of decency and of indecency—of probity and of knavery. In short, he learns to form a judgment of men, to estimate their actions—to distinguish the various sentiments excited in

them, according to the diversity of those effects which they make him experience.

It is upon the necessary diversity of these effects that is founded the discrimination between good and evil—between virtue and vice; distinctions which do not rest, as some thinkers have believed, on the conventions made between men; still less upon the chimerical will of a supernatural being, but upon the invariable, the eternal relations that subsist between beings of the human species congregated together, and living in society—relations which will have existence as long as man shall remain, and as long as society shall continue to exist.

Thus *virtue* is every thing that is truly and constantly useful to the individuals of the human race living together in society; *vice*, every thing that is injurious to them. The greatest virtues are those which procure for man the most durable and solid advantages: the greatest vices, are those which most disturb his tendency to happiness, and which most interrupt the necessary order of society. The *virtuous man* is he whose actions tend uniformly to the welfare of his fellow creatures. The *vicious man* is he whose conduct tends to the misery of those with whom he lives; from whence his own peculiar misery most commonly results. Every thing that procures for man a true and a permanent happiness, is reasonable; every thing that disturbs his individual felicity, or that of the beings necessary to his happiness, is foolish or unreasonable. The man who injures others, is wicked—the man who injures himself, is an imprudent being, who neither has a knowledge of reason, of his own peculiar interests, nor of truth.

Man's *duties* are the means pointed out to him by experience and reason, by which he is to arrive at that goal he proposes to himself: these duties are the necessary consequence of the relations subsisting between mortals who equally desire happiness, and who are equally anxious to preserve their existence. When it is said, these duties *compel him*, it signifies nothing more than that, without taking these means, he could not reach the end proposed to him by his nature. Thus, *moral obli-*

gation is the necessity of employing the natural means to render the beings with whom he lives happy, to the end that he may determine them in turn, to contribute to his own individual happiness: his obligation towards himself is the necessity he is under to take those means without which he would be incapable to conserve himself, and render his existence solidly happy. Morals, like the universe, are founded upon necessity, or upon the eternal relation of things.

Happiness, is a mode of existence of which man naturally wishes the duration, or in which he is willing to continue. It is measured by its duration and its vivacity. The greatest happiness is that which has the longest continuance: transient happiness, or that which has only a short duration, is called *pleasure;* the more lively it is, the more fugitive, because man's senses are only susceptible of a certain quantum of motion. When pleasure exceeds this given quantity, it is changed into *anguish,* or into that painful mode of existence of which he ardently desires the cessation: this is the reason why pleasure and pain frequently so closely approximate each other as scarcely to be discriminated. Immoderate pleasure is the forerunner of regret. It is succeeded by ennui and weariness, and it ends in disgust: transient happiness frequently converts itself into durable misfortune. According to these principles, it will be seen that man, who in each moment of his duration seeks necessarily after happiness, ought, when he is reasonable, to regulate his pleasures, and to refuse himself to all those of which the indulgence would be succeeded by regret or pain ; whilst he should endeavour to procure for himself the most permanent felicity. .

Happiness cannot be the same for all the beings of the human species ; the same pleasures cannot equally affect men whose confirmation is different, whose modification is diverse. This, no doubt, is the true reason why the greater number of moral philosophers are so little in accord upon those objects in which they have made man's happiness consist, as well as on the means by which it may be obtained. Nevertheless, in general happiness ap-

pears to be a state, whether momentary or durable, in which man readily acquiesces, because he finds it conformable to his being. This state results from the accord which is found between himself and those circumstances in which he has been placed by nature : or, if it be preferred, *happiness is the co-ordination of man with the causes that give him impulse.*

The ideas which man forms to himself of happiness, depend not only on his temperament, on his individual conformation, but also upon the habits he has contracted. *Habit,* is in man a mode of existence—of thinking—of acting, which his organs, as well interior as exterior, contract by the frequent reiteration of the same motion, from whence results the faculty of performing these actions with promptitude and with facility.

If things be attentively considered, it will be found that almost the whole conduct of man, the entire system of his actions, his occupations, his connexions, his studies, his amusements, his manners, his customs, his very garments, even his aliments, are the effect of habit. He owes equally to habit the facility with which he exercises his mental faculties of thought, of judgment, of wit, of reason, of taste, &c. It is to habit he owes the greater part of his inclinations, of his desires, of his opinions, of his prejudices, of the ideas, true or false, he forms to himself of his welfare. In short, it is to habit, consecrated by time, that he owes those errours into which every thing strives to precipitate him, and to prevent him from emancipating himself. It is habit that attaches him either to virtue or to vice.*

Man is so much modified by habit, that it is frequently confounded with his nature : from hence results, as will presently be seen, those opinions, or those ideas which he has called *innate,*

---

* Experience proves that the first crime is always accompanied by more pangs of remorse than the second; this again, by more than the third, and so on to those that follow. A first action is the commencement of a habit ; those which succeed confirm it : by force of combating the obstacles that prevent the commission of criminal actions, man arrives at the power of vanquishing them with ease and with facility. Thus he frequently becomes wicked from habit.

because he has been unwilling to recur back to the source from whence they sprung, which has, as it were, identified itself with his brain. However this may be, he adheres with great strength of attachment to all those things to which he is habituated; his mind experiences a sort of violence, or incommodious revulsion, when it is endeavoured to make him change the course of his ideas: a fatal predilection frequently conducts him back to the old track in despite of reason.

It is by a pure mechanism that may be explained the phenomena of habit, as well physical as moral; the soul, notwithstanding its pretended spirituality, is modified exactly in the same manner as the body. Habit, in man, causes the organs of voice to learn the mode of expressing quickly the ideas consigned to his brain, by means of certain motion, which, during his infancy, the tongue acquires the power of executing with facility : his tongue, once habituated to move itself in a certain manner, finds much trouble to move itself after another mode; the throat yields with difficulty to those inflections which are exacted by a language different from that to which he has been accustomed. It is the same with his ideas; his brain, his interior organ, his soul, inured to a given manner of modification, accustomed to attach certain ideas to certain objects, long used to form to itself a system connected with certain opinions, whether true or false, experiences a painful sensation whenever he undertakes to give it a new impulse, or alter the direction of its habitual motion. It is nearly as difficult to make him change his opinions as his language.*

Here then, without doubt, is the cause of that almost invincible attachment which man displays to those customs, those prejudices, those institutions of which it is in vain that reason, experience, good sense, prove to him the inutility, or even the danger.

Habit opposes itself to the clearest demonstrations; these can avail nothing against those passions and those vices which time has rooted in him—against the most ridiculous systems—against the strangest customs—especially when he has learned to attach to them the ideas of utility—of common interest— of the welfare of society. Such is the source of that obstinacy which man evinces for his religion—for ancient usages—for unreasonable customs—for laws, so little accordant with justice— for abuses, which so frequently make him suffer—for prejudices of which he sometimes acknowledges the absurdity, although unwilling to divest himself of them. Here is the reason why nations contemplate the most useful novelties as mischievous innovations, and believe they would be lost if they were to remedy those evils which they have learned to consider as necessary to their repose, and which they have been taught to consider dangerous to be cured.*

Education, is the only art of making man contract in early life, that is to say, when his organs are extremely flexible, the habits, the opinions, and the modes of existence adopted by the society in which he is placed. The first moments of his infancy are employed in collecting experience; those who are charged with the care of bringing him up, teach him how to apply it: it is they who develop reason in him: the first impulse they give him commonly decides of his condition, his passions, the ideas he forms to himself of happiness, and the means he shall employ to procure it—of his virtues and his vices. Under the eyes of his masters, the infant acquires ideas, and learns to associate them—to think in a certain manner—to judge well or ill. They point out to him various objects, which they accustom him either to love or to hate, to desire or to avoid, to esteem or to despise. It is thus opinions are transmitted from fathers, from mothers, from nurses, and from masters, to man in his infantile state. It is thus that his mind by

---

* Hobbes says that, "It is the nature of all corporeal beings, who have been frequently moved in the same manner, to continually receive a greater aptitude, or to produce the same motions with more facility." It is this which constitutes habit as well in morals as in physics. V. Hobbes's Essay on Human Nature.

---

* Assiduitate quotidiana et consuetudine oculorum assuescunt animi, neque admirantur, neque requirunt rationes earum rerum quas vident. Cicero de Natur : Deorum Lib. ii. Cap. 2.

degrees saturates itself with truth, or fills itself with errour, and as either of them regulates his conduct, it renders him either happy or miserable, virtuous or vicious, estimable or hateful. It is thus he becomes either contented or discontented with his destiny, according to the objects towards which they have directed his passions, and bent the energies of his mind; that is to say, in which they have shown him his interest, or taught him to place his felicity: in consequence he loves and seeks after that which they have instructed him to revere, which they have made the object of his research : he has those tastes, those inclinations, those phantasms, which, during the whole course of his life, he is forward to indulge, which he is eager to satisfy, in proportion to the activity they have excited in him, and the capacity with which he has been provided by nature.

Politics ought to be the art of regulating the passions of man, and of directing them to the welfare of society ; but too frequently it is nothing more than the detestable art of arming the passions of the various members of society against each other, to accomplish their mutual destruction, and fill with rancorous animosities that association, from which, if properly managed, man ought to derive his felicity. Society is commonly so vicious because it is not founded upon nature, upon experience, upon general utility, but on the contrary, upon the passions, the caprices, the particular interests of those by whom it is governed.

Politics, to be useful, should found its principles upon nature; that is to say, should conform itself to the essence of man, and to the great end of society : and society being a whole, formed by the union of a great number of families or individuals, assembled from a reciprocity of interest in order that they may satisfy with greater facility their reciprocal wants, and procure the advantages they desire ; that they may obtain mutual succours ; above all, that they may gain the faculty of enjoying in security those benefits with which nature and industry may furnish them ; it follows, of course, that politics, destined to maintain society, ought to enter into its views, facilitate the means of giving them efficiency,

and remove all those obstacles that have a tendency to counteract the intention with which man entered into association.

Man in approximating to his fellow man to live with him in society, has made, either formally or tacitly, a covenant, by which he engages to render mutual services, and to do nothing that can be prejudicial to his neighbour. But as the nature of each individual impels him constantly to seek after his own welfare, which he has mistaken to consist in the gratification of his passions, in the indulgence of his transitory caprices, without any regard to the convenience of his fellows ; there needed a power to conduct him back to his duty, to oblige him to conform himself to his obligations, and to recall him to engagements which the hurry of his passions frequently make him forget. This power is the *law;* it is the collection of the will of society, reunited to fix the conduct of its members, and to direct their action in such a mode that it may concur to the great end of his association.

But as society, more especially when very numerous, cannot assemble itself unless with great difficulty, and without tumult make known its intentions, it is obliged to choose citizens in whom it places confidence ; whom it makes the interpreter of its will ; whom it constitutes the depositaries of the power requisite to carry it into execution. Such is the origin of all *government,* which to be legitimate can only be founded on the free consent of society— without which it is violence, usurpation, robbery. Those who are charged with the care of governing, call themselves *sovereigns, chiefs, legislators,* and, according to the form which society has been willing to give to its government, these sovereigns are styled *monarchs, magistrates, representatives,* &c. Government only borrows its power from society : being established for no other purpose than its welfare, it is evident society can revoke this power whenever its interest shall exact it—change the form of its government— extend or limit the power which it has confided to its chiefs, over whom, by the immutable laws of nature, it always conserves a supreme authority ; because these laws enjoin, that the part shall always remain subordinate to the whole.

Thus sovereigns are the ministers of society—its interpreters—the depositaries of a greater or of a less portion of its power, but they are not its absolute masters, neither are they the proprietors of nations. By a *corenant*, either expressed or implied, they engage themselves to watch over the maintenance, and to occupy themselves with the welfare, of society; it is only upon these conditions that society consents to obey them. The price of obedience is protection.* No society upon earth was ever willing or competent to confer irrevocably upon its chiefs the right of doing it injury. Such a compact would be annulled by nature; because she wills that each society, the same as each individual of the human species, shall tend to its own conservation; it has not, therefore, the capacity to consent to its permanent misery.

Laws, in order that they may be just, ought invariably to have for their end the general interest of society; that is to say, to assure to the greater number of citizens those advantages for which man originally associated. These advantages are, *liberty, property, security.*

Liberty, to man, is the faculty of doing, for his own peculiar happiness, every thing which does not injure or diminish the happiness of his associates: in associating, each individual renounced the exercise of that portion of his natural liberty, which would be able to prejudice or injure the liberty of his fellows. The exercise of that liberty which is injurious to society is called *licentiousness. Property* is the faculty of enjoying those advantages which spring from labour—those benefits which industry or talent has procured to each member of society. *Security* is the certitude that each individual ought to have, of enjoying in his person and his property, the protection of the laws, as long as he shall faithfully perform his engagements with society. *Justice* assures to all the members of society,

the possession of those advantages or rights which belong to them. From this it will appear, that, without justice, society is not in a condition to procure the happiness of any man. Justice is also called *equity,* because, by the assistance of the laws, made to command the whole, she reduces all its members to a state of equality; that is to say, she prevents them from prevailing one over the other by the inequality which nature·or industry may have made between their respective powers. *Rights* are every thing which society, by equitable laws, permits each individual to do for his own peculiar felicity. These rights are evidently limited by the invariable end of all association; society has, on its part, rights over all its members, by virtue of the advantages which it procures for them; all its members, in turn, have a right to claim from society, or secure from its ministers, those advantages for the procuring of which they congregated, and renounced a portion of their natural liberty. A society of which the chiefs, aided by the laws, do not procure any good for its members, evidently loses its right over them: those chiefs who injure society, lose the right of commanding. It is not our country without it secures the welfare of its inhabitants; a society without equity contains only enemies; a society oppressed is composed only of tyrants and slaves; slaves are incapable of being citizens; it is liberty—property — security, that render our country dear to us; and it is the true love of his country that forms the citizen.†

For want of having a proper knowledge of these truths, or for want of applying them when known, some nations have become unhappy—have contained nothing but a vile heap of slaves, separated from each other, and detached from society, which neither procures for them any good, nor secures to them any one advantage. In consequence of the imprudence of some nations, or of the craft, the cunning, the violence of those to whom they have confided the power of making laws, and of carrying them into execution, their sovereigns have rendered themselves absolute masters

* There ought to be a reciprocity of interest between the governed and the governor: whenever this reciprocity is wanting, society is in that state of confusion, spoken of in the fifth chapter,—it is verging on destruction.

† An ancient poet has justly said, *Servorum nulla est unquam civitas.*

of society. These, mistaking the true source of their power, pretended to hold it from heaven; to be accountable for their actions to God alone; to owe nothing to society, in a word, to be Gods upon earth, and to possess the right of governing arbitrarily, as the God or Gods above. From thence politics became corrupted, they were only a mockery. Such nations, disgraced and grown contemptible, did not dare resist the will of their chiefs—their laws were nothing more than the expression of the caprice of these chiefs; public welfare was sacrificed to their peculiar interests—the force of society was turned against itself—its members withdrew to attach themselves to its oppressors, to its tyrants; these, to seduce them, permitted them to injure it with impunity, to profit by its misfortunes. Thus liberty, justice, security, virtue, were banished from many nations—politics was no longer any thing more than the art of availing itself of the forces of a people, of the treasure of society, of dividing it on the subject of its interest, in order to subjugate it by itself: at length a stupid and mechanical habit made them love their chains.

Man, when he has nothing to fear, presently becomes wicked; he who believes he has not occasion for his fellow, persuades himself he may follow the inclinations of his heart, without caution or discretion. Thus, fear is the only obstacle society can effectually oppose to the passions of its chiefs: without it they will quickly become corrupt, and will not scruple to avail themselves of the means society has placed in their hands to make them accomplices in their iniquity. To prevent these abuses it is requisite society should set bounds to its confidence; should limit the power which it delegates to its chiefs; should reserve to itself a sufficient portion of authority to prevent them from injuring it; it must establish prudent checks; it must cautiously divide the powers it confers, because united it will be infallibly oppressed. The slightest reflection will make men feel, that the burden of governing is too ponderous to be borne by an individual—that the scope and the multiplicity of his duties must always render him negligent—that the

extent of his power has ever a tendency to render him mischievous. In short, the experience of all ages will convince nations that man is continually tempted to the abuse of power: that therefore the sovereign ought to be subject to the law, not the law to the sovereign.

Government has necessarily an equal influence over the philosophy as over the morals of nations. In the same manner that its care produces labour, activity, abundance, salubrity, justice, and its negligence induces idleness, sloth, discouragement, penury, contagion, injustice, vices and crimes. It depends upon government either to foster industry, mature genius, give a spring to talents, or to stifle them. Indeed, government, the distributer of dignities, of riches, of rewards, of punishments—the master of those objects in which man from his infancy has learned to place his felicity—acquires a necessary influence over his conduct; it kindles his passions; gives them direction; makes him instrumental to whatever purpose it pleases: it modifies him; determines his manners; which in a whole people, as in the individual, is nothing more than the conduct, or the general system of wills and of actions that necessarily result from his education, his government, his laws, his religious opinions, his institutions, whether rational or irrational. In short, manners are the habits of a people: these are good whenever society draws from them true and solid happiness; they are detestable in the eye of reason, when the happiness of society does not spring from them, and when they have nothing more in their favour than the suffrage of time, or the countenance of prejudice, which rarely consults experience and good sense. If experience be consulted, it will be found there is no action, however abominable, that has not received the applause of some people. Parricide — the sacrifice of children- robbery—usurpation—cruelty—intolerance—prostitution, have all in their turn been licensed actions, and have been deemed laudable and meritorious deeds with some nations of the earth. Above all, Religion has consecrated the most unreasonable, the most revolting customs.

Man's passions depending on the motion of attraction and of repulsion of which he is rendered susceptible by nature, who enables him, by his peculiar essence, to be attracted by those objects which appear useful to him, to be repelled by those which he considers prejudicial; it follows that government, by holding the magnet, has the power either of restraining them, or of giving them a favourable or an unfavourable direction. All his passions are constantly limited by either loving or hating—seeking or avoiding—desiring or fearing. These passions, so necessary to the conservation of man, are a consequence of his organization, and display themselves with more or less energy, according to his temperament: education and habit develop them, and government conducts them towards those objects which it believes itself interested in making desirable to its subjects. The various names which have been given to these passions are relative to the different objects by which they are excited, such as pleasure—grandeur—riches, which produce voluptuousness—ambition—vanity—avarice. If the source of those passions which predominate in nations be attentively examined, it will be commonly found in their governments. It is the impulse received from their chiefs that renders them sometimes warlike—sometimes superstitious—sometimes aspiring after glory—sometimes greedy after wealth—sometimes rational—sometimes unreasonable. If sovereigns, in order to enlighten and to render happy their dominions, were to employ only the *tenth* part of the vast expenditures which they lavish, and only a *tithe* of the pains which they employ to stupify them—to deceive them—to afflict them, their subjects would presently be as wise and as happy, as they are now remarkable for being blind, ignorant, and miserable.

Let the vain project of destroying passions from the heart of man be abandoned; let an effort be made to direct them towards objects that may be useful to himself and to his associates. Let education, let government, let the laws, habituate him to restrain his passions within those just bounds which experience and reason prescribe. Let the ambitious have honours, titles, distinctions, power, when they shall have usefully served their country; let riches be given to those who covet them, when they shall have rendered themselves necessary to their fellow citizens; let eulogies encourage those who shall be actuated by the love of glory. In short, let the passions of man have a free course, whenever there shall result from their exercise real and durable advantages to society. Let education kindle only those which are truly beneficial to the human species; let it favour those alone which are really necessary to the maintenance of society. The passions of man are dangerous, only because every thing conspires to give them an evil direction.

Nature does not make man either good or wicked;* she combines machines more or less active, mobile, and energetic; she furnishes him with organs, with temperament, of which his passions, more or less impetuous, are the necessary consequence; these passions have always his happiness for their object; therefore they are legitimate and natural, and they can only be called bad or good, relatively to the influence they have on the beings of his species. Nature gives man legs proper to sustain his weight, necessary to transport him from one place to another; the care of those who rear them, strengthens them; habituates him to avail himself of them; accustoms him to make either a good or a bad use of them. The arm which he has received from nature is neither good nor bad; it is necessary to a great number of the actions of life; nevertheless the use of this arm becomes criminal if he has contracted the habit of using it to rob or to assassinate, with a view to obtain that money which he has been taught from his infancy to desire; which the society in which he lives renders necessary to him, but which his industry will enable him to obtain without doing injury to his fellow man.

The heart of man is a soil which nature has made equally suitable to the production of brambles or of useful grain—of deleterious poison or of refreshing fruit, by virtue of the seeds which may be sown in it—by the cultivation that may be bestowed upon it. In his infancy those objects are pointed

---

* Seneca has said with great reason.—Erras si existimes vitia nobiscum nasci; supervenerunt, ingesta sunt. V. *Senec. Epist.* 91, 95, 124

out to him which he is to estimate or to despise—to seek after or to avoid—to love or to hate. It is his parents and his instructers who render him either virtuous or wicked—wise or unreason-able—studious or dissipated—steady or trifling—solid or vain. Their example and their discourse modify him through his whole life, teaching him what are the things he ought either to desire or to avoid: he desires them in conse-quence; and he imposes on himself the task of obtaining them according to the energy of his temperament, which ever decides the force of his passions. It is thus that education, by inspiring him with opinions and ideas either true or false, gives him those primitive im-pulsions after which he acts in a man-ner either advantageous or prejudicial, both to himself and to others. Man, at his birth, brings with him into the world nothing but the necessity of con-serving himself and of rendering his existence happy: instruction, example, the customs of the world, present him with the means, either real or imagi-nary, of achieving it: habit procures for him the facility of employing these means; and he attaches himself strongly to those he judges best calculated to secure to him the possession of those objects which he has learned to desire as the preferable good attached to his existence. Whenever his education, whenever the examples which have been afforded him, whenever the means with which he has been provided, are approved by reason, are the result of experience, every thing concurs to ren-der him virtuous: habit strengthens these dispositions in him; and he be-comes, in consequence, a useful mem-ber of society, to the interests of which every thing ought to prove to him that his own permanent well-being is neces-sarily allied. If, on the contrary, his education—his institutions—the exam-ples which are set before him—the opinions which are suggested to him in his infancy, are of a nature to exhibit to his mind virtue as useless and repug-nant, and vice as useful and congenial to his own individual happiness, he will become vicious; he will believe him-self interested in injuring society; he will be carried along by the general current: he will renounce virtue, which to him will no longer be any thing more

No. III.—10

than a vain idol, without attractions to induce him to follow it; without charms to tempt his adoration, because it will appear to exact that he should immolate at its shrine all those objects which he has been constantly taught to consider the most dear to himself and as benefits the most desirable.

In order that man may become vir-tuous, it is absolutely requisite that he should have an interest or should find advantages in practising virtue. For this end, it is necessary that education should implant in him reasonable ideas; that public opinion should lean towards virtue as the most desirable good; that example should point it out as the object most worthy esteem; that government should faithfully reward it; that honour should always accompany its practice; that vice and crime should invariably be despised and punished. Is virtue in this situation amongst men? Does the education of man infuse into him just ideas of happiness; true notions of virtue; dispositions really favourable to the beings with whom he is to live? The examples spread before him, are they suitable to innocence of manners? are they calculated to make him respect decency—to cause him to love probity—to practise honesty—to value good faith—to esteem equity—to revere conjugal fidelity—to observe exactitude in ful-filling his duties? Religion, which alone pretends to regulate his manners, does it render him sociable—does it make him pacific—does it teach him to be humane? The arbiters of society, are they faithful in rewarding those who have best served their country, in punishing those who have plundered, divided, and ruined it? Justice, does she hold her scales with an even hand between all the citizens of the state? The laws, do they never support the strong against the weak; favour the rich against the poor; uphold the happy against the miserable? In short, is it an uncommon spectacle to behold crime frequently justified, or crowned with success, insolently triumphing over that merit which it disdains, over that virtue which it outrages? Well, then, in societies thus constituted, virtue can only be heard by a very small number of peaceable citizens, who know how to estimate its value, and who enjoy it in secret. For the others, it is only a

disgusting object, as they see in it
nothing but the supposed enemy to
their happiness, or the censor of their
individual conduct.

If man, according to his nature, is
necessitated to desire his welfare, he
is equally obliged to cherish the means
by which he believes it is to be acquired:
it would be useless, and perhaps unjust,
to demand that a man should be virtuous, if he could not be so without rendering himself miserable. Whenever
he thinks vice renders him happy, he
must necessarily love vice; whenever
he sees inutility or crime rewarded and
honoured, what interest will he find in
occupying himself with the happiness
of his fellow creatures, or in restraining
the fury of his passions ? In fine, whenever his mind is saturated with false
ideas and dangerous opinions, it follows
of course that his whole conduct will
become nothing more than a long chain
of errours, a series of depraved actions.

We are informed, that the savages,
in order to flatten the heads of their
children, squeeze them between two
boards, by that means preventing them
from taking the shape designed for
them by nature. It is pretty nearly the
same thing with the institutions of
man ; they commonly conspire to counteract nature—to constrain—to divert
—to extinguish the impulse nature has
given him, to substitute others which
are the source of all his misfortunes.
In almost all the countries of the earth
man is bereft of truth. is fed with falsehoods, is amused with marvellous chimeras : he is treated like those children
whose members are, by the imprudent
care of their nurses, swathed with little
fillets, bound up with rollers, which
deprive them of the free use of their
limbs, obstruct their growth, prevent
their activity, and oppose themselves
to their health.

Most of the religious opinions of man
have for their object only to display to
him his supreme felicity in those illusions for which they kindle his passions:
but as the phantoms which are presented to his imagination are incapable of
being considered in the same light by all
who contemplate them, he is perpetually
in dispute concerning these objects ; he
hates and persecutes his neighbour—
his neighbour in turn persecutes him—
he believes in doing this he is doing

well; that in committing the greatest
crimes to sustain his opinions he is acting right. It is thus religion infatuates
man from his infancy, fills him with
vanity and fanaticism : if he has a
heated imagination it drives him on to
fury ; if he has activity, it makes him
a madman, who is frequently as cruel
to himself, as he is dangerous and incommodious to others: if, on the contrary, he be phlegmatic or of a slothful
habit, he becomes melancholy and is
useless to society.

Public opinion every instant offers to
man's contemplation false ideas of honour and wrong notions of glory : it
attaches his esteem not only to frivolous advantages, but also to prejudicial
and injurious actions, which example
authorizes—which prejudice consecrates—which habit precludes him from
viewing with disgust, from eying with
the horrour they merit. Indeed, habit
familiarizes his mind with the most
absurd ideas—with the most unreasonable customs—with the most blameable actions—with prejudices the most
contrary to his own interests, the most
detrimental to the society in which he
lives. He finds nothing strange, nothing singular, nothing despicable, nothing ridiculous, except those opinions
and those objects to which he is himself unaccustomed. There are countries in which the most laudable actions
appear very blameable and extremely
ridiculous, and where the foulest, the
most diabolical actions, pass for very
honest and perfectly rational.*

Authority commonly believes itself
interested in maintaining the received
opinions ; those prejudices and those
errours which it considers requisite to
the maintenance of its power, are sustained by force, which is never ration-

---

* In some nations they kill the old men ; in
some the children strangle their fathers. The
Phenicians and the Carthagenians immolated
their children to their Gods. Europeans approve duels; and those who refuse to blow
out the brains of another are contemplated by
them as dishonoured. The Spaniards, the
Portuguese, think it meritorious to burn a
heretic. Christians deem it right to cut the
throats of those who differ from them in opinion. In some countries women prostitute
themselves without dishonour ; in others it is
the height of hospitality for man to present
his wife to the embraces of the stranger: the
refusal to accept this, elicits his scorn, calls
forth his resentment.

al. Princes filled with deceptive images of happiness; with mistaken notions of power; with erroneous opinions of grandeur; with false ideas of glory, are surrounded with flattering courtiers, who are interested in keeping up the delusion of their masters: these contemptible men have acquired ideas of virtue only that they may outrage it: by degrees they corrupt the people, these become depraved, lend themselves to their debaucheries, pander to the vices of the great, then make a merit of imitating them in their irregularities. A court is the true focus of the corruption of a people.

This is the true source of moral evil. It is thus that every thing conspires to render man vicious, to give a fatal impulse to his soul; from whence results the general confusion of society, which becomes unhappy from the misery of almost every one of its members. The strongest motive-powers are put in action to inspire man with a passion for futile or indifferent objects, which make him become dangerous to his fellow man by the means which he is compelled to employ in order to obtain them. Those who have the charge of guiding his steps, either impostors themselves, or the dupes to their own prejudices, forbid him to hearken to reason; they make truth appear dangerous to him, and exhibit errour as requisite to his welfare, not only in this world but in the next. In short, habit strongly attaches him to his irrational opinions—to his perilous inclinations—to his blind passion for objects either useless or dangerous. Here then is the reason why for the most part man finds himself necessarily determined to evil; the reason why the passions, inherent in his nature and necessary to his conservation, become the instruments of his destruction, the bane of that society which they ought to preserve. Here, then, the reason why society becomes a state of warfare, and why it does nothing but assemble enemies, who are envious of each other and always rivals for the prize. If some virtuous beings are to be found in these societies, they must be sought for in the very small number of those, who, born with a phlegmatic temperament, have moderate passions, who therefore either do not desire at all, or desire very feebly, those objects with which their associates are continually inebriated.

Man's nature diversely cultivated, decides upon his faculties, as well corporeal as intellectual—upon his qualities, as well moral as physical. The man who is of a sanguine, robust constitution, must necessarily have strong passions: he who is of a bilious, melancholy habit, will as necessarily have fantastical and gloomy passions: the man of a gay turn, of a sprightly imagination, will have cheerful passions; while the man, in whom phlegm abounds, will have those which are gentle, or which have a very slight degree of violence. It appears to be upon the equilibrium of the humours that depends the state of the man who is called *virtuous:* his temperament seems to be the result of a combination, in which the elements or principles are balanced with such precision, that no one passion predominates over another, or carries into his machine more disorder than its neighbour. Habit, as we have seen, is man's nature modified: this latter furnishes the matter; education, domestic example, national manners, give it the form: these acting on his temperament, make him either reasonable or irrational, enlightened or stupid, a fanatic or a hero, an enthusiast for the public good, or an unbridled criminal, a wise man smitten with the advantages of virtue or a libertine plunged into every kind of vice. All the varieties of the moral man depend on the diversity of his ideas, which are themselves arranged and combined in his brain by the intervention of his senses. His temperament is the produce of physical substances; his habits are the effect of physical modifications; the opinions, whether good or bad, injurious or beneficial, true or false, which form themselves in his mind, are never more than the effect of those physical impulsions which the brain receives by the medium of the senses.

———

CHAPTER X.

The Soul does not derive its Ideas from itself.
It has no innate Ideas.

WHAT has preceded suffices to prove that the interior organ of man, which is called his *soul*, is purely material. He will be enabled to convince himself of this truth, by the manner in which he acquires his ideas; from those impressions, which material objects successively make on his organs, which are themselves acknowledged to be material. It has been seen that the faculties which are called *intellectual*, are to be ascribed to that of feeling; the different qualities of those faculties, which are called moral, have been explained after the necessary laws of a very simple mechanism : it now remains to reply to those who still obstinately persist in making the soul a substance distinguished from the body, or who insist on giving it an essence totally distinct. They seem to found their distinction upon this, that this interior organ has the faculty of drawing its ideas from within itself; they will have it that man, at his birth, brings with him ideas into the world, which according to this wonderful notion, they have called *innate.** They have believed, then, that the soul, by a special privilege, in a nature where every thing is connected, enjoyed the faculty of moving itself without receiving any impulse; of creating to itself ideas, of thinking on a subject, without being determined to such action by any exterior object, which, by moving its organs, should furnish it with an image of the subject of its thoughts. In consequence of these gratuitous suppositions, which it is only requisite to expose in order to confute, some very able speculators, who were prepossessed by their superstitious prejudices, have ventured the length to assert, that, without model, without prototype, to act on the senses, the soul is competent

* Some ancient philosophers have held, that the soul originally contains the principles of several notions or doctrines: the Stoics designated this by the term Πρωληψις, *anticipated opinions*; the Greek mathematicians Κοιναι Εννοιαι, *universal ideas*. The Jews have a similar doctrine which they borrowed from the Chaldeans; their Rabbins taught that each soul, before it was united to the seed that must form an infant in the womb of a woman, is confided to the care of an angel, which causes him to behold heaven, earth, and hell: this, they pretend, is done by the assistance of a lamp which extinguishes itself, as soon as the infant comes into the world. See *Gaulmin. De vita et morte Mosis.* .

to delineate to itself the whole universe, with all the beings it contains. Descartes and his disciples have assured us, that the body went absolutely for nothing in the sensations or ideas of the soul; that it can feel—that it can perceive, understand, taste, and touch, even when there should exist nothing that is corporeal or material exterior to ourselves.

But what shall be said of a Berkeley, who has endeavoured to prove to man, that every thing in this world is nothing more than a chimerical illusion, and that the universe exists nowhere but in himself: that it has no identity but in his imagination ; who has rendered the existence of all things problematical by the aid of sophisms, insolvable even to those who maintain the doctrine of the spirituality of the soul.†

To justify such monstrous opinions, they assert that ideas are only the objects of thought. But according to the last analysis, these ideas can only reach man from exterior objects, which in giving impulse to his senses, modify his brain ; or from the material beings contained within the interior of his machine, who make some parts of his body experience those sensations which he perceives, and which furnish him with ideas, which he relates, faithfully or otherwise, to the cause that moves him. Each idea is an effect, but however difficult it may be to recur to the cause, can we possibly suppose it is not ascribable to a cause? If we can only form ideas of material substances,

† Extravagant as this doctrine of the bishop of Cloyne may appear, it cannot well be more so than that of Malebranche, the champion of innate ideas, who makes the divinity the common bond between the soul and the body : or than that of those metaphysicians who maintain, that the soul is a substance heterogeneous to the body, and, who, by ascribing to this soul the thoughts of man, have, in fact, rendered the body superfluous. They have not perceived, they were liable to one solid objection, which is, that if the ideas of man are innate, if he derives them from a superior being, independent of exterior causes, if he sees every thing in God ; how comes it that so many false ideas are afloat, that so many errours prevail with which the human mind is saturated ? From whence come those opinions which, according to the theologians, are so displeasing to God ? Might it not be a question to the Malebranchists, was it in the Divinity that Spinosa beheld his system ?

how can we suppose the cause of our ideas can possibly be immaterial? To pretend that man, without the aid of exterior objects, without the intervention of his senses, is competent to form ideas of the universe, is to assert, that a blind man is in a capacity to form a true idea of a picture that represents some fact of which he has never heard any one speak.

It is very easy to perceive the source of those errours into which men, otherwise extremely profound and very enlightened, have fallen, when they have been desirous to speak of the soul and of its operations. Obliged, either by their own prejudices, or by the fear of combating the opinions of an imperious theology, they have become the advocates of the principle, that the soul was a *pure spirit*, an immaterial substance, of an essence directly different from that of the body, or from every thing we behold: this granted, they have been incompetent to conceive how material objects could operate, or in what manner gross and corporeal organs were enabled to act on a substance that had no kind of analogy with them, and how they were in a capacity to modify it by conveying it ideas; in the impossibility of explaining this phenomenon, at the same time perceiving that the soul had ideas, they concluded that it must draw them from itself, and not from those beings, which, according to their own hypothesis, were incapable of acting on it; they therefore imagined that all the modifications of this soul, sprung from its own peculiar energy, were imprinted on it from its first formation by the author of nature —an immaterial being like itself; and that these did not in any manner depend upon the beings of which we have a knowledge, or which act upon it by the gross means of our senses.

There are, however, some phenomena which, considered superficially, appear to support the opinion of these philosophers, and to announce a faculty in the human soul of producing ideas within itself, without any exterior aid; these are *dreams*, in which the interior organ of man, deprived of objects that move it visibly, does not, however, cease to have ideas, to be set in activity, and to be modified in a manner that is sufficiently sensible to have an influence upon his body. But if a little reflection be called in, the solution to this difficulty will be found : it will be perceived, that, even during sleep, his brain is supplied with a multitude of ideas, with which the eye or time before has stocked it ; these ideas were communicated to it by exterior and corporeal objects, by which it has been modified : it will be found that these modifications renew themselves, not by any spontaneous or voluntary motion on its part, but by a chain of involuntary movements which take place in his machine, which determine or excite those that give play to the brain; these modifications renew themselves with more or less fidelity, with a greater or lesser degree of conformity to those which it has anteriorly experienced. Sometimes in dreaming he has memory, then he retraces to himself the objects which have struck him faithfully ; at other times, these modifications renew themselves without order, without connexion, or very differently from those which real objects have before excited in his interior organ. If in a dream he believe he sees a friend, his brain renews in itself the modifications or the ideas which this friend had formerly excited, in the same order that they arranged themselves when his eyes really beheld him ; this is nothing more than an effect of memory. If, in his dream, he fancy he sees a monster which has no model in nature, his brain is then modified in the same manner that it was by the particular or detached ideas with which it then does nothing more than compose an ideal whole, by assembling and associating, in a ridiculous manner, the scattered ideas that were consigned to its keeping; it is then, that in dreaming he has imagination.

Those dreams that are troublesome, extravagant, whimsical, or unconnected, are commonly the effect of some confusion in his machine; such as painful indigestion, an overheated blood, a prejudicial fermentation, &c. —these material causes excite in his body a disorderly motion, which precludes the brain from being modified in the same manner it was on the day before ; in consequence of this irregular motion, the brain is disturbed, it only represents to itself confused ideas that want connexion.. When in a dream he believes he sees a sphinx,* either he has seen the representation of one when he was awake, or else the disorderly motion of the brain is such, that it causes it to combine ideas, to connect parts, from which there results a whole without model, of which the parts were not formed to be united. It is thus, that his brain combines the head of a woman, of which it already has the idea, with the body of a lioness, of which it also has the image. In this his head acts in the same manner as when, by any defect in the interior organ, his disordered imagination paints to him some objects, notwithstanding he is awake. He frequently dreams without being asleep: his dreams never produce any thing so strange but that they have some resemblance with the objects which have anteriorly acted on his senses, or have already communicated ideas to his brain. The crafty theologians have composed at their leisure, and in their waking hours, those phantoms of which they avail themselves to terrify man; they have done nothing more than assemble the scattered traits which they have found in the most terrible beings of their own species; by exaggerating the powers and the rights claimed by tyrants, they have formed Gods before whom man trembles.

Thus it is seen that dreams, far from proving that the soul acts by its own peculiar energy, or draws its ideas from its own recesses, prove, on the contrary, that in sleep it is entirely passive, that it does not even renew its modifications, but according to the involuntary confusion, which physical causes produce in the body, of which every thing tends to show the identity and the consubstantiality with the soul. What appears to have led those into a mistake, who maintained that the soul drew its ideas from itself, is this, they have contemplated these ideas as if they were real beings, when, in point of fact, they are nothing more than the modifications produced in the brain of man by objects to which this brain is

a stranger; they are these objects, who are the true models or archetypes to which it is necessary to recur: here is the source of their errours.   .

In the individual who dreams, the soul does not act more from itself than it does in the man who is drunk, that is to say, who is modified by some spirituous liquor; or than it does in the sick man when he is delirious, that is to say, when he is modified by those physical causes which disturb his machine in the performance of its functions; or than it does in him whose brain is disordered: dreams, like these various states, announce nothing more than a physical confusion in the human machine, under the influence of which the brain ceases to act after a precise and regular manner: this disorder may be traced to physical causes, such as the aliments, the humours, the combinations, the fermentations, which are but little analogous to the salutary state of man; from which it will appear, that his brain is necessarily confused whenever his body is agitated in an extraordinary manner.

Do not let him, therefore, believe that his soul acts by itself, or without a cause, in any one moment of his existence; it is, conjointly with the body, submitted to the impulse of beings who act on him necessarily, and according to their various properties. Wine, taken in too great a quantity, necessarily disturbs his ideas, causes confusion in his corporeal functions, occasions disorder in his mental faculties.

If there really existed a being in nature with the capability of moving itself by its own peculiar energies, that is to say, able to produce motion independent of all other causes, such a being would have the power of arresting itself, or of suspending the motion of the universe, which is nothing more than an immense chain of causes linked one to the other, acting and reacting by necessary and by immutable laws, which cannot be changed or suspended,—unless the essences of every thing in it were changed—nay, annihilated. In the general system of the world, nothing more can be perceived than a long series of motion, received and communicated in succession by beings capacitated to give impulse to each other: it is thus that each body is moved, by the collision

* A being supposed by the poets to have a head and face like a woman, a body like a dog, wings like a bird, and claws like a lion, who put forth riddles and killed those who could not expound them.

of some other body. The invisible motion of his soul is to be attributed to causes concealed within himself; he believes that it is moved by itself, because he does not see the springs which put it in motion, or because he conceives those motive-powers are incapable of producing the effects he so much admires: but, does he more clearly conceive how a spark in exploding gunpowder is capable of producing the terrible effects he witnesses? The source of his errours arises from this, that he regards his body as gross and inert, whilst this body is a sensible machine, which has necessarily an instantaneous conscience the moment it receives an impression, and which is conscious of its own existence by the recollection of impressions successively experienced; memory, by resuscitating an impression anteriorly received, by detaining it, or by causing an impression which it receives to remain, whilst it associates it with another, then with a third, gives all the mechanism of *reasoning.*

An idea, which is only an imperceptible modification of the brain, gives play to the organ of speech, which displays itself by the motion it excites in the tongue: this, in its turn, breeds ideas, thoughts, passions, in those beings who are provided with organs susceptible of receiving analogous motion; in consequence of which, the wills of a great number of men are influenced, who, combining their efforts, produce a revolution in a state, or even have an influence over the entire globe. It is thus that an Alexander decided the fate of Asia ; it is thus that a Mahomet changed the face of the earth ; it is thus that imperceptible causes produce the most terrible, the most extended effects, by a series of necessary motion imprinted on the brain of man.

The difficulty of comprehending the effects produced on the soul of man, has made him attribute to it those incomprehensible qualities which have been examined. By the aid of imagination, by the power of thought, this soul appears to quit his body, to transport itself with the utmost facility towards the most distant objects; to run over and to approximate in the twinkling of an eye all the points of the universe: he has therefore believed that a being,

who is susceptible of such rapid motion, must be of a nature very distinguished from all others ; he has persuaded himself that this soul in reality does travel, that it actually springs over the immense space necessary to meet these various objects ; he did not perceive, that to do it in an instant, it had only to run over itself, and approximate the ideas consigned to its keeping by means of the senses.

Indeed, it is never by any other means than by his senses, that beings become known to man, or furnish him with ideas ; it is only in consequence of the impulse given to his body, that his brain is modified ; or that his soul thinks, wills, and acts. If, as Aristotle asserted more than two thousand years ago, " *nothing enters the mind of man, but through the medium of his senses* ;" it follows as a consequence, that every thing that issues from it, must find some sensible object to which it can attach its ideas, whether immediately, as a man, a tree, a bird, &c., or in the last analysis or decomposition, such as pleasure, happiness, vice, virtue, &c.* Whenever, therefore, a word or its idea, does not connect itself with some sensible object, to which it can be related, this word, or this idea, is unmeaning, is void of sense : it were better for man that the idea was banished from his mind, struck out of his language. This principle is only the converse of the axiom of Aristotle ; if the direct be evident, the inverse must be so likewise.

How has it happened, that the profound Locke, who, to the great mortification of the metaphysicians, has placed this principle of Aristotle in the clearest point of view; how is it that all those who, like him, have recognised the absurdity of the system of innate ideas, have not drawn the immediate and necessary consequences ? How has it come to pass, that they have not had sufficient courage to apply so clear a principle to all those fanciful chimeras with which the human mind has for such a length of time been so vainly occupied ? Did they not perceive, that their principle sapped the very founda

---

* This principle, so true, so luminous, so important in its consequence, has been set forth in all its lustre by a great number of philosophers; among the rest, by the great Locke.

tions of that theology, which never occupies man but with those objects, of which, as they are inaccessible to his senses, he, consequently, can never form to himself any accurate idea? But prejudice, particularly when it is held sacred, prevents him from seeing the most simple application of the most self-evident principles; in religious matters, the greatest men are frequently nothing more than children, who are incapable of either foreseeing or deducing the consequence of their own data.

Locke, as well as all those who have adopted his system, which is so demonstrable, or the axiom of Aristotle, which is so clear, ought to have concluded from it, that all those wonderful things with which theologians have amused themselves, are mere chimeras; that an immaterial spirit or substance, without extent, without parts, is nothing more than an absence of ideas; in short, they ought to have felt, that the ineffable intelligence which they have supposed to preside at the helm of the world, is nothing more than a being of their own imagination, of which it is impossible his senses can ever prove either the existence or the qualities.

For the same reason moral philosophers ought to have concluded, that what is called *moral sentiment, moral instinct,* that is, innate ideas of virtue, anterior to all experience of the good or bad effects resulting from its practice, are mere chimerical notions, which, like a great many others, have for their guarantee and base only theological speculation.* Before man can judge, he must feel; before he can distinguish good from evil, he must compare.

To undeceive him with respect to innate ideas or modifications imprinted on his soul at the moment of his birth, it is simply requisite to recur to their source; he will then see, that those

---

* Morals is a science of facts: to found it, therefore, on an hypothesis inaccessible to his senses, of which he has no means of proving the reality, is to render it uncertain; it is to cast the log of discord into his lap; to cause him unceasingly to dispute upon that which he can never understand. To assert that the ideas of morals are *innate,* or the effect of *instinct,* is to pretend that man knows how to read before he has learned the letters of the alphabet; that he is acquainted with the laws of society, before they are either made or promulgated.

with which he is familiar, which have, as it were, identified themselves with his existence, have all come to him through the medium of some of his senses; that they are sometimes engraven on his brain with great difficulty; that they have never been permanent, and that they have perpetually varied in him : he will see that these pretended inherent ideas of his soul, are the effect of education, of example, above all, of habit, which, by reiterated motion, has taught his brain to associate his ideas, either in a confused or perspicuous manner; to familiarize itself with systems, either rational or absurd. In short, he takes those for innate ideas, of which he has forgotten the origin; he no longer recalls to himself either the precise epoch or the successive circumstances when these ideas were first consigned to his brain: arrived at a certain age, he believes he has always had the same notions; his memory, crowded with experience and a multitude of facts, is no longer able to distinguish the particular circumstances which have contributed to give his brain its present modifications, its instantaneous mode of thinking, its actual opinions. For example, not one of his race recollects the first time the word God struck his ears, the first ideas that it formed in him, the first thoughts that it produced in him; nevertheless, it is certain that from thence he has searched for some being with whom to connect the idea which he has either formed to himself, or which has been suggested to him: accustomed to hear God continually spoken of, he has, when in other respects most enlightened, regarded this idea as if it were infused into him by nature; whilst it is clearly to be attributed to those delineations of it which his parents or his instructers have made to him, and which he has afterwards modified according to his own particular organization, and the circumstances in which he has been placed: it is thus that each individual forms to himself a God of which he is himself the model, or which he modifies after his own fashion.†

His ideas of morals, although more real than those of metaphysics, are not, however, *innate :* the moral sentiments he forms on the will, or the judgment

---

† See Vol. II., Chapter iv.

he passes on the actions of man, are founded on experience, which, alone, can enable him to discriminate those which are either useful or prejudicial, virtuous or vicious, honest or dishonest, worthy his esteem or deserving his censure. His moral sentiments are the fruit of a multitude of experience, frequently very long and very complicated. He gathers it with time: it is more or less faithful, by reason of his particular organization, and the causes by which he is modified; he ultimately applies this experience with greater or lesser facility, and on this depends his habit of judging. The celerity with which he applies his experience, when he judges of the moral actions of his fellow man, is what has been termed *moral instinct.* ·┼·

That which in natural philosophy is called *instinct*, is only the effect of some want of the body, the consequence of some attraction, or some repulsion, in man or animals. The child that is newly born, sucks for the first time: the nipple of the breast is put into his mouth: the natural analogy that is found between the conglomerate glands which line his mouth, and the milk which flows from the bosom of the nurse through the medium of the nipple, causes the child to press it with his mouth, in order to express the fluid appropriate to nourish his tender age; from all this the infant gathers experience; by degrees the ideas of a nipple, of milk, of pleasure, associate themselves in his brain, and every time he sees the nipple, he seizes it, promptly conveys it to his mouth, and applies it to the use for which it is designed.

What has been said will enable us to judge of those prompt and sudden sentiments, which have been designated *the force of blood.* Those sentiments of love, which fathers and mothers have for their children; those feelings of affection, which children, with good inclinations, bear towards their parents, are by no means innate sentiments; they are nothing more than the effect of experience, of reflection, of habit, in souls of sensibility. These sentiments do not even exist in a great number of human beings. We but too often witness tyrannical parents, occupied with making enemies of their children, who appear to have been formed only to

No. III.—11

be the victims of their irrational caprices.

From the instant in which man commences, until that in which he ceases to exist, he feels, he is moved either agreeably or unpleasantly, he collects facts, he gathers experience, which produce ideas in his brain that are either cheerful or gloomy. Not one individual has this experience present to his memory at the same time, nor does it ever represent to him the whole clew at once: it is however this experience that mechanically, and without his knowledge, directs him in all his actions; it was to designate the rapidity with which he applied this experience, of which he so frequently loses the connexion, of which he is so often at a loss to render himself an account, that he imagined the word *instinct:* it appears to be the effect of a magical and supernatural power to the greater number of individuals; but it is a word devoid of sense to many others; however, to the philosopher it is the effect of a very lively feeling, which, to him, consists in the faculty of combining promptly a multitude of experiences and a long and numerous train of extremely complicated ideas. It is want that causes the inexplicable instinct we behold in animals, which have been denied souls without reason; whilst they are susceptible of an infinity of actions that prove they think, they judge, have memory, are capable of experience, can combine ideas, can apply them with more or less facility to satisfy the wants engendered by their particular organization; in short, that prove they have passions, and that these are capable of being modified.*

The embarrassments which animals have thrown in the way of the partisans of the doctrine of spirituality is well known: they have been fearful, if they allowed them to have a spiritual soul, of elevating them to the condition of human creatures; on the other hand, in not allowing them to have a soul, they have furnished their adversaries with authority to deny it in like manner

---

* Nothing but the height of folly can refuse intellectual faculties to animals; they feel, choose, deliberate, express love, show hatred; in many instances their senses are much keener than those of man. Fish will return periodically to the spot where it is the custom to throw them bread.

OF THE SOUL.

to man, who thus finds himself debased to the condition of the animal. Theologians have never known how to extricate themselves from this difficulty. Descartes fancied he solved it by saying that beasts have no souls, are mere machines. Nothing can be nearer the surface than the absurdity of this principle. Whoever contemplates nature without prejudice, will readily acknowledge, that there is no other difference between the man and the beast than that which is to be attributed to the diversity of his organization.

In some beings of the human species, who appear to be endowed with a greater sensibility of organs than others, may be seen *an instinct*, by the assistance of which they very promptly judge of the concealed dispositions of their fellows, simply by inspecting the lineaments of their face. Those who are denominated *physiognomists*, are only men of very acute feelings, who have gathered an experience of which others, whether from the coarseness of their organs, from the little attention they have paid, or from some defect in their senses, are totally incapable: these last do not believe in the science of physiognomy, which appears to them perfectly ideal. Nevertheless, it is certain that the action of this soul, which has been made spiritual, makes impressions that are extremely marked upon the exterior of the body; these impressions continually reiterated, their image remains: thus, the habitual passions of man paint themselves on his countenance, by which the attentive observer, who is endowed with acute feeling, is enabled to judge with great rapidity of his mode of existence, and even to foresee his actions, his inclinations, his desires, his predominant passions, &c. Although the science of physiognomy appears chimerical to a great number of persons, yet there are few who have not a clear idea of a tender regard, of a cruel eye, of an austere aspect, of a false and dissimulating look, of an open countenance, &c. Keen and practised optics acquire, without doubt, the faculty of penetrating the concealed motion of the soul, by the visible traces it leaves upon features that it has continually modified. Above all, the eyes of man very quickly undergo changes, according to the motion which is excited in

him: these delicate organs are visibly altered by the smallest shock communicated to his brain. Serene eyes announce a tranquil soul; wild eyes indicate a restless mind; fiery eyes portray a choleric and sanguine temperament; fickle or inconstant eyes give room to suspect a soul either alarmed or dissimulating. It is the study of this variety of shades that renders man practised and acute: upon the spot he combines a multitude of acquired experience, in order to form his judgment of the person he beholds. His judgment partakes in nothing of the supernatural or the wonderful: such a man is only distinguished by the fineness of his organs, and by the celerity with which his brain performs its functions.

It is the same with some beings of the human species, in whom may be discovered an extraordinary sagacity, which to the uninformed appears Divine and miraculous.* Indeed, we see men who are capable of appreciating in the twinkling of an eye a multitude of circumstances, and who have sometimes the faculty of foreseeing the most distant events, yet this species of *prophetic* talent has nothing in it of the supernatural; it indicates nothing more than great experience, with an extremely delicate organization, from which they derive the faculty of judging with extreme facility of causes, and of foreseeing their very remote effects. . This faculty is also found in animals, who foresee much better than man the variations of the atmosphere, with the various changes of the weather. Birds have long been the prophets and even the guides of several nations who pretend to be extremely enlightened.

It is, then, to their organization, exercised after a particular manner, that must be attributed those wondrous faculties which distinguish some beings. To have *instinct* only signifies to judge quickly, without requiring to make a long reasoning on the subject. Man's ideas upon vice and upon virtue are by no means innate; they are, like

* It appears that the most skilful practitioners in medicine have been men endowed with very acute feelings, similar to those of the physiognomists, by the assistance of which they judged with great facility of diseases, and very promptly drew their prognostics.

all others, acquired; the judgment he forms is founded upon experience, whether true or false: this depends upon his conformation, and upon the habits that have modified him. The infant has no ideas either of the Divinity or of virtue: it is from those who instruct him that he receives these ideas: he makes more or less use of them, according to his natural organization, or as his dispositions have been more or less exercised. Nature gives man legs, the nurse teaches him their use, his agility depends upon their natural conformation, and the manner in which he exercises them.

What is called *taste* in the fine arts, is to be attributed, in the same manner, only to the acuteness of man's organs practised by the habit of seeing, of comparing, and of judging certain objects: from whence results, to some of his species, the faculty of judging with great rapidity, or in the twinkling of an eye, the whole with its various relations. It is by the force of seeing, of feeling, of experiencing objects, that he attains to a knowledge of them; it is in consequence of reiterating this experience, that he acquires the power and the habit of judging with celerity. But this experience is by no means *innate*, for he did not possess it before he was born; he is neither able to think, to judge, nor to have ideas, before he has feeling; he is neither in a capacity to love nor to hate; to approve nor to blame, before he has been moved either agreeably or disagreeably. This is, however, what must be supposed by those who are desirous to make man admit *innate ideas*, or opinions infused by nature, whether in morals, theology, or in any science. That his mind should have the faculty of thought, and should occupy itself with an object, it is requisite it should be acquainted with its qualities; that it may have a knowledge of these qualities, it is necessary that some of his senses should have been struck by them: those objects, therefore, of which he does not know any of the qualities are nullities, or at least they do not exist for him.

It will be asserted, perhaps, that the universal consent of man upon certain propositions, such as *the whole is greater than its part*, and upon all geometrical demonstrations, appear to warrant the supposition of certain primary notions that are innate, or not acquired. It may be replied, that these notions are always acquired, and that they are the fruit of an experience more or less prompt; that it is requisite to have compared the whole with its part before conviction can ensue that the whole is the greater of the two. Man, when he is born, does not bring with him the idea that two and two make four; but he is, nevertheless, very speedily convinced of its truth. Before forming any judgment whatever, it is absolutely necessary to have compared facts.

It is evident that those who have gratuitously supposed innate ideas, or notions inherent in man, have confounded his organization, or his natural dispositions, with the habit by which he is modified, and with the greater or less aptitude he has of making experiments, and of applying them in his judgment. A man who has taste in painting, has, without doubt, brought with him into the world eyes more acute and more penetrating than another; but these eyes would by no means enable him to judge with promptitude if he had never had occasion to exercise them; much less, in some respects, can those dispositions which are called *natural* be regarded as innate. Man is not at twenty years of age the same as he was when he came into the world; the physical causes that are continually acting upon him, necessarily have an influence upon his organization, and so modify it, that his natural dispositions themselves are not at one period what they are at another.* Every day may be seen children who, to a certain age, display a great deal of ingenuity, a strong aptitude for the sciences, and who finish by falling into stupidity. Others may be observed, who, during their infancy, have shown dispositions but little favourable to improvement, yet develop themselves in the end, and astonish us

* "We think," says La Motte Le Vayer, "quite otherwise of things at one time than at another: when young than when old—when hungry than when our appetite is satisfied—in the night than in the day—when peevish than when cheerful; thus varying every hour, by a thousand other circumstances which keep us in a state of perpetual inconstancy and instability."

by an exhibition of those qualities of which we judged them deficient: there arrives a moment in which the mind makes use of a multitude of experience which it has amassed without its having been perceived, and, if I may be allowed the expression, without their own knowledge.

Thus, it cannot be too often repeated, all the ideas, all the notions, all the modes of existence, all the thoughts of man are acquired. His mind cannot act and exercise itself but upon that of which it has knowledge; it can understand either well or ill only those things which it has previously felt. Such of his ideas that do not suppose some exterior material object for their model, or one to which he is able to relate them, which are therefore called *abstract ideas*, are only modes in which his interior organ considers its own peculiar modifications, of which it chooses some without respect to others. The words which he uses to designate these ideas, such as *bounty, beauty, order, intelligence, virtue,* &c., do not offer any one sense if he does not relate them to, or if he does not explain them by those objects which his senses have shown him to be susceptible of those qualities, or of those modes of existence and of acting, which are known to him. What is it that points out to him the vague idea of *beauty,* if he does not attach it to some object that has struck his senses in a particular manner, to which, in consequence, he attributes this quality? What is it that represents the word *intelligence,* if he does not connect it with a certain mode of being and of acting? Does the word *order* signify any thing, if he does not relate it to a series of actions, to a chain of motion, by which he is affected in a certain manner? Is not the word *virtue* void of sense, if he does not apply it to those dispositions of his fellows which produce known effects, different from those which result from contrary inclinations? What do the words *pain* and *pleasure* offer to his mind in the moment when his organs neither suffer nor enjoy, if it be not the modes in which he has been affected, of which his brain conserves the remembrance or the impressions, and which experience has shown him to be either useful or prejudicial? But

when he hears the words *spirituality, immateriality, incorporeality, divinity,* &c., pronounced, neither his senses nor his memory afford him any assistance: they do not furnish him with any means by which he can form an idea of their qualities, nor of the objects to which he ought to apply them: in that which is not matter, he can only see vacuum and emptiness, which cannot be susceptible of any one quality.

All the errours and all the disputes of men, have their foundation in this, that they have renounced experience and the evidence of their senses, to give themselves up to the guidance of notions which they have believed *infused* or *innate,* although in reality they are no more than the effect of a distempered imagination; of prejudices in which they have been instructed from their infancy; with which habit has familiarized them; and which authority has obliged them to conserve. Languages are filled with abstract words, to which are attached confused and vague ideas; of which, when they come to be examined, no model can be found in nature; no object to which they can be related. When man gives himself the trouble to analyze things, he is quite surprised to find that those words which are continually in the mouths of men, never present any fixed and determinate idea: he hears them unceasingly speaking of *spirits*—of the *soul* and its faculties—of *God* and his attributes —of *duration*—of *space*—of *immensity*—of *infinity*—of *perfection*—of *virtue*—of *reason*—of *sentiment*—of *instinct*—of *taste,* &c., without his being able to tell precisely what they themselves understand by these words. And yet words appear to have been invented but for the purpose of representing the images of things, or to paint, by the assistance of the senses, those known objects on which the mind is able to meditate, which it is competent to appreciate, to compare, and to judge. •

For man to think of that which has not acted on any of his senses, is to think on words: it is a dream of sounds; it is to seek in his own imagination for objects to which he can attach his wandering ideas. To assign qualities to these objects is, unquestionably, to redouble his extravagance. The word

God is destined to represent to him an object that has not the capacity to act on any one of his organs, of which, consequently, it is impossible for him to prove either the existence or the qualities; still, his imagination, by dint of racking itself, will in some measure supply him with the ideas he wants, and compose some kind of a picture with the images or colours he is always obliged to borrow from those objects of which he has a knowledge: thus the Divinity has been represented under the character of a venerable old man, or under that of a puissant monarch, &c. It is evident, however, that man with some of his qualities has served for the model of this picture. But if he be informed that this God is a pure spirit; that has neither body nor extent; that he is not contained in space; that he is beyond nature; here then he is plunged into emptiness; his mind no longer has any ideas: it no longer knows upon what it meditates. This, as will be seen in the sequel, is the source of those unformed notions which men have formed of the divinity; they themselves annihilate him, by assembling incompatible and contradictory attributes.* In giving him moral and known qualities, they make him a man; in assigning him the negative attributes of theology, they destroy all antecedent ideas; they make him a mere nothing—a chimera. From this it will appear that those sublime sciences which are called *theology, psychology, metaphysics*, have been mere sciences of words: morals and politics, which they too often infect, have, in consequence, become inexplicable enigmas, which nothing short of the study of nature can enable us to expound.

Man has occasion for truth; it consists in a knowledge of the true relations he has with those things which can have an influence on his welfare: these relations are to be known only by experience: without experience there can be no reason; without reason man is only a blind creature who conducts himself by chance. But how is he to acquire experience upon ideal objects, which his senses neither enable him to know nor to examine? How is he to assure himself of the existence and

the qualities of beings he is not able to feel? How can he judge whether these objects be favourable or prejudicial to him? How is he to know what he ought to love, what he should hate, what to seek after, what to shun, what to do, what to leave undone? Yet it is upon this knowledge that his condition in this world rests—the only world of which he knows any thing; it is upon this knowledge that morals is founded. From whence it may be seen, that, by causing him to blend vague theological notions with morals, or the science of the certain and invariable relations which subsist between mankind, or by weakly establishing them upon chimerical beings, which have no existence but in his imagination, this science, upon which the welfare of society so much depends, is rendered uncertain and arbitrary, is abandoned to the caprices of fancy, is not fixed upon any solid basis.

Beings essentially different by their natural organization, by the modifications they experience, by the habits they contract, by the opinions they acquire, must of necessity think differently. His temperament, as we have seen, decides the mental qualities of man; this temperament itself, is diversely modified in him; from whence it consecutively follows, his imagination cannot possibly be the same, neither can it create to him the same images. Each individual is a connected whole, of which all the parts have a necessary correspondence. Different eyes must see differently, must give extremely varied ideas of the objects they contemplate, even when these objects are real. What, then, must be the diversity of these ideas if the objects meditated upon do not act upon the senses? Mankind have pretty nearly the same ideas, in the gross, of those substances that act on his organs with vivacity; he is sufficiently in unison upon some qualities which he contemplates very nearly in the same manner; I say *very nearly*, because the intelligence, the notion, the conviction of any one proposition, however simple, however evident, however clear it may be supposed, is not, nor cannot be strictly the same in any two men. Indeed, one man not being another man, the first cannot, for example, have rigorously and mathematically the same notion of unity as the second, seeing that an identical effect cannot be the result of two different causes. Thus when men agree in their ideas, in their modes of thinking, in their judgment, in their passions, in their desires, and in their tastes, their consent does not arise from their seeing or feeling the same objects precisely in the same manner, but pretty nearly, for language is not, nor cannot be, sufficiently copious to designate the vast variety of shades, the multiplicity of imperceptible differences which are to be found in their modes of seeing and thinking. Each man has, I may say, a language which is peculiar to himself alone, and this language is incommunicable to others. What harmony, then, can possibly exist between them when they discourse with each other upon objects only known to their imagination? Can this imagination in one individual, ever be the same as in another? How can they possibly understand each other when they assign to these objects qualities that can only be attributed to the particular manner in which their brain is affected.

For one man to exact from another that he shall think like himself, is to insist that he shall be organized precisely in the same manner, that he shall have been modified exactly the same in every moment of his existence; that he shall have received the same temperament, the same nourishment, the same education; in a word, that he shall require that other to be himself. Wherefore is it not exacted that all men shall have the same features? Is man more the master of his opinions? Are not his opinions the necessary consequence of his nature, and of those peculiar circumstances which, from his infancy, have necessarily had an influence upon his mode of thinking and manner of acting? If man be a connected whole, whenever a single feature differs from his own, ought he not to conclude that it is not possible his brain can either think, associate ideas, imagine, or dream precisely in the same manner with that other.

The diversity in the temperament of man is the natural and necessary source of the diversity of his passions, of his taste, of his ideas of happiness,

of his opinions of every kind. Thus the same diversity will be the fatal source of his disputes, of his hatreds, and of his injustice, every time he shall reason upon unknown objects, but to which he shall attach the greatest importance. He will never understand either himself or others in speaking of a spiritual soul, or of an immaterial God distinguished from nature; he will, from that moment, cease to speak the same language, and he will never attach the same ideas to the same words. What, then, shall be the common standard that shall decide which is the man that thinks most correctly? What is the scale by which to measure who has the best regulated imagination? what balance shall be found sufficiently exact to determine whose knowledge is most certain when he agitates subjects which experience cannot enable him to examine; that escape all his senses; that have no model; that are above reason? Each individual, each legislator, each speculator, each nation, has ever formed to himself different ideas of these things, and each believes that his own peculiar reveries ought to be preferred to those of his neighbours; which always appear to him as absurd, as ridiculous, as false as his own can possibly have appeared to his fellow Each clings to his own opinion, because each retains his own peculiar mode of existence, and believes his happiness depends upon his attachment to his prejudices, which he never adopts but because he believes them beneficial to his welfare. Propose to a man to change his religion for yours, he will believe you a madman; you will only excite his indignation, elicit his contempt; he will propose to you, in his turn, to adopt his own peculiar opinions; after much reasoning, you will treat each other as absurd beings, ridiculously opiniated and stubborn; and he will display the least folly who shall first yield. But if the adversaries become heated in the dispute, which always happens when they suppose the matter important, or when they would defend the cause of their own self-love, then their passions sharpen, they grow angry, quarrels are provoked, they hate each other, and end by reciprocal injury. It is thus, that for opinions which no man can demonstrate, we see the Brahmin despised; the Mohammedan hated; the Pagan held in contempt; and that they oppress and disdain each other with the most rancorous animosity: the Christian burns the Jew because he clings to the faith of his fathers; the Roman Catholic condemns the Protestant to the flames, and makes a conscience of massacring him in cold blood; this reacts in his turn; again the various sects of Christians have leagued together against the incredulous, and for a moment suspended their own bloody disputes, that they might chastise their enemies: then, having glutted their revenge, they returned with redoubled fury to wreak over again their infuriated vengeance on each other.

If the imaginations of men were the same, the chimeras which they bring forth would be everywhere the same; there would be no disputes among them on this subject if they all dreamt in the same manner; great numbers of human beings would be spared, if man occupied his mind with objects capable of being known, of which the existence was proved, of which he was competent to discover the true qualities by sure and reiterated experience. *Systems of philosophy* are subject to dispute only when their principles are not sufficiently proved; by degrees experience, in pointing out the truth, terminates these quarrels. There is no variance among *geometricians* upon the principles of their science; it is only raised when their suppositions are false, or their objects too much complicated. Theologians find so much difficulty in agreeing among themselves, simply because in their contests they divide without ceasing, not known and examined propositions, but prejudices with which they have been imbued in their youth, in the schools, in their books, &c. They are perpetually reasoning, not upon real objects, of which the existence is demonstrated, but upon imaginary systems, of which they have never examined the reality; they found these disputes not upon averred experience nor upon constant facts, but upon gratuitous suppositions, which each endeavours to convince the other are without solidity. Finding these ideas of long standing, and that few people refuse to admit them, they

take them for incontestable truths, that ought to be received merely upon being announced; whenever they attach great importance to them, they irritate themselves against the temerity of those who have the audacity to doubt, or even to examine them.

If prejudice had been laid aside, it would perhaps have been discovered that many of those objects which have given birth to the most shocking, the most sanguinary disputes among men, were mere phantoms which a little examination would have shown to be unworthy their notice. The most trifling reflection would have shown him the necessity of this diversity in his notions, of this contrariety in his imagination, which depends upon his natural conformation diversely modified, and which necessarily has an influence over his thoughts, over his will, and over his actions. In short, if he had consulted morals and reason, every thing would have proved to him, that beings who call themselves rational, were made to think variously, without or that account, ceasing to live peaceably with each other, love each other; and lend each other mutual succours; and that whatever might be their opinions upon subjects either impossible to be known or to be contemplated under the same point of view: every thing would have joined in evidence to convince him of the unreasonable tyranny, of the unjust violence, and of the useless cruelty of those men of blood, who persecute mankind in order that they may mould others to their own peculiar opinions: every thing would have conducted mortals to *mildness*, to *indulgence*, to *toleration ;* virtues unquestionably of more real importance to the welfare of society than the marvellous speculations by which it is divided, and by which it is frequently hurried on to sacrifice the pretended enemies to these revered opinions.

From this it must be evident of what importance it is to *morals* to examine the ideas to which it has been agreed to attach so much worth, and to which man, at the irrational command of fanatical and cruel guides, is continually sacrificing his own peculiar happiness and the tranquillity of nations. Let him return to experience, to nature, and to reason; let him consult those objects.

that are real and useful to his permanent felicity; let him study nature's laws; let him study himself; let him consult the bonds which unite him to his fellow mortals; let him tear asunder the fictitious bonds that enchain him to a mere phantom. If his imagination must always feed itself with illusions, if he remains steadfast in his own opinions, if his prejudices are dear to him, let him at least permit others to ramble in their own manner or seek after truth as best suits their inclination; but let him always recollect, that all the opinions, all the ideas, all the systems, all the wills, all the actions of man, are the necessary consequence of his nature, of his temperament, of his organization, and of those causes, either transitory or constant, which modify him: in short, that *man is not more a free agent to think than to act:* a truth that will be again proved in the following chapter.

CHAPTER XI.

Of the System of Man's Free Agency.

THOSE who have pretended that the *soul* is distinguished from the body, is immaterial, draws its ideas from its own peculiar source, acts by its own energies, without the aid of any exterior object, have, by a consequence of their own system, enfranchised it from those physical laws according to which all beings of which we have a knowledge are obliged to act. They have believed that the soul is mistress of its own conduct, is able to regulate its own peculiar operations, has the faculty to determine its will by its own natural energy; in a word, they have pretended that man is a *free agent.*

It has been already sufficiently proved that the soul is nothing more than the body considered relatively to some of its functions more concealed than others: it has been shown that this soul, even when it shall be supposed immaterial, is continually modified conjointly with the body, is submitted to all its motion, and that without this it would remain inert and dead: that, consequently, it is subjected to the influence of those material and physical causes which give impulse to the body; of which the mode of existence, whether habitual

or transitory, depends upon the material elements by which it is surrounded, that form its texture, constitute its temperament, enter into it by means of the aliments, and penetrate it by their subtility. The faculties which are called *intellectual*, and those qualities which are styled *moral*, have been explained in a manner purely physical and natural. In the last place it has been demonstrated that all the ideas, all the systems, all the affections, all the opinions, whether true or false, which man forms to himself, are to be attributed to his physical and material senses. Thus man is a being purely physical; in whatever manner he is considered, he is connected to universal nature, and submitted to the necessary and immutable laws that she imposes on all the beings she contains, according to their peculiar essences or to the respective properties with which, without consulting them, she endows each particular species. Man's life is a line that nature commands him to describe upon the surface of the earth, without his ever being able to swerve from it, even for an instant. He is born without his own consent; his organization does in nowise depend upon himself; his ideas come to him involuntarily; his habits are in the power of those who cause him to contract them; he is unceasingly modified by causes, whether visible or concealed, over which he has no control, which necessarily regulate his mode of existence, give the hue to his way of thinking, and determine his manner of acting. He is good or bad, happy or miserable, wise or foolish, reasonable or irrational, without his will being for any thing in these various states. Nevertheless, in despite of the shackles by which he is bound, it is pretended he is a free agent, or that independent of the causes by which he is moved, he determines his own will, and regulates his own condition.

However slender the foundation of this opinion, of which every thing ought to point out to him the errour, it is current at this day and passes for an incontestable truth with a great number of people, otherwise extremely enlightened; it is the basis of religion, which, supposing relations between man and the unknown being she has placed above nature, has been incapable of imagining how man could either merit reward or deserve punishment from this being, if he was not a free agent. Society has been believed interested in this system; because an idea has gone abroad, that if all the actions of man were to be contemplated as necessary, the right of punishing those who injure their associates would no longer exist. At length human vanity accommodated itself to a hypothesis which, unquestionably, appears to distinguish man from all other physical beings, by assigning to him the special privilege of a total independence of all other causes, but of which a very little reflection would have shown him the impossibility.

As a part subordinate to the great whole, man is obliged to experience its influence. To be a free agent, it were needful that each individual was of greater strength than the entire of nature; or that he was out of this nature, who, always in action herself, obliges all the beings she embraces to act, and to concur to her general motion; or, as it has been said elsewhere, to conserve her active existence by the motion that all beings produce in consequence of their particular energies, submitted to fixed, eternal, and immutable laws. In order that man might be a free agent, it were needful that all beings should lose their essences; it would be equally necessary that he himself should no longer enjoy physical sensibility; that he should neither know good nor evil, pleasure nor pain; but if this were the case, from that moment he would no longer be in a state to conserve himself, or render his existence happy; all beings would become indifferent to him; he would no longer have any choice; he would cease to know what he ought to love, what it was right he should fear; he would not have any acquaintance with that which he should seek after, or with that which it is requisite he should avoid. In short, man would be an unnatural being, totally incapable of acting in the manner we behold. It is the actual essence of man to tend to his well being, or to be desirous to conserve his existence; if all the motion of his machine spring as a necessary consequence from this primitive impulse; if pain warn him of that which

he ought to avoid; if pleasure announce to him that which he should desire; if it be in his essence to love that which either excites delight, or that from which he expects agreeable sensations; to hate that which either makes him fear contrary impressions or that which afflicts him with uneasiness; it must necessarily be that he will be attracted by that which he deems advantageous; that his will shall be determined by those objects which he judges useful; that he will be repelled by those beings which he believes prejudicial, either to his habitual or to his transitory mode of existence. It is only by the aid of experience that man acquires the faculty of understanding what he ought to love or to fear. Are his organs sound? his experience will be true; are they unsound? it will be false: in the first instance he will have reason, prudence, foresight; he will frequently foresee very remote effects; he will know that what he sometimes contemplates as a good, may possibly become an evil by its necessary or probable consequences; that what must be to him a transient evil, may by its result procure him a solid and durable good. It is thus experience enables him to foresee, that the amputation of a limb will cause him painful sensation, he consequently is obliged to fear this operation, and he endeavours to avoid the pain; but, if experience has also shown him that the transitory pain this amputation will cause him may be the means of saving his life; the preservation of his existence being of necessity dear to him, he is obliged to submit himself to the momentary pain, with a view to procuring a permanent good by which it will be overbalanced.

The will, as we have elsewhere said, is a modification of the brain, by which it is disposed to action, or prepared to give play to the organs. This will is necessarily determined by the qualities, good or bad, agreeable or painful, of the object or the motive that acts upon his senses, or of which the idea remains with him, and is resuscitated by his memory. In consequence, he acts necessarily, his action is the result of the impulse he receives either from the motive, from the object, or from the idea which has modified his brain, or disposed his will. When he does not act according to this impulse. it is because there comes some new cause, some new motive, some new idea, which modifies his brain in a different manner, gives him a new impulse, determines his will in another way, by which the action of the former impulse is suspended: tnus, the sight of an agreeable object, or its idea, determines his will to set him in action to procure it; but if a new object or a new idea more powerfully attracts him, it gives a new direction to his will, annihilates the effect of the former, and prevents the action by which it was to be procured. This is the mode in which reflection, experience, reason, necessarily arrests or suspends the action of man's will: without this he would of necessity have followed the anterior impulse which carried him towards a then desirable object. In all this he always acts according to necessary laws, from which he has no means of emancipating himself.

If when tormented with violent thirst, he figures to himself in idea, or really perceives a fountain, whose limpid streams might cool his feverish want, is he sufficient master of himself to desire or not to desire the object competent to satisfy so lively a want? It will no doubt be conceded, that it is impossible he should not be desirous to satisfy it; but it will be said—if at this moment it is announced to him that the water he so ardently desires is poisoned, he will, notwithstanding his vehement thirst, abstain from drinking it: and it has, therefore, been falsely concluded that he is a free agent. The fact, however, is, that the motive in either case is exactly the same: his own conservation. The same necessity that determined him to drink before he knew the water was deleterious, upon this new discovery equally determines him not to drink; the desire of conserving himself either annihilates or suspends the former impulse ; the second motive becomes stronger than the preceding, that is, the fear of death, or the desire of preserving himself, necessarily prevails over the painful sensation caused by his eagerness to drink: but, it will be said, if the thirst is very parching, an inconsiderate man without regarding the danger will risk swallowing the water. Nothing is

gained by this remark : in this case, the anterior impulse only regains the ascendency; he is persuaded that life may possibly be longer preserved, or that he shall derive a greater good by drinking the poisoned water than by enduring the torment, which, to his mind, threatens instant dissolution : thus the first becomes the strongest and necessarily urges him on to action. Nevertheless, in either case, whether he partakes of the water, or whether he does not, the two "actions will be equally necessary ; they will be the effect of that motive which finds itself most puissant; which consequently acts in the most coercive manner upon his will.

This example will serve to explain the whole phenomena of the human will. This will, or rather the brain, finds itself in the same situation as a bowl, which, although it has received an impulse that drives it forward in a straight line, is deranged in its course whenever a force superior to the first obliges it to change its direction. The man who drinks the poisoned water appears a madman ; but the actions of fools are as necessary as those of the most prudent individuals. The motives that determine the voluptuary and the debauchee to risk their health, are as powerful, and their actions are as necessary, as those which decide the wise man to manage his. But, it will be insisted, the debauchee may be prevailed on to change his conduct: this does not imply that he is a free agent; but that motives may be found sufficiently powerful to annihilate the effect of those that previously acted upon him; then these new motives determine his will to the new mode of conduct he may adopt as necessarily as the former did to the old mode.

Man is said to *deliberate*, when the action of the will is suspended; this happens when two opposite motives act alternately upon him. *To deliberate*, is to hate and to love in succession; it is to be alternately attracted and repelled ; it is to be moved, sometimes by one motive, sometimes by another. Man only deliberates when he does not distinctly understand the quality of the objects from which he receives impulse, or when experience has not sufficiently apprised him of the effects, more or less remote, which his actions will produce. He would take the air, but the weather is uncertain; he deliberates in consequence; he weighs the various motives that urge his will to go out or to stay at home; he is at length determined by that motive which is most probable; this removes his indecision, which necessarily settles his will, either to remain within or to go abroad: this motive is always either the immediate or ultimate advantage he finds, or thinks he finds, in the action to which he is persuaded.

Man's will frequently fluctuates between two objects, of which either the presence or the ideas move him alternately: he waits until he has contemplated the objects, or the ideas they have left in his brain which solicit him to different actions; he then compares these objects or ideas; but even in the time of deliberation, during the comparison, pending these alternatives of love and hatred which succeed each other, sometimes with the utmost rapidity, he is not a free agent for a single instant; the good or the evil which he believes he finds successively in the objects, are the necessary motives of these momentary wills; of the rapid motion of desire or fear, that he experiences as long as his uncertainty continues. From this it will be obvious that deliberation is necessary; that uncertainty is necessary; that ·whatever part he takes, in consequence of this deliberation, it will always necessarily be that which he has judged, whether well or ill, is most probable to turn to his advantage.

When the soul is assailed by two motives that act alternately upon it, or modify it successively, it deliberates; the brain is in a sort of equilibrium, accompanied with perpetual oscillations, sometimes towards one object, sometimes towards the other, until the most forcible carries the point, and thereby extricates it from this state of suspense, in which consists the indecision of his will. But when the brain is simultaneously assailed by causes equally strong that move it in opposite directions, agreeable to the general law of all bodies when they are struck equally by contrary powers, it stops, it is in *nisu;* it is neither capable to will nor to act; it waits until one of

the two causes has obtained sufficient force to overpower the other; to determine its will; to attract it in such a manner that it may prevail over the efforts of the other cause.

This mechanism, so simple, so natural, suffices to demonstrate why uncertainty is painful, and why suspense is always a violent state for man. The brain, an organ so delicate and so mobile, experiences such rapid modifications that it is fatigued; or when it is urged in contrary directions, by causes equally powerful, it suffers a kind of compression, that prevents the activity which is suitable to the preservation of the whole, and which is necessary to procure what is advantageous to its existence. This mechanism will also explain the irregularity, the indecision, the inconstancy of man, and account for that conduct which frequently appears an inexplicable mystery, and which is, indeed, the effect of the received systems. In consulting experience, it will be found that the soul is submitted to precisely the same physical laws as the material body. If the will of each individual, during a given time, was only moved by a single cause or passion, nothing would be more easy than to foresee his actions; but his heart is frequently assailed by contrary powers, by adverse motives, which either act on him simultaneously or in succession; then his brain, attracted in opposite directions, is either fatigued, or else tormented by a state of compression, which deprives it of activity. Sometimes it is in a state of incommodious inaction; sometimes it is the sport of the alternate shocks it undergoes. Such, no doubt, is the state in which man finds himself when a lively passion solicits him to the commission of crime, whilst fear points out to him the danger by which it is attended: such, also, is the condition of him whom remorse, by the continued labour of his distracted soul, prevents from enjoying the objects he has criminally obtained.

If the powers or causes, whether exterior or interior, acting on the mind of man, tend towards opposite points, his soul, as well as all other bodies, will take a mean direction between the two; and in consequence of the violence with which his soul is urged, his condition becomes sometimes so painful that his existence is troublesome: he has no longer a tendency to his own peculiar conservation; he seeks after death as a sanctuary against himself, and as the only remedy to his despair: it is thus we behold men, miserable and discontented, voluntarily destroy themselves whenever life becomes insupportable. Man cannot cherish his existence any longer than life holds out charms to him: when he is wrought upon by painful sensations, or drawn by contrary impulsions, his natural tendency is deranged; he is under the necessity to follow a new route; this conducts him to his end, which it even displays to him as the most desirable good. In this manner may be explained the conduct of those melancholy beings, whose vicious temperaments, whose tortured consciences, whose chagrin, whose *ennui* sometimes determine them to renounce life.*

The various powers, frequently very complicated, that act either successively or simultaneously upon the brain of man, which modify him so diversely in the different periods of his existence, are the true causes of that obscurity in morals, of that difficulty which is found, when it is desired to unravel the concealed springs of his enigmatical conduct. The heart of man is a labyrinth, only because it very rarely happens that we possess the necessary gift of judging it; from whence it will appear, that his circumstances, his indecision, his conduct, whether ridiculous or unexpected, are the necessary consequences of the changes operated in him; are nothing but the effect of motives that successively determine his will; which are dependant on the frequent variations experienced by his machine. According to these variations the same motives have not always the same influence over his will; the same objects no longer enjoy the faculty of pleasing him; his temperament has changed, either for the moment, or for ever: it follows as a consequence, that his taste, his desires, his passions, will change;

there can be no kind of uniformity in his conduct; nor any certitude in the effects to be expected.

Choice by no means proves the free agency of man: he only deliberates when he does not yet know which to choose of the many objects that move him, he is then in an embarrassment, which does not terminate until his will is decided by the greater advantage he believes he shall find in the object he chooses, or the action he undertakes. From whence it may be seen, that choice is necessary, because he would not determine for an object, or for an action, if he did not believe that he should find in it some direct advantage. That man should have free agency it were needful that he should be able to will or choose without motive, or that he could prevent motives coercing his will. Action always being the effect of his will once determined, and as his will cannot be determined but by a motive which is not in his own power, it follows that he is never the master of the determination of his own peculiar will; that consequently he never acts as a free agent. It has been believed that man was a free agent because he had a will with the power of choosing; but attention has not been paid to the fact that even his will is moved by causes independent of himself; is owing to that which is inherent in his own organization, or which belongs to the nature of the beings acting on him.† Is he the master of willing not to withdraw his hand from the fire when he fears it will be burnt? Or has he the power to take away from fire the property which makes him fear it? Is he the master of not choosing a dish of meat, which he knows to be agreeable, or analogous to his palate;

---

* See *Chapter* xiv.—Man is oftener induced to destroy himself by mental than by bodily pains. A thousand things may cause him to forget his bodily sufferings, whilst in those of the mind his brain is wholly absorbed; and this is the reason why intellectual pleasures are superior to all others.

† Man passes a great portion of his life without even willing. His will depends on the motive by which he is determined. If he were to render an exact account of every thing he does in the course of each day—from rising in the morning to lying down at night—he would find that not one of his actions have been in the least voluntary; that they have been mechanical, habitual, determined by causes which he was not able to foresee; to which he was either obliged to yield, or with which he was allured to acquiesce: he would discover, that all the motives of his labours, of his amusements, of his discourses, of his thoughts, have been necessary; that they have evidently either seduced him or drawn him along.

of not preferring it to that which he knows to be disagreeable or dangerous? It is always according to his sensations, to his own peculiar experience, or to his suppositions, that he judges of things, either well or ill; but whatever may be his judgment, it depends necessarily on his mode of feeling, whether habitual or accidental, and the qualities he finds in the causes that move him, which exist in despite of himself.

All the causes by which his will is actuated, must act upon him in a manner sufficiently marked to give him some sensation, some perception, some idea; whether complete or incomplete, true or false: as soon as his will is determined, he must have felt either strongly or feebly; if this was not the case he would have determined without motive: thus, to speak correctly, there are no causes which are truly indifferent to the will: however faint the impulse he receives, whether on the part of the objects themselves, or on the part of their images or ideas, as soon as his will acts, the impulse has been competent to determine him. In consequence of a slight or feeble impulse, the will is weak; it is this weakness in his will, that is called *indifference*. His brain with difficulty perceives the sensation it has received; it consequently acts with less vigour, either to obtain or to remove the object or the idea that has modified it. If the impulse is powerful, the will is strong, it makes him act vigorously to obtain or to remove the object which appears to him either very agreeable or very incommodious.

It has been believed that man was a free agent, because it has been imagined that his soul could at will recall ideas which sometimes suffice to check his most unruly desires.[*] Thus, the idea of a remote evil, frequently prevents him from enjoying a present and actual good: thus remembrance, which is an almost insensible or slight modification of his brain, annihilates, at each instant, the real objects that act upon his will. But he is not master of recalling to himself his ideas at pleasure; their association is independent of him; they

are arranged in his brain in despite of him and without his own knowledge, where they have made an impression more or less profound; his memory itself depends upon his organization; its fidelity depends upon the habitual or momentary state in which he finds himself; when his will is vigorously determined to some object or idea that excites a very lively passion in him, those objects or ideas that would be able to arrest his action, no longer present themselves to his mind; in those moments his eyes are shut to the dangers that menace him; of which the idea ought to make him forbear; he marches forwards headlong towards the object by whose image he is hurried on; reflection cannot operate upon him in any way; he sees nothing but the object of his desires; the salutary ideas which might be able to arrest his progress disappear, or else display themselves either too faintly or too late to prevent his acting. Such is the case with all those who, blinded by some strong passion, are not in a condition to recall to themselves those motives, of which the idea alone, in cooler moments, would be sufficient to deter them from proceeding; the disorder in which they are, prevents their judging soundly; renders them incapable of foreseeing the consequence of their actions; precludes them from applying to their experience; from making use of their reason; natural operations which suppose a justness in the manner of associating their ideas, but to which their brain is then not more competent, in consequence of the momentary delirium it suffers, than their hand is to write whilst they are taking violent exercise.

Man's mode of thinking is necessarily determined by his manner of being; it must therefore depend on his natural organization, and the modification his system receives independently of his will. From this, we are obliged to conclude, that his thoughts, his reflections, his manner of viewing things, of feeling, of judging, of combining ideas, is neither voluntary nor free. In a word, that his soul is neither mistress of the motion excited in it, nor of representing to itself, when wanted, those images or ideas that are capable of counterbalancing the impulse it re-

---

[*] St. Augustine says: " Non enim cuiquam in potestate est quid veniat in mentem."

ceives. This is the reason, why man, when in a passion, ceases to reason; at that moment reason is as impossible to oe heard, as it is during an ecstacy, or in a fit of drunkenness. The wicked are never more than men who are either drunk or mad ; if they reason, it is not until tranquillity is re-establish-ed in their machine; then, and not till then, the tardy ideas that present them-selves to their mind enable them to see the consequence of their actions, and give birth to ideas that bring on them that trouble, which is designated *shame,* *regret, remorse.*

The errours of philosophers on the free agency of man, have arisen from their regarding his will as the *primum mobile,* the original motive of his ac-tions; for want of recurring back, they have not perceived the multiplied, the complicated causes which, independ-ently of him, give motion to the will itself; or which dispose and modify his brain, whilst he himself is purely passive in the motion he receives. Is he the master of desiring or not desir-ing an object that appears desirable to him? Without doubt it will be an-swered, no: but he is the master of re-sisting his desire, if he reflects on the consequences. But, I ask, is he capa-ble of reflecting on these consequences, when his soul is hurried along by a very lively passion, which entirely de-pends upon his natural organization, and the causes by which he is modi-fied ? Is it in his power to add to these consequences all the weight necessary to counterbalance his desire? Is he the master of preventing the qualities which render an object desirable from residing in it? I shall be told: he ought to have learned to resist his pas-sions, to contract a habit of putting a a curb on his desires. I agree to it without any difficulty. But in reply, I again ask, is his nature susceptible of this modification? Does his boiling blood, his unruly imagination, the ig-neous fluid that circulates in his veins, permit him to make, enable him to ap-ply true experience in the moment when it is wanted ? And even when his temperament has capacitated him, has his education, the examples set be-fore him, the ideas with which he has been inspired in early life, been suitable to make him contract this habit of re-

pressing his desires? Have not all these things rather contributed to in-duce him to seek with avidity, to make him actually desire those objects which you say he ought to resist.

The *ambitious man* cries out: you will have me resist my passion; but have they not unceasingly repeated to me that rank, honours, power, are the most desirable advantages in life? Have I not seen my fellow citizens envy them, the nobles of my country sacrifice every thing to obtain them ? In the society in which I live, am I not obliged to feel, that if I am deprived of these advantages, I must expect to languish in contempt; to cringe under the rod of oppression ?

The *miser* says : you forbid me to love money, to seek after the means of acquiring it : alas ! does not every thing tell me that, in this world, money is the greatest blessing ; that it is amply sufficient to render me happy ? In the country I inhabit, do I not see all my fellow citizens covetous of riches ? but do I not also witness that they are littl scrupulous in the means of obtaining wealth ? As soon as they are enrich-ed by the means which you censure, are they not cherished, considered and respected ? By what authority, then, do you defend me from amassing trea-sure ? what right have you to prevent my using means, which, although you call them sordid and criminal, I see ap-proved by the sovereign ? Will you have me renounce my happiness ?

The *voluptuary* argues : you pre tend that I should resist my desires ; but was I the maker of my own tem-perament, which unceasingly invites me to pleasure? You call my plea-sures disgraceful; but in the country in which I live, do I not witness the most dissipated men enjoying the most distinguished rank ? Do I not behold that no one is ashamed of adultery but the husband it has outraged ? do not I see men making trophies of their de-baucheries, boasting of their libertinism, rewarded with applause ?

The *choleric man* vociferates: you advise me to put a curb on my passions, and to resist the desire of avenging myself: but can I conquer my nature? Can I alter the received opinions of the world ? Shall I not be for ever dis-graced, infallibly dishonoured in society, if I do not wash out in the blood of my fellow creature the injuries I have received?

The *zealous enthusiast* exclaims: you recommend me mildness; you advise me to be tolerant; to be indulgent to the opinions of my fellow men; but is not my temperament violent? Do I not ardently love my God? Do they not assure me, that zeal is pleasing to him; that sanguinary inhuman persecutors have been his friends? As I wish to render myself acceptable in his sight, I therefore adopt the same means.

In short, the actions of man are never free; they are always the necessary consequence of his temperament, of the received ideas, and of the notions, either true or false, which he has formed to himself of happiness; of his opinions, strengthened by example, by education, and by daily experience. So many crimes are witnessed on the earth only because every thing conspires to render man vicious and criminal; the religion he has adopted, his government, his education, the examples set before him, irresistibly drive him on to evil: under these circumstances, morality preaches virtue to him in vain. In those societies where vice is esteemed, where crime is crowned, where venality is constantly recompensed, where the most dreadful disorders are punished only in those who are too weak to enjoy the privilege of committing them with impunity, the practice of virtue is considered nothing more than a painful sacrifice of happiness. Such societies chastise, in the lower orders, those excesses which they respect in the higher ranks; and frequently have the injustice to condemn those in the penalty of death, whom public prejudices, maintained by constant example, have rendered criminal. Man, then, is not a free agent in any one instant of his life; he is necessarily guided in each step by those advantages, whether real or fictitious, that he attaches to the objects by which his passions are roused: these passions themselves are necessary in a being who unceasingly tends towards his own happiness; their energy is necessary, since that depends on his temperament; his temperament is necessary, because it depends on the physical ele-

ments which enter into his composition; the modification of this temperament is necessary, as it is the infallible and inevitable consequence of the impulse he receives from the incessant action of moral and physical beings.

In despite of these proofs of the want of free agency in man, so clear to unprejudiced minds, it will, perhaps, be insisted upon with no small feeling of triumph, that if it be proposed to any one, to move or not to move his hand, an action in the number of those called *indifferent*, he evidently appears to be the master of choosing; from which it is concluded that evidence has been offered of his free agency. The reply is, this example is perfectly simple; man in performing some action which he is resolved on doing, does not by any means prove his free agency: the very desire of displaying this quality, excited by the dispute, becomes a necessary motive, which decides his will either for the one or the other of these actions: what deludes him in this instance, or that which persuades him he is a free agent at this moment, is, that he does not discern the true motive which sets him in action, namely, the desire of convincing his opponent: if in the heat of the dispute he insists and asks, "Am I not the master of throwing myself out of the window?" I shall answer him, no; that whilst he preserves his reason there is no probability that the desire of proving his free agency, will become a motive sufficiently powerful to make him sacrifice his life to the attempt: if, notwithstanding this, to prove he is a free agent, he should actually precipitate himself from the window, it would not be a sufficient warranty to conclude he acted freely, but rather that it was the violence of his temperament which spurred him on to this folly. Madness is a state, that depends upon the heat of the blood, not upon the will. A fanatic or a hero, braves death as necessarily as a more phlegmatic man or a coward flies from it.*

---

* There is, in point of fact, no difference between the man that is cast out of the window by another, and the man who throws himself out of it, except that the impulse in the first instance comes immediately from without, whilst that which determines the fall in the second case, springs from within his own peculiar machine, having its more re-

It is said that free agency is the absence of those obstacles competent to oppose themselves to the actions of man, or to the exercise of his faculties: it is pretended that he is a free agent whenever, making use of these faculties, he produces the effect he has proposed to himself. In reply to this reasoning, it is sufficient to consider that it in nowise depends upon himself to place or remove the obstacles that either determine or resist him; the motive that causes his action is no more in his own power than the obstacle that impedes him, whether this obstacle or motive be within his own machine or exterior of his person: he is not master of the thought presented to his mind, which determines his will; this thought is excited by some cause independent of himself.

To be undeceived on the system of his free agency, man has simply to recur to the motive by which his will is determined; he will always find this motive is out of his own controul. It is said: that in consequence of an idea to which the mind gives birth, man acts freely if he encounters no obstacle. But the question is, what gives birth to this idea in his brain? was the mind master either to prevent it from presenting itself, or from renewing itself in his brain? Does not this idea depend either upon objects that strike him exteriorly and in despite of himself, or upon causes, that without his knowledge, act within himself and modify his brain? Can he prevent his eyes, cast without design upon any object whatever, from giving him an idea of this object, and from moving his brain? He is not more master of the obstacles; they are the necessary effects of either interior or exterior causes, which al-

mote cause also exterior. When Mutius Scævola held his hand in the fire, he was as much acting under the influence of necessity (caused by interior motives) that urged him to this strange action, as if his arm had been held by strong men: pride, despair, the desire of braving his enemy, a wish to astonish him, an anxiety to intimidate him, &c., were the invisible chains that held his hand bound to the fire. The love of glory, enthusiasm for their country, in like manner caused Codrus and Decius to devote themselves for their fellow-citizens. The Indian Colanus and the philosopher Peregrinus were equally obliged to burn themselves, by desire of exciting the astonishment of the Grecian assembly

ways act according to their given properties. A man insults a coward, this necessarily irritates him against his insulter, but his will cannot vanquish the obstacle that cowardice places to the object of his desire, because his natural conformation, which does not depend upon himself, prevents his having courage. In this case, the coward is insulted in despite of himself; and against his will is obliged patiently to brook the insult he has received.

The partisans of the system of free agency appear ever to have confounded constraint with necessity. Man believes he acts as a free agent, every time he does not see any thing that places obstacles to his actions; he does not perceive that the motive which causes him to will, is always necessary and independent of himself. A prisoner loaded with chains is compelled to remain in prison; but he is not a free agent in the desire to emancipate himself; his chains prevent him from acting, but they do not prevent him from willing; he would save himself if they would loose his fetters; but he would not save himself as a free agent; fear or the idea of punishment would be sufficient motives for his action.

Man may, therefore, cease to be restrained, without, for that reason, becoming a free agent: in whatever manner he acts, he will act necessarily, according to motives by which he shall be determined. He may be compared to a heavy body that finds itself arrested in its descent by any obstacle whatever: take away this obstacle, it will gravitate or continue to fall; but who shall say this dense body is free to fall or not? Is not its descent the necessary effect of its own specific gravity? The virtuous Socrates submitted to the laws of his country, although they were unjust; and though the doors of his jail were left open to him, he would not save himself; but in this he did not act as a free agent: the invisible chains of opinion, the secret love of decorum, the inward respect for the laws, even when they were iniquitous, the fear of tarnishing his glory, kept him in his prison; they were motives sufficiently powerful with this enthusiast for virtue, to induce him to wait death with tranquillity; it was not in his power to save himself, because he could find no potential motive to bring him to depart, even for an instant, from those principles to which his mind was accustomed.

Man, it is said, frequently acts against his inclination, from whence it is falsely concluded he is a free agent; but when he appears to act contrary to his inclination, he is always determined to it by some motive sufficiently efficacious to vanquish this inclination. A sick man, with a view to his cure, arrives at conquering his repugnance to the most disgusting remedies: the fear of pain, or the dread of death, then becomes necessary motives; consequently this sick man cannot be said to act freely.

When it is said, that man is not a free agent, it is not pretended to compare him to a body moved by a simple impulsive cause: he contains within himself causes inherent to his existence; he is moved by an interior organ, which has its own peculiar laws, and is itself necessarily determined in consequence of ideas formed from perceptions resulting from sensations which it receives from exterior objects. As the mechanism of these sensations, of these perceptions, and the manner they engrave ideas on the brain of man, are not known to him; because he is unable to unravel all these motions; because he cannot perceive the chain of operations in his soul, or the motive principle that acts within him, he supposes himself a free agent; which, literally translated, signifies, that he moves himself by himself; that he determines himself without cause: when he rather ought to say, that he is ignorant how or for why he acts in the manner he does. It is true the soul enjoys an activity peculiar to itself: but it is equally certain that this activity would never be displayed, if some motive or some cause did not put it in a condition to exercise itself: at least it will not be pretended that the soul is able either to love or to hate without being moved, without knowing the objects, without having some idea of their qualities. Gunpowder has unquestionably a particular activity, but this activity will never display itself, unless fire be applied to it; this, however, immediately sets it in motion.

It is the great complication of motion in man, it is the variety of his action, it is the multiplicity of causes that move

him, whether simultaneously or in continual succession, that persuades him, he is a free agent: if all his motions were simple, if the causes that move him did not confound themselves with each other, if they were distinct, if his machine were less complicated, he would perceive that all his actions were necessary, because he would be enabled to recur instantly to the cause that made him act. A man who should be always obliged to go towards the west, would always go on that side; but he would feel that, in so going, he was not a free agent: if he had another sense, as his actions or his motion, augmented by a sixth, would be still more varied and much more complicated, he would believe himself still more a free agent than he does with his five senses.

It is, then, for want of recurring to the causes that move him; for want of being able to analyze, from not being competent to decompose the complicated motion of his machine, that man believes himself a free agent: it is only upon his own ignorance that he founds the profound yet deceitful notion he has of his free agency; that he builds those opinions which he brings forward as a striking proof of his pretended freedom of action. If, for a short time, each man was willing to examine his own peculiar actions, search out their true motives to discover their concatenation, he would remain convinced that the sentiment he has of his natural free agency, is a chimera that must speedily be destroyed by experience.

Nevertheless it must be acknowledged that the multiplicity and diversity of the causes which continually act upon man, frequently without even his knowledge, render it impossible, or at least extremely difficult for him to recur to the true principles of his own peculiar actions, much less the actions of others: they frequently depend upon causes so fugitive, so remote from their effects, and which, superficially examined, appear to have so little analogy, so slender a relation with them, that it requires singular sagacity to bring them into light. This is what renders the study of the moral man a task of such difficulty; this is the reason why his heart is an abyss, of which it is frequently impossible for him to fathom the depth. He is then obliged to content himself with a knowledge of the general and necessary laws by which the human heart is regulated : for the individuals of his own species these laws are pretty nearly the same ; they vary only in consequence of the organization that is peculiar to each, and of the modification it undergoes : this, however, cannot be rigorously the same in any two. It suffices to know, that by his essence, man tends to conserve himself, and to render his existence happy : this granted, whatever may be his actions, if he recur back to this first principle, to this general, this necessary tendency of his will, he never can be deceived with regard to his motives. Man, without doubt, for want of cultivating reason and experience, frequently deceives himself upon the means of arriving at this end ; sometimes the means he employs are unpleasant to his fellows, because they are prejudicial to their interests ; or else those of which he avails himself appear irrational, because they remove him from the end to which he would approximate : but whatever may be these means, they have always necessarily and invariably for object either an existing or imaginary happiness, directed to preserve himself in a state analogous to his mode of existence, to his manner of feeling, to his way of thinking, whether durable or transitory. It is from having mistaken this truth, that the greater number of moral philosophers have made rather the romance than the history of the human heart ; they have attributed the actions of man to fictitious causes ; at least they have not sought out the necessary motives of his conduct. Politicians and legislators have been in the same state of ignorance, or else impostors have found it much shorter to employ imaginary motive-powers, than those which really have existence : they have rather chosen to make him tremble under incommodious phantoms, than guide him to virtue by the direct road to happiness, notwithstanding the conformity of the latter with the natural desires of his heart.

However this may be, man either sees or believes he sees much more distinctly the necessary relation of effects with their causes in natural philosophy than in the human heart : at least he sees in the former sensible causes constantly produce sensible effects, ever the same, when the circumstances are alike. After this he hesitates not to look upon physical effects as necessary ; whilst he refuses to acknowledge necessity in the acts of the human will : these he has, without any just foundation, attributed to a motive-power that acts independently by its own peculiar energy, which is capable of modifying itself without the concurrence of exterior causes, and which is distinguished from all material or physical beings. Agriculture is founded upon the assurance, afforded by experience, that the earth, cultivated and sown in a certain manner, when it has otherwise the requisite qualities, will furnish grain, fruit and flowers, either necessary for subsistence or pleasing to the senses. If things were considered without prejudice, it would be perceived, that in morals, education is nothing more than *the agriculture of the mind ;* that, like the earth, by reason of its natural disposition, of the culture bestowed upon it, of the seeds with which it is sown, of the seasons, more or less favourable that conduct it to maturity, we may be assured that the soul will produce either virtue or vice—*moral fruit,* that will be either salubrious for man or baneful to society. *Morals* is the science of the relations that subsist between the minds, the wills, and the actions of men, in the same manner that geometry is the science of the relations that are found between bodies. Morals would be a chimera and would have no certain principles, if it was not founded upon the knowledge of the motives which must necessarily have an influence upon the human will, and which must necessarily determine the actions of human beings.

If, in the moral as well as in the physical world, a cause, of which the action is not interrupted, be necessarily followed by a given effect, it flows consecutively that a reasonable education, grafted upon truth, and founded upon wise laws ; that honest principles instilled during youth ; virtuous examples continually held forth ; esteem attached solely to merit and good actions ; contempt and shame and chastisements regularly visiting vice and falsehood and crime, are causes that would necessarily act on the will of man, and would determine the greater number of his species to exhibit virtue. But if, on the contrary, religion, politics, example, public opinion, all labour to countenance wickedness and to train man viciously; if instead of fanning his virtues, they stifle good principles; if instead of directing his studies to his advantage, they render his education either useless or unprofitable; if this education itself, instead of grounding him in virtue, only inoculates him with vice; if, instead of inculcating reason it imbues him with prejudice; if, instead of making him enamoured of truth, it furnishes him with false notions and with dangerous opinions; if, instead of fostering mildness and forbearance, it kindles in his breast only those passions which are incommodious to himself and hurtful to others; it must be of necessity that the will of the greater number shall determine them to evil.* Here, without doubt, is the real source from whence springs that universal corruption of which moralists, with great justice, so loudly complain, without, however, pointing out those causes of the evil, which are as true as they are necessary. Instead of this, they search for it in human nature; say it is corrupt;† blame

---

* Many authors have acknowledged the importance of a good education, and that youth was the season to feed the human heart with wholesome diet; but they have not felt that a good education is incompatible, nay impossible, with the superstition of man, since this commences with giving his mind a false bias; that it is equally inconsistent with arbitrary government, because this always dreads, lest he should become enlightened, and is ever sedulous to render him servile, mean, contemptible, and cringing; that it is incongruous with laws that are too frequently bottomed on injustice; that it cannot obtain with those received customs that are opposed to good sense; that it cannot exist whilst public opinion is unfavourable to virtue; above all, that it is absurd to expect it from incapable instructers, from masters with weak minds, who have only the ability to infuse into their scholars those false ideas with which they are themselves infected.

† We can scarcely conceive a more baneful doctrine than that which inculcates the natural corruption of man, and the absolute need of the grace of God to make him good. Such a doctrine tends necessarily to discourage him; it either makes him sluggish or drives him to despair whilst waiting for this grace. What a strange system of morals is that of theologians, who attribute all moral evil to an

man for loving himself; stigmatize him for seeking after his own happiness; insist that he must have *supernatural assistance* to enable him to become good; yet, notwithstanding the supposed free agency of man, it is insisted that nothing less than the author of nature himself, is necessary to destroy the wicked desires of his heart: but, alas! this powerful agent himself is found inefficacious to controul those unhappy propensities, which, under the fatal constitution of things, the most vigorous motives, as has been before observed, are continually infusing into the will of man. He is indeed incessantly exhorted to resist these passions; to stifle and root them out of his heart: but is it not evident they are necessary to his welfare, and inherent in his nature? Does not experience prove them to be useful to his conservation, since they have for object, only to avoid that which may be injurious and to procure that which may be advantageous? In short, is it not easy to be seen, that these passions well directed, that is to say, carried towards objects that are truly useful, that are really interesting to himself, which would embrace the happiness of others, would necessarily contribute to the substantial and permanent well-being of society? The passions of man are like fire, at once necessary to the wants of life, and equally capable of producing the most terrible ravages.‡

Every thing becomes an impulse to the will: a single word frequently suffices to modify a man for the whole course of his life; to decide for ever his propensities; an infant, who has burned his finger by having approached it too near to the flame of a lighted taper, is warned that he ought to abstain from indulging a similar temptation; a man once punished and despised for having committed a dishonest action, is not often tempted to continue so

original sin, and all moral good to the pardon of it! But it ought certainly not to excite surprise that a moral system, founded upon such ridiculous hypotheses, is of no efficacy.— *See* Vol. II. chap. viii.

‡ Theologians themselves, have felt, they have acknowledged, the necessity of the passions: many of the fathers of the church have broached this doctrine; among the rest Father Senault has written a book expressly on the subject. entitled, *Of the Use of the Passions*

unfavourable a course. Under whatever point of view man is considered, he never acts but after the impulse given to his will, whether it be by the will of others, or by more perceptible physical causes. The particular organization decides the nature of the impulse; souls act upon souls that are analogous; fiery imaginations act with facility upon strong passions, and upon imaginations easy to be inflamed: the surprising progress of enthusiasm, the hereditary propagation of superstition, the transmission of religious errours from race to race, the excessive ardour with which man seizes on the marvellous, are effects as necessary as those which result from the action and reaction of bodies.

In despite of the gratuitous ideas which man has formed to himself on his pretended free agency; in defiance of the illusions of this supposed intimate sense, which, maugre his experience, persuades him that he is master of his will; all his institutions are really founded upon necessity: on this, as on a variety of other occasions, practice throws aside speculation. Indeed, if it was not believed that certain motives embraced the power requisite to determine the will of man, to arrest the progress of his passions; to direct them towards an end, to modify him, of what use would be the faculty of speech? What benefit could arise from education, from legislation, from morals, even from religion itself? What does education achieve, save give the first impulse to the human will; make man contract habits; oblige him to persist in them; furnish him with motives, whether true or false, to act after a given manner? When the father either menaces his son with punishment, or promises him a reward, is he not convinced these things will act upon his will? What does legislation attempt except it be to present to the citizens of a state those motives which are supposed necessary to determine them to perform some actions that are considered worthy; to abstain from committing others that are looked upon as unworthy? What is the object of morals, if it be not to show man that his interest exacts he should suppress the momentary ebullition of his passions, with a view to promote a more certain happiness, a more lasting well-

being, than can possibly result from the gratification of his transitory desires? Does not the religion of all countries suppose the human race, together with the entire of nature, submitted to the irresistible will of a necessary being who regulates their condition after the eternal laws of immutable wisdom? Is not this God, which man adores, the absolute master of their destiny? Is it not this divine being who chooses and who rejects? The anathemas fulminated by religion, the promises it holds forth, are they not founded upon the idea of the effects these chimeras will necessarily produce upon ignorant and timid people? Is not man brought into existence by this kind Divinity without his own knowledge? Is he not obliged to play a part against his will? Does not either his happiness or his misery depend on the part he plays?*

Education, then, is only necessity shown to children: legislation, is necessity shown to the members of the body politic: morals, is the necessity of the relations subsisting between men, shown to reasonable beings: in short, man grants necessity in every thing for which he believes he has certain unerring experience: that of which he does not comprehend the necessary connexion of causes with their effects he styles probability: he would not act as he does, if he was not convinced, or, at least, if he did not presume that certain effects will necessarily follow his actions. The moralist preaches reason, because he believes it necessary

* Every religion is evidently founded upon fatalism. Among the Greeks they supposed men were punished for their *necessary* faults—as may be seen in Orestes, in Œdipus, etc., who only committed crimes predicted by the oracles. Christians have made vain efforts to justify God Almighty in throwing the faults of men on their *free will*, which is opposed to *Predestination*, another name for *fatalism.* However, their system of *Grace* will by no means obviate the difficulty, for God gives grace only to those whom he pleases. In all countries religion has no other foundation than the fatal decrees of an irresistible being who arbitrarily decides the fate of his creatures. All theological hypotheses turn upon this point; and yet those theologians who regard the system of fatalism as false or dangerous, do not see that the Fall of Angels, Original Sin, Predestination, the System of Grace, the small number of the Elect, etc. incontestably prove that religion is a true system of fatalism

to man: the philosopher writes, because he believes truth must sooner or later prevail over falsehood: theologians and tyrants necessarily hate truth and despise reason, because they believe them prejudicial to their interests: the sovereign, who strives to terrify crime by the severity of his laws, but who, nevertheless, oftener renders it useful and even necessary to his purposes, presumes the motives he employs will be sufficient to keep his subjects within bounds. All reckon equally upon the power or upon the necessity of the motives they make use of, and each individual flatters himself, either with or without reason, that these motives will have an influence on the conduct of mankind. The education of man is commonly thus defective or inefficacious, only because it is regulated by prejudice: even when this education is good, it is but too often speedily counteracted and annihilated by every thing that takes place in society. Legislation and politics are very frequently iniquitous, and serve no better purpose than to kindle passions in the bosom of man, which, once set afloat, they are no longer competent to restrain. The great art of the moralist should be to point out to man and to those who are intrusted with the office of regulating his will, that their interests are identified; that their reciprocal happiness depends upon the harmony of their passions; that the safety, the power, the duration of empires, necessarily depend on the good sense diffused among the individual members; on the truth of the notions inculcated in the mind of the citizens; on the moral goodness that is sown in their hearts; on the virtues that are cultivated in their breasts. Religion should not be admissible unless it truly fortified and strengthened these motives, and unless it were possible for falsehood to lend real assistance to truth. But in the miserable state into which errour has plunged a considerable portion of the human species, man, for the most part, is obliged to be wicked or to injure his fellow creature; the strongest motives invite him to the commission of evil. Religion renders him a useless being; makes him an abject slave; causes him to tremble under its terrours; or else turns him into a furious fanatic, who

is at once cruel, intolerant and inhuman: arbitrary power crushes him and obliges him to become cringing and vicious: law visits crime with punishment only in those who are too feeble to oppose its course, or when it has become incapable of restraining the violent excesses to which a bad government gives birth. In short, education neglected and despised, depends either upon priests, who are impostors, or else upon parents without understanding and devoid of morals, who impress on the ductile mind of their scholars those vices with which they are themselves tormented, and who transmit to them the false opinions which they have an interest in making them adopt.

All this proves the necessity of recurring to the primitive source of man's wanderings, if it be seriously intended to furnish him with suitable remedies. It is useless to dream of correcting his mistakes, until the true causes that move his will are unravelled, or until more real, more beneficial, more certain motives, are substituted for those which are found so inefficacious and so dangerous both to society and to himself. It is for those who guide the human will who regulate the condition of nations, to seek after these motives with which reason will readily furnish them; even a good book, by touching the heart of a great prince, may become a very powerful cause that shall necessarily have an influence over the conduct of a whole people; that shall decide upon the felicity of a portion of the human race.

From all that has been advanced in this chapter, it results, that in no one moment of his existence is man a free agent. He is not the architect of his own conformation, which he holds from nature; he has no controul over his own ideas, or over the modification of his brain; these are due to causes, that, in despite of him, and without his own knowledge, unceasingly act upon him; he is not the master of not loving or coveting that which he finds amiable or desirable; he is not capable of refusing to deliberate, when he is uncertain of the effects certain objects will produce upon him; he cannot avoid choosing that which he believes will be most advantageous to him; in the moment when his will is determined oy his choice he is not competent to act otherwise than he does. In what instance, then, is he the master of his own actions? In what moment is he a free agent?*

That which a man is about to do, is always a consequence of that which he has been—of that which he is—of that which he has done up to the moment of the action: his total and actual existence, considered under all its possible circumstances, contains the sum of all the motives to the action he is about to commit; this is a principle the truth of which no thinking being will be able to refuse accrediting: his life is a series of necessary moments; his conduct, whether good or bad, virtuous or vicious, useful or prejudicial, either to himself or to others, is a concatenation of action, as necessary as all the moments of his existence. To *live*, is to exist in a necessary mode during the points of that duration which succeed each other necessarily: *to will*, is to acquiesce or not in remaining such as he is: *to be free*, is to yield to the necessary motives he carries within himself.

If he understood the play of his organs, if he was able to recall to himself all the impulsions they have received, all the modifications they have undergone, all the effects they have produced, he would perceive that all his actions are submitted to that *fatality*, which regulates his own particular system, as it does the entire system of the universe: no one effect in him, any more than in nature, produces itself by *chance;* this, as has been before proved, is a word void of sense. All that passes in him; all that is done by him; as well as all that happens in nature, or that is attributed

to her, is derived from necessary causes, which act according to necessary laws, and which produce necessary effects from whence necessarily flow others.

Fatality, is the eternal, the immutable, the necessary order, established in nature; or the indispensable connexion of causes that act, with the effects they operate. Conforming to this order, heavy bodies fall; light bodies rise; that which is analogous in matter reciprocally attracts; that which is heterogeneous mutually repels; man congregates himself in society, modifies each his fellow; becomes either virtuous or wicked; either contributes to his mutual happiness, or reciprocates his misery; either loves his neighbour, or hates his companion necessarily, according to the manner in which the one acts upon the other. From whence it may be seen, that the same necessity which regulates the physical, also regulates the moral world, in which every thing is in consequence submitted to fatality. Man, in running over, frequently without his own knowledge, often in despite of himself, the route which nature has marked out for him, resembles a swimmer who is obliged to follow the current that carries him along: he believes himself a free agent, because he sometimes consents, sometimes does not consent, to glide with the stream, which, notwithstanding, always hurries him forward; he believes himself the master of his condition, because he is obliged to use his arms under the fear of sinking.

Volentem ducunt fata, nolentem trahunt.
                                        Senec.

The false ideas he has formed to himself upon free agency, are in general thus founded: there are certain

---

* The question of *Free Will* may be reduced to this:—Liberty, or Free Will, cannot be associated with any known functions of the soul; for the soul, at the moment in which it acts, deliberates, or wills, cannot act, deliberate, or will otherwise than it does, because a thing cannot exist and not exist at the same time. Now, it is my will, such as it is, that makes me deliberate; my deliberation, that makes me choose; my choice that makes me act; my determination that makes me execute that which my deliberation has made me choose, and I have only deliberated because I have had motives which rendered it impossible for me not to be willing to deliberate. Thus liberty is not found either in the will, in the

deliberation, in the choice, or in the action. Theologians must not, therefore, connect liberty with these operations of the soul, otherwise there will be a contradiction of ideas. If the soul is not free when it wills, deliberates, chooses, or acts, will theologians tell us when it can exercise its liberty?

It is evident that the system of liberty, or free will, has been invented to exonerate God from the evil that is done in this world. But is it not from God man received this liberty? Is it not from God he received the faculty of choosing evil and rejecting the good? If so, God created him with a determination to sin, else liberty is essential to man and independent of God.—*See* " *Treatise of Systems,*" p. 124.

events which he judges *necessary;* either because he sees that they are effects constantly and invariably linked to certain causes, which nothing seems to prevent; or because he believes he has discovered the chain of causes and effects that is put in play to produce those events: whilst he contemplates as *contingent* other events of whose causes he is ignorant, and with whose mode of acting he is unacquainted: but in nature, where every thing is connected by one common bond, there exists no effect without a cause. In the moral as well as in the physical world, every thing that happens is a necessary consequence of causes, either visible or concealed, which are of necessity obliged to act after their peculiar essences. *In man, free agency is nothing more than necessity contained within himself.*

CHAPTER XII.

An Examination of the Opinion which pretends that the System of Fatalism is Dangerous.

For a being whose essence obliges him to have a constant tendency to his own conservation and to render himself happy, experience is indispensable: without it he cannot discover truth, which is nothing more, as has been already said, than a knowledge of the constant relations which subsist between man and those objects that act upon him; according to his experience he denominates those that contribute to his permanent welfare, useful and salutary; those that procure him pleasure, more or less durable, he calls agreeable. Truth itself becomes the object of his desires, only when he believes it is useful; he dreads it whenever he presumes it will injure him. But has truth the power to injure him? Is it possible that evil can result to man from a correct understanding of the relations he has with other beings? Can it be true that he can be harmed by becoming acquainted with those things of which, for his own happiness, he is interested in having a knowledge? No! unquestionably not: it is upon its utility that truth founds its worth and its rights: sometimes it may be disagreeable to individuals, it may even

appear contrary to their interests; but it will always be useful to the whole human species, whose interests must for ever remain distinct from those of men who, duped by their own peculiar passions, believe their advantage consists in plunging others into errour.

Utility, then, is the touchstone of the systems, the opinions, and the actions of man; it is the standard of the esteem and the love he owes to truth itself: the most useful truths are the most estimable: those truths which are most interesting for his species, he styles eminent; those of which the utility limits itself to the amusement of some individuals who have not correspondent ideas, similar modes of feeling, wants analogous to his own, he either disdains, or else calls them barren.

It is according to this standard that the principles laid down in this work ought to be judged. Those who are acquainted with the immense chain of mischief produced on the earth by erroneous systems of superstition, will acknowledge the importance of opposing to them systems more accordant with truth, drawn from nature, and founded on experience. Those who are, or believe they are, interested in maintaining the established errours, will contemplate with horrour the truths here presented to them: in short, those infatuated mortals, who only feel very faintly the enormous load of misery brought upon mankind by theological prejudices, will regard all our principles as useless, or, at most, as steril truths, calculated to amuse the idle hours of a few speculators.

No astonishment, therefore, need be excited at the various judgments formed by man: his interests never being the same, any more than his notions of utility, he condemns or disdains every thing that does not accord with his own peculiar ideas. This granted, let us examine if, in the eyes of the disinterested man, who is not entangled by prejudice, who is sensible to the happiness of his species, *the doctrine of fatalism* be useful or dangerous? Let us see if it be a barren speculation, that has not any influence upon the felicity of the human race? It has been already shown that it will furnish morals with efficacious arguments, with real motives to determine the will, supply politics with the true lever to raise the proper activity in the mind of man. It will also be seen that it serves to explain in a simple manner the mechanism of man's actions, and the most striking phenomena of the human heart: on the other hand, if his ideas are only the result of unfruitful speculations, they cannot interest the happiness of the human species. Whether he believes himself a free agent, or whether he acknowledges the necessity of things, he always equally follows the desires imprinted on his soul. A rational education, honest habits, wise systems, equitable laws, rewards uprightly distributed, punishments justly inflicted, will render man virtuous; while thorny speculations, filled with difficulties. can, at most, only have an influence over persons accustomed to think.

After these reflections it will be very easy to remove the difficulties that are unceasingly opposed to the system of fatalism; which so many persons, blinded by their religious systems, are desirous to have considered as dangerous; as deserving of punishment; as calculated to disturb public tranquillity; as tending to unchain the passions, and to confound ideas of vice and of virtue. The opposers of necessity say: that if all the actions of man are necessary, no right whatever exists to punish bad ones, or even to be angry with those who commit them; that nothing ought to be imputed to them; that the laws would be unjust, if they should decree punishment for necessary actions; in short, that under this system, man could neither have merit nor demerit. In reply it may be argued, that, to impute an action to any .one, is to attribute that action to him—to acknowledge him for the author: thus, when even an action was supposed to be the effect of an agent, and that agent *necessity*, the imputation would still lie: the merit or demerit that is ascribed to an action are ideas originating in the effects, whether favourable or pernicious, that result to those who experience its operation: when, therefore, it should be conceded that the agent was necessity, it is not less certain that the action would be either good or bad; estimable or contemptible, to those who must feel its influence; in short, that it would be capable of either

eliciting their love, or exciting their anger. Love and anger are modes of existence suitable to modify beings of the human species: when, therefore, man irritates himself against his fellow, he intends to excite his fear, or even to punish him. Moreover, his anger is necessary; it is the result of his nature and of his temperament. The painful sensation produced by a stone that falls on the arm, does not displease the less because it comes from a cause deprived of will, and which acts by the necessity of its nature. In contemplating man as acting necessarily, it is impossible to avoid distinguishing that mode of action or being which is agreeable, which elicits approbation, from' that which is afflicting, which irritates, which nature obliges him to blame and to prevent. From this it will be seen that the system of fatalism does not in any manner change the actual state of things, and is by no means calculated to confound man's ideas of virtue and vice.*

Laws are made with a view to maintain society, and to prevent man associated from injuring his neighbour; they are therefore competent to punish those who disturb its harmony, or those who commit actions that are injurious to their fellows; whether these associates may be the agents of necessity, or whether they are free agents, it suffices to know that they are susceptible of modification, and are therefore submitted to the operation of law. Penal laws are those motives which experience has shown capable of restraining or of annihilating the impulse passions give to man's will: from whatever necessary cause man may derive these passions, the legislator proposes to arrest their effect, and when he takes suitable means he is certain of success. The jurisconsult,

---

* Man's nature always revolts against that which opposes it: there are men so choleric, that they infuriate themselves even against insensible and inanimate objects; reflection on their own impotence to modify these objects ought to conduct them back to reason. Parents are frequently very much to be blamed for correcting their children with anger: they should be contemplated as beings who are not yet modified, or who have, perhaps, been very badly modified by themselves: nothing is more common in life, than to see men punish faults of which they are themselves the cause.

in decreeing to crime, gibbets, tortures, or any other chastisement whatever, does nothing more than is done by the architect, who in building a house places gutters to carry off the rain, and prevent it from sapping the foundation.

Whatever may be the cause that obliges man to act, society possesses the right to crush the effects: as much as the man whose land would be ruined by a river, has to restrain its waters by a bank, or even, if he is able, to turn its course. It is by virtue of this right, that society has the power to intimidate and to punish, with a view to its own conservation, those who may be tempted to injure it; or those who commit actions which are acknowledged really to interrupt its repose, to be inimical to its security, or repugnant to his happiness.

It will perhaps be argued, that society does not usually punish those faults in which the will has no share; that it punishes the will alone; that this it is which decides the nature of the crime, and the degree of its atrocity: that if this will be not free, it ought not to be punished. I reply, that society is an assemblage of sensible beings, susceptible of reason, who desire their own welfare, who fear evil, and seek after good. These dispositions enable their will to be so modified or determined, that they are capable of holding such a conduct as will conduce to the end they have in view. Education, the laws, public opinion, example, habit, fear, are the causes that must modify associated man, influence his will, regulate his passions, restrain the actions of him who is capable of injuring the end of his association, and thereby make him concur to the general happiness. These causes are of a nature to make impressions on every man whose organization and whose essence place him in a capacity to contract the habits, the modes of thinking, and the manner of acting, with which society is willing to inspire him. All the individuals of the human species are susceptible of fear; from whence it flows as a natural consequence,·that the fear of punishment, or the privation of the happiness he desires, are motives that must necessarily more or less influence his will, and regulate his actions. If the man is to be found, who is so badly constituted as to resist or to be

insensible to those motives which operate upon all his fellows, he is not fit to live in society; he would contradict the very end of his association; he would be its enemy; he would place obstacles to its natural tendency; his rebellious disposition, his unsociable will, not being susceptible of that modification which is convenient to his own true interests and to the interests of his fellow citizens, these would unite themselves against such an enemy; and the law which is the expression of the general will, would visit with condign punishment that refractory individual upon whom the motives presented to him by society had not the effect which it had been induced to expect: in consequence such an unsociable man would be chastised; he would be rendered miserable; and according to the nature of his crime he would be excluded from society, as a being but little calculated to concur in its views.

If society has the right to conserve itself, it has also the right to take the means: these means are the laws which present to the will of man those motives which are most suitable to deter him from committing injurious actions. If these motives fail of the proper effect, if they are unable to influence him, society, for its own peculiar good, is obliged to wrest from him the power of doing it farther injury. From whatever source his actions may arise, whether they are the result of free agency, or whether they are the offspring of necessity, society coerces him, if after having furnished him with motives sufficiently powerful to act upon reasonable beings, it perceives that these motives have not been competent to vanquish his depraved nature. It punishes him with justice, when the actions from which it dissuades him are truly injurious to society; it has an unquestionable right to punish, when it only commands or defends those things that are conformable to the end proposed by man in his association. But, on the other hand, the law has not acquired the right to punish him, if it has failed to present to him the motives necessary to have an influence over his will; it has not the right to coerce him, if the negligence of society has deprived him of the means of subsisting, of exercising his talents, of exerting his·

industry, and of labouring for its welfare. It is unjust, when it punishes those to whom it has neither given an education, nor honest principles; whom it has not enabled to contract habits necessary to the maintenance of society: it is unjust, when it punishes them for faults which the wants of their nature, or the constitution of society has rendered necessary to them: it is unjust and irrational, whenever it chastises them for having followed those propensities which example, which public opinion, which the institutions, which society itself conspires to give them. In short, the law is defective when it does not proportion the punishment to the real evil which society has sustained. The last degree of injustice and folly is, when society is so blinded as to inflict punishment on those citizens who have served it usefully.

Thus penal laws in exhibiting terrifying objects to man who must be supposed susceptible of fear, present him with motives calculated to have an influence over his will. The idea of pain, the privation of liberty, the fear of death, are, to a being well constituted and in the full enjoyment of his faculties, very puissant obstacles that strongly oppose themselves to the impulse of his unruly desires: when these do not coerce his will, when they fail to arrest his progress, he is an irrational being, a madman, a being badly organized, against whom society has the right to guaranty itself and to take measures for its own security. Madness is, without doubt, an involuntary and a necessary state; nevertheless, no one feels it unjust to deprive the insane of their liberty, although their actions can only be imputed to the derangement of their brain. The wicked are men whose brain is either constantly or transitorily disturbed; still they must be punished by reason of the evil they commit; they must always be placed in the impossibility of injuring society; if no hope remains of bringing them back to a reasonable conduct, and to adopt a mode of action conformable to the great end of association, they must be for ever excluded its benefits.

It will not be requisite to examine here how far the punishments, which society inflicts upon those who offend against it, may be reasonably carried.

Reason should seem to indicate, that the law ought to show to the necessary crimes of man all the indulgence that is compatible with the conservation of society. The system of fatalism, as we have seen, does not leave crime unpunished; but it is at least calculated to moderate the barbarity with which a number of nations punish the victims to their anger. This cruelty becomes still more absurd when experience has shown its inutility: the habit of witnessing ferocious punishments, familiarizes criminals with the idea. If it be true that society possesses the right of taking away the life of its members; if it be really a fact that the death of a criminal, thenceforth useless, can be advantageous for society, (which it will be necessary to examine,) humanity at least exacts, that this death should not be accompanied with useless tortures with which laws too frequently seem to delight in overwhelming their victim. This cruelty defeats its own end, as it only serves to make the culprit, who is immolated to the public vengeance, suffer without any advantage to society: it moves the compassion of the spectator, and interests him in favour of the miserable offender who groans under its weight: it impresses nothing upon the wicked; whilst the sight of those cruelties destined for himself but too frequently renders him more ferocious, more cruel, and more the enemy of his associates : if the example of death were less frequent, even without being accompanied with tortures, it would be more efficacious.*

_____

* The greater number of criminals only look upon death as a *bad quarter of an hour.* A thief seeing one of his comrades display a want of firmness under the punishment, said to him: "*Is not this what I hate often told you, that in our business we have one evil more than the rest of mankind?*" Robberies are daily committed even at the foot of the scaffolds where criminals are punished: In those nations, where the penalty of death is so lightly inflicted, has sufficient attention been paid to the fact, that society is yearly deprived of a great number of individuals who would be able to render it very useful service, if made to work, and thus indemnify the community for the injuries they have committed ? The facility with which the lives of men are taken away, proves the tyranny and incapacity of legislators : they find it a much shorter road to destroy the citizens, than to seek after the means to render them better.

What shall be said for the unjust cruelty of some nations, in which the law, that ought to have for its object the advantage of the whole, appears to be made only for the security of the most powerful ; in which punishments the most disproportionate to the crime, unmercifully take away the lives of men, whom the most urgent necessity has obliged to become criminal ? It is thus, that in a great number of civilized nations, the life of the citizen is placed in the same scales with money ; that the unhappy wretch, who is perishing from hunger and misery, is put to death for having taken a pitiful portion of the superfluity of another whom he beholds rolling in abundance ? It is this, that in many otherwise very enlightened societies, is called *justice*, or making the punishment commensurate with the crime.

This dreadful iniquity becomes yet more crying, when the laws decree the most cruel tortures for crimes to which the most irrational customs give birth; which bad institutions multiply. Man, as it cannot be too frequently repeated, is so prone to evil, only because every thing appears to urge him on to the commission, by too frequently showing him vice triumphant : his education is void in most states ; he receives from society no other principles, save those of an unintelligible religion, which make but a feeble barrier against his propensities : in vain the law cries out to him : " abstain from the goods of thy neighbour ;" his wants, more powerful, loudly declare to him that he must live at the expense of a society who has done nothing for him, and who condemns him to groan in misery and in indigence ; frequently deprived of the common necessaries, he compensates himself by theft, and by assassination ; he becomes a plunderer by profession, a murderer by trade, and seeks, at the risk of his life, to satisfy those wants, whether real or imaginary, to which every thing around him conspires to give birth. Deprived of education, he has not been taught to restrain the fury of his temperament. Without ideas of decency, destitute of the true principles of honour, he engages in criminal pursuits that injure his country, which has been to him nothing more than a step-

mother. In the paroxysm of his rage, he only sees the gibbet that awaits him ; his unruly desires have become too potent ; they have given an invete-racy to his habits which preclude him from changing them ; laziness has made him torpid ; despair has rendered him blind ; he rushes on to death ; and society punishes him rigorously for those fatal and necessary dispositions, which it has itself engendered in his heart, or which at least it has not taken the pains seasonably to root out and to oppose by motives calculated to give him honest principles. Thus society frequently punishes those propensities of which it is itself the author, or which its negligence has suffered to spring up in the mind of man : it acts like those unjust fathers, who chastise their children for vices which they have themselves made them contract.

However unjust and unreasonable this conduct may be, or appear to be, it is not the less necessary : society, such as it is, whatever may be its corruption, whatever vices may pervade its institutions, like every thing else in nature, tends to subsist and to conserve itself : in consequence it is obliged to punish those excesses which its own vicious constitution has produced : in despite of its peculiar prejudices and vices, it feels cogently that its own immediate security demands, that it should destroy the conspiracies of those who make war against its tranquillity : if these, hurried on by necessary propensities, disturb its repose and injure its interests, this following the natural law, which obliges it to labour to its own peculiar conservation, removes them out of its road, and punishes them with more or less rigour, according to the objects to which it attaches the greatest importance, or which it supposes best suited to further its own peculiar welfare : without doubt it deceives itself frequently, but it deceives itself necessarily, for want of the knowledge calculated to. enlighten it with regard to its true interests, or for want of those, who regulate its movements, possessing proper vigilance, suitable talents, and the requisite virtue. From this it will appear, that the injustice of a society badly constituted, and blinded by its prejudices, is as neces- ·

sary as the crimes of those by whom
it is hostily attacked and distracted.*
The body politic, when in a state of
insanity, cannot act more consistently
with reason than one of its members
whose brain is disturbed by madness.
It will still be said that these maxims, by submitting every thing to necessity, must confound, or even destroy, the notions man forms of justice
and injustice, of good and evil, of
merit and demerit. I deny it: although
man, in every thing he does, acts necessarily, his actions are good, just, and
meritorious, every time they tend to the
real utility of his fellows, and of the
society of which he makes a part: they
are, of necessity, distinguished from
those which are really prejudicial to
the welfare of his associates. Society
is just, good, and worthy our reverence,
when it procures for all its members
their physical wants, affords them protection, secures their liberty, and puts
them in possession of their natural
rights. It is in this that consists all
the happiness of which the social compact is susceptible. Society is unjust,
and unworthy our esteem, when it is
partial to a few, and cruel to the greater number: it is then that it multiplies its enemies, and obliges them to
revenge themselves by criminal actions
which it is under the necessity to punish. It is not upon the caprices of
political society that depend the true
notions of justice and injustice, the
right ideas of moral good and evil, a
just appreciation of merit and demerit;
it is upon *utility*—upon the necessity
of things—which always forces man
to feel that there exists a mode of acting which he is obliged to venerate and
approve, either in his fellows or in
society: whilst there is another mode
which his nature makes him hate,
which his feelings compel him to condemn. It is upon his own peculiar
essence that man founds his ideas of
pleasure and of pain, of right and of
wrong, of vice and of virtue: the only
difference between these is, that pleasure and pain make them instantane-

ously felt in his brain; whilst the advantages that accrue to him from justice and virtue, frequently do not display themselves but after a long train
of reflections, and after multiplied experiences, which many, either from a
defect in their conformation or from the
peculiarity of the circumstances under
which they are placed, are prevented
from making, or, at least, from making
correctly.

By a necessary consequence of this
truism, the system of fatalism, although
it has frequently been so accused, does
not tend to encourage man in crime,
and to make remorse vanish from his
mind. His propensities are to be ascribed to his nature; the use he makes
of his passions depends upon his habits
upon his opinions, upon the ideas he
has received in his education, and upon
the examples held forth by the society
in which he lives. These things are
what necessarily decide his conduct.
Thus when his temperament renders
him susceptible of strong passions, he
is violent in his desires, whatever may
be his speculations.

Remorse is the painful sentiment
excited in him by grief caused either
by the immediate or probable future
effect of his passions: if these effects
were always useful to him, he would
not experience remorse; but, as soon
as he is assured that his actions render
him hateful or contemptible; or as soon
as he fears he shall be punished in
some mode or other, he becomes restless and discontented with himself: he
reproaches himself with his own conduct; he feels ashamed; he fears the
judgment of those beings whose affection he has learned to esteem; in
whose good will he finds his own comfort deeply interested. His experience
proves to him, that the wicked man is
odious to all those upon whom his actions have any influence: if these actions are concealed at the moment, he
knows it very rarely happens they
remain so forever. The smallest reflection convinces him, that there is no
wicked man who is not ashamed of
his own conduct; who is truly contented with himself; who does not
envy the condition of the good man;
who is not obliged to acknowledge,
that he has paid very dearly for those
advantages he is never able to enjoy

* A society punishing excesses to which it
has itself given birth, may be compared to a
man attacked with the *lousy* disorder, who is
obliged to kill the insects, although it is his
own diseased constitution which every moment produces them.

without making the most bitter reproaches against himself: then he feels ashamed, despises himself, hates himself, his conscience becomes alarmed, remorse follows in its train. To be convinced of the truth of this principle, it is only requisite to cast our eyes on the extreme precautions that tyrants and villains, who are otherwise sufficiently powerful not to dread the punishment of man, take to prevent exposure; to what lengths they push their cruelties against some, to what meanness they stoop to others, of those who are able to hold them up to public scorn. Have they not then a consciousness of their own iniquities? Do they not know, that they are hateful and contemptible? Have they not remorse? Is their condition happy? Persons well brought up acquire these sentiments in their education; which are either strengthened or enfeebled by public opinion, by habit, by the examples set before them. In a depraved society, remorse, either does not exist, or presently disappears: because in all his actions, it is ever the judgment of his fellow man that man is obliged necessarily to regard. He never feels either shame or remorse for actions he sees approved, that are practised by all the world. Under corrupt governments, venal souls, avaricious beings, mercenary individuals, do not blush, either at meanness, robbery, or rapine, when it is authorized by example: in licentious nations no one blushes at adultery; in superstitious countries, man does not blush to assassinate his fellow for his opinions. It will be obvious, therefore, that his remorse, as well as the ideas, whether right or wrong, which man has of decency, virtue, justice, &c. are the necessary consequence of his temperament, modified by the society in which he lives: assassins and thieves, when they live only among themselves, have neither shame nor remorse.

Thus, I repeat, all the actions of man, are necessary; those which are always useful, which constantly contribute to the real, tend to the permanent happiness of his species, are called *virtues*, and are necessarily pleasing to all who experience their influence—at least, if their passions or false opinions do not oblige them to judge

in that manner which is but little accordant with the nature of things: each man acts, each individual judges necessarily according to his own peculiar mode of existence, and after the ideas, whether true or false, which he has formed with regard to his happiness. There are necessary actions, which man is obliged to approve; there are others, that in despite of himself, he is compelled to censure, of which the idea generates shame, when his reflection permits him to contemplate them under the same point of view that they are regarded by his associates. The virtuous man and the wicked act from motives equally necessary; they differ simply in their organization, and in the ideas they form to themselves of happiness: we love the one, necessarily, we detest the other from the same necessity. The law of his nature which wills that a sensible being shall constantly labour to preserve himself, has not left to man the power to choose, or the free agency to prefer pain to pleasure, vice to utility, crime to virtue. It is then the essence of man himself, that obliges him to discriminate those actions which are advantageous to him, from those which are prejudicial.

This distinction subsists even in the most corrupt societies, in which the ideas of virtue, although completely effaced from their conduct, remain the same in their mind. Let us suppose a man, who had decidedly determined for villany, who should say to himself: "It is folly to be virtuous in a society that is depraved, in a community that is debauched." Let us suppose also that he has sufficient address and good fortune to escape censure or punishment during a long series of years; I say, that despite of all these circumstances, apparently so advantageous for himself, such a man has neither been happy nor contented with his own conduct. He has been in continual agonies; ever at war with his own actions; in a state of constant agitation. How much pain, how much anxiety, has he not endured in this perpetual conflict with himself? how many precautions, what excessive labour, what endless solicitude, has he not been compelled to employ in this continued struggle; how many embarrassments, how many cares, has he not experienced in this eternal wrestling with his associates, whose penetration he dreads? Demand· of him what he thinks of himself, he will shrink from the question. Approach the bedside of this villain at the moment he is dying, ask him ·if he would be willing to recommence, at the same price, a life of similar agitation? If he is ingenuous, he will avow that he has tasted neither repose nor happiness; that each crime filled him with inquietude; that reflection prevented him from sleeping; that the world has been to him only one continued scene of alarm and an everlasting anxiety of mind; that to live peaceably upon bread and water, appears to him to be a much happier, a more easy condition, than to possess riches, credit, reputation, honours, on the same terms that he has himself acquired them. If this villain, maugre all his success, finds his condition so deplorable, what must be thought of the feelings of those who have neither the ·same resources, nor the same advantages, to succeed in their criminal projects?

Thus the system of necessity, is a truth not only founded upon certain experience, but, again, it establishes morals upon an immoveable basis. Far from sapping the foundations of virtue, it points out its necessity; it clearly shows the invariable sentiments it must excite—sentiments so necessary, so strong, that all the prejudices and all the vices of man's institutions, have never been able entirely to eradicate them from his mind. When he mistakes the advantages of virtue, it ought to be ascribed to the errours that are infused into him; to the irrationality of his institutions. All his· wanderings are the fatal and necessary consequences of errour and of prejudices which have identified themselves with his existence. Let it not therefore any longer be imputed to his nature that he has become wicked, but to those baneful opinions he has imbibed with his mother's milk which have rendered him ambitious, avaricious, envious, haughty, arrogant, debauched, intolerant, obstinate, prejudiced, incommodious to his fellows, and mischievous to himself. It is education that carries into his system the germ of those vices,

which necessarily torment him during the whole course of his life.

Fatalism is reproached with discouraging man, damping the ardour of his soul, plunging him into apathy, and with destroying the bonds that should connect him with society. Its opponents say: "If every thing is necessary, we must let things go on, and not be disturbed at any thing." But does it depend on man to be sensible or not? Is he master of feeling, or not feeling pain? If nature has endowed him with a humane and tender soul, is it possible he should not interest himself in a very lively manner in the welfare of beings whom he knows are necessary to his own peculiar happiness? His feelings are necessary; they depend on his own peculiar nature, cultivated by education. His imagination, prompt to concern itself with the felicity of his race, causes his heart to be oppressed at the sight of those evils his fellow creature is obliged to endure: makes his soul tremble in the contemplation of the misery arising from the despotism that crushes him; from the superstition that leads him astray; from the passions that distract him; from the follies that are perpetually ranking him in a state of warfare against his neighbour. Although he knows that death is the fatal and necessary period to the form of all beings; his soul is not affected in a less lively manner at the loss of a beloved wife—at the demise of a child calculated to console his old age—at the final separation from an esteemed friend, who had become dear to his heart. Although he is not ignorant that it is the essence of fire to burn, he does not believe he is dispensed from using his utmost efforts to arrest the progress of a conflagration. Although he is intimately convinced that the evils to which he is a witness are the necessary consequence of primitive errours with which his fellow citizens are imbued, yet he feels he ought to display truth to them, (if nature has given him the necessary courage,) under the conviction that if they listen to it, it will by degrees become a certain remedy for their sufferings—that it will produce those necessary effects which it is of its essence to operate.

If the speculations of man modify

his conduct, if they change his temperament, he ought not to doubt that the system of necessity would have the most advantageous influence over him: not only is it suitable to calm the greater part of his inquietude, but it will also contribute to inspire him with a useful submission, a rational resignation to the decrees of a destiny, with which his too great sensibility frequently causes him to be overwhelmed. This happy apathy without doubt would be desirable to those whose souls, too tender to brook the inequalities of life, frequently render them the deplorable sport of their fate; or whose organs, too weak to make resistance to the buffetings of fortune, incessantly expose them to be dashed in pieces under the rude blows of adversity.

But, of all the important advantages the human race would be enabled to derive from the doctrine of fatalism if man was to apply it to his conduct, none would be of greater magnitude, none of more happy consequence, none that would more efficaciously corroborate his happiness, than that general indulgence, that universal toleration, that must necessarily spring from the opinion that *all is necessary.* In consequence of the adoption of this principle, the fatalist, if he had a sensible soul, would commiserate the prejudices of his fellow man, would lament over his wanderings, would seek to undeceive him, without ever irritating himself against his weakness—without ever insulting his misery. Indeed, what right have we to hate or despise man for his opinions? His ignorance, his prejudices, his imbecility, his vices, his passions, his weakness, are they not the inevitable consequence of vicious institutions? Is he not sufficiently punished by the multitude of evils that afflict him on every side? Those despots who crush him with an iron sceptre, are they not continual victims to their own peculiar restlessness, and eternal slaves to their suspicions? Is there one wicked individual who enjoys a pure, an unmixed, a real happiness? Do not nations unceasingly suffer from their follies? Are they not the incessant dupes to their prejudices? Is not the ignorance of chiefs, the ill-will they bear to reason, the hatred they have for truth, punished by the imbecility of their citizens, and by the ruin of the states they govern? In short, the fatalist would grieve to witness necessity each moment exercising its severe decrees upon mortals who are ignorant of its power, or who feel its castigation, without being willing to acknowledge the hand from whence it proceeds; he will perceive, that ignorance is necessary, that credulity is the necessary result of ignorance, that slavery and bondage are necessary consequences of ignorant credulity; that corruption of manners springs necessarily from slavery; that the miseries of society and of its members, are the necessary offspring of this corruption.

The fatalist, in consequence of these ideas, will neither be a gloomy misanthrope, nor a dangerous citizen. He will pardon in his brethren those wanderings which their nature vitiated by a thousand causes, has rendered necessary; he will offer them consolation; he will endeavour to inspire them with courage; he will be sedulous to undeceive them in their idle notions; in their chimerical ideas; but he will never show them that rancorous animosity which is more suitable to make them revolt from his doctrines than to attract them to reason. He will not disturb the repose of society; he will not raise the people to insurrection against the sovereign authority; on the contrary, he will feel that the miserable blindness and perverseness of so many conductors of the people, are the necessary consequence of that flattery administered to them in their infancy; of the depraved malice of those who surround them, and who wickedly corrupt them, that they may profit by their folly: in short, that these things are the inevitable effect of that profound ignorance of their true interest, in which every thing strives to keep them.

The fatalist has no right to be vain of his peculiar talents or of his virtues: he knows that these qualities are only the consequence of his natural organization, modified by circumstances that have in nowise depended upon himself. He will neither have hatred nor feel contempt for those whom nature and circumstances have not favoured in a similar manner. It is the fatalist who ought to be humble and modest from principle: is he not obliged to acknow ledge that he possesses nothing that he has not previously received?

In fact, every thing will conduct to indulgence the fatalist whom experience has convinced of the necessity of things. He will see with pain that it is the essence of a society badly constituted, unwisely governed, enslaved to prejudice, attached to unreasonable customs, submitted to irrational laws, degraded under despotism, corrupted by luxury, inebriated with false opinions, to be filled with trifling members; to be composed of vicious citizens; to be made up of cringing slaves, who are proud of their chains; of ambitious men, without ideas of true glory; of misers and prodigals; of fanatics and libertines! Convinced of the necessary connexion of things, he will not be surprised to see that the supineness of their chiefs carries discouragement into their country; or that the influence of their governors stirs up bloody wars by which it is depopulated; causes useless expenditures that empoverish it; and that all these excesses united is the reason why so many nations contain only men wanting happiness, who are devoid of morals, destitute of virtue. In all this, he will contemplate nothing more than the necessary action and reaction of physics upon morals, of morals upon physics. In short, all who acknowledge fatality, will remain persuaded that a nation badly governed is a soil very abundant in poisonous plants; that these have such a plentiful growth as to crowd each other and choke themselves. It is in a country cultivated by the hands of a Lycurgus, that he will witness the production of intrepid citizens, of noble-minded individuals, of disinterested men, who are strangers to irregular pleasures. In a country cultivated by a Tiberius, he will find nothing but villains, with depraved hearts, men with mean contemptible souls, despicable informers, and execrable traitors. It is the soil, it is the circumstances in which man finds himself placed, that renders him either a useful object or a prejudicial being: the wise man avoids the one, as he would those dangerous reptiles whose nature it is to sting and communicate their deadly venom; he attaches himself to the other, esteems him, loves him, as he does those deli-

cious fruits, with whose rich maturity his palate is pleasantly gratified, and with whose cooling juices he finds himself agreeably refreshed: he sees the wicked without anger; he cherishes the good with pleasure; he delights in the bountiful; he knows full well that the tree which is languishing without culture in the arid, sandy desert; that is stunted for want of attention; leafless for want of moisture; that has grown crooked from neglect; become barren from want of loam; would perhaps have expanded far and wide its verdant boughs, brought forth delectable fruit, afforded an umbrageous refreshing shelter, if its seed had been fortunately sown in a more fertile soil, or if it had experienced the fostering cares of a skilful cultivator.

Let it not then be said, that it is degrading man to reduce his functions to a pure mechanism; that it is shamefully to undervalue him, to compare him to a tree—to an abject vegetation. The philosopher devoid of prejudice, does not understand this language invented by those who are ignorant of what constitutes the true dignity of man. A tree is an object which, in its station, joins the useful with the agreeable; it merits our approbation when it produces sweet and pleasant fruit, and when it affords a favourable shade. All machines are precious, whenever they are truly useful, and when they faithfully perform the functions for which they are designed. Yes, I speak it with courage, the honest man, when he has talents and possesses virtue, is, for the beings of his species, a tree that furnishes them with delicious fruit, and affords them refreshing shelter: the honest man is a machine, of which the springs are adapted to fulfil its functions in a manner that must gratify the expectation of all his fellows. No, I should not blush to be a machine of this sort; and my heart would leap with joy if I could foresee that the fruit of my reflections would one day be useful and consoling to my fellow man.

Is not nature herself a vast machine, of which the human species is but a very feeble spring? I see nothing contemptible either in her or in her productions: all the beings who come out of her hands are good, are noble, are sublime, whenever they co-operate to the production of order; to the maintenance of harmony in the sphere where they must act. Of whatever nature the soul may be, whether mortal or immortal; whether it be regarded as a spirit, or whether it be looked upon as a portion of the body; it will be found noble, great, and sublime, in a Socrates, in an Aristides, in a Cato: it will be thought abject, it will be viewed as despicable and corrupt in a Claudius, in a Sejanus, in a Nero: its energies will be admired in a Shakspeare, in a Corneille, in a Newton, in a Montesquieu: its baseness will be lamented when we behold mean men, who flatter tyranny, or who servilely cringe at the foot of superstition.

All that has been said in the course of this work, proves clearly that every thing is necessary; that every thing is always in order relatively to nature, where all beings do nothing more than follow the laws that are imposed on their respective classes. It is part of her plan, that certain portions of the earth shall bring forth delicious fruits, whilst others shall only furnish brambles and noxious vegetables: she has been willing that some societies should produce wise men and great heroes, that others should only give birth to contemptible men, without energy, and destitute of virtue. Winds, tempests, hurricanes, volcanoes, wars, plagues, famine, diseases, death, are as necessary to her eternal march, as the beneficent heat of the sun, the serenity of the atmosphere, the gentle showers of spring, plentiful years, peace, health, harmony, life: vice and virtue, darkness and light, ignorance and science, are equally necessary; the one are not benefits, the other are not evils, except for those beings whose happiness they influence, by either favouring or deranging their peculiar mode of existence. *The whole cannot be miserable, but it may contain unhappy individuals.*

Nature, then, distributes with the same hand that which is called *order*, and that which is called *disorder;* that which is called *pleasure*, and that which is called *pain;* in short, she diffuses, by the necessity of her existence, good and evil, in the world we inhabit. Let not man therefore either arraign her bounty, or tax her with malice; let him not imagine that his vociferations or his supplications, can

No. IV.—15

ever arrest her colossal power, always acting after immutable laws. Let him submit silently to his condition; and when he suffers, let him not seek a remedy by recurring to chimeras that his own distempered imagination has created; let him draw from the stores of nature herself the remedies which she offers for the evil she brings upon him: if she send him diseases, let him search in her bosom for those salutary productions to which she has given birth. If she gives him errours, she also furnishes him with experience and truth to counteract and destroy their fatal effects. If she permits man to groan under the pressure of his vices, beneath the load of his follies, she also shows him in virtue a sure remedy for his infirmities: if the evils that some societies experience are necessary, when they shall have become too incommodious, they will be irresistibly obliged to search for those remedies which nature will always point out to them. If this nature has rendered existence insupportable to some unfortunate beings whom she may appear to have selected for her victims, still death is a door that will surely be opened to them, and will deliver them from their misfortunes, although they may be deemed impossible of cure.

Let not man, then, accuse nature with being inexorable to him; since there does not exist an evil for which she has not furnished the remedy to those who have the courage to seek and apply it. Nature follows general and necessary laws in all her operations; physical and moral evil are not to be ascribed to her want of kindness, but to the necessity of things. Physical calamity is the derangement produced in man's organs by physical causes which he sees act: moral evil is the derangement produced in him by physical causes, of which the action is to him a secret. These causes always terminate by producing sensible effects, which are capable of striking his senses; neither the thoughts nor the will of man ever show themselves but by the marked effects they produce either in himself or upon those beings whom nature has rendered susceptible of feeling their impulse. He suffers, because it is of the essence of some beings to derange the economy of his machine; he enjoys because the properties of some beings are analogous to his own mode of existence; he is born, because it is of the nature of some matter to combine itself under a determinate form; he lives, he acts, he thinks, because it is of the essence of certain combinations to maintain themselves in existence for a season; at length he dies, because a necessary law prescribes that all the combinations which are formed, shall either be destroyed or dissolve themselves. From all this it results, that nature is impartial to all its productions; she submits man, like all other beings, to those eternal laws from which she has not been able to exempt herself: if she was to suspend these laws, even for an instant, from that moment disorder would reign in her system, and her harmony would be disturbed.

Those who wish to study nature, must take experience for their guide; this, and this only, can enable them to dive into her secrets, and to unravel by degrees the frequently imperceptible woof of those slender causes of which she avails herself to operate the greatest phenomena: by the aid of experience, may often discovers in her new properties, perceives modes of action entirely unknown to the ages which have preceded him; those effects which his grandfathers contemplated as marvellous, which they regarded as supernatural efforts, looked upon as miracles, have become familiar to him in the present day; are at this moment contemplated as simple and natural consequences of which he comprehends the mechanism and the cause. Man, in fathoming nature, has arrived at discovering the true causes of earthquakes, of the periodical motion of the sea, of subterraneous conflagrations, of meteors, of the electrical fluid, the whole of which were considered by his ancestors, and are still so by the ignorant, as indubitable signs of heaven's wrath. His posterity, in following up, in rectifying the experience already made, will go still farther, and discover effects and causes which are totally veiled from present eyes. The united efforts of the human species, will one day perhaps penetrate even into the sanctuary of nature, and throw into light many of those mysteries, which, up to the present time, she seems to have refused to all his researches.

In contemplating man under his true aspect; in quitting authority to follow experience; in laying aside errour to consult reason; in submitting every thing to physical laws, from which his imagination has vainly exerted its utmost power to withdraw them; it will be found, that the phenomena of the moral world follow exactly the same general rules as those of the physical, and that the greater part of those astonishing effects, which ignorance aided by his prejudices, makes him consider as inexplicable and as wonderful, are natural consequences flowing from simple causes. He will find, that the eruption of a volcano and the birth of a Tamerlane are to nature the same thing; in recurring to the primitive causes of those striking events which he beholds with consternation, of those terrible revolutions, those frightful convulsions that distract mankind, lay waste the fairest works of nature, and ravage nations, he will find the wills that compassed the most surprising changes, that operated the most extensive alterations in the state of things, were moved by physical causes, whose exility made him treat them as contemptible, and as utterly incapable to give birth to the phenomena, whose magnitude strikes him with awe and amazement.

If man was to judge of causes by their effects, there would be no small causes in the universe. In a nature where every thing is connected; where every thing acts and reacts, moves and changes, composes and decomposes, forms and destroys, there is not an atom which does not play an important and necessary part; there is not an imperceptible particle, however minute, which, placed in convenient circumstances, does not operate the most prodigious effects. If man was in a capacity to follow the eternal chain, to pursue the concatenated links that connect with their causes all the effects he witnesses, without losing sight of any one of its rings, if he could unravel the ends of those insensible threads that give impulse to the thoughts, decision to the will, direction to the passions of those men who are called mighty, according to their actions; he would find that they are true atoms which nature employs to move the moral world; that it is the unexpected but necessary junction of these indiscernible particles of matter, it is their aggregation, their combination, their proportion, their fermentation, which modifying the individual by degrees, in despite of himself, and frequently without his own knowledge, make him think, will and act in a determinate but necessary mode. If the will and the actions of this individual have an influence over a great number of other men, here is the moral world in a state of the greatest combustion. Too much acrimony in the bile of a fanatic, blood too much inflamed in the heart of a conqueror, a painful indigestion in the stomach of a monarch, a whim that passes in the mind of a woman, are sometimes causes sufficient to bring on war, to send millions of men to the slaughter, to root out an entire people, to overthrow walls, to reduce cities into ashes, to plunge nations into slavery, to put a whole people into mourning, to breed famine in a land, to engender pestilence, to propagate calamity, to extend misery, to spread desolation far and wide upon the surface of our globe, through a long series of ages.

The dominant passion of an individual of the human species, when it disposes of the passions of many others, arrives at combining their will, at uniting their efforts, and thus decides the condition of man. It is after this manner that an ambitious, crafty, and voluptuous Arab gave to his countrymen an impulse, of which the effect was the subjugation and desolation of vast countries in Asia, in Africa, and in Europe; whose consequences were sufficiently potential to give a novel system of religion to millions of human beings; to overturn the altars of their former gods; in short, to alter the opinions, to change the customs of a considerable portion of the population of the earth. But in examining the primitive sources of this strange revolution, what were the concealed causes that had an influence over this man, that excited his peculiar passions, that modified his temperament? What was the matter from the combination of which resulted a crafty, ambitious, enthusiastic, and eloquent man; in short, a personage

competent to impose on his fellow creatures, and capable of making them concur in his views. They were the insensible particles of his blood, the imperceptible texture of his fibres, the salts, more or less acrid, that stimulated his nerves, the proportion of igneous fluid that circulated in his system. From whence came these elements? It was from the womb of his mother, from the aliments which nourished him, from the climate in which he had his birth, from the ideas he received, from the air which he respired, without reckoning a thousand inappreciable and transitory causes, that, in the instance given, had modified, had determined the passions of this important being, who had thereby acquired the capacity to change the face of this mundane sphere.

To causes so weak in their principles, if in the origin the slightest obstacle had been opposed, these wonderful events, which have astounded man, would never have been produced. The fit of an ague, the consequence of bile a little too much inflamed, had sufficed, perhaps, to have rendered abortive all the vast projects of the legislator of the Mussulmen. Spare diet, a glass of water, a sanguinary evacuation, would sometimes have been sufficient to have saved kingdoms.

It will be seen, then, that the condition of the human species, as well as that of each of its individuals, every instant depends on insensible causes, to which circumstances, frequently fugitive, give birth; that opportunity develops, and convenience puts in action: man attributes their effects to chance, whilst these causes operate necessarily and act according to fixed rules: he has frequently neither the sagacity, nor the honesty, to recur to their true principles; he regards such feeble motives with contempt, because he has been taught to consider them as incapable of producing such stupendous events. They are, however, these motives, weak as they may appear to be, these springs, so pitiful in his eyes, which, according to her necessary laws, suffice in the hands of nature, to move the universe. The conquests of a Gengis-khan have nothing in them that is more strange to the eye of a philosopher than the explosion of a mine, caused in its principle by a feeble spark, which commences with setting fire to a single grain of powder; this presently communicates itself to many millions of other contiguous grains, of which the united and multiplied powers, terminate by blowing up mountains, overthrowing fortifications, or converting populous cities into heaps of ruins.

Thus, imperceptible causes, concealed in the bosom of nature until the moment their action is displayed, frequently decide the fate of man. The happiness or the wretchedness, the prosperity or the misery of each individual, as well as that of whole nations, are attached to powers which it is impossible for him to foresee, to appreciate, or to arrest the action. Perhaps, *at this moment*, atoms are amassing, insensible particles are combining, of which the assemblage shall form a sovereign, who will be either the scourge or the saviour of a mighty empire.* Man cannot answer for his own destiny one single instant; he has no cognizance of what is passing within himself; he is ignorant of the causes which act in the interior of his machine; he knows nothing of the circumstances that will give them activity and develop their energy; it is, nevertheless, on these causes, impossible to be unravelled by him, that depends his condition in life. Frequently an unforeseen rencounter gives birth to a passion in his soul, of which the consequences shall necessarily have an influence over his felicity. It is thus that the most virtuous man, by a whimsical combination of unlooked for circumstances, may become in an instant the most criminal of his species.

This truth, without doubt, will be found frightful and terrible: but at bottom, what has it more revolting than that which teaches him that an infinity of accidents, as irremediable as they are unforeseen, may every instant wrest from him that life to which he is so strongly attached? Fatalism reconciles the good man easily to death: it makes him contemplate it as a certain means of withdrawing himself from wickedness; this system shows death, even to the happy man himself, as a medium between him and those mis-

fortunes which frequently terminate by poisoning his happiness, and with im bittering the most fortunate existence.

Let man then submit to necessity in despite of himself it will always hurry him forward: let him resign himself to nature; let him accept the good with which she presents him; let him oppose to the necessary evil which she makes him experience, those necessary remedies which she consents to afford him: let him not disturb his mind with useless inquietude; let him enjoy with moderation, because he will find that pain is the necessary companion of excess: let him follow the paths of virtue, because every thing will prove to him, even in this world of perverseness, that it is absolutely necessary to render him estimable in the eyes of others, and to make him contented with himself.

Feeble, and vain mortal, thou pretendest to be a free agent; alas, dost not thou see all the threads which enchain thee? Dost thou not perceive that they are atoms which form thee; that they are atoms which move thee; that they are circumstances independent of thyself that modify thy being, and rule thy destiny? In the puissant nature that environs thee, shalt thou pretend to be the only being who is able to resist her power? Dost thou really believe, that thy weak prayers will induce her to stop in her eternal march or change her everlasting course?

CHAPTER XIII.

Of the Immortality of the Soul;—Of the Doctrine of a Future State;—Of the Fear of Death.

THE reflections presented to the reader in this work, tend to show, what ought to be thought of the human soul, as well as of its operations and faculties: every thing proves, in the most convincing manner, that it acts and moves according to laws similar to those prescribed to the other beings of nature; that it cannot be distinguished from the body; that it is born with it; that it grows up with it; that it is modified in the same progression; in short, every thing ought to make man conclude that it perishes with it. This soul, as well

---

* By a strange coincidence, Napoleon Buonaparte was born the same year in which the System of Nature was first published.

as the body, passes through a state of weakness and infancy; it is in this stage of its existence that it is assailed by a multitude of modifications and of ideas which it receives from exterior objects through the medium of the organs; that it amasses facts; that it collects experience, whether true or false; that it forms to itself a system of conduct, according to which it thinks and acts, and from whence results either its happiness or its misery, its reason or its delirium, its virtues or its vices; arrived with the body at its full powers; having in conjunction with it reached maturity, it does not cease for a single instant to partake in common of its sensations, whether these are agreeable or disagreeable; in consequence it conjointly approves or disapproves its state; like it, it is either sound or diseased, active or languishing, awake or asleep. In old age, man extinguishes entirely, his fibres become rigid, his nerves lose their elasticity, his senses are obtunded, his sight grows dim, his ears lose their quickness, his ideas become unconnected, his memory fails, his imagination cools; what, then, becomes of his soul? Alas! it sinks down with the body; it gets benumbed as this loses its feeling, becomes sluggish as this decays in activity; like it, when enfeebled by years it fulfils its functions with pain; and this substance, which is deemed spiritual or *immaterial*, undergoes the same revolutions, and experiences the same vicissitudes as does the body itself.

In despite of this convincing proof of the materiality of the soul, and of its identity with the body, some thinkers have supposed that although the latter is perishable, the former does not perish; that this portion of man enjoys the especial privilege of *immortality*; that it is exempt from dissolution and free from those changes of form all the beings in nature undergo: in consequence of this, man has persuaded himself that this privileged soul does not die: its immortality above all appears indubitable to those who suppose it spiritual: after having made it a simple being, without extent, devoid of parts, totally different from any thing of which he has a knowledge, he pretended that it was not subjected to the laws of decomposition common to all

beings, of which experience shows him the continual operation.

Man, feeling within himself a concealed force that insensibly produced action, that imperceptibly gave direction to the motion of his machine, believed that the entire of nature, of whose energies he is ignorant, with whose modes of acting he is unacquainted, owed its motion to an agent analogous to his own soul, who acted upon the great macrocosm in the same manner that this soul acted upon his body. Man having supposed himself double, made nature double also: he distinguished her from her own peculiar energy; he separated her from her mover, which by degrees he made spiritual. Thus this being distinguished from nature was regarded as the soul of the world, and the soul of man was considered as portions emanating from this universal soul. This notion upon the origin of the soul, is of very remote antiquity. It was that of the Egyptians, of the Chaldeans, of the Hebrews, of the greater number of the *wise men of the east.* It was in these schools that Pherecydes, Pythagoras, Plato, drew up a doctrine so flattering to the vanity of human nature—so gratifying to the imagination of mortals. Thus man believed himself a portion of the Divinity; immortal, like the Godhead, in one part of himself; nevertheless, religions subsequently invented have renounced these advantages, which they

---

* It appears that Moses believed, with the Egyptians, the divine emanation of souls: according to him, " *God formed man of the dust of the ground, and breathed into his nostrils the breath of life; and man became a living soul:*" Gen. ii. 7.—nevertheless Christians at this day reject this system of *Divine emanation*, seeing that it supposes the Divinity divisible; besides, their religion having need of a Hell to torment the souls of the damned, it would have been necessary to send a portion of the Divinity to Hell, conjointly with the souls of those victims that were sacrificed to his own vengeance. Although Moses, in the above quotation, seems to indicate that the soul was a portion of the Divinity, it does not appear that the doctrine of the *immortality of the soul* was established in any one of the books attributed to him. It was during the Babylonish captivity that the Jews learned the doctrine of future rewards and punishments, taught by Zoroaster to the Persians, but which the Hebrew Legislator did not understand, or at least he left his people ignorant on the subject.

judged incompatible with the other parts of their systems: they held forth that the sovereign of nature, or her contriver, was not the soul of man, but that in virtue of his omnipotence, he created human souls in proportion as he produced the bodies which they must animate; and they taught, that these souls once produced, by an effect of the same omnipotence, enjoyed immortality.

However it may be with these variations upon the origin of souls, those who supposed them emanating from the Divinity, believed that after the death of the body, which served them for an envelope, they returned by refunding to their first source. Those who, without adopting the opinion of divine emanation, admired the spirituality and the immortality of the soul, were under the necessity to suppose a region, to find out an abode for these souls, which their imagination painted to them each according to his fears, his hopes, his desires, and his prejudices.

Nothing is more popular than the doctrine of the *immortality of the soul;* nothing is more universally diffused than the expectation of another life. Nature having inspired man with the most ardent love for his existence, the desire of preserving himself for ever was a necessary consequence : this desire was presently converted into certainty; from that desire of existing eternally, which nature has implanted in him, he made an argument to prove that man would never cease to exist. Abbadie says: "Our soul has no useless desires, it desires naturally an eternal life;" and by a very strange logic he concludes, that this desire could not fail to be fulfilled.* However this may be, man, thus disposed, listened with avidity to those who announced to him systems so conformable with his wishes. Nevertheless, he ought not to regard as supernatural the desire of existing, which always was, and always will be,

of the essence of man; it ought not to excite surprise if he received with eagerness an hypothesis that flattered his hopes, by promising that his desire would one day be gratified; but let him beware how he concludes, that this desire itself is an indubitable proof of the reality of this future life, with which, for his present happiness, he seems to be far too much occupied. The passion for existence, is in man only a natural consequence of the tendency of a sensible being, whose essence it is to be willing to conserve himself: in the human being, it follows the energy of his soul or keeps pace with the force of his imagination, always ready to realize that which he strongly desires. He desires the life of the body, nevertheless this desire is frustrated; wherefore should not the desire for the life of the soul be frustrated like the other?†

The most simple reflection upon the nature of his soul, ought to convince man that the idea of its immortality is only an illusion of the brain. Indeed, what is his soul, save the principle of sensibility? What is it to think, to enjoy, to suffer; is it not to feel? What is life, except it be the assemblage of modifications, the congregation of motion, peculiar to an organized being? Thus, as soon as the body ceases to live, its sensibility can no longer exercise itself; therefore it can no longer have ideas, nor in consequence thoughts. Ideas, as we have proved, can only reach man through his senses; now, how will they have it, that once deprived of his senses, he is yet capable of receiving sensations, of having perceptions, of forming ideas? As they have made the soul of man a being separated from the animated body, wherefore have they not made life a being distinguished from the living body? Life in a body is the totality of its motion; feeling and thought make a part of this motion: thus, in the dead man, these motions will cease like all the others.

Indeed, by what reasoning will it be proved, that this soul, which cannot

* Cicero before Abbadie had declared the immortality of the soul to be an innate idea in man ; yet, strange to tell, in another part of his works he considers Pherecydes as the inventor of the doctrine —Naturam ipsam de immortalitate animarum tacitam judicare; nescio quomodo inhæret in mentibus quasi sæculorum quodam augurium. Permanere animos arbitramur consensu nationum omnium.— *Tusculam Disputat.* lib. i.

† The partisans of the doctrine of the immortality of the soul, reason thus: "All men desire to live for ever; therefore they will live for ever." Suppose the argument retorted on them : "All men naturally desire to be rich; therefore, all men will one day be rich."

feel, think, will, or act, but by aid of man's organs, can suffer pain, be susceptible of pleasure, or even have a consciousness of its own existence, when the organs which should warn it of their presence, are decomposed or destroyed? Is it not evident that the soul depends on the arrangement of the various parts of the body, and on the order with which these parts conspire to perform their functions or motions? Thus the organic structure once destroyed, can it be doubted the soul will be destroyed also? Is it not seen, that during the whole course of human life, this soul is stimulated, changed, deranged, disturbed, by all the changes man's organs experience? And yet it will be insisted that this soul acts, thinks, subsists, when these same organs have entirely disappeared!

An organized being may be compared to a clock, which, once broken, is no longer suitable to the use for which it was designed. To say, that the soul shall feel, shall think, shall enjoy, shall suffer, after the death of the body, is to pretend, that a clock, shivered into a thousand pieces, will continue to strike the hour, and have the faculty of marking the progress of time. Those who say, that the soul of man is able to subsist notwithstanding the destruction of the body, evidently support the position, that the modification of a body will be enabled to conserve itself, after the subject is destroyed: but this is completely absurd.

It will be said, that the conservation of the soul after the death of the body, is an effect of the divine omnipotence: but this is supporting an absurdity by a gratuitous hypothesis. It surely is not meant by divine omnipotence, of whatever nature it may be supposed, that a thing shall exist and not exist at the same time: that a soul shall feel and think without the intermediates necessary to thought.

Let them, then, at least forbear asserting, that reason is not wounded by the doctrine of the immortality of the soul, or by the expectation of a future life. These notions, formed to flatter man, or to disturb the imagination of the uninformed who do not reason, cannot appear either convincing or probable to enlightened minds. Reason, exempted from the illusions of prejudice, is, without doubt, wounded by the supposition of a soul that feels, that thinks, that is afflicted, that rejoices, that has ideas, without having organs; that is to say, destitute of the only known and natural means by which it is possible for it to feel sensations, have perceptions, or form ideas. If it be replied, that other means are able to exist, which are *supernatural* or *unknown*; it may be answered, that these means of transmitting ideas to the soul separated from the body, are not better known to, or more within the reach of those who suppose it than they are of other men. It is at least very certain, that all those who reject the system of innate ideas, cannot, without contradicting their own principles, admit the groundless doctrine of the immortality of the soul.

In defiance of the consolation that so many persons pretend to find in the notion of an eternal existence; in despite of that firm persuasion, which such numbers of men assure us they have, that their souls will survive their bodies, they seem so very much alarmed at the dissolution of this body, that they do not contemplate their end, which they ought to desire as the period of so many miseries, but with the greatest inquietude: so true it is, that the real, the present, even accompanied with pain, has much more influence over mankind, than the most beautiful chimeras of the future, which he never views but through the clouds of uncertainty. Indeed the most religious men, notwithstanding the conviction they express of a blessed eternity, do not find these flattering hopes sufficiently consoling to repress their fears and trembling when they think on the necessary dissolution of their bodies. Death was always for mortals the most frightful point of view; they regard it as a strange phenomenon, contrary to the order of things, opposed to nature; in a word, as an effect of the celestial vengeance, as the *wages of sin.* Although every thing proves to man that death is inevitable, he is never able to familiarize himself with its idea; he never thinks on it without shuddering, and the assurance of possessing an immortal soul, but feebly indemnifies him for the grief he feels in the deprivation of his perishable body. Two causes contribute to strengthen and nourish his alarm; the one is, that this death, commonly accompanied with pain, wrests from him an existence that pleases him, with which he is acquainted, to which he is accustomed; the other is the uncertainty of the state that must succeed his actual existence.

The illustrious Bacon has said: that "Men fear death, for the same reason that children dread being alone in darkness."* Man naturally challenges every thing with which he is unacquainted; he is desirous to see clearly, to the end that he may guaranty himself against those objects which may menace his safety, or that he may be enabled to procure for himself those which may be useful to him. The man who exists, cannot form to himself any idea of non-existence; as this circumstance disturbs him, for want of experience his imagination sets to work; this points out to him, either well or ill, this uncertain state: accustomed to think, to feel, to be stimulated into activity, to enjoy society, he contemplates as the greatest misfortune a dissolution that will strip him of these objects, and deprive him of those sensations which his present nature has rendered necessary to him; that will prevent his being warned of his own existence; that shall bereave him of his pleasures to plunge him into nothing. In supposing it even exempt from pain, he always looks upon this nothing as an afflicting solitude, as a heap of profound darkness; he sees himself in a state of general desolation, destitute of all assistance, and feeling the rigour of this frightful situation. But does not a profound sleep help to give him a true idea of this nothing? Does not that deprive him of every thing? Does it not appear to annihilate the universe to him, and him to the universe? Is death any thing more than a profound and permanent sleep? Is it for want of being able to form an idea of death, that man dreads it; if he could figure to himself a true image of this state of annihilation, he would from thence cease to fear it; but he is not able to conceive a state in which there is no feeling; he therefore believes, that

when he shall no longer exist, he will have the same feelings and the same consciousness of things which during his existence appear to his mind in such gloomy colours: imagination pictures to him his funeral pomp; the grave they are digging for him; the lamentations that will accompany him to his last abode; he persuades himself that these melancholy objects will affect him as painfully, even after his decease, as they do in his present condition in which he is in full possession of his senses.†

Mortal, led astray by fear! after thy death thine eyes will see no more; thine ears will hear no longer; in the depth of thy grave, thou wilt no more be witness to this scene which thine imagination. at present represents to thee under such dismal colours; thou wilt no longer take part in what shall be done in the world; thou wilt no more be occupied with what may befall thine inanimate remains, than thou wast able to be the day previous to that which ranked thee among the beings of thy species. To die, is to cease to think, to feel, to enjoy, to suffer; thy sorrows will not follow thee to the silent tomb. Think of death, not to feed thy fears and to nourish thy melancholy, but to accustom thyself to look upon it with a peaceable eye, and to cheer thee up against those false terrours with which the enemies to thy repose labour to inspire thee!

The fears of death are vain illusions, that must disappear as soon as we learn to contemplate this necessary event under its true point of view. A great man has defined philosophy to be *a meditation on death*;‡ he is not desirous by that to have it understood that man ought to occupy himself sorrowfully with his end, with a view to nourish his fears; on the contrary he wishes to invite him to familiarize himself with an object that nature has rendered necessary to him, and to accustom himself to expect it with a serene countenance. If life is a benefit, if it be necessary to love it, it is no less necessary to quit it, and reason ought

---

* Nam veluti pueri trepidant, atque omnia cæcis
In tenebris metuunt: sic nos in luce timemus
Interdum, nihilo quæ sunt metuenda magis quam
Quæ pueri in tenebris pavitant, finguntque futura.
Lucretius, Lib. III. v. 87, et seq.

† Nec videt in vera nullum fore morte alium, *se:*
Qui possit vivus sibi *se* lugere peremptum,
Stansque jacentum, nec lacerari.-urire dolore.
Lucret. Lib. III.

‡ Μελετη τε θανατε. And Lucan has said:
Scire mori sors prima viris.

to teach him a calm resignation to the decrees of fate: his welfare exacts that he should contract the habit of contemplating without alarm an event that his essence has rendered inevitable: his interest demands that he should not by continual dread imbitter his life, the charms of which he must inevitably destroy, if he can never view its termination but with trepidation. Reason and his interest concur to assure him against those vague terrours with which his imagination inspires him in this respect. If he was to call them to his assistance, they would reconcile him to an object that only startles him because he has no knowledge of it, or because it is only shown to him with those hideous accompaniments with which it is clothed by superstition. Let him, then, endeavour to despoil death of these vain illusions, and he will perceive that it is only the sleep of life; that this sleep will not be disturbed with disagreeable dreams, and that an unpleasant awakening will never follow it. To die, is to sleep; it is to reenter into that state of insensibility in which he was previous to his birth; before he had senses, before he was conscious of his actual existence. Laws, as necessary as those which gave him birth, will make him return into the bosom of nature from whence he was drawn, in order to reproduce him afterwards under some new form, which it would be useless for him to know: without consulting him, nature places him for a season in the order of organized beings; without his consent, she will oblige him to quit it to occupy some other order.

Let him not complain, then, that nature is callous; she only makes him undergo a law from which shè does not exempt any one being she contains.* If all are born and perish; if every thing is changed and destroyed; if the birth of a being is never more than the first step towards its end; how is it possible to expect that man, whose machine is so frail, of which the parts are so com-

plicated, the whole of which possesses such extreme mobility, should be exempted from the common law which decrees that even the soild earth he inhabits shall experience change, shall undergo alteration — perhaps be destroyed! Feeble, frail mortal! thou pretendest to exist for ever; wilt thou, then, that for thee alone, eternal nature shall change her undeviating course? Dost thou not behold in those eccentric comets, with which 'thine eyes are sometimes astonished, that the planets themselves are subject to death? Live then in peace, for the season that nature permits thee; and if thy mind be enlightened by reason, thou wilt die without terrour!

Notwithstanding the simplicity of these reflections, nothing is more rare than the sight of men truly fortified against the fears of death: the wise man himself turns pale at its approach; he has occasion to collect the whole force of his mind to expect it with serenity. It cannot then furnish matter for surprise, if the idea of death is so revolting to the generality of mortals; it terrifies the young; it redoubles the chagrin and sorrow of the old, who are worn down with infirmity: indeed the aged although enfeebled by time, dread it much more than the young who are in the full vigour of life; the man of many lustres is more accustomed to live; the powers of his mind are weakened; he has less energy: at length disease consumes him; yet the unhappy wretch thus plunged into misfortune, and labouring under excruciating tortures, has scarcely ever dared to contemplate death which he ought to consider as the period to all his anguish.

If the source of this pusillanimity be sought, it will be found in his nature, which attaches him to life, and in that deficiency of energy in his soul, which hardly any thing tends to corroborate, but which every thing strives to enfeeble and bruise. All human institutions, all the opinions of man, conspire to augment his fears, and to render his ideas of death more terrible and revolting. Indeed, superstition pleases itself with exhibiting death under the most frightful traits; as a dreadful moment, which not only puts an end to his pleasures, but gives him up without

* Quid de rerum natura querimur, illa se bene gessit; vita si scias uti, longa est.— V. *Senec. de Brevitate Vitæ.* Man complains of the short duration of life—of the rapidity with which time flies away; yet the greater number of men do not know how to employ either time or life.

defence to the strange rigour of a pitiless despot, which nothing can soften. According to this superstition, the most virtuous man is never sure of pleasing him; but has reason to tremble for the severity of his judgments; to fear the dreadful torments and endless punishments which await the victims of his caprice, for involuntary weakness or the necessary faults of a short-lived existence. This implacable tyrant will avenge himself* of man's infirmities, his momentary offences, of the propensities that have been planted in his heart, of the errours of his mind, the opinions he has imbibed in the society in which he was born without his own consent, the ideas he has formed, the passions he has indulged, and above all, his not being able to comprehend an inconceivable being, and all the extravagant dogmas offered to his acceptance.*

Such, then, are the afflicting objects with which religion occupies its unhappy and credulous disciples; such are the fears, which the tyrant of human thoughts points out to them as *salutary.* In defiance of the exility of the effect which these notions produce on the greater number of those who say they are, or who believe themselves persuaded, they are held forth as the most powerful rampart that can be opposed to the irregularities of man. Nevertheless, as will be seen presently, it will be found that these systems, or rather these chimeras so terrible to behold, operate little or nothing on the larger portion of mankind, who think

* Those who dare to think for themselves— those who have refused to listen to their enthusiastic guides—those who have no reverence for the Bible—those who have had the audacity to consult their reason—those who have boldly ventured to detect impostors— those who have doubted the divine mission of Jesus Christ—those who believe that Jehovah violated decency in his visit to the carpenter's wife—those who look upon Mary as no better than a strolling wench—those who think that St Paul was an arch knave,—are to smart everlastingly in flaming oceans of burning sulphur, are to float to all eternity in the most excruciating agonies, on seas of liquid brimstone, wailing and gnashing their teeth: what wonder, then, if man dreads to be cast into these hideous gulfs—if his mind loathes the horrific picture—if he wishes to defer for a season these dreadful punishments—if he clings to an existence, painful as it may be, rather than encounter such revolting cruelties.

of them but seldom, and never in the moment that passion, interest, pleasure, or example, hurries them along. If these fears act, it is commonly on those who have but little occasion to abstain from evil: they make honest hearts tremble, but fail of effect on the perverse. They torment sensible souls, but leave those that are hardened in repose; they disturb tractable and gentle minds, but cause no trouble to rebellious spirits: thus they alarm none but those who are already sufficiently alarmed; they coerce only those who are already restrained.

These notions, then, impress nothing on the wicked; when by accident they do act on them, it is only to redouble the wickedness of their natural character, to justify them in their own eyes, to furnish them with pretexts to exercise it without fear, and to follow it without scruple. Indeed, the experience of a great number of ages has shown to what excess of wickedness, to what lengths the passions of man have carried him, when they have been authorized and unchained by religion; or, at least, when he has been enabled to cover himself with its mantle. Man has never been more ambitious, never more covetous, never more crafty, never more cruel, never more seditious, than when he has persuaded himself that religion permitted or commanded him to be so: thus religion did nothing more than lend an invincible force to his natural passions, which, under its sacred auspices, he could exercise with impunity and without remorse; still more, the greatest villains, in giving free vent to the detestable propensities of their natural wickedness, have believed that by displaying an over-heated zeal they merited well of heaven; that they exempted themselves by crimes from that chastisement at the hand of their God, which they thought their anterior conduct had richly merited.

These, then, are the effects which the *salutary* notions of theology produce on mortals. These reflections will furnish an answer to those who say, that, " if religion promised heaven equally to the wicked as to the righteous, there would be found none incredulous of another life." We reply that, in point of fact, religion does accord heaven to the wicked, since it frequently places in this happy abode the most useless and the most depraved of men.*

Thus religion, as we have seen, sharpens the passions of evil disposed men, by legitimating those crimes, at which, without this sanction, they would shudder to commit; or for which, at least, they would feel shame and experience remorse. In short, the ministers of religion furnish to the most profligate men the means of diverting from their own heads the thunderbolt that should strike their crimes, with the promise of a never-fading happiness.

With respect to the incredulous, without doubt there may be amongst them wicked men, as well as amongst the most credulous; but incredulity no more supposes wickedness than credulity supposes righteousness. On the contrary, the man who thinks, who meditates, knows far better the true motives to goodness, than he who suffers himself to be blindly guided by uncertain motives, or by the interest of others. Sensible men have the greatest advantage in examining opinions which it is pretended must have an influence over their eternal happiness: if these are found false or injurious to their present life, they will not therefore conclude that they have not another life either to fear or to hope; that they are permitted to deliver themselves up with impunity to vices which would do an injury to themselves, or would draw upon them the contempt and anger of society: the man who does not expect another life, is the more interested in prolonging his existence in this, and in rendering himself dear to his fellows in the only life of which he has any knowledge: he has made a great stride towards felicity, in disengaging himself from those terrors which afflict others.†

Superstition, in fact, takes a pride in rendering man slothful, credulous, and pusillanimous! It is its principle to afflict him without intermission; to redouble in him the horrours of death: ever ingenious in tormenting him, it has extended his inquietudes beyond even his known existence; and its ministers, the more securely to dispose of him in this world, invented future regions, reserving to themselves the privilege of awarding recompenses to those who yielded most implicitly to their arbitrary laws, and of having their God decree punishments to those refractory beings who rebelled against their power.‡

Thus, far from holding forth consolation to mortals, far from cultivating man's reason, far from teaching him to yield under the hands of necessity, religion strives to render death still more bitter to him, to make its yoke sit heavy, to fill up its retinue with a multitude of hideous phantoms, and to render its approach terrible. By this means it has crowded the world with enthusiasts, whom it seduces by vague promises; with contemptible slaves, whom it coerces with the fear of imaginary evils. It has at length persuaded man, that his actual existence is only a journey by which he will arrive at a more important life. This irrational doctrine of a future life prevents him from occupying himself with his true happiness; from thinking of ameliorating his institutions, of improving his laws, of advancing the progress of science, and of perfecting his morals. Vain and gloomy ideas have absorbed his attention: he consents to groan under religious and political tyranny; to live in errour, to languish in mis-

* Such were Moses, Samuel, and David, among the Jews; Mahomet amongst the Mussulmen; amongst the Christians, Constantine, St. Cyril, St. Athanasius, St. Dominic, and a great many more pious robbers and zealous persecutors, *whom the Church reveres!* We may also add to this list the Crusaders, Leaguers, Puritans, and our modern heterodox Saints, the *Unitarian Inquisitors* of Massachusetts, who, if they had had the power, would have condemned Abner Kneeland to the devouring flames.

† A virtuous and good man has nothing to fear, but every thing to hope; for, if contrary to what he is able to judge, there should be a hereafter existence, will not his actions have

been so regulated by virtue, will he not have so comported himself in his present existence, as to stand a fair chance of enjoying in their fullest extent those felicities prepared for his species?

‡ Let us review the history of Priestcraft in all ages, and we shall invariably find it the same crafty and contemptible system. Tantalus, for divulging their secrets, must eternally fear, engulfed in burning sulphur, the stone ready to fall on his devoted head; whilst Romulus was beatified and worshipped as a God under the name of Quirinus. The same system of Priestcraft caused the philosopher Callisthenes to be put to death, for opposing the worship of Alexander, and elevated the monk Athanasius to be a saint in heaven!

fortune, in the hope, when he shall be no more, of being one day happier; in the firm confidence, that his calamities, his stupid patience, will conduct him to a never-ending felicity: he has believed himself submitted to a cruel God, who is willing to make him purchase his future welfare, at the expense of every thing most dear and most valuable to his existence here below: they have pictured their God as irritated against him, as disposed to appease itself by punishing him eternally for any efforts he should make to withdraw himself from their power. It is thus that the doctrine of a future life has been most fatal to the human species: it plunged whole nations into sloth, made them languid, filled them with indifference to their present welfare; or else precipitated them into the most furious enthusiasm, which hurried them on to tear each other in pieces in order to merit heaven.

It will be asked, perhaps, by what road has man been conducted, to form to himself these strange and gratuitous ideas of another world? I reply, that it is a truth man has no idea of a future life, which does not exist for him; the ideas of the past and the present furnish his imagination with the materials of which he constructs the edifice of the regions of futurity; and Hobbes says, " We believe that that which is, will always be," and that the same causes will have the same effects." Man in his actual state, has two modes of feeling, one that he approves, another that he disapproves: thus, persuaded that these two modes of feeling must accompany him, even beyond his present existence, he placed in the regions of eternity two distinguished abodes; one destined to felicity, the other to misery: the one will contain the friends of his God; the other is a prison, destined to avenge Him on all those who shall not faithfully believe the doctrines promulgated by the ministers of a vast variety of superstitions.*

---

* Has sufficient attention been paid to the fact that results as a necessary consequence from this reasoning, which on examination will be found to have rendered the first place entirely useless, seeing that by the number and contradiction of these various systems, let man believe which ever he may, let him follow it in the most faithful manner, still he

Such is the origin of the ideas upon a future life, so diffused among mankind. Every where may be seen an *Elysium* and a *Tartarus; a Paradise* and a *Hell;* in a word, two distinguished abodes, constructed according to the imagination of the knaves or enthusiasts who have invented them, and who have accommodated them to the peculiar prejudices, to the hopes, to the fears, of the people who believe in them. The Indian figures the first of these abodes as one of inaction and of permanent repose, because, being the inhabitant of a hot climate, he has learned to contemplate rest as the extreme of felicity: the Mussulman promises himself corporeal pleasures, similar to those that actually constitute the object of his research in this life: the Christian hopes for ineffable and spiritual pleasures—in a word, for a happiness of which he has no idea.

Of whatever nature these pleasures may be, man perceived that a body was needful, in order that his soul might be enabled to enjoy the pleasures. or to experience the pains in reserve for him by the Divinity: from hence the doctrine of the *resurrection;* but as he beheld this body putrify, as he saw it dissolve, as he witnessed its decomposition after death, he therefore had recourse to the divine omnipotence, by whose interposition he now believes it will be formed anew. This opinion, so incomprehensible, is said to have originated in Persia, among the Magi, and finds a great number of adherents, who have never given it a serious examination.† Others, incapable of ele-

---

must be ranked as an infidel, as a rebel to the Divinity, because he cannot believe in all; and those from which he dissents, by a consequence of their own creed, condemn him to the prison-house?

† The doctrine of the *resurrection* appears perfectly useless to all those who believe in the existence of a soul, that feels, thinks, suffers, and enjoys after a separation from the body: indeed, there are already sects who begin to maintain, that the body is not necessary, that therefore it will never be resurrected.— Like Berkeley, they conceive that "the soul has need neither of body nor any exterior being, either to experience sensations, or to have ideas." The *Malebranchists*, in particular, must suppose that the rejected souls will see hell in the Divinity, and will feel themselves burn without having occasion for bodies for that purpose.

rating themselves to these sublime notions, believed, that under divers forms, man animated successively different animals, of various species, and that he never ceased to be an inhabitant of the earth ; such was the opinion of those who adopted the doctrine of *Metempsychosis.*

As for the miserable abode of souls, the imagination of fanatics, who were desirous of governing the people, strove to assemble the most frightful images to render it still more terrible. Fire is of all beings that which produces in man the most pungent sensation ; it was therefore supposed that God could not invent any thing more cruel to punish his enemies : then fire was the point at which their imagination was obliged to stop; and it was agreed pretty generally, that fire would one day avenge the offended divinity :* thus they painted the victims to his anger as confined in fiery dungeons ; as perpetually rolling in a vortex of bituminous flames ; as plunged in unfathomed gulfs of liquid sulphur ; and making the infernal caverns resound with their useless groanings, and with their unavailing gnashing of teeth.

But it will perhaps be inquired, how could man reconcile himself to the belief of an existence accompanied with eternal torments ; above all, as many according to their own religious systems had reason to fear it for themselves ? Many causes have concurred to make him adopt so revolting an opinion. In the first place, very few thinking men have ever believed such an absurdity, when they have deigned to make use of their reason ; or, when they have accredited it, this notion was always counterbalanced by the idea of the goodness, by a reliance on the mercy, which they attributed to their God.†

In the second place, those who were blinded by their fears, never rendered to themselves any account of these strange doctrines, which they either received with awe from their legislators, or which were transmitted to them by their fathers. In the third place each sees the object of his terrours only at a favourable distance ; moreover superstition promises him the means of escaping the tortures he believes he has merited. At length, like those sick people whom we see cling with fondness even to the most painful life, man preferred the idea of an unhappy though unknown existence, to that of non-existence, which he looked upon as the most frightful evil that could befall him. either because he could form no idea of it, or, because his imagination painted to him this non-existence, this nothing, as the confused assemblage of all evils. A known evil, of whatever magnitude, alarmed him less, above all when there remained the hope of being able to avoid it, than an evil of which he knew nothing, upon which consequently his imagination was painfully employed, but to, which he knew not how to oppose a remedy.

It will be seen, then, that superstition, far from consoling man upon the necessity of death, only redoubles his terrours, by the evils which it pretends his decease will be followed : these terrours are so strong, that the miserable wretches who believe strictly in these formidable doctrines, pass their days in affliction, bathed in the most bitter tears. What shall be said of an opinion, so destructive to society, yet adopted by so many nations, which announces to them, that a severe God, may at each instant, *like a thief,* take them unprovided ; that at each moment they are liable to pass under the most rigorous judgment ? What idea can

---

* It is no doubt to this we owe the atonements by fire used by a great number of oriental nations, and practised at this very day by the priests of the *God of Peace*, who are so cruel as to consign to the flames all those who differ from them in their ideas of the Divinity. As a consequence of this absurd system, the civil magistrates condemn to the fire the sacrilegious and the blasphemer—that is to say, persons who do no harm to any one ; whilst they are content to punish more mildly those who do a real injury to society. So much for religion and its effects !

† If, as Christians assume, the torments in hell are to be infinite in their duration and in-

tenseness, we must conclude that man, who is a finite being, cannot suffer infinitely. God himself, in despite of the efforts he might make to punish eternally for faults which are limited by time, cannot communicate infinity to man. The same may be said of the joys of Paradise, where a finite being will no more comprehend an infinite God, than he does in this world. On the other hand, if God perpetuates the existence of the damned, as Christianity teaches, he perpetuates the existence of sin, which is not very consistent with his supposed love of order.

be bettei suited to terrify man, what
more likely to discourage him, what
more calculated to damp the desire of
ameliorating his condition, than the
afflicting prospect of a world always
on the brink of dissolution, and of a
divinity seated upon the ruins of nature, ready to pass judgment on the
human species? Such are, nevertheless, the fatal opinions with which the
mind of nations has been fed for thousands of years; they are so dangerous,
that if by a happy want of just inference, he did not derogate in his conduct from these afflicting ideas, he
would fall into the most abject stupidity. How could man occupy himself
with a perishable world, ready every
moment to crumble into atoms? How
think of rendering himself happy on
earth, when it is only the porch to an
eternal kingdom? Is it, then, surprising that the superstitions to which such
doctrines serve for a basis, have prescribed to their disciples a total detach-
ment from things below: an entire renunciation of the most innocent pleasures; and have given birth to a sluggishness, to a pusillanimity, to an abjection of soul, to an insociability, that
renders him useless to himself and
dangerous to others? If necessity did
not oblige man to depart in his practice from these irrational systems; if
his wants did not bring him back to
reason, in despite of his religious doctrines, the whole world would presently become a vast desert, inhabited by
some few isolated savages, who would
not even have courage to multiply
themselves. What kind of notions
are those which must necessarily be
put aside, in order that human association may subsist?

Nuvurtholoss, tho doctrine of a future life, accompanied with rewards
and punishments, has been regarded
for a great number of ages as the most
powerful, or even as the only motive
capable of coercing the passions of
man—as the sole means that can oblige
him to be virtuous. By degrees, this
doctrine has become the basis of almost
all religious and political systems, so
much so, that at this day it is said this
prejudice cannot be attacked without
absolutely rending asunder the bonds
of society. The founders of religions
have made use of it to attach their

credulous disciples; legislators have
looked at it as the curb best calculated
to keep mankind under discipline.
Many philosophers themselves have
believed with sincerity, that this doctrine was requisite to terrify man, and
thus divert him from crime.*

It must indeed be allowed, that this
doctrine has been of the greatest utility
to those who have given religions to
nations and made themselves its ministers: it was the foundation of their
power; the source of their wealth; the
permanent cause of that blindness, the
solid basis of those terrours, which it
was their interest to nourish in the human race. It was by this doctrine that
priest became first the rival, then the
master of kings: it is by this dogma
that nations are filled with enthusiasts
inebriated with religion, always more
disposed to listen to its menaces than
to the counsels of reason, to the orders
of the sovereign, to the cries of nature,
or to the laws of society. Politics itself, was enslaved to the caprice of the
priest; the temporal monarch was
obliged to bend under the yoke of the
eternal monarch; the one only disposed of this perishable world; the other
extended his power into the world to
come, much more important for man
than the earth, on which he is only a
pilgrim, a mere passenger. Thus the
doctrine of another life, placed the government itself in a state of dependance
upon the priest; the monarch was nothing more than his first subject, and
he was never obeyed, but when the two
were in accord to oppress the human
race. Nature in vain cried out to man,
to be careful of his present happiness;
the priest ordered him to be unhappy.
in the expectation of future felicity.
Rwasou in vain exhorted him to be
peaceable, the priest breathed forth fanaticism and fury, and obliged him to
disturb the public tranquillity, every
time there was a question of the inte-

* When the doctrine of the immortality of the
soul first came out of the school of Plato, and
first diffused itself among the Greeks, it caused
the greatest ravages; it determined a multitude
of men, who were discontented with their condition, to terminate their existence. Ptolemy
Philadelphus, king of Egypt, seeing the effect
this doctrine, which at the present day is looked upon as so salutary, produced on the brains
of his subjects, defended the teaching of it,
under the penalty of death.

• rests of the invisible monarch of another life, or the real interests of his ministers in this.

Such is the fruit that politics has gathered from the doctrine of a future life. The regions of the world to come, have enabled the priesthood to conquer the present world. The expectation of celestial happiness, and the dread of future tortures, only served to prevent man from seeking after the means to render himself happy here below. Thus errour, under whatever aspect it is considered, will never be more than a source of evil for mankind. The doctrine of another life, in presenting to mortals an ideal happiness, will render them enthusiasts; in overwhelming them with fears, it will make useless beings, generate cowards, form atrabilarious or furious men, who will lose sight of their present abode, to occupy themselves with the pictured regions of a world to come, and with those dreadful evils which they must fear after their death.

If it be insisted, that the doctrine of future rewards and punishments is the most powerful curb to restrain the passions of man; we shall reply by calling in daily experience. If we only cast our eyes around, we shall see this assertion contradicted; and we shall find that these marvellous speculations do not in any manner diminish the number of the wicked, because they are incapable of changing the temperament of man, of annihilating those passions which the vices of society engender in his heart. In those nations who appear the most thoroughly convinced of this future punishment, may be seen assassins, thieves, crafty knaves, oppressors, adulterers, voluptuaries; all these pretend they are firmly persuaded of the reality of an hereafter; yet in the whirlwind of dissipation, in the vortex of pleasure, in the fury of their passions, they no longer behold this formidable future existence, which in those moments has no kind of influence over their earthly conduct.

In short, in many of those countries where the doctrine of another life is so firmly established that each individual irritates himself against whoever may have the temerity to combat the opinion, or even to doubt it, we see that it is utterly incapable of impressing any

thing on rulers who are unjust, who are negligent of the welfare of their people, who are debauched; on courtesans who are lewd in their habits; on covetous misers; on flinty extortioners, who fatten on the substance of a nation; on women without modesty; on a vast multitude of drunken, intemperate and vicious men; on great numbers even amongst those priests, whose function it is to announce the vengeance of heaven. If it be inquired of them, how they dare to give themselves up to such scandalous actions, which they ought to know are certain to draw upon them eternal punishment? They will reply: that the madness of their passions, the force of their habits, the contagion of example, or even the power of circumstances, have hurried them along, and have made them forget the dreadful consequences in which their conduct is likely to involve them; besides, they will say that the treasures of the divine mercy are infinite, and that repentance suffices to efface the foulest transgressions, the blackest guilt, and the most enormous crimes.* In this multitude of wretched beings, who, each after his own manner, desolates society with his criminal pursuits, you will find only a small number who are sufficiently intimidated by the fears of a miserable hereafter to resist their evil propensities. What did I say? these propensities are in themselves too weak to carry them forward, and without the aid of the doctrine of another life, the law and the fear of censure would have been motives sufficient to prevent them from rendering themselves criminal.

It is, indeed, fearful, timorous souls, upon whom the terrors of another life make a profound impression: human beings of this sort come into the world

---

* The idea of Divine Mercy cheers up the wicked, and makes him forget Divine Justice. And indeed, these two attributes, supposed to be equally infinite in God, must counterbalance each other in such a manner, that neither the one nor the other are able to act. Yet, the wicked reckon upon an *immoveable* God, or at least flatter themselves to escape from the effects of his justice by means of his mercy. The highwayman, who knows that sooner or later he must perish on the gallows, says, that he has nothing to fear, as he will then have an opportunity of *making a good end*. Every Christian believes that *true repentance* blots out all their sins. The East Indian attributes the same virtues to the waters of the Ganges.

with moderate passions, a weakly organization, and a cool imagination; it is not therefore surprising that in such men, who are already restrained by their nature, the fear of future punishment counterbalances the weak efforts of their feeble passions; but it is by no means the same with those hardened criminals, with those men who are habitually vicious, whose unseemly excesses nothing can arrest, and who, in their violence, shut their eyes to the fear of the laws of this world, despising still more those of the other.

Nevertheless, how many persons say they are, and even believe themselves restrained by the fears of the life to come! But, either they deceive us, or they impose upon themselves, by attributing to these fears that which is only the effect of motives much nearer at hand, such as the feebleness of their machine, the mildness of their temperament, the slender energy of their souls, heir natural timidity, the ideas imbibed n their education, the fear of consequences immediately resulting from criminal actions, the physical evils attendant on unbridled irregularities: these are the true motives that restrain them, and not the notions of a future life, which men who say they are most firmly persuaded of its existence, forget whenever a powerful interest solicits them to sin. If for a time man would pay attention to what passes before his eyes, he would perceive that he ascribes to the fear of his God that which is in reality only the effect of peculiar weakness, of pusillanimity, of the small interest found to commit evil: these men would not act otherwise than they do if they had not this fear before them; if therefore he reflected, he would feel that it is always necessity that makes men act as they do.

Man cannot be restrained, when he does not find within himself motives sufficiently powerful to conduct him back to reason. There is nothing, either in this world or in the other, that can render him virtuous when an untoward organization, a mind badly cultivated, a violent imagination, inveterate habits, fatal examples, powerful interests, invite him from every quarter to the commission of crime. No speculations are capable of restraining the man who braves public opinion,

who despises the law, who is careless of its censure, who turns a deaf ear to the cries of conscience, whose power in this world places him out of the reach of punishment.* In the violence of his transports he will fear still less a distant futurity, of which the idea always recedes before that which he believes necessary to his immediate and present happiness. All lively passions blind man to every thing that is not its immediate object; the terrours of a future life, of which his passions always possess the secret to diminish to him the probability, can effect nothing upon the wicked man who does not fear even the much nearer punishment of the law—who sets at naught the assured hatred of those by whom he is surrounded. Man, when he delivers himself up to crime, sees nothing certain except the supposed advantage which attends it; the rest always appear to him either false or problematical.

If man would but open his eyes, he would clearly perceive, that to effect any thing upon hearts hardened by crime, he must not reckon upon the chastisement of an avenging Divinity, which the self-love natural to man always shows him as pacified in the long run. He who has arrived at persuading himself that he cannot be happy without crime, will always readily deliver himself up to it notwithstanding the menaces of religion. Whoever is sufficiently blind, not to read his infamy in his own heart, to see his own vileness in the countenances of his associates, his own condemnation in the anger of his fellow men, his own unworthiness in the indignation of the judges established to punish the offences he may commit; such a man. I say, will never feel the impression his crimes

* It will be said, that the fear of another life is a curb useful at least to restrain princes and nobles, who have no other; and that this curb, such as it is, is better than none. But it has been sufficiently proved that the belief in a future life does not controul the actions of sovereigns. The only way to prevent sovereigns from injuring society, is, to make them subservient to the laws, and to prevent their ever having the right or power of enslaving and oppressing nations according to the whim or caprice of the moment. Therefore, a good political constitution, founded upon natural rights and a sound education, is the only efficient check to the malpractices of the rulers of nations.

shall make on the features of a judge that is either hidden from his view, or that he only contemplates at a distance. The tyrant, who with dry eyes can hear the cries of the distressed, who with callous heart can behold the tears of a whole people of whose misery he is the cause, will not see the angry countenance of a more powerful master. When a haughty, arrogant monarch, pretends to be accountable for his actions to the Divinity alone, it is because he fears his nation more than he does his God.

On the other hand, does not religion itself annihilate the effects of those fears which it announces as salutary? Does it not furnish its disciples with the means of extricating themselves from the punishments with which it has so frequently menaced them? Does it not tell them, that a steril repentance will, even at the moment of death, disarm the celestial wrath; that it will purify the filthy souls of sinners? Do not even the priests, in some superstitions, arrogate to themselves the right of remitting to the dying, the punishment due to the crimes committed during the course of a disorderly life? In short, do not the most perverse men, encouraged in iniquity, debauchery, and crime, reckon, even to the last moment, upon the aid of a religion that promises them the infallible means of reconciling themselves to the Divinity whom they have irritated, and of avoiding his rigorous punishments?

In consequence of these notions, so favourable to the wicked, so suitable to tranquillize their fears, we see that the hope of an easy expiation, far from correcting man, engages him to persist until death in the most crying disorders. Indeed, in despite of the numberless advantages which he is assured flows from the doctrine of a life to come, in defiance of its pretended efficacy to repress the passions of men, do not the priests themselves, although so interested in the maintenance of this system, every day complain of its insufficiency? They acknowledge, that mortals, whom from their infancy they have imbued with these ideas, are not less hurried forward by their evil propensities, less suuk in the vortex of dissipation, less the slaves to their pleasures, less captivated by bad habits, less driven along

No. V.—17

by the torrent of the world, less seduced by their present interest, which make them forget equally the recompense and the chastisement of a future existence. In a word, the ministers of Heaven allow, that their disciples, for the greater part, conduct themselves in this world as if they had nothing either to hope or to fear in another.

But let it be supposed for a moment that the doctrine of eternal punishments was of some utility, and that it really restrained a small number of individuals; what are these feeble advantages compared to the numberless evils that flow from it? Against one timid man, whom this idea restrains, there are thousands upon whom it operates nothing; there are millions whom it makes irrational; whom it renders savage persecutors; whom it converts into useless and wicked fanatics; there are millions whose mind it disturbs, and whom it diverts from their duties towards society; there are an infinity whom it grievously afflicts and troubles, without producing any real good for their associates.*

* Many persons, convinced of the utility of the belief in another life, consider those who do not fall in with this doctrine as the enemies of society. However, it will be found on examination that the wisest and the most enlightened men of antiquity have believed, not only that the soul is material and perishes with the body, but also that they have attacked without hesitation and without subterfuge the opinion of future punishments. This sentiment was not peculiar to the Epicureans, but was adopted by philosophers of all sects, by Pythagoreans, by Stoics, by Peripatetics, by Academics; in short, by the most godly and the most virtuous men of Greece and Rome. Pythagoras, according to Ovid, speaks thus:—

O Genus attonitum gelidæ formidine Mortis,
Quid stiga, quid tenebras, et nomina vana timetis
Matericm vatum, falsique pericula mundi?

Timæus of Locris, who was a Pythagorean, admits that the doctrine of future punishments was fabulous, solely destined for the imbecility of the uninformed, and but little calculated for those who cultivate their reason.

Aristotle expressly says, that, "Man has neither good to hope, nor evil to fear after death."

The Platonists, who made the soul immortal, could not have any idea of future punishments, because the soul according to them was a portion of the Divinity, which, after the dissolution of the body, it returned to rejoin. Now, a portion of the Divinity could not be subject to suffer.

Zeno, according to Cicero, supposed the soul to be an igneous substance, from whence

CHAPTER XIV.

Education, Morals, and the Laws, suffice to restrain Man.— Of the Desire of Immortality.— Of Suicide.

It is not then in an ideal world, existing no where but in the imagination of man, that he must seek to collect motives calculated to make him act properly in this; it is in the visible world that will be found incitements to divert him from crime and to rouse him to virtue. It is in nature, in experience,* in truth, that he must search out remedies for the evils of his species, and for motives suitable to infuse into the human heart propensities truly useful for society.

If attention has been paid to what has been said in the course of this work, it will be seen, that above all it is education that will best furnish the true means of rectifying the wanderings of mankind. It is this that should scatter the seeds in his heart; cultivate the tender shoots; make a profitable use

he concluded it destroyed itself.—*Zenoni* Stoico animus ignis videtur. Si sit ignis, extinguetur; interibit cum reliquo corpore.

This philosophical orator, who was of the sect of the Academics, is not always in accord with himself; however, on several occasions he treats openly as fables the torments of Hell, and looks upon death as the end of every thing for man.—*Vide Tusculan.*, C. 38.

Seneca is filled with passages which contemplate death as a state of total annihilation:— Mors est non esse. Id quale sit jam scio; hoc erit post me quod ante me fuit. Si quid in hac re tormenti est, necesse est et fuisse antequam prodiremus in lucem; atqui nullam sensimus tunc vexationem. Speaking of the death of his brother, he says:—Quid itaque ejus desiderio maceror, qui aut beatus, aut nullus est? But nothing can be more decisive than what he writes to Marcia to console him. (chap. 19.)— Cogita nullis defunctum malis affici: illa quæ nobis inferos faciunt terribiles, fabulam esse: nullas imminere mortuis tenebras, nec carcerem, nec flumina flagrantia igne, nec oblivionis amnem, nec tribunalia, et reos et in illa libertate tam laxa iterum tyrannos: luserunt ista poetæ et vanis nos agitavere terroribus. Mors omnium dolorum est solutio est et finis: ultra quam mala nostra non exeunt, quæ nos in illam tranquilitatem, in qua antequam nasceremur, jacuimus, reponit.

Here is also another conclusive passage from this philosopher, which is deserving of the attention of the reader:—Si animus fortuita contempsit; si deorum hominumque formidinem ejecit, et scit non multum ab homine timendum, a deo nihil; si contemptor omnium quolium torquetur vita eo perduotus est ut illi liquent mortem nullius mali esse materiam, multorum finem.—*V. De Beneficiis, VII.* i.

Seneca, the tragedian, explains himself in the same manner as the philosopher:—

Post mortem nihil est, ipsaque mors nihil.
Velocis spatii meta novissima.
Quæris quo jaceas post obitum loco?
Quo non nata jacent.
Mors individua est noxia corpori,
Nec parcens animæ.
                                    Troades.

Epictetus has the same idea. In a passage reported by Arrian, he says:—" But where are you going? It cannot be to a place of suffering: you will only return to the place from whence you came; you are about to be again peace-

ably associated with the elements from whence you are derived. That which in your composition is of the nature of fire, will return to the element of fire; that which is of the nature of earth, will rejoin itself to the earth; that which is air, will reunite itself with air; that which is water, will resolve itself into water; there is no Hell, no Acheron, no Cocytus, no Phlegethon."—*See Arrian. in Epictet. lib.* iii. *cap.* 13. In another place he says: " The hour of death approaches; but do not aggravate your evil, nor render things worse than they are: represent them to yourself under their true point of view. The time is come when the materials of which you are composed, go to resolve themselves into the elements from whence they were originally borrowed. What is there that is terrible or grievous in that? Is there any thing in the world, that perishes totally?"— *See Arrian. lib.* iv. *cap.* 7. § 1.

The sage and pious Antoninus says: " He who fears death, either fears to be deprived of all feeling, or dreads to experience different sensations. If you lose all feeling, you will no longer be subject either to pain or to misery. If you are provided with other senses of a different nature, you will become a creature of a different species." This great emperor further says: " that we must expect death with tranquillity, seeing that it is only a dissolution of the elements of which each animal is composed."—*See the Moral Reflections of Marcus Antoninus, lib.* ii.

To the evidence of so many great men of Pagan antiquity, may be joined that of the author of Ecclesiastes, who speaks of death and of the condition of the human soul, like an Epicurean: he says: " For that which befalleth the sons of men befalleth beasts; even one thing befalleth them: as the one dieth, so dieth the other; yea, they have all one breath; so that a man hath no præeminence above a beast; for all is vanity. All go unto one place; all are of the dust, and all turn to dust again." And further, " wherefore I perceive, that there is nothing better, than that a man should rejoice in his own works; for that is his portion: for who shall bring him to see what shall be after him?"

In short, how can Christians reconcile the utility or the necessity of this doctrine with the fact, that the legislator of the Jews, inspired by the Divinity, remained silent on a subject that is said to be of so much importance?

of his dispositions; turn to account those faculties which depend on his organization; which should cherish the fire of his imagination, kindle it for useful objects; damp it, or extinguish it for others; in short, it is this which should make sensible souls contract habits that are advantageous for society, and beneficial to the individual. Brought up in this manner, man would not have occasion for celestial punishments to teach him the value of virtue; he would not need to behold burning gulfs of brimstone under his feet, to induce him to feel horrour for crime; nature, without these fables, would teach him much better what he owes to himself, and the law would point out to him what he owes to the body politic of which he is a member. It is thus that education would form valuable citizens to the state; the depositaries of power would distinguish those whom education should have thus formed, by reason of the advantages which they would procure for their country; they would punish those who should be found injurious to it; it would make the citizens see, that the promises of reward which education and morals held forth, are by no means vain; and that in a state well constituted, virtue is the true and only road to happiness; talents the way to gain respect; and that inutility and crime lead to contempt and misfortune.

A just, enlightened, virtuous, and vigilant government, who should honestly propose the public good, would have no occasion either for fables or for falsehoods to govern reasonable subjects; it would blush to make use of imposture to deceive citizens who, instructed in their duties, would find their interest in submitting to equitable laws; who would be capable of feeling the benefit these have the power of conferring on them; it would know, that public esteem has more power over men of elevated minds than the terrour of the laws; it would feel, that habit is sufficient to inspire them with horrour, even for those concealed crimes that escape the eyes of society; it would understand, that the visible punishments of this world impose much more on the ignorant than those of an uncertain and distant futurity: in short, it would ascertain that the sensible benefits within the compass of the sovereign power to distribute, touch the imagination of mortals more keenly than those vague recompenses which are held forth to them in a future existence.

Man is almost every where so wicked, so corrupt, so rebellious to reason, only because he is not governed according to his nature, nor properly instructed in her necessary laws: he is every where fed with useless chimeras; every where submitted to masters who neglect his instruction, or who only seek to deceive him. On the face of this globe we only see unjust sovereigns, enervated by luxury, corrupted by flattery, depraved by licentiousness, made wicked by impunity, devoid of talents, without morals, destitute of virtue, and incapable of exerting any energy for the benefit of the states they govern; they are consequently but little occupied with the welfare of their people, and indifferent to their duties, of which indeed they are often ignorant. Stimulated by the desire of continually finding means to feed their insatiable ambition, they engage in useless, depopulating wars, and never occupy their mind with those objects which are the most important to the happiness of their nation: interested in maintaining the received prejudices, they never wish to consider the means of curing them: in short, deprived themselves of that understanding which teaches man that it is his interest to be kind, just, and virtuous, they ordinarily reward only those crimes which their imbecility makes them imagine as useful to them, and they generally punish those virtues which are opposed to their own imprudent passions. Under such masters, is it surprising that society should be ravaged by perverse men who emulate each other in oppressing its members, in sacrificing its dearest interests. The state of society is a state of hostility of the sovereign against the whole, of each of its members the one against the other.* Man is wicked,

* It must be observed I do not say here, like Hobbes, that the state of nature is a state of war, but that men, by their nature, are neither good nor wicked; in fact, man will be either good or bad, according as he is modified. If men are so ready to injure one another, it is only because every thing conspires to give them different interests. Each one, if I may say so, lives isolated in society, and their chiefs avail themselves of their divisions to subdue

not because he is born so, but because he is rendered so; the great, the powerful, crush with impunity the indigent and the unhappy; these, at the risk of their lives, seek to retaliate the evil they have received: they attack either openly or in secret a country who to them is a stepmother, who gives all to some of her children, and deprives the others of every thing: they punish it for its partiality, and clearly show that the motives borrowed from a life hereafter are impotent against the fury of those passions to which a corrupt administration has given birth in this life; that the terrour of the punishments in this world are too feeble against necessity, against criminal habits; against a dangerous organization uncorrected by education.

In all countries the morals of the people are neglected, and the government is occupied only with rendering them timid and miserable. Man is almost every where a slave; it must then follow, of necessity, that he is base, interested, dissimulating, without honour; in a word, that he has the vices of the state of which he is a citizen. Every where he is deceived, encouraged in ignorance, and prevented from cultivating his reason; of course he must every where be stupid, irrational, and wicked; every where he sees vice and crime applauded and honoured; thence he concludes vice to be a good; virtue only a useless sacrifice of himself: every where he is miserable, therefore he injures his fellow men to relieve his own anguish: it is in vain to show him heaven, in order to restrain him; his views presently descend again to the earth, where he is willing to be happy at any price; therefore the laws, which have neither provided for his instruction, for his morals, nor his happiness, menace him uselessly, and punish him for the unjust negligence of his legislators. If politics, more enlightened, did seriously occupy itself with the instruction and with the welfare of the people; if laws were more equitable; if each society, less partial, bestowed on its members the care, the education, and the assist-

the whole. `Divide et impera` is the maxim that all bad governments follow by instinct. Tyrants would be badly off if they had to rule over virtuous men only.

ance which they have a right to expect if governments less covetous, and more vigilant, were sedulous to render their subjects more happy, there would not be seen such numbers of malefactors, of robbers, of murderers, who every where infest society; they would not be obliged to destroy life, in order to punish a wickedness, which is commonly ascribable to the vices of their own institutions: it would be unnecessary to seek in another life for fanciful chimeras, which always prove abortive against the infuriate passions, and against the real wants of man. In short, if the people were better instructed and more happy, politics would no longer be reduced to the exigency of deceiving them in order to restrain them; nor to destroy so many unfortunates for having procured necessaries at the expense of their hardhearted fellow citizens.

When it shall be desired to enlighten man, let him always have truth laid before him. Instead of kindling his imagination by the idea of those pretended goods that a future state has in reserve for him, let him be solaced, let him be succoured; or, at least, let him be permitted to enjoy the fruit of his labour; let not his substance be ravaged from him by cruel imposts; let him not be discouraged from work, by finding all his labour inadequate to support his existence, let him not be driven into that idleness that will surely lead him on to crime: let him consider his present existence, without carrying his views to that which may attend him after his death: let his industry be excited; let his talents be rewarded; let him be rendered active, laborious, beneficent, and virtuous, in the world he inhabits; let it be shown to him that his actions are capable of having an influence over his fellow men, but not on those imaginary beings located in an ideal world. Let him not be menaced with the tortures of a God when he shall be no more; let him behold society armed against those who disturb its repose; let him see the consequence of the hatred of his associates; let him learn to feel the value of their affection; let him be taught to esteem himself; let him understand, that to obtain the esteem of others he must have virtue; above all, that the virtue us man in a well constituted society has nothing to fear either from his fellow citizens or from the Gods.

If it be desired to form honest, courageous, industrious citizens, who may be useful to their country, let them beware of inspiring man from his infancy with an ill-founded dread of death—of amusing his imagination with marvellous fables—of occupying his mind with his destiny in a future life, quite useless to be known, and which has nothing in common with his real felicity. Let them speak of immortality to intrepid and noble souls; let them show it as the price of their labours to energetic minds, who, springing forward beyond the boundaries of their actual existence, are little satisfied with eliciting the admiration and with gaining the love of their contemporaries, but are determined also to wrest the homage, to secure the affection of future races. Indeed, there is an immortality to which genius, talents, virtue, have a just right to pretend; do not therefore let them censure or endeavour to stifle so noble a passion in man, which is founded upon his nature, and from which society gathers the most advantageous fruits.

The idea of being buried in total oblivion; of having nothing in common after his death, with the beings of his species; of losing all possibility of again having any influence over them, is a thought extremely painful to man; it is above all afflicting to those who possess an ardent imagination. The desire of immortality, or of living in the memory of his fellow men, was always the passion of great souls; it was the motive to the actions of all those who have played a great part on the earth. I Heroes, whether virtuous or criminal, philosophers as well as conquerors, men of genius, and men of talents, those sublime personages who have done honour to their species, as well as those illustrious villains who have debased and ravaged it, have had an eye to posterity in all their enterprises, and have flattered themselves with the hope of acting upon the souls of men, even when they themselves should no longer exist. If man in general does not carry his views so far, he is at least sensible to the idea of seeing himself regenerated in his children, whom he knows are destined to

survive him, to transmit his name, to preserve his memory, and to represent him in society; it is for them that he rebuilds his cottage; it is for them that he plants the tree which his eyes will never behold in its vigour; it is that they may be happy that he labours. The sorrow which imbitters the life of those rich men, frequently so useless to the world, when they have lost the hope of continuing their race, has its source in the fear of being entirely forgotten: they feel, that the useless man dies entirely. The idea that his name will be in the mouths of men; the thought that it will be pronounced with tenderness, that it will be recollected with kindness, that it will excite in their hearts favourable sentiments, is an illusion that is useful and suitable to flatter even those who know that nothing will result from it. Man pleases himself with dreaming that he shall have power; that he shall pass for something in the universe, even after the term of his human existence; he partakes by imagination in the projects, in the actions, in the discussions of future ages, and would be extremely unhappy if he believed himself entirely excluded from their society. The laws in all countries have entered into these views; they have so far been willing to console their citizens for the necessity of dying, by giving them the means of exercising their will, even for a long time after their death: this condescension goes to that length, that the dead frequently regulate the condition of the living during a long series of years.

Every thing serves to prove the desire in man of surviving himself. Pyramids, mausoleums, monuments, epitaphs, all show that he is willing to prolong his existence, even beyond his decease. He is not insensible to the judgment of posterity; it is for him the philosopher writes; it is to astonish him that the monarch erects sumptuous edifices, it is his praises that the great man already hears echo in his ears; it is to him that the virtuous citizen appeals from prejudiced or unjust contemporaries. Happy chimera! Sweet illusion! that realizes itself to ardent imaginations, and which is calculated to give birth to, and to nurture the enthusiasm of genius, courage, grandeur of soul, and talent; its influence is sometimes able to restrain the excesses of the most powerful men, who are frequently very much disquieted for the judgment of posterity, from a conviction that this will, sooner or later, avenge the living of the foul injustice which they have made them suffer.

No man, therefore, can consent to be entirely effaced from the remembrance of his fellows; some men have not the temerity to place themselves above the judgment of the future human species, to degrade themselves in its eyes. Where is the being who is insensible to the pleasure of exciting the tears of those who shall survive him; of again acting upon their souls; of once more occupying their thoughts; of exercising upon them his power, even from the bottom of his grave? Let, then, eternal silence be imposed upon those superstitious and melancholy men who censure a sentiment from which society derives so many real advantages; let not mankind listen to those passionless philosophers, who are willing to smother this great, this noble spring of his soul; let him not be seduced by the sarcasms of those voluptuaries, who pretend to despise an immortality towards which they lack the power to set forward. The desire of pleasing posterity and of rendering his name agreeable to generations yet to come, is a laudable motive, when it causes him to undertake those things of which the utility may have an influence over men and nations who have not yet an existence. Let him not treat as irrational the enthusiasm of those beneficent and mighty geniuses, whose keen eyes have foreseen him even in their day; who have occupied themselves of him for his welfare; who have desired his suffrage; who have written for him; who have enriched him by their discoveries; who have cured him of his errours. Let him render them the homage which they have expected at his hands; let him at least reverence their memory for the benefits he has derived from them; let him treat their mouldering remains with respect for the pleasure he receives from their labours; let him pay to their ashes a tribute of grateful recollection for the happiness they have been sedulous to procure for him. Let him sprinkle with his tears the urns of Socrates, of Phocion; let him wash out the stain that their punishment has made on the human species; let him expiate by his regret the Athenian ingratitude; let him learn by their example to dread religious and political fanaticism; let him fear to harass merit and virtue, in persecuting those who may happen to differ from him in his prejudices.

Let him strew flowers over the tombs of a Homer, of a Tasso, of a Milton; let him revere the immortal shades of those happy geniuses, whose harmonious lays excite in his soul the most tender sentiments; let him bless the memory of all those benefactors to the people, who were the delight of the human race; let him adore the virtues of a Titus, of a Trajan, of an Antoninus, of a Julian; let him merit, in his sphere, the eulogies of future ages; and let him always remember, that to carry with him to the grave the regret of his fellow man, he must display talents and practise virtue. The funeral ceremonies of the most powerful monarchs, have rarely been wetted with the tears of the people—they have commonly drained them while living. The names of tyrants excite the horrour of those who hear them pronounced. Tremble, then, cruel kings! ye who plunge your subjects into misery—who bathe them with bitter tears; who ravage nations, who change the fruitful earth into a barren cemetery; tremble for the sanguinary traits under which the future historian will paint you to generations yet unborn: neither your splendid monuments, your imposing victories, your innumerable armies, nor your sycophant courtiers, can prevent posterity from insulting your odious manes, and from avenging their grandfathers of your transcendent crimes.

Not only man sees his dissolution with pain, but again he wishes his death may be an interesting event for others. But, as we have already said, he must have talents, he must have beneficence, he must have virtue, in order that those who surround him may interest themselves in his condition, and may give regret to his ashes. Is it, then, surprising if the greater number of men, occupied entirely with themselves, completely absorbed by their own vanity, devoted to their own puerile objects, for ever busied with the care of gratifying their vile passions, at the expense of their family happiness, unheedful of the wants of a wife, unmindful of the necessity of their children, careless of the calls of friendship, regardless of their duty to society, do not by their death excite the sensibilities of their survivors, or that they should be presently forgotten? There is an infinity of monarchs of whom history does not tell us any thing, save that they have lived. In despite of the inutility in which men for the most part pass their existence; maugre the little care they bestow to render themselves dear to the beings who environ them; notwithstanding the numerous actions they commit to displease their associates, the self-love of each individual persuades him that his death must be an interesting occurrence: shows him, we may say, the order of things as overturned at his decease. O mortal, feeble and vain! Dost thou not know the Sesostrises, the Alexanders, the Cesars, are dead? Yet the course of the universe is not arrested: the demise of those famous conquerors, afflicting to some few favoured slaves, was a subject of delight for the whole human race. Dost thou, then, foolishly believe, that thy talents ought to interest thy species, and put it into mourning at thy decease? Alas! the Corneilles, the Lockes, the Newtons, the Boyles, the Harveys, the Montesquieus, are no more! Regretted by a small uumber of friends, who have presently consoled themselves by their necessary vocations, their death was indifferent to the greater number· of their fellow citizens. Darest thou, then, flatter thyself, that thy reputation, thy titles, thy riches, thy sumptuous repasts, thy diversified pleasures, will make thy funeral a memorable event! It will be spoken of by some few for two days, and do not be at all surprised: learn that there have died in former ages, in Babylon, in Sardis, in Carthage, in Athens, in Rome, millions of citizens, more illustrious, more powerful, more opulent, more voluptuous than thou art, of whom, however, no one has taken care to transmit to thee even the names. Be then virtuous, O man! in whatever station thy destiny assigns thee, and thou shalt be happy in thy lifetime; do thou good, and thou shalt be cherished: acquire talents, and thou shalt

be respected; posterity shall admire thee, if those talents, by becoming beneficial to their interests, shall bring them acquainted with the name under which they formerly designated thy annihilated being. But the universe will not be disturbed by thy loss; and when thou comest to die, whilst thy wife, thy children, thy friends, fondly leaning over thy sickly couch, shall be occupied with the melancholy task of closing thine eyes, thy nearest neighbour shall, perhaps, be exulting with joy!

Let not then man occupy himself with his future condition, but let him sedulously endeavour to make himself useful to those with whom he lives; let him, for his own peculiar happiness, render himself dutiful to his parents, attentive to his children, kind to his relations, true to his friends, lenient to his servants; let him strive to become estimable in the eyes of his fellow citizens; let him faithfully serve a country which assures to him his welfare; let the desire of pleasing posterity excite him to those labours that shall elicit their eulogies; let a legitimate self-love, when he shall be worthy of it, make him taste in advance those commendations which he is willing to deserve; let him learn to love and esteem himself; but never let him consent that concealed vices, that secret crimes, shall degrade him in his own eyes, and oblige him to be ashamed of his own conduct.

Thus disposed, let him contemplate his own decease with the same indifference that it will be looked upon by the greater number of his fellows; let him expect death with constancy, and wait for it with calm resignation; let him learn to shake off those vain terrours, with which superstition would overwhelm him; let him leave to the enthusiast his vague hopes; to the fanatic his madbrained speculations; to the bigot those fears with which he ministers to his own melancholy; but let his heart, fortified by reason, no longer dread a dissolution that will destroy all feeling.

Whatever may be the attachment man has to life, whatever may be his fear of death, it is every day seen that habit, that opinion, that prejudice, are motives sufficiently powerful to annihilate these passions in his breast, to make him brave danger, to cause him to hazard his existence. Ambition, pride, jealousy, love, vanity, avarice, the desire of glory, that deference to opinion which is decorated with the sounding title of a *point of honour*, have the efficacy to make him shut his eyes to danger, and to push him on to death; vexation, anxiety of mind, disgrace, want of success, softens to him its hard features, and makes him regard it as a door that will afford him shelter from the injustice of mankind: indigence, trouble, adversity, familiarizes him with this death, so terrible to the happy. The poor man, condemned to labour, inured to privations, deprived of the comforts of life, views its approach with indifference; the unfortunate, when he is unhappy, when he is without resource, embraces it in despair, and accelerates its march as soon as he sees that happiness is no longer within his grasp.

Man in different ages, and in different countries, has formed opinions extremely various upon the conduct of those who have had the courage to put an end to their own existence. His ideas upon this subject, as upon all others, have taken their tone from his religious and political institutions. The Greeks, the Romans, and other nations, which every thing conspired to render courageous and magnanimous, regarded as heroes and as Gods, those who voluntarily cut the thread of life. In Hindostan, the Brahmin yet knows how to inspire even women with sufficient fortitude to burn themselves upon the dead bodies of their husbands. The Japanese upon the most trifling occasion makes no kind of difficulty in plunging a dagger into his bosom.

Among the people of our own country religion renders man less prodigal of life; it teaches him that his God, who is willing he should suffer, and who is pleased with his torments, readily consents to his being put to a lingering death, but not that he should free himself from a life of misery by at once cutting the thread of his days. Some moralists, abstracting the height of religious ideas, have held that it never is permitted to man to break the conditions of the covenant that he has made with society. Others have looked upon suicide as cowardice, they have

thought that it was weakness, that it to hazard his existence. displayed pusillanimity, to suffer himself to be overwhelmed with the shafts of his destiny, and have held, that there would be much more courage and elevation of soul, in supporting his afflictions and in resisting the blows of fate.

If nature be consulted upon this point, it will be found, that all the actions of man, that feeble plaything in the hands of necessity, are indispensable; that they depend on causes which move him in despite of himself, and that without his knowledge make him accomplish at each moment of his existence some one of its decrees. If the same power that obliges all intelligent beings to cherish their existence, renders that of man so painful and so cruel that he finds it insupportable, he quits his species; order is destroyed for him, and he accomplishes a decree of nature that wills he shall no longer exist. This nature has laboured during thousands of years to form in the bowels of the earth the iron that must number his days.

If the relation of man with nature be examined, it will be found that his engagement was neither voluntary on his part, nor reciprocal on the part of nature or God. The volition of his will had no share in his birth; it is commonly against his will that he is obliged to finish life; and his actions are, as we have proved, only the necessary effects of unknown causes which determine his will. He is, in the hands of nature, that which a sword is in his own hands; he can fall upon it without its being able to accuse him with breaking his engagements, or of stamping with ingratitude the hand that holds it: man can only love his existence on condition of being happy; as soon as the entire of nature refuses him this happiness; as soon as all that surrounds him becomes incommodious to him; as soon as his melancholy ideas offer nothing but afflicting pictures to his imagination, he already exists no longer; he is suspended in the void; and he may quit a rank which no longer suits him; in which he finds no one interest; which offers him no protection; and in which he can no more be useful either to himself or to others.

If the covenant which unites man to society, be considered, it will be obvious that every contract is conditional, must be reciprocal; that is to say, supposes mutual advantages between the contracting parties. The citizen cannot be bound to his country, to his associates, but by the bonds of happiness. Are these bonds cut asunder? he is restored to liberty. Society, or those who represent it, do they use him with harshness, do they treat him with injustice, do they render his existence painful? Does disgrace hold him out to the finger of scorn; does indigence menace him, in an obdurate world? Perfidious friends, do they forsake him in adversity? An unfaithful wife, does she outrage his heart? Rebellious, ungrateful children, do they afflict his old age? Has he placed his happiness exclusively on some object which it is impossible for him to procure? Chagrin, remorse, melancholy, despair, have they disfigured to him the spectacle of the universe? In short, for whatever cause it may be, if he is not able to support his evils, let him quit a world which from thenceforth is for him only a frightful desert: let him remove himself for ever from a country he thinks no longer willing to reckon him amongst the number of her children: let him quit a house that to his mind is ready to bury him under its ruins: let him renounce a society to the happiness of which he can no longer contribute; which his own peculiar felicity alone can render dear to him. And could the man be blamed, who finding himself useless, who being without resources in the town where destiny gave him birth, should quit it in his chagrin to plunge himself in solitude? Death is to the wretched the only remedy for despair; the sword is then the only friend—the only comfort that is left to the unhappy: as long as hope remains the tenant of his bosom; as long as his evils appear to him at all supportable; as long as he flatters himself with seeing them brought to a termination; as long as he finds some comfort in existence however slender, he will not consent to deprive himself of life: but when nothing any longer sustains in him the love of this existence, then to live, is to him the greatest of evils; to die, the only mode by which he can avoid the excess of despair.*

* This has been the opinion of many great

That society who has not the ability, or who is not willing to procure man any one benefit, loses all its rights over him; nature, when it has rendered his existence completely miserable, has in fact ordered him to quit it: in dying he does no more than fulfil one of her decrees, as he did when he first drew his breath. To him who is fearless of death, there is no evil without a remedy; for him who refuses to die, there yet exist benefits which attach him to the world; in this case let him rally his powers, let him oppose courage to a destiny that oppresses him; let him call forth those resources with which nature yet furnishes him; she cannot have totally abandoned him whilst she yet leaves him the sensation of pleasure, and the hopes of seeing a period to his pains. As to the superstitious, there is no end to his sufferings, for he is not allowed to abridge them.† His religion bids him to continue to groan, and forbids his recurring to death, which would lead him to a miserable state of existence: he would be eternally punished for daring to anticipate the tardy orders of a cruel God, who takes pleasure in

men : Seneca, the moralist, whom Lactantius calls the divine Pagan, who has been praised equally by St. Austin and by St. Augustine, endeavours by every kind of argument to make death a matter of indifference to man:—Malum est in necessitate vivere : sed in necessitate vivere, necessitas nulla est. Quidni nulla sit? Patent undique ad libertatem viœ multœ, breves, faciles. Aganus Deo gratias, quod nemo in vita teneri possit.—V. Senec. Epist. xii. Cato has always been commended, because he would not survive the cause of liberty,—for that he would not live a slave. Curtius, who rode voluntarily into the gap to save his country, has always been held forth as a model of heroic virtue. Is it not evident that those martyrs who have delivered themselves up to punishment, have preferred quitting the world, to living in it contrary to their own ideas of happiness? When the fabulous Samson wished to be revenged on the Philistines, did he not consent to die with them as the only means? If our country is attacked, do we not voluntarily sacrifice our lives in its defence?

† Christianity, and the civil laws of Christians, are very inconsistent in censuring *suicide*. The Old Testament furnishes examples in Samson and Eleazar—that is to say, in men who stood very high with God. The *Messiah*, or the son of the Christians' God, if it be true that he died of his own accord, was evidently a *suicide*. The same may be said of those penitents who have made it a merit of gradually destroying themselves.

seeing him reduced to despair, and who wills that man should not have the audacity to quit, without his consent, the post assigned to him.

Man regulates his judgment on his fellows only by his own peculiar mode of feeling; he deems as folly, he calls delirium, all those violent actions which he believes but little commensurate with their causes, or which appear to him calculated to deprive him of that happiness towards which he supposes a being, in the enjoyment of his senses, cannot cease to have a tendency: he treats his associate as a weak creature when he sees him affected with that which touches him but lightly, or when he is incapable of supporting those evils which his self-love flatters him he would himself be able to endure with more fortitude. He accuses of madness whoever deprives himself of life, for objects that he thinks unworthy so dear a sacrifice; he taxes him with phrensy, because he has himself learned to regard this life as the greatest blessing. It is thus that he always erects himself into a judge of the happiness of others, of their mode of seeing, and of their manner of feeling. A miser who destroys himself after the loss of his treasure, appears a fool in the eyes of him who is less attached to riches; he does not feel, that without money life to this miser is only a continued torture, and that nothing in the world is capable of diverting him from his painful sensations: he will proudly tell you, that in his place he had not done so much; but to be exactly in the place of another man, it is needful to have his organization, his temperament, his passions, his ideas; it is in fact needful to be that other—to be placed exactly in the same circumstances, to be moved by the same causes; and in this case all men, like the miser, would sacrifice their life after being deprived of the only source of their happiness.

He who deprives himself of his existence, does not adopt this extremity, so repugnant to his natural tendency, but when nothing in this world has the faculty of rejoicing him—when no means are left of diverting his affliction. His misfortune, whatever it may be, for him is real; his organization, be it strong, or be it weak, is his own, not that of another; a man who is sick only in imagination, really suffers, and

even troublesome dreams place him in a very uncomfortable situation. Thus when a man kills himself, it ought to be concluded, that life, in the room of being a benefit, had become a very great evil to him; that existence had lost all its charms in his eyes; that the entire of nature was to him destitute of attraction; that it no longer contained any thing that could seduce him; that after the comparison which his disturbed imagination had made of existence with non-existence, the latter appeared to him preferable to the first.

Many persons will not fail to consider as dangerous these maxims, which, in spite of the received prejudices, authorize the unhappy to cut the thread of life; but *maxims* will never induce a man to adopt such a violent resolution: it is a temperament soured by chagrin, a bilious constitution, a melancholy habit, a defect in the organization, a derangement in the whole machine, it is in fact necessity, and not reasonable speculations, that breed in man the design of destroying himself. Nothing invites him to this step so long as reason remains with him, or whilst he yet possesses hope—that sovereign balm for every evil. As for the unfortunate, who cannot lose sight of his sorrows, who cannot forget his pains, who has his evils always present to his mind; he is obliged to take counsel from these alone. Besides, what assistance or what advantage can society promise to itself from a miserable wretch reduced to despair, from a misanthrope overwhelmed with grief, from a wretch tormented with remorse, who has no longer any motive to render himself useful to others, who has abandoned himself, and who finds no more interest in preserving his life? Those who destroy themselves are such, that had they lived, the offended laws must have ultimately been obliged to remove them from a society which they disgraced.

As life is, commonly, the greatest blessing for man, it is to be presumed that he who deprives himself of it, is impelled thereto by an invincible force. It is the excess of misery, the height of despair, the derangement of his brain, caused by melancholy, that urges man on to destroy himself. Agitated by contrary impulsions, he is, as we have before said, obliged to follow a middle course, that conducts him to his death; if man be not a free agent, in any one instant of his life, he is again much less so in the act by which it is terminated.*

It will be seen, then, that he who kills himself, does not, as it is pretended, commit an outrage on nature or its author. He follows an impulse of that nature, and thus adopts the only means left him to quit his anguish; he goes out of a door which she leaves open to him; he cannot offend her in accomplishing a law of necessity; the iron hand of this having broken the spring that renders life desirable to him, and which urged him to self-conservation, shows him he ought to quit a rank or system where he finds himself too miserable to have the desire of remaining. His country or his family have no right to complain of a member whom it has no means of rendering happy, and from whom consequently they have nothing more to hope. To be useful to either, it is necessary he should cherish his own peculiar existence; that he should have an interest in conserving himself; that he should love the bonds by which he is united to others; that he should be capable of occupying himself with their felicity. That the suicide should be punished in another world, and should repent of his precipitancy, he should outlive himself, and should carry with him into his future residence his organs, his senses, his memory, his ideas, his actual mode of existing, his determinate manner of thinking.

In short, nothing is more useful for society than to inspire man with a contempt for death, and to banish from his mind the false ideas he has of its consequences. The fear of death can never do more than make cowards; the fear of its pretended consequences will make nothing but fanatics or melancholy beings, who are useless to themselves and unprofitable to others. Death is a resource that ought not to be taken away from oppressed virtue, which the injustice of man frequently reduces to despair. If man feared death less, he would neither be a slave nor super-

stitious; truth would find defenders more zealous; the rights of mankind would be more hardily sustained; errour would be more powerfully opposed; tyranny would be banished from nations: cowardice nourishes it, fear perpetuates it. In fact, man can neither be contented nor happy, whilst his opinions shall oblige him to tremble.

. CHAPTER XV.

Of Man's true Interest, or of the Ideas he forms to himself of Happiness.—Man cannot be Happy without Virtue.

UTILITY, as has been before observed, ought to be the only standard of the judgment of man. To be useful, is to contribute to the happiness of his fellow creatures; to be prejudicial, is to further their misery. This granted, let us examine if the principles we have hitherto established be prejudicial or advantageous, useful or useless, to the human race. If man unceasingly seeks after his happiness, he can only approve of that which procures for him his object, or furnishes him the means by which it is to be obtained.

What has been already said will serve in fixing our ideas upon what constitutes this happiness: it has been already shown, that it is only continued pleasure;† but in order that an object may please, it is necessary that the impressions it makes, the perceptions it gives, the ideas which it leaves, in short, that the motion it excites in man should be analogous to his organization, conformable to his temperament, assimilated to his individual nature: modified as it is by habit, determined as it is by an infinity of circumstances, it is necessary that the action of the object by which he is moved, or of which the idea remains with him, far from enfeebling him, far from annihilating his feelings should tend to strengthen him; it is necessary, that without fatiguing his mind, exhausting his faculties or deranging his organs, this object should impart to his machine that degree of activity for which it continually has occasion. What is the object that unites all these qualities? Where is the man whose organs are susceptible

* Suicide is said to be very common in England, whose climate produces melancholy in its inhabitants. In that country those who kill themselves are looked upon as *lunatics*;— their disease does not seem more blameable than any other delirium.

† See Chapter IX.

of continual agitation without being fatigued, without experiencing a painful sensation, without sinking? Man is always willing to be warned of his existence in the most lively manner, as long as he can be so without pain. What do I say? He consents frequently to suffer, rather than not feel. He accustoms himself to a thousand things, which at first must have affected nim in a disagreeable manner, and which frequently end, either by converting themselves into wants, or by no longer affecting him any way.* Where, indeed, can he always find objects in nature capable of continually supplying the stimulus requisite to keep him in an activity that shall be ever proportioned to the state of his own organization, which his extreme mobility renders subject to perpetual variation? The most lively pleasures are always the least durable, seeing they are those which exhaust him most.

That man should be uninterruptedly happy, it would be requisite that his powers were infinite; it would require, that, to his mobility he joined a vigour, a solidity, which nothing could change; or else it is necessary that the objects from which he receives impulse should either acquire or lose properties, according to the different states through which his machine is successively obliged to pass; it would need that the essences of beings should be changed in the same proportion as his dispositions, and should be submitted to the continual influence of a thousand causes, which modify him without his knowledge, and in despite of himself. If, at each moment his machine undergoes changes, more or less marked, which are ascribable to the different degrees of elasticity, of density, of serenity of the atmosphere, to the portion of igneous fluid circulating through his blood, to

the harmony of his organs, to the order that exists between the various parts of his body; if, at every period of his existence, his nerves have not the same tensions, his fibres the same elasticity, his mind the same activity, his imagination the same ardour, &c., it is evident, that the same causes in preserving to him only the same qualities, cannot always affect him in the same manner. Here is the reason why those objects that please him in one season displease him in another: these objects have not themselves sensibly changed, but his organs, his dispositions, his ideas, his mode of seeing, his manner of feeling, have changed; such is the source of man's inconstancy.

If the same objects are not constantly in that state competent to form the happiness of the same individual, it is easy to perceive, that they are yet less in a capacity to please all men; or that the same happiness cannot be suitable to all. Beings already various by their temperament, their faculties, their organization, their imagination, their ideas, of distinct opinions, of contrary habits, which an infinity of circumstances, whether physical or moral, have variously modified, must necessarily form very different notions of happiness. Those of a miser cannot be the same as those of a prodigal; those of the voluptuary, the same as those of one who is phlegmatic; those of an intemperate, the same as those of a rational man who husbands his health. The happiness of each is in consequence composed of his natural organization, and of those circumstances, of those habits, of those ideas, whether true or false, that have modified him: this organization and these circumstances never being the same in any two men, it follows that what is the object of one man's views, most be indifferent or even displeasing to another; thus, as we have before said, no one can be capable of judging of that which may contribute to the felicity of his fellow man.

Interest, is the object to which each individual, according to his temperament and his own peculiar ideas, attaches his welfare; from which it will be perceived, that this *interest* is never more than that which each contemplates as necessary to his happiness.

---

* Of this truth, tobacco, coffee, and above all, brandy, furnish examples. It was this last which enabled the Europeans to enslave the negro and to subdue the savage. This is also the reason man runs to see tragedies and to witness the execution of criminals. In short, the desire of feeling, or of being powerfully moved, appears to be the principle of curiosity— of that avidity with which we seize on the marvellous, the supernatural, the incomprehensible, and on every thing that excites the imagination. Men cling to their religions as the savage does to brandy.

It must, therefore, be concluded, that no man is totally without interest. That of the miser, is to amass wealth; that of the prodigal, to dissipate it; the interest of the ambitious, is to obtain power; that of the modest philosopher, to enjoy tranquillity: the interest of the debauchee, is to give himself up without reserve to all sorts of pleasure; that of the prudent man, to abstain from those which may injure him: the interest of the wicked, is to gratify his passions at any price: that of the virtuous, to merit by his conduct the love and the approbation of others; to do nothing that can degrade himself in his own eyes.

Thus, when it is said, that *interest is the only motive of human actions,* it is meant to indicate, that each man labours after his own manner to his own peculiar happiness, which he places in some object, either visible or concealed, either real or imaginary, and that the whole system of his conduct is directed to its attainment. This granted, no man can be called disinterested; this appellation is only applied to those of whose motives we are ignorant, or whose interest we approve. Thus, the man who finds a greater pleasure in assisting his friends in misfortune, than preserving in his coffers useless treasure, is called generous, faithful, and disinterested: in like manner all men are denominated disinterested, who feel their glory far more precious than their fortune. In short, all men are designated disinterested, who place their happiness in making sacrifices which man considers costly, because he does not attach the same value to the object for which the sacrifice is made.

Man frequently judges very erroneously of the interest of others, either because the motives that animate them are too complicated for him to unravel; or, because to be enabled to judge of them fairly, it is needful to have the same eyes, the same organs, the same passions, the same opinions: nevertheless, obliged to form his judgment of the actions of mankind by their effect on himself, he approves the interest that actuates them, whenever the result is advantageous for his species: thus, he admires valour, generosity, the love of liberty, great talents, virtue, &c., he then only approves of the objects, in which the beings he applauds, have placed their happiness; he approves these dispositions even when he is not in a capacity to feel their effects; but in this judgment he is not himself disinterested; experience, reflection, habit, reason, have given him a taste for morals, and he finds as much pleasure in being witness to a great and generous action, as the man of *virtu* finds in the sight of a fine picture of which he is not the proprietor. He who has formed to himself a habit of practising virtue, is a man who has unceasingly before his eyes the interest that he has in meriting the affection, in deserving the esteem, in securing the assistance of others, as well as to love and esteem himself: impressed with these ideas, which have become habitual to him, he abstains even from concealed crimes, since these would degrade him in his own eyes · he resembles a man, who having from his infancy contracted a habit of cleanliness, would be painfully affected at seeing himself dirty, even when no one should witness it. The honest man is he to whom truth has shown his interest or his happiness in a mode of acting that others are obliged to love and to approve for their own peculiar interest.

These principles, duly developed, are the true basis of morals; nothing is more chimerical than those which are founded upon imaginary motives, placed out of nature; or upon innate sentiments, which some speculators have regarded as anterior to man's experience, and as wholly independent of those advantages which result to him from its use: it is the essence of man to love himself: to tend to his own conservation; to seek to render his existence happy :* thus interest, or the desire of happiness, is the only real motive of all his actions; this interest depends upon his natural organization, his wants, his acquired ideas, the habits he has contracted; he is without doubt in errour, when either a vitiated organization or false opinions show him his welfare in objects either useless or injurious to himself, as well as to others; he marches steadily in the paths of vir-

---

* Seneca says: Modus ergo diligendi præcipiendus est homini, id est quomodo se diligat aut prosit sibi; quin autem diligat aut pro sit sibi, dubitare dementis est.

tue, when true ideas have made him rest his happiness on a conduct useful to his species, approved by others, and which renders him an interesting object to his associates. Morals would be a vain science, if it did not incontestably prove to man that *his interest consists in being virtuous.* Obligation, of whatever kind, can only be founded upon the probability or the certitude of either obtaining a good or avoiding an evil.

Indeed, in no one instant of his duration, can a sensible and intelligent being either lose sight of his own preservation or forget his own welfare; he owes happiness to himself; but experience quickly proves to him, that bereaved of assistance, he cannot alone procure all those objects which are requisite to his felicity: he lives with sensible, with intelligent beings, occupied like himself with their own peculiar happiness, but capable of assisting him in obtaining those objects he most desires; he discovers that these beings will not be favourable to his views, but when they find their interest involved; from which he concludes, that his own happiness demands that he should conduct himself at all times in a manner suitable to conciliate the attachment, to obtain the approbation, to elicit the esteem, to secure the assistance of those beings who are most capacitated to further his designs. He perceives that it is man who is most necessary to the welfare of man, and that to induce him to join in his interests, he ought to make him find real advantages in seconding his projects: but to procure real advantages to the beings of the human species, is to have virtue; the reasonable man, therefore, is obliged to feel that it is his interest to be virtuous. Virtue is only the art of rendering himself happy, by the felicity of others. The virtuous man is he who communicates happiness to those beings who are capable of rendering his own condition happy, who are necessary to his conservation, who have the ability to procure him a felicitous existence.

Such, then, is the true foundation of all morals; merit and virtue are founded upon the nature of man; have their dependance upon his wants. It is virtue, alone, that can render him truly happy:[*] without virtue, society can neither be useful nor indeed subsist; it can only have real utility when it assembles beings animated with the desire of pleasing each other, and disposed to labour to their reciprocal advantage: there exists no comfort in those families whose members are not in the happy disposition to lend each other mutual succours; who have not a reciprocity of feeling that stimulates them to assist one the other; that induces them to cling to each other, to support the sorrows of life; to unite their efforts to put away those evils to which nature has subjected them. The conjugal bonds are sweet only in proportion as they identify the interest of two beings, united by the want of legitimate pleasure, from whence results the maintenance of political society, and the means of furnishing it with citizens. Friendship has charms, only when it more particularly associates two virtuous beings; that is to say, two beings animated with the sincere desire of conspiring to their reciprocal happiness. In short, it is only by displaying virtue that man can merit the benevolence, the confidence, the esteem, of all those with whom he has relation; in a word, no man can be independently happy.

Indeed, the happiness of each human individual depends on those sentiments to which he gives birth, on those feelings which he nourishes in the beings amongst whom his destiny has placed him; grandeur may dazzle them; power and force may wrest from them an involuntary homage; opulence may seduce mean and venal souls; but it is humanity, it is benevolence, it is compassion, it is equity, that, unassisted by these, can without efforts obtain for him those delicious sentiments of attachment, of tenderness, of esteem, of which all reasonable men feel the necessity. To be virtuous, then, is to place his interest in that which accords with the interest of others; it is to enjoy those benefits and that pleasure which he himself diffuses over his fellows. He, whom his nature, his education, his reflections, his habits, have rendered sus-

---

[*] Est autem virtus nihil aliud quam in se perfecta et ad summum perducta natura..— *Cicero. De Legibus I.* He says elsewhere Virtus rationis absol itio definitur.

ceptible of these dispositions, and to whom his circumstances have given him the faculty of gratifying them, becomes an interesting object to all those who approach him: he enjoys every instant; he reads with satisfaction the contentment and the joy which he has diffused over all countenances: his wife, his children, his friends, his servants, greet him with gay and serene faces, indicative of that content and of that peace which he recognises for his own work: every thing that environs him is ready to partake his pleasures and to share his pains; cherished, respected, looked up to by others, every thing conducts him to agreeable reflections: he knows the rights he has acquired over their hearts; he applauds himself for being the source of a felicity that captivates all the world; his own condition, his sentiments of self-love, become a hundred times more delicious when he sees them participated by all those with whom his destiny has connected him. The habit of virtue creates for him no wants but those which virtue itself suffices to satisfy; it is thus that virtue is always its own peculiar reward, that it remunerates itself with all the advantages it incessantly procures for others.

It will be said, and perhaps even proved, that under the present constitution of things, virtue, far from procuring the welfare of those who practise it, frequently plunges man into misfortune, and often places continual obstacles to his felicity; that almost every where it is without recompense. What do I say? A thousand examples could be adduced as evidence that in almost every country it is hated, persecuted, obliged to lament the ingratitude of human nature. I reply, with avowing, that by a necessary consequence of the wanderings and errours of his race, virtue rarely conducts man to those objects in which the uninformed make their happiness consist. The greater number of societies, too frequently ruled by those whose ignorance makes them abuse their power, whose prejudices render them the enemies of virtue, who flattered by sycophants, secure in the impunity their actions enjoy. commonly lavish their esteem, bestow their kindness on none but the most unworthy objects, reward only the most frivolous, recompense none but the most prejudicial qualities: and hardly ever accord that justice to merit which is unquestionably its due. But the truly honest man is neither ambitious of remuneration, nor sedulous of the suffrages of a society thus badly constituted: contented with domestic happiness, he seeks not to augment relations which would do no more than increase his danger; he knows that a vitiated community is a whirlwind, with which an honest man cannot co-order himself; he therefore steps aside, quits the beaten path, by continuing in which he would infallibly be crushed. He does all the good of which he is capable in his sphere; he leaves the road free to the wicked, who are willing to wade through its mire; he laments the heavy strokes they inflict on themselves; he applauds the mediocrity that affords him security; he pities those nations made miserable by their errours; rendered unhappy by those passions which are the fatal but necessary consequence; he sees they contain nothing but wretched citizens, who far from cultivating their true interest, far from labouring to their mutual felicity, far from feeling the real value of virtue, unconscious how dear it ought to be to them, do nothing but either openly attack or secretly injure it; in short, who detest a quality which would restrain their disorderly propensities.

In saying that virtue is its own peculiar reward, it is simply meant to announce, that, in a society whose views were guided by truth, by experience, and by reason, each individual would be acquainted with his real interests, would understand the true end of association, would have sound motives to perform his duties, and find real advantages in fulfilling them; in fact, would be convinced that to render himself solidly happy, he should occupy his actions with the welfare of his fellows, and by their utility, merit their esteem, their kindness, and their assistance. In a well constituted society, the government, the laws, education, example, would all conspire to prove to the citizen, that the nation of which he forms a part is a whole that cannot be happy that cannot subsist without virtue; experience would, at each step, convince him that the welfare of its parts can only result from that of the whole body corporate; justice would make him feel, that no society can be advantageous to its members where the volition of wills in those who act, is not so conformable to the interests of the whole, as to produce an advantageous reaction.

But, alas! by the confusion which the errours of man have carried into his ideas, virtue, disgraced, banished and persecuted, finds not one of those advantages it has a right to expect; man is indeed shown those pretended rewards for it in a future life, of which he is almost always deprived in his actual existence. It is thought necessary to deceive, to seduce, to intimidate him, in order to induce him to follow that virtue which every thing renders incommodious to him; he is fed with distant hopes, in order to solicit him to practise virtue, while contemplation of the world makes it hateful to him; he is alarmed by remote terrours to deter him from committing evil, which all conspires to render amiable and necessary. It is thus that politics and superstition, by the formation of chimeras, by the creation of fictitious interests, pretend to supply those true and real motives which nature furnishes, which experience points out, which an enlightened government should hold forth, which the law ought to enforce, which instruction should sanction, which example should encourage, which rational opinions would render pleasant. Man, blinded by his passions, not less dangerous than necessary, led away by precedent, authorized by custom, enslaved by habit, pays no attention to these uncertain promises and menaces; the actual interest of his immediate pleasures, the force of his passions, the inveteracy of his habits, always rise superior to the distant interests pointed out in his future welfare, or the remote evils with which he is threatened, which always appear doubtful whenever he compares them with present advantages.

Thus superstition, far from making man virtuous by principle, does nothing more than impose upon him a yoke as severe as it is useless: it is borne by none but enthusiasts, or by the pusillanimous, who, without becoming better, tremblingly champ the feeble bit put into their mouth. Indeed, experience incontestably proves, that religion is a dike inadequate to restrain the torrent of corruption to which so many accumulated causes give an irresistible force: nay more, does not this religion itself augment the public disorder, by the dangerous passions which it lets loose and consecrates? Virtue, in almost every climate, is confined to some few rational beings, who have sufficient strength of mind to resist the stream of prejudice; who are contented by remunerating themselves with the benefits they diffuse over society; whose temperate dispositions are gratified with the suffrages of a small number of virtuous approvers: in short, who are detached from those frivolous advantages which the injustice of society but too commonly accords only to baseness, to intrigue, and to crime.

In despite of the injustice that reigns in the world, there are, however, some virtuous men; in the bosom even of the most degenerate nations, there are some benevolent beings, still enamoured of virtue, who are fully acquainted with its true value, who are sufficiently enlightened to know that it exacts homage even from its enemies; who are at least satisfied with those concealed pleasures and recompenses, of which no earthly power is competent to deprive them. The honest man acquires a right to the esteem, the veneration, the confidence, the love, even of those whose conduct is exposed by a contrast with his own. In short, vice is obliged to cede to virtue, of which it blushingly acknowledges the superiority. Independent of this ascendency so gentle, so grand, so infallible, if even the whole universe should be unjust to him, there yet remains to the honest man the advantage of loving his own conduct, of esteeming himself, of diving with satisfaction into the recesses of his own heart, of contemplating his own actions with that delicious complacency that others ought to do, if they were not hoodwinked. No power is adequate to ravish from him the merited esteem of himself; no authority is sufficiently potent to give it to him when he deserves it not; but when it is not well founded it is then a ridiculous sentiment: it ought to be censured when it displays itself in a mode that is mortifying and troublesome to others; it is then called *arrogance;* if it rest itself upon frivolous actions, it is called *vanity*; but when it cannot be condemned, when it is known for legitimate, when it is discovered to have a solid foundation, when it bottoms itself upon talents, when it rises upon great actions that are useful to the community, when it erects its edifice upon virtue, even though society should not set these merits at their just price, it is noble pride, elevation of mind, grandeur of soul.

Let us not, then, listen to the preaching of those superstitions which, enemies to man's happiness, have been desirous of destroying it, even in the inmost recesses of his heart; which have prescribed to him hatred of his fellows and contempt for himself; which pretend to wrest from the honest man that self-respect which is frequently the only reward that remains to virtue in a perverse world. To annihilate in him this sentiment so full of justice, this love of himself, is to break the most powerful spring that urges him to act right. What motive, indeed, except it be this, remains for him in the greater part of human societies? Is not virtue discouraged and contemned? Is not audacious crime and cunning vice rewarded? Is not love of the public weal taxed as folly; exactitude in fulfilling duties looked upon as a bubble? Is not compassion, sensibility, tenderness, conjugal fidelity, sincerity, inviolable friendship, treated with ridicule? Man must have motives for action: he neither acts well nor ill, but with a view to his own happiness—to that which he thinks his interest; he does nothing gratuitously; and when reward for useful actions is withheld from him, he is reduced either to become as abandoned as others, or else to remunerate himself with his own applause.

This granted, the honest man can never be completely unhappy; he can never be entirely deprived of the recompense which is his due; virtue can amply make up to him all the happiness denied him by public opinion; but nothing can compensate to him the want of virtue. It does not follow that the honest man will be exempted from afflictions: like the wicked, he is subjected to physical evils; he may be worn down with disease; he may frequently be the subject of calumny, of injustice, of ingrat-

itude, of hatred; but in the midst of all his misfortunes, of his sorrows, he finds support in himself, he is contented with his own conduct, he respects himself, he feels his own dignity, he knows the equity of his rights, and consoles himself with the confidence inspired by the justness of his cause. These supports are not calculated for the wicked. Equally liable with the honest man to infirmities and to the caprices of his destiny, he finds the recesses of his own heart filled with dreadful alarms, cares, solicitude, regret, and remorse; he dies within himself; his conscience sustains him not, but loads him with reproach; and his mind, overwhelmed, sinks under the storm. The honest man is not an insensible stoic; virtue does not procure impassibility, but if wretched, it enables him to cast off despair; if infirm, he has less to complain of than the vicious being who is oppressed with sickness; if indigent, he is less unhappy in his poverty; if in disgrace, he is not overwhelmed by its pressure, like the wretched slave to crime.

Thus the happiness of each individual depends on the cultivation of his temperament; nature makes both the happy and the unhappy; it is culture that gives value to the soil nature has formed, and instruction and reflection make it useful. For man to be happily born, is to have received from nature a sound body, organs that act with precision, a just mind, a heart whose passions and desires are analogous and conformable to the circumstances in which his destiny has placed him. Nature, then, has done every thing for him, when she has joined to these faculties the quantum of vigour and energy sufficient to enable him to obtain those things, which his station, his mode of thinking, his temperament, have rendered desirable. Nature has made him a fatal present, when she has filled his sanguinary vessels with an overheated fluid, given him an imagination too active, desires too impetuous after objects either impossible or improper to be obtained under his circumstances; or which at least he cannot procure without those incredible efforts that either place his own welfare in danger or disturb the repose of society. The most happy man is commonly he who possesses a peaceable mind, who only desires those things which he can procure by labour suitable to maintain his activity, without causing shocks that are either too violent or troublesome. A philosopher, whose wants are easily satisfied, who is a stranger to ambition, who is contented with the limited circle of a small number of friends, is, without doubt, a being much more happily constituted than an ambitious conqueror, whose greedy imagination is reduced to despair by having only one world to ravage. He who is happily born, or whom nature has rendered susceptible of being conveniently modified, is not a being injurious to society: it is generally disturbed by men who are unhappily born, whose organization renders them turbulent, who are discontented with their destiny, who are inebriated with their own licentious passions, who are smitten with difficult enterprises, who set the world in combustion to gather imaginary benefits, in which they make their own happiness consist. An Alexander requires the destruction of empires, nations to be deluged with blood, cities to be laid in ashes, its inhabitants to be exterminated, to content that passion for glory of which he has formed to himself a false idea, but which his too ardent imagination anxiously thirsts after : 'for a Diogenes there needs only a tub, with the liberty of appearing whimsical : a Socrates wants nothing but the pleasure of forming disciples to virtue.

Man by his organization is a being to whom motion is always necessary, he must therefore always desire it ; this is the reason why too much facility in procuring the objects of his search, renders them quickly insipid. To feel happiness, it is necessary to make efforts to obtain it ; to find charms in its enjoyment, it is necessary that the desire should be whetted by obstacles; he is presently disgusted with those benefits which have cost him but little pains. The expectations of happiness, the labour requisite to procure it, the varied and multiplied pictures which his imagination forms to him, supply his brain with that motion for which it has occasion ; this gives impulse to his organs, puts his whole machine into activity, exercises his faculties, sets all his springs in play ; in a word, puts him into that agreeable activity, for the want

of which the enjoyment of happiness itself cannot compensate him. Action is the true element of the human mind ; as soon as it ceases to act, it sinks into lassitude. His mind has the same occasion for ideas his stomach has for aliment.*

Thus the impulse given him by desire is itself a great benefit; it is to the mind what exercise is to the body; without it he would not derive any pleasure in the aliments presented to him; it is thirst that renders the pleasure of drinking so agreeable. Life is a perpetual circle of regenerated desires and wants satisfied : repose is only a pleasure to him who labours; it is a source of weariness, the cause of sorrow, the spring of vice to him who has nothing to do. To enjoy without interruption is not to enjoy any thing : the man who has nothing to desire is certainly more unhappy than he who suffers.

These reflections, grounded upon experience, ought to prove to man that good as well as evil depends on the essence of things. Happiness to be felt cannot be continued. Labour is necessary to make intervals between his pleasures ; his body has occasion for exercise to co-order him with the beings who surround him ; his heart must have desires ; trouble alone can give him the right relish of his welfare; it is this which puts in the shadows to the picture of human life. By an irrevocable law of his destiny, man is obliged to be discontented with his present condition ; to make efforts to change it; to reciprocally envy that felicity which no individual enjoys perfectly. Thus the poor man envies the opulence of the rich, although this one is frequently more unhappy than his needy neighbour; thus the rich man views with pain the advantages of a

---

* The advantage which philosophers and men of letters have over the ignorant and the idle, or over those that neither think nor study, is owing to the variety as well as quantity of ideas furnished to the mind by study and reflection. The mind of a man who thinks finds more delight in a good book than can be obtained by all the riches at the command of the ignorant. To study is to amass ideas ; and the number and combination of ideas make that difference between man and man which we observe, besides giving him an advantage over all other animals.

poverty which he sees active, healthy, and frequently jocund even in the bosom of penury.

If man were perfectly contented, there would no longer be any activity in the world; it is necessary that he should desire, act, labour, in order that he may be happy: such is the course of nature, of which the life consists in action. Human societies can only subsist by the continual exchange of those things in which man places his happiness. The poor man is obliged to desire and to labour, that he may procure what he knows is requisite to the preservation of his existence; the primary wants given to him by nature, are to nourish himself, clothe himself, lodge himself, and propagate his species; has he satisfied these? he is quickly obliged to create others entirely new; or rather, his imagination only refines upon the first; he seeks to diversify them; he is willing to give them fresh zest; arrived at opulence, when he has run over the whole circle of wants, when he has completely exhausted their combinations, he falls into disgust. Dispensed from labour, his body amasses humours; destitute of desires, his heart feels a languor; deprived of activity, he is obliged to divide his riches with beings more active, more laborious than himself: these, following their own peculiar interests, take upon themselves the task of labouring for his advantage, of procuring for him means to satisfy his wants, of ministering to his caprices in order to remove the languor that oppresses him. It is thus the great, the rich, excite the energies, the activity, the industry of the indigent; these labour to their own peculiar welfare by working for others: thus the desire of ameliorating his condition, renders man necessary to his fellow man; thus wants, always regenerating, never satisfied, are the principles of life, of activity, the source of health, the basis of society. If each individual were competent to the supply of his own exigencies, there would be no occasion for him to congregate in society, but his wants, his desires, his whims, place him in a state of dependance on others: these are the causes that each individual, in order to further his own peculiar interest, is obliged to be useful to those who have the capability of procuring for him the objects which he himself has not. A nation is nothing more than the union of a great number of individuals, connected with each other by the reciprocity of their wants, or by their mutual desire of pleasure; the most happy man is he who has the fewest wants, and the most numerous means of satisfying them.*

In the individuals of the human species, as well as in political society, the progression of wants, is a thing absolutely necessary; it is founded upon the essence of man; it is requisite that the natural wants once satisfied, should be replaced by those which he calls *imaginary*, or *wants of the fancy;* these become as necessary to his happiness as the first. Custom, which permits the native American to go quite naked, obliges the more civilized inhabitant of Europe to clothe himself; the poor man contents himself with very simple attire, which equally serves him for winter and for summer; the rich man desires to have garments suitable to each season; he would experience pain if he had not the convenience of changing his raiment with every variation of his climate; he would be unhappy if the expense and variety of his costume did not display to the surrounding multitude his opulence, mark his rank, announce his superiority. It is thus habit multiplies the wants of the wealthy; it is thus that vanity itself becomes a want, which sets a thousand hands in motion, who are all eager to gratify its cravings; in short, this very vanity procures for the necessitous man the means of subsisting at the expense of his opulent neighbour. He who is accustomed to pomp, who is used to ostentatious splendour, whose habits are luxurious, whenever he is deprived of these insignia of opulence to which he has attached the idea of happiness, finds himself just as unhappy as the needy wretch who has not wherewith to cover his nakedness. The civilized nations of the present day were in their origin savages composed of erratick tribes, mere wanderers who were occupied with war and the chase, obliged to seek a precarious subsistence by hunting in

---

* The man who would be truly rich, has no need to increase his fortune, it suffices he should diminish his wants.

those woods: in time they have become stationary; they first applied themselves to agriculture, afterwards to commerce; by degrees they have refined on their primitive wants, extended their sphere of action, given birth to a thousand new wants, imagined a thousand new means to satisfy them; this is the natural and necessary progression of active beings, who cannot live without feeling; who, to be happy, must of necessity diversify their sensations.

In proportion as man's wants multiply, the means to satisfy them becomes more difficult; he is obliged to depend on a greater number of his fellow creatures; his interest obliges him to rouse their activity to engage them to concur with his views, consequently he is obliged to procure for them those objects by which they can be excited. The savage need only put forth his hand to gather the fruit he finds sufficient for his nourishment. The opulent citizen of a flourishing society is obliged to set numerous hands to work to produce the sumptuous repast and to procure the far-fetched viands become necessary to revive his languishing appetite, or to flatter his inordinate vanity. From this it will appear, that in the same proportion the wants of man are multiplied, he is obliged to augment the means to satisfy them. Riches are nothing more than the measure of a convention, by the assistance of which man is enabled to make a greater number of his fellows concur in the gratification of his desires; by which he is capacitated to invite them, for their own peculiar interests, to contribute to his pleasures. What, in fact, does the rich man do, except announce to the needy that he can furnish him with the means of subsistence if he consents to lend himself to his will? What does the man in power except show to others that he is in a state to supply the requisites to render them happy? Sovereigns, nobles, men of wealth, appear to be happy only because they possess the ability, are masters of the motives, sufficient to determine a great number of individuals to occupy themselves with their respective felicity.

The more things are considered, the more man will be convinced that his false opinions are the true source of his misery; and the clearer it will appear

to him that happiness is so rare only because he attaches it to objects either indifferent or useless to his welfare, or which, when enjoyed, convert themselves into real evils.

Riches are indifferent in themselves, it is only by their application that they either become objects of utility to man, or are rendered prejudicial to his welfare. Money, useless to the savage, who understands not its value, is amassed by the miser, (to whom it is useless) lest it should be squandered by the prodigal or by the voluptuary, who makes no other use of it than to purchase infirmities and regret. Pleasures are nothing for the man who is incapable of feeling them; they become real evils when they are too freely indulged; when they are destructive to his health; when they derange the economy of his machine; when they make him neglect his duties, and when they render him despicable in the eyes of others. Power is nothing in itself; it is useless to man if he does not avail himself of it to promote his own peculiar felicity: it becomes fatal to him as soon as he abuses it; it becomes odious whenever he employs it to render others miserable. For want of being enlightened on his true interest, the man who enjoys all the means of rendering himself completely happy, scarcely ever discovers the secret of making those means truly subservient to his own peculiar felicity. The art of enjoying is that which of all others is least understood: man should learn this art before he begins to desire; the earth is covered with individuals who only occupy themselves with the care of procuring the means, without ever being acquainted with the end. All the world desire fortune and power, yet very few indeed are those whom these objects render truly happy.

It is quite natural in man, it is extremely reasonable, it is absolutely necessary, to desire those things which can contribute to augment the sum of his felicity. Pleasure, riches, power, are objects worthy his ambition, and deserving his most strenuous efforts, when he has learned how to employ them to render his existence really more agreeable. It is impossible to censure him who desires them, to despise him who commands them, to hate him who possesses them, but when to obtain them he employs odious means, or when after he has obtained them he makes a pernicious use of them, injurious to himself, prejudicial to others. Let him wish for power, let him seek after grandeur, let him be ambitious of reputation, when he can obtain them without making the purchase at the expense of his own repose, or that of the beings with whom he lives: let him desire riches, when he knows how to make a use of them that is truly advantageous for himself, really beneficial for others; but never let him employ those means to procure them with which he may be obliged to reproach himself, or which may draw upon him the hatred of his associates. Let him always recollect, that his solid happiness should rest its foundations upon his own esteem, and upon the advantages he procures for others; and above all, that of all the objects to which his ambition may point, the most impracticable for a being who lives in society, is that of attempting to render himself exclusively happy.

CHAPTER XVI.

The Errours of Man, upon what constitutes Happiness, the true Source of his Evil.— Remedies that may be applied.

REASON by no means forbids man from forming capacious desires; ambition is a passion useful to his species, when it has for its object the happiness of his race. Great minds are desirous of acting on an extended sphere; geniuses who are powerful, enlightened, beneficent, distribute very widely their benign influence; they must necessarily, in order to promote their own peculiar felicity, render great numbers happy. So many princes fail to enjoy true happiness, only because their feeble, narrow souls, are obliged to act in a sphere too extensive for their energies: it is thus that by the supineness, the indolence, the incapacity of their chiefs, nations frequently pine in misery, and are often submitted to masters whose exility of mind is as little calculated to promote their own immediate happiness, as it is to further that of their miserable subjects. On the other hand, minds too vehement, too much

inflamed, too active, are themselves tormented by the narrow sphere that confines them, and their misplaced ardour becomes the scourge of the human race.* Alexander was a monarch, who was as injurious to the earth, as discontented with his condition, as the indolent despot whom he dethroned.— The souls of neither were by any means commensurate with their sphere of action.

The happiness of man will never be more than the result of the harmony that subsists between his desires and his circumstances. The sovereign power, to him who knows not how to apply it to the advantage of his citizens, is as nothing; if it renders him miserable, it is a real evil; if it produces the misfortune of a portion of the human race, it is a detestable abuse. The most powerful princes are ordinarily such strangers to happiness, their subjects are commonly so unfortunate only because they first possess all the means of rendering themselves happy, without ever giving them activity, or because the only knowledge they have of them is their abuse. A wise man, seated on a throne, would be the most happy of mortals. A monarch is a man for whom his power, let it be of whatever extent, cannot procure other organs, other modes of feeling, than the meanest of his subjects; if he has an advantage over them, it is by the grandeur, the variety, the multiplicity of the objects with which he can occupy himself, which, by giving perpetual activity to his mind, can prevent it from decay and from falling into sloth. If his mind is virtuous and expansive, his ambition finds continual food in the contemplation of the power he possesses to unite by gentleness and kindness the will of his subjects with his own; to interest them in his own conservation, to merit their affections, to draw forth the respect of strangers, and to elicit the eulogies of all nations. Such are the conquests that reason proposes to all those whose destiny it is to govern the fate of empires: they are sufficiently grand to satisfy the

---

* Æstuat infelix augusto limite mundi.— Seneca says of Alexander, Post Darium and Indos pauper est Alexander; inventus est qui concupisceret aliquid post omnia. *V. Senec. Epist.* 120.

most ardent imagination, to gratify the most capacious ambition. Kings are the most happy of men only because they have the power of making a great number of other men happy, and thus of multiplying the causes of legitimate content with themselves.

The advantages of the sovereign power are participated by all those who contribute to the government of states. Thus grandeur, rank, reputation, are desirable for all who are acquainted with all the means of rendering them subservient to their own peculiar. felicity; they are useless to those ordinary men, who have neither the energy nor the capacity to employ them in a mode advantageous to themselves; they are detestable whenever to obtain them man compromises his own happiness and the welfare of society: this society itself is in an errour every time it respects men who only employ to its destruction a power, the exercise of which it ought never to approve but when it reaps from it substantial benefits.

Riches, useless to the miser, who is no more than their miserable jailer, prejudicial to the debauchee, for whom they only procure infirmities, disgust, and satiety, can, in the hands of the honest man, produce unnumbered means of augmenting the sum of his happiness; but before man covets wealth, it is proper he should know how to employ it; money is only a representative of happiness: to enjoy it so as to make others happy, this is the reality. Money, according to the compact of man, procures for him all those benefits he can desire; there is only one which it will not procure, that is, the knowledge how to apply it properly. For man to have money, without the true secret how to enjoy it, is to possess the key of a commodious palace to which he is interdicted entrance; to lavish it prodigally, is to throw the key into the

homage of all those who surround him; he will restrain himself in his pleasures, in order that he may be enabled truly to enjoy them; he will know that money cannot re-establish a mind worn out with enjoyment, enfceebled by excess; cannot invigorate a body enervated by debauchery, from thenceforth become incapable of sustaining him, except by the necessity of privations; he will know that the licentiousness of the voluptuary stifles pleasure in its source, and that all the treasure in the world cannot renew his senses.

From this it will be obvious, that nothing is more frivolous than the declamations of a gloomy philosophy against the desire of power, the pursuit of grandeur, the acquisition of riches, the enjoyment of pleasure.—These objects are desirable for man, whenever his condition permits him to make pretensions to them, or whenever he has acquired the knowledge of making them turn to his own real advantage; reason cannot either censure or despise him, when to obtain them he wounds no one's interest: his associates will esteem him when he employs their agency to secure his own happiness, and that of his fellows. Pleasure is a benefit, it is of the essence of man to love it; it is even rational, when it renders his existence really valuable to himself, when its consequences are not grievous to others. Riches are the symbols of the great majority of the benefits of this life; they become a reality in the hands of the man who has the clew to their just application. Power is the most sterling of all benefits, when he who is its depositary has received from nature a mind sufficiently noble, elevated, benevolent, and energetic, which enables him to extend his happy influence over whole nations, which, by this means, he places in a state of legitimate is only upon the faculty of rendering him happy that legitimate authority builds its structure. No man derives from nature the right of commanding another; but it is voluntarily accorded to those from whom he expects his welfare. Government is the right of commanding conferred on the sovereign, only for the advantage of those who are governed. Sovereigns are the defenders of the persons, the guardians of the property, the protectors of the liberty of their subjects: it is only on this condition these consent to obey; government would not be better than a robbery whenever it availed itself of the powers confided to it to render society unhappy. The empire of religion is founded on the opinion man entertains of its having power to render nations happy; and the Gods are horrible phantoms if they do render man unhappy.* Government and religion, could be reasonable institutions only inasmuch as they equally contributed to the felicity of man: it would be folly in him to submit himself to a yoke from which these resulted nothing but evil: it would be rank injustice to oblige him to renounce his rights, without some corresponding advantage.

The authority which a father exercises over his family, is only founded on the advantages which he is supposed to procure for it. Rank, in political society, has only for its basis the real or imaginary utility of some citizens, for which the others are willing to distinguish, respect, and obey them. The rich acquire rights over the indigent, only by virtue of the welfare they are able to procure them. Genius, talents, science, arts, have rights over man, only in consequence of their utility, of the delight they confer, of the advantages they procure for society. In a word, it is happiness, it is the expectation of happiness, it is its image, that man cherishes, esteems, and unceasingly adores. Gods and monarchs, the rich and the great, may easily impose

on him, may dazzle him, may intimidate him, but they will never be able to obtain the voluntary submission of his heart, which alone can confer upon them legitimate rights, without they make him experience real benefits and display virtue. Utility is nothing more than true happiness; to be useful is to be virtuous; to be virtuous is to make others happy.

The happiness which man derives from them, is the invariable and necessary standard of his sentiments for the beings of his species, for the objects he desires, for the opinions he embraces, for those actions on which he decides; he is the dupe of his prejudices every time he ceases to avail himself of this standard to regulate his judgment. He will never run the risk of deceiving himself, when he shall examine strictly what is the real utility resulting to his species from the religion, from the laws, from the institutions, from the inventions and the various actions of all mankind. A superficial view may sometimes seduce him; but experience, aided by reflection, will re-conduct him to reason, which is incapable of deceiving him. This teaches him that pleasure is a momentary happiness, which frequently becomes an evil; that evil is a fleeting trouble, that frequently becomes a good: it makes him understand the true nature of objects, and enables him to foresee the effects he may expect; it makes him distinguish those desires to which his welfare permits him to lend himself from those to whose seduction he ought to make resistance. In short, it will always convince him, that the true interest of intelligent beings, who love happiness, who desire to render their own existence felicitous, demands that they should root out all those phantoms, abolish all those chimerical ideas, destroy all those prejudices, which obstruct their felicity in this world.

If he consults experience, he will perceive that it is in illusions and opinions looked upon as sacred, that he ought to search out the source of that multitude of evils, which almost every where overwhelms mankind. From ignorance of natural causes, man has created Gods; imposture rendered these Gods terrible to him; and these fatal

---

* Cicero says—Nisi homini placuerit, Deus non erit.—" God cannot oblige men to obey him, unless he proves to them that he has the power of rendering them happy or unhappy." *See the Defence of Religion,* Vol. I. p. 433. From this we must conclude that we are right in judging of religion and of the Gods by the advantages or disadvantages they procure to society.

ideas haunted him without rendering him better, made him tremble without either benefit to himself or to others; filled his mind with chimeras, opposed themselves to the progress of his reason, prevented him from seeking after his happiness. His fears rendered him the slave of those who have deceived him under pretence of consulting his welfare; he committed evil whenever they told him his Gods demanded crimes; hé lived in misfortune, because they made him believe these Gods condemned him to be miserable; the slave of these Gods, he never dared to disentangle himself from his chains, because the artful ministers of these Divinities gave him to understand, that stupidity, the renunciation of reason, sloth of mind, abjection of soul, were the sure means of obtaining eternal felicity.

Prejudices, not less dangerous, have blinded man upon the true nature of government; nations are ignorant of the true foundations of authority; they dare not demand happiness from those kings who are charged with the care of procuring it for them: they have believed that their sovereigns were Gods disguised, who received with their birth, the right of commanding the rest of mankind; that they could at their pleasure dispose of the felicity of the people, and that they were not accountable for the misery they engendered. By a necessary consequence of these opinions, politics have almost every where degenerated into the fatal art of sacrificing the interests of the many, either to the caprice of an individual, or to some few privileged rascals. In despite of the evils which assailed them, nations fell down in adoration before the idols they themselves had made, and foolishly respected the instruments of their misery; obeyed their unjust will: lavished their blood, exhausted their treasure, sacrificed their lives, to glut the ambition, the cupidity, the never-ending caprices of these men; they bent the knee to established opinion, bowed to rank, yielded to title, to opulence, to pageantry, to ostentation: at length, victims to their prejudices, they in vain expected their welfare at the hands of men who were themselves unhappy from their own vices, whose neglect of virtue had rendered them incapable of enjoying

true felicity, who were but little disposed to occupy themselves with their prosperity: under such chiefs their physical and moral happiness were equally neglected or even annihilated.

The same blindness may be perceived in the science of morals. Religion, which never had any thing but ignorance for its basis, and imagination for its guide, did not found ethics upon man's nature, upon his relations with his fellows, upon those duties which necessarily flow from these relations, it preferred founding them upon imaginary relations, which it pretended subsisted between him and some invisible powers it had gratuitously imagined, and had falsely been made to speak.*

It was these invisible Gods which religion always paints as furious tyrants, who were declared the arbiters of man's destiny—the models of his conduct; when he was willing to imitate these tyrannical Gods, when he was willing to conform himself to the lessons of their interpreters, he became wicked, was an unsociable creature, a useless being, or else a turbulent maniac and a zealous fanatic. It was these alone who profited by religion, who advantaged themselves by the darkness in which it involved the human mind; nations were ignorant of nature, they knew nothing of reason, they understood not truth; they had only a gloomy religion, without one certain idea of either morals or virtue. When man committed evil against his fellow creature, he believed he had offended his God; but he also believed himself forgiven, as soon as he had prostrated himself before him; as soon as he had made him costly presents, and gained over the priest to his interest. Thus religion, for from giving a sure, a natural, and a known basis to morals, only rested it on an unsteady foundation, made it consist in ideal duties, impossible to

* Thus Trophonius, from his cave, made affrighted mortals tremble, shook the stoutest nerves, made them turn pale with fear; his miserable, deluded supplicants, who were obliged to sacrifice to him, anointed their bodies with oil, bathed in certain rivers, and after they had offered their cake of honey and received their destiny, became so dejected, so wretchedly forlorn, that to this day their descendants, when they behold a melancholy man, exclaim, "*He has consulted the oracle of Trophonius.*"

be accurately understood. What did I say? It first corrupted him, and his expiations finished by ruining him. Thus when religion was desirous to combat the unruly passions of man, it attempted it in vain; always enthusiastic, and deprived of experience, it knew nothing of the true remedies; those which it applied were disgusting, only suitable to make the sick revolt against them; it made them pass for divine, because they were not made of man; they were inefficacious, because chimeras could effectuate nothing against those substantive passions to which motives more real and more powerful concurred to give birth, which every thing conspired to nourish in his heart. The voice of religion, or of the Gods, could not make itself heard amidst the tumult of society, where all cried out to man, that he could not render himself happy without injuring his fellow creature; these vain clamours only made virtue hateful to him, because they always represented it as the enemy to his happiness—as the bane of human pleasures. He consequently failed in the observation of his duties, because real motives were never held forth to induce him to make the requisite sacrifice: the present prevailed over the future, the visible over the invisible, the known over the unknown; and man became wicked, because every thing informed him he must be so in order to obtain happiness.

Thus, the sum of human misery was never diminished; on the contrary, it was accumulating either by his religion, by his government, by his education, by his opinions, or by the institutions he adopted under the idea of rendering his condition more pleasant. It cannot be too often repeated, it is in errour that man will find the true spring of those evils with which the human race is afflicted; it is not nature that renders him miserable and unhappy; it is not an irritated Divinity, who is desirous he should live in tears; it is not hereditary depravation that has caused him to be wicked and miserable, it is to errour that these deplorable effects are to be ascribed.

The sovereign good, so much sought after by some philosophers, announced with so much emphasis by others, may be considered as a chimera, like unto

that marvellous *panacea*, which some adepts have been willing to pass upon mankind for a universal remedy. All men are diseased; the moment of their birth delivers them over to the contagion of errour; but individuals are variously affected by it, by a consequence of their natural organization and of their peculiar circumstances. If there is a sovereign remedy which can be indiscriminately applied to the diseases of man, there is without doubt only *one*, and this remedy is *truth*, which he must draw from nature.

At the sight of those errours which blind the greater number of mortals—of those delusions which man is doomed to suck in with his mother's milk; at the sight of those desires, of those propensities, by which he is perpetually agitated, of those passions which torment him, of those inquietudes which gnaw his repose, of those evils, as well physical as moral, which assail him on every side, the contemplator of humanity would be tempted to believe that happiness was not made for this world, and that any effort to cure those minds which every thing unites to poison, would be a vain enterprise. When he considers those numerous superstitions by which man is kept in a continual state of alarm, that divide him from his fellow, that render him irrational; when he beholds the many despotic governments that oppress him; when he examines those multitudinous, unintelligible, contradictory laws that torture him; the manifold injustice under which he groans; when he turns his mind to the barbarous ignorance in which he is steeped, almost over the whole surface of the earth; when he witnesses those enormous crimes that debase society, and render it so hateful to almost every individual; he has great difficulty to prevent his mind from embracing the idea, that misfortune is the only appendage of the human species; that this world is made solely to assemble the unhappy; that human felicity is a chimera, or at least a point so fugitive, that it is impossible it can be fixed.

Thus superstitious and atrabilious mortals, nourished in melancholy, unceasingly see either nature or its author exasperated against the human race; they suppose that man is the constant object of heaven's wrath; that he irrtates it even by his desires, and renders himself criminal by seeking a felicity which is not made for him. Struck with beholding that those objects which he covets in the most lively manner, are never competent to content his heart, they have decried them as abominations, as things prejudicial to his interest, as odious; they prescribe him that he should entirely shun them; they have endeavoured to put to the rout all his passions, without any distinction even of those which are the most useful to himself, the most beneficial to those beings with whom he lives: they have been willing that man should render himself insensible—should become his own enemy—that he should separate himself from his fellow creatures—that he should renounce all pleasure—that he should refuse happiness; in short, that he should cease to be a man; that he should become unnatural. "Mortals!" have they said, "ye were born to be unhappy; the author of your existence has destined ye for misfortune; enter then into his views, and render yourselves miserable. Combat those desires which have felicity for their object; renounce those pleasures which it is your essence to love; attach yourselves to nothing in this world; fly a society that only serves to inflame your imagination, to make you sigh after benefits you ought not to enjoy; break up the spring of your souls; repress that activity that seeks to put a period to your sufferings; suffer, afflict yourselves, groan, be wretched; such is for you the true road to happiness."

Blind physicians! who have mistaken for a disease the natural state of man! they have not seen that his desires and his passions were essential to him; that to defend him from loving and desiring, is to deprive him of that activity, which is the vital principle of society; that to tell him to hate and despise himself, is to take from him the most substantive motive that can conduct him to virtue. It is thus, that, by its supernatural remedies, religion, far from curing evils, has only increased them, and made them more desperate; in the room of calming his passions, it gives them inveteracy, makes them more dangerous, renders them more venomous, turns that into a curse which

nature has given him for his preservation and happiness. It is not by extinguishing the passions of man that he is to be rendered happier, it is by directing them towards useful objects, which, by being truly advantageous to himself, must of necessity be beneficial to others.

In despite of the errours which blind the human race; in despite of the extravagance of man's religious and political institutions, notwithstanding the complaints and murmurs he is continually breathing forth against his destiny, there are yet happy individuals on the earth. Man has sometimes the felicity to behold sovereigns animated by the noble passion to render nations flourishing and happy; now and then he encounters an Antoninus, a Trajan, a Julian, an Alfred, a Henri IV.;* he meets with elevated minds, who place their glory in encouraging merit, who rest their happiness in succouring indigence, who think it honourable to lend a helping hand to oppressed virtue: he sees genius, occupied with the desire of eliciting the admiration of his fellow-citizens by serving them usefully, and satisfied with enjoying that happiness he procures for others.

Let it not be believed that the man of poverty himself, is excluded from happiness. Mediocrity and indigence frequently procure for him advantages that opulence and grandeur are obliged to acknowledge. The soul of the needy man, always in action, never ceases to form desires, whilst the rich and the powerful are frequently in the afflicting embarrassment of either not knowing what to wish for, or else of desiring those objects which it is impossible for them to obtain.† The poor man's body, habituated to labour, knows the sweets of repose; this repose of the body is the most troublesome fatigue to him who is wearied with his idleness. Exercise and frugality procure for the one vigour, health, and contentment; the intemperance and sloth of the other furnish him only with disgust and infirmities. Indigence sets all the springs of the soul to work; it is the mother of industry; from its bosom arise genius,

---

* To this scanty list may now be added the names of George Washington and Thomas Jefferson.

† Petronius says: Nescio quomodo bonæ mentis soror est paupertas.

talents, and merit, to which opulence and grandeur pay their homage. In short, the blows of fate find in the poor man a flexible reed, who bends without breaking.

Thus nature is not a stepmother to the greater number of her children. He whom fortune has placed in an obscure station, is ignorant of that ambition which devours the courtier; knows nothing of the inquietude which deprives the intriguer of his rest; is a stranger to the remorse, disgust, and weariness of the man, who, enriched with the spoils of a nation, does not know how to turn them to his profit. The more the body labours, the more the imagination reposes itself; it is the diversity of the objects man runs over that kindles it; it is the satiety of those objects that causes him disgust; the imagination of the indigent is circumscribed by necessity: he receives but few ideas, he is acquainted with but few objects; in consequence he has but little to desire; he contents himself with that little, whilst the entire of nature with difficulty suffices to satisfy the insatiable desires, to gratify the imaginary wants of the man plunged in luxury, who has run over and exhausted all common objects. Those, whom prejudice contemplates as the most unhappy of men, frequently enjoy advantages more real and much greater those who oppress them, who despise them, but who are nevertheless often reduced to the misery of envying them. Limited desires are a real benefit: the man of meaner condition, in his humble fortune, desires only bread: he obtains it by the sweat of his brow; he would eat it with pleasure if injustice did not almost always render it bitter to him. By the delirium of governments, those who roll in abundance, without for that reason being more happy, dispute with the cultivator even the fruits which the earth yields to the labour of his hands. Princes sacrifice their true happiness, as well as that of their states, to these passions, to those caprices, which discourage the people, which plunge their provinces in misery, which make millions unhappy without any advantage to themselves. Tyrants oblige their subjects to curse their existence, to abandon labour, and take from them the courage of propagating

a progeny who would be as unhappy as their fathers: the excess of oppression sometimes obliges them to revolt and to avenge themselves by wicked outrages of the injustice it has heaped on their devoted heads. Injustice, by reducing indigence to despair, obliges it to seek in crime resources against its misery. An unjust government produces discouragement; its vexations depopulate a country; the earth remains without culture; from thence is bred frightful famine, which gives birth to contagion and plague. The misery of a people produce revolutions: soured by misfortunes their minds get into a state of fermentation, and the overthrow of an empire is the necessary effect. It is thus that physics and morals are always connected, or rather are the same thing.

If the bad morals of chiefs do not always produce such marked effects, at least they generate slothfulness, of which the effect is to fill society with mendicants and malefactors, whose vicious course neither religion nor the terrour of the laws can arrest; which nothing can induce to remain the unhappy spectators of a welfare they are not permitted to participate. They seek a fleeting happiness at the expense even of their lives, when injustice has shut up to them the road of labour and industry, which would have rendered them both useful and honest.

Let it not then be said, that no government can render all its subjects happy: without doubt it cannot flatter itself with contenting the capricious humours of some idle citizens, who are obliged to rack their imagination to appease the disgust arising from lassitude: but it can, and it ought to occupy itself with ministering to the real wants of the multitude. A society enjoys all the happiness of which it is susceptible, whenever the greater number of its members are wholesomely fed, decently clothed, comfortably lodged; in short, when they can without an excess, of toil beyond their strength procure wherewith to satisfy those wants which nature has made necessary to their existence. Their minds rest contented as soon as they are convinced no power can ravish from them the fruits of their industry, and that they labour for themselves. By a consequence of human folly, whole nations are obliged to toil incessantly, to waste their strength, to sweat under their burdens, to drench the earth with their tears, in order to maintain the luxury, to gratify the whims, to support the corruption of a small number of irrational beings, of some few useless men, to whom happiness has become impossible, because their bewildered imaginations no longer know any bounds. It is thus that religious and political errours have changed the fair face of nature into a valley of tears.

For want of consulting reason, for want of knowing the value of virtue, for want of being instructed in their true interests, for want of being acquainted with what constitutes solid and real felicity, the prince and the people, the rich and the poor, the great and the little, are unquestionably frequently very far removed from content; nevertheless if an impartial eye be glanced over the human race, it will be found to comprise a greater number of benefits than of evils. No man is entirely happy, but he is so in detail. Those who make the most bitter complaints of the rigour of their fate, are, however, held in existence by threads frequently imperceptible, which prevent the desire of quitting it. In short, habit lightens to man the burden of his troubles; grief suspended becomes true enjoyment; every want is a pleasure in the moment when it is satisfied; freedom from chagrin, the absence of disease, is a happy state which he enjoys secretly and without even perceiving it; hope, which rarely abandons him entirely, helps him to support the most cruel disasters. The prisoner laughs in his irons; the wearied villager returns singing to his cottage; in short, the man who calls himself the most unfortunate, never sees death approach without dismay, at least if despair has not totally disfigured nature in his eyes.*

As long as man desires the continuation of his being, he has no right to call himself completely unhappy; whilst hope sustains him, he still enjoys a great benefit. If man was more just in rendering to himself an account of his pleasures and of his pains, he would

acknowledge that the sum of the first exceeds by much the amount of the last; he would perceive that he keeps a very exact leger of the evil, but a very unfaithful journal of the good: indeed he would avow, that there are but few days entirely unhappy during the whole course of his existence. His periodical wants procure for him the pleasure of satisfying them: his mind is perpetually moved by a thousand objects, of which the variety, the multiplicity, the novelty, rejoices, him, suspends his sorrows, diverts his chagrin. His physical evils, are they violent? They are not of long duration; they conduct him quickly to his end: the sorrows of his mind conduct him to it equally. At the same time that nature refuses him every happiness, she opens to him a door by which he quits life: does he refuse to enter it? it is that he yet finds pleasure in existence. Are nations reduced to despair? Are they completely miserable? They have recourse to arms; and, at the risk of perishing, they make the most violent efforts to terminate their sufferings.

Thus, as he sees so many of his fellows cling to life, man ought to conclude they are not so unhappy as he thinks. Then let him not exaggerate the evils of the human race: let him impose silence on that gloomy humour, that persuades him these evils are without remedy; let him diminish by degrees the number of his errours, and his calamities will vanish in the same proportion. He is not to conclude himself infelicitous, because his heart never ceases to form new desires. Since his body daily requires nourishment, let him infer that it is sound, that it fulfils its functions. As long as he has desire, the proper deduction ought to be, that his mind is kept in the necessary activity; he should also gather from all this that passions are essential to him, that they constitute the happiness of a being who feels, who thinks, who receives ideas, who must necessarily love and desire that which promises him a mode of existence analogous to his natural energies. As long as he exists, as long as the spring of his mind maintains its elasticity, this mind desires; as long as it desires, he experiences the activity which is necessary to him; as long as he acts, so long he lives. Human life may be compared to a river, of which the waters succeed each other, drive each other forward, and flow on without interruption;. these waters obliged to roll over an unequal bed, encounter at intervals those obstacles which prevent their stagnation; they never cease to undulate, recoil, and to rush forward, until they are restored to the ocean of nature.

CHAPTER XVII.

Those Ideas which are true, or founded upon Nature, are the only Remedies for the Evils of Man.—Recapitulation.—Conclusion of the First Part.

WHENEVER man ceases to take experience for his guide, he falls into errour. His errours become yet more dangerous and assume a more determined inveteracy, when they are clothed with the sanction of religion: it is then that he hardly ever consents to return into the paths of truth; he believes himself deeply interested in no longer seeing clearly that which lies before him; he fancies he has an essential advantage in no longer understanding himself, and that his happiness exacts that he should shut his eyes to truth. If the majority of moral philosophers have mistaken the human heart; if they have deceived themselves upon its diseases and the remedies that are suitable; if the remedies they have administered have been inefficacious or even dangerous, it is because they have abandoned nature, have resisted experience, and have not had sufficient steadiness to consult their reason; because, having renounced the evidence of their senses, they have only followed the caprices of an imagination either dazzled by enthusiasm or disturbed by fear, and have preferred the illusions it has held forth to the realities of nature, who never deceives.

It is for want of having felt, that an intelligent being cannot for an instant lose sight of his own peculiar conservation—of his particular interests, either real or fictitious—of his own welfare, whether permanent or transitory; in short, of his happiness, either true or false; it is for want of having considered that desires and passions are essential and natural, that both the one and the other are motions necessary to the mind of man, that the physicians of the human mind have supposed supernatural causes for his wanderings, and have only applied to his evils topical remedies, either useless or dangerous. Indeed, in desiring him to stifle his desires, to combat his propensities, to annihilate his passions, they have done no more than give him steril precepts, at once vague and impracticable; these vain lessons have influenced no one; they have at most restrained some few mortals, whom a quiet imagination but feebly solicited to evil; the terrours with which they have accompanied them, have disturbed the tranquillity of those persons, who were moderate by their nature, without ever arresting the ungovernable temperament of those who were inebriated by their passions, or hurried along by the torrent of habit. In short, the promises of superstition, as well as the menaces it holds forth, have only formed fanatics and enthusiasts, who are either dangerous or useless to society, without ever making man truly virtuous, that is to say, useful to his fellow creatures.

These empirics, guided by a blind routine, have not seen that man, as long as he exists, is obliged to feel, to desire, to have passions, and to satisfy them in proportion to the energy which his organization has given him; they have not perceived that education planted these desires in his heart, that habit rooted them, that his government, frequently vicious, corroborated their growth, that public opinion stamped them with its approbation, that experience rendered them necessary, and that to tell men thus constituted to destroy their passions, was either to plunge them into despair, or else to order them remedies too revolting for their temperament. In the actual state of opulent societies, to say to a man who knows by experience that riches procure every pleasure, that he must not desire them, that he must not make any efforts to obtain them, that he ought to detach himself from them, is to persuade him to render himself miserable. To tell an ambitious man not to desire grandeur and power, which every thing conspires to point out to him as the height of felicity, is to order him to overturn at one blow the habitual system of his ideas; it is to speak to a deaf man. To tell a lover of an impetuous temperament, to stifle his passion for the object that enchants him, is to make him understand that he ought to renounce his happiness. To oppose religion to such puissant interests, is to combat realities by chimerical speculations.

Indeed, if things were examined without prepossession, it would be found that the greater part of the precepts inculcated by religion, or which fanatical and supernatural morals give to man, are as ridiculous as they are impossible to be put into practice. To interdict passion to man, is to desire of him not to be a human creature; to counsel an individual of violent imagination to moderate his desires, is to advise him to change his temperament —to request his blood to flow more sluggishly. To tell a man to renounce his habits, is to be willing that a citizen, accustomed to clothe himself, should consent to walk quite naked; it would avail as much to desire him to change the lineament of his face, to destroy his configuration, to extinguish his imagination, to alter the course of his fluids, as to command him not to have passions analogous with his natural energy, or to lay aside those which habit and his circumstances have converted into wants.* Such are, however, the so much boasted remedies which the greater number of moral philosophers apply to human depravity. Is it then surprising they do not produce the desired effect, or that they only reduce man to a state of despair, by the effervescence that results from the continual conflict which they excite between the passions of his heart, between his vices and his virtues, between his habits and those chimerical fears with which superstition is at all times ready to overwhelm him? The vices of society, aided by the objects of which it avails itself to whet the desires of man, the pleasures, the riches, the grandeur, which his government holds forth to him as so many seductive magnets, the advantage which education, the benefits, example, public opinion render dear to him, attract him on one side; whilst a gloomy morality vainly solicits him on the other; thus, religion plunges him into misery—holds a violent struggle with his heart, without ever gaining the victory; when by accident it does prevail against so many united forces, it renders him unhappy —it completely destroys the spring of his mind.

Passions are the true counterpoise to passions; then, let him not seek to destroy them, but let him endeavour to direct them; let him balance those which are prejudicial, by those which are useful to society. Reason, the fruit of experience, is only the art of choosing those passions to which, for his own peculiar happiness, he ought to listen. Education is the true art of disseminating, the proper method of cultivating advantageous passions in the heart of man. Legislation is the art of restraining dangerous passions, and of exciting those which may be conducive to the public welfare. Religion is only the art of planting and of nourishing in the mind of man those chimeras, those illusions, those impostures, those incertitudes, from whence spring passions fatal to himself as well as to others: it is only by bearing up with fortitude against these, that he can place himself on the road to happiness.†

Reason and morals cannot effect any thing on mankind, if they do not point out to each individual, that his true interest is attached to a conduct useful to others and beneficial to himself; this conduct to be useful must conciliate for him the favour of those beings who are necessary to his happi-

---

* It is evident that these counsels, extravagant as they are, have been suggested to man by all religions. The Indian, the Japanese, the Mahometan, the Christian, the Jew, each, according to his superstition, has made perfection to consist in fasting, mortification, abstinence from the most rational pleasures, retirement from the busy world, and in labouring without ceasing to counteract nature. Among the Pagans the priests of the Syrian Goddess were not more rational—their piety led them to mutilate themselves.

† To these we may add philosophy, which is the art of advocating truth, of renouncing errour, of contemplating reality, of drawing wisdom from experience, of cultivating man's nature to his own felicity, by teaching him to contribute to that of his associates; in short, it is reason, education, and legislation, united to further the great end of human existence, by causing the passions of man to flow in a current genial to his own happiness.

ness: it is then for the interest of mankind, for the happiness of the human race, it is for the esteem of himself, for the love of his fellows, for the advantages which ensue, that education in early life should kindle the imagination of the citizen; this is the true means of obtaining those happy results with which habit should familiarize him, which public opinion should render dear to his heart, for which example ought continually to rouse his faculties. Government, by the aid of recompenses, ought to encourage him to follow this plan; by visiting crime with punishment, it ought to deter those who are willing to interrupt it. Thus the hope of a true welfare, the fear of real evil, will be passions suitable to countervail those which, by their impetuosity, would injure society; these last will at least become very rare, if instead of feeding man's mind with unintelligible speculations, in lieu of vibrating on his ears words void of sense, he is only spoken to of realities, only shown those interests which are in unison with truth.

Man is frequently so wicked, only because he almost always feels himself interested in being so; let him be more enlightened and more happy, and he will necessarily become better. An equitable government, a vigilant administration will presently fill the state with honest citizens; it will hold forth to them present reasons, real and palpable, to be virtuous; it will instruct them in their duties; it will foster them with its cares; it will allure them by the assurance of their own peculiar happiness; its promises and its menaces faithfully executed, will, unquestionably, have much more weight than those of superstition, which never exhibits to their view other than illusory benefits, fallacious punishments, which the man hardened in wickedness will doubt every time he finds an interest in questioning them; present motives will tell more home to his heart, than those which are distant and at best uncertain. The vicious and the wicked are so common upon the earth, so pertinacious in their evil courses, so attached to their irregularities, only because there are but few governments that make man feel the advantage of being just, honest, and benevolent; on the

contrary, there is hardly any place where the most powerful interests do not solicit him to crime by favouring the propensities of a vicious organization, which nothing has attempted to rectify or lead towards virtue.* A savage, who in his horde, knows not the value of money, certainly would not commit a crime; if transplanted into civilized society, he will presently learn to desire it, will make efforts to obtain it, and, if he can without danger, finish by stealing it, above all if he had not been taught to respect the property of the beings who environ him. The savage and the child are precisely in the same state; it is the negligence of society, of those intrusted with their education, that render both the one and the other wicked. The son of a noble, from his infancy learns to desire power, at a riper age he becomes ambitious; if he has the address to insinuate himself into favour, he becomes wicked, and he may be so with impunity. It is not therefore nature that makes man wicked, they are his institutions which determine him to vice. The infant brought up amongst robbers, can generally become nothing but a malefactor; if he had been reared with honest people, the chance is he would have been a virtuous man.

If the source be traced of that profound ignorance in which man is with respect to his morals, to the motives that can give volition to his will, it will be found in those false ideas which the greater number of speculators have formed to themselves of human nature. The science of morals has become an enigma, which it is impossible to unravel, because man has made himself double, has distinguished his mind from his body, supposed it of a nature different from all known beings, with modes of action, with properties distinct from all other bodies; because he has emancipated this mind from physical laws, in order to submit it to capricious laws derived from imaginary regions. Metaphysicians, seized upon these gratuitous suppositions, and by dint of subtilizing them, have rendered

* Sallust says, Nemo gratuito malus est. We can say in the same manner. Nemo gratuito bonus est

them completely unintelligible. These moralists have not perceived, that motion is essential to the mind as well as to the living body; that both the one and the other are never moved but by material, by physical objects; that the wants of each regenerate themselves unceasingly; that the wants of the mind, as well as those of the body, are purely physical; that the most intimate, the most constant connexion subsists between the mind and the body, or rather they have been unwilling to allow, that they are only the same thing considered under different points of view. Obstinate in their supernatural or unintelligible opinions, they have refused to open their eyes, which would have convinced them, that the body in suffering rendered the mind miserable; that the mind afflicted undermined the body and brought it to decay; that both the pleasures and agonies of the mind, have an influence over the body, either plunge it into sloth or give it activity: they have rather chosen to believe, that the mind draws its thoughts, whether pleasant or gloomy, from its own peculiar sources; while the fact is, that it derives its ideas only from material objects, that strike on the physical organs; that it is neither determined to gayety nor led on to sorrow, but by the actual state, whether permanent or transitory, in which the fluids and solids of the body are found. In short, they have been loath to acknowledge, that the mind, purely passive, undergoes the same changes which the body experiences; that it is only moved by its intervention, acts only by its assistance, receives its sensations, its perceptions, forms its ideas, derives either its happiness, or its misery, from physical objects, through the medium of the organs of which the body is composed, frequently without its own cognizance, and often in despite of itself.

By a consequence of these opinions, connected with marvellous systems, and systems invented to justify them, they have supposed the human mind to be a free agent; that is to say that it has the faculty of moving itself—that it enjoys the privilege of acting independent of the impulse received from exterior objects through the organs of the body; that regardless of these impulsions, it can even resist them, and follow its own direction by its own energies; that it is not only different in its nature from all other beings, but has also a separate mode of action; in other words, that it is an isolated point, which is not submitted to that uninterrupted chain of motion, which bodies communicate to each other in nature whose parts are always in action.—Smitten with their sublime notions, these speculators were not aware, that in thus distinguishing the soul or mind, from the body and from all known beings, they rendered it an impossibility to form any true idea of it; they were unwilling to perceive the perfect analogy which is found between the manner of the mind's action, and that by which the body is affected; they shut their eyes to the necessary and continual correspondence which is found between the mind and the body; they would not see that like the body it is subjected to the motion of attraction and repulsion, which is ascribable to qualities inherent in those physical substances which give play to the organs of the body; that the volition of its will, the activity of its passions, the continual regeneration of its desires, are never more than consequences of that activity which is produced on the body by material objects which are not under its controul, and that these objects render it either happy or miserable, active or languishing, contented or discontented, in despite of itself and of all the efforts it is capable of making to render it otherwise: they have rather chosen to seek in the heavens for fictitious powers to set it in motion; they have held forth to man only imaginary interests: under the pretext of procuring for him an ideal happiness, he has been prevented from labouring to his true felicity, which has been studiously withheld from his knowledge: his regards have been fixed upon the heavens, that he might lose sight of the earth: truth has been concealed from him, and it has been pretended he would be rendered happy by dint of terrours, by means of phantoms, and of chimeras. In short, hoodwinked and blind, he was only guided through the flexuous paths of life by men as blind as himself, where both the one and the other were lost in the maze.

From every thing which has been hitherto said, it evidently results that all the errours of mankind, of whatever nature they may be, arise from man's having renounced reason, quitted experience, and refused the evidence of his senses, that he might be guided by imagination, frequently deceitful, and by authority, always suspicious. Man will ever mistake his true happiness, as long as he neglects to study nature, to investigate her immutable laws, to seek in her alone the remedies for those evils which are the consequence of his present errours: he will be an enigma to himself, as long as he shall believe himself double, and that he is moved by an inconceivable power, of the laws and nature of which he is ignorant. His intellectual as well as his moral faculties will remain unintelligible to him if he does not contemplate them with the same eyes as he does his corporeal qualities, and does not view them as submitted in every thing to the same regulations. The system of his pretended free agency is without support; experience contradicts it every instant, and proves that he never ceases to be under the influence of necessity in all his actions; this truth, far from being dangerous to man, far from being destructive of his morals, furnishes him with their true basis, by making him feel the necessity of those relations which subsist between sensible beings united in society, who have congregated with a view of uniting their common efforts for their reciprocal felicity. From the necessity of these relations, spring the necessity of his duties; these point out to him the sentiments of love which he should accord to virtuous conduct, or that aversion he should have for what is vicious. From hence the true foundation of *moral obligation*, will be obvious, which is only the necessity of taking means to obtain the end man proposes to himself by uniting in society, in which each individual for his own peculiar interest, his own particular happiness, his own personal security, is obliged to display and to hold a conduct suitable to the preservation of the community, and to contribute by his actions to the happiness of the whole. In a word, it is upon the necessary action and reaction of the hu-

man will, upon the necessary attraction and repulsion of man's mind, that all his morals are bottomed: it is the unison of his will, the concert of his actions, that maintain society: it is rendered miserable by his discordance; it is dissolved by his want of union.

From what has been said it may be concluded, that the names under which man has designated the concealed causes acting in nature, and their various effects, are never more than *necessity* considered under different points of view. It will be found, that what he calls *order*, is a necessary consequence of causes and effects, of which he sees, or believes he sees, the entire connexion, the complete routine, and which pleases him as a whole when he finds it conformable to his existence. In like manner it will be seen that what he calls *confusion*, is a consequence of like necessary causes and effects, which he thinks unfavourable to himself, or but little suitable to his being. He has designated by the name of *intelligence*, those necessary causes that necessarily operate the chain of events which he comprises under the term *order*. He has called *divinity*, those necessary but invisible causes which give play to nature, in which every thing acts according to immutable and necessary laws: *destiny* or *fatality*, the necessary connexion of those unknown causes and effects which he beholds in the world: *chance*, those effects which he is not able to foresee, or of which he ignores the necessary connexion with their causes. Finally, *intellectual* and *moral faculties*, those effects and those modifications necessary to an organized being, whom he has supposed to be moved by an inconceivable agent, that he has believed distinguished from his body, of a nature totally different from it, and which he has designated by the word *soul.* In consequence, he has believed this agent immortal, and not dissoluble like the body.

It has been shown that the marvellous doctrine of another life, is founded upon gratuitous suppositions, contradicted by reflection. It has been proved, that the hypothesis is not only useless to man's morals, but again, that it is calculated to palsy his exertions, to divert him from actively pursuing the true road to his own happiness, to fill him with romantic caprices, and to inebriate him with opinions prejudicial to his tranquillity; in short, to lull to slumber the vigilance of legislators, by dispensing them from giving to education, to the institutions, to the laws of society, all that attention which it is the duty and for his interest they should bestow. It must have been felt, that politics has unaccountably rested itself upon opinions little capable of satisfying those passions which every thing conspires to kindle in the heart of man, who ceases to view the future, while the present seduces and hurries him along. It has been shown, that contempt of death is an advantageous sentiment, calculated to inspire man's mind with courage to undertake that which may be truly useful to society. In short, from what has preceded, it will be obvious what is competent to conduct man to happiness, and also what are the obstacles that errour opposes to his felicity.

Let us not then be accused of demolishing without rebuilding, with combating errour without substituting truth, with sapping at one and the same time the foundations of religion and of sound morals. The last is necessary to man; it is founded upon his nature; its duties are certain, they must last as long as the human race remains; it imposes obligations on him, because, without it, neither individuals nor society could be able to subsist, either obtain or enjoy those advantages which nature obliges them to desire.

Listen then, O man! to those morals which are established upon experience and upon the necessity of things; do not lend thine ear to those superstitions founded upon revorion, imposture, and the capricious whims of a disordered imagination. Follow the lessons of those humane and gentle morals, which conduct man to virtue by the path of happiness: turn a deaf ear to the inefficacious cries of religion which renders man really unhappy; which can never make him reverence virtue, which it paints in hideous and hateful colours; in short, let him see if reason, without the assistance of a rival who prohibits its use, will not more surely conduct him towards that great end which is the object and tendency of all his views.

Indeed, what benefit has the human race hitherto drawn from those sublime and supernatural notions, with which theology has fed mortals during so many ages? All those phantoms conjured up by ignorance and imagination; all those hypotheses, as subtile as they are irrational, from which experience is banished; all those words devoid of meaning with which languages are crowded; all those fantastical hopes and panic terrours, which have been brought to operate on the will of man, have they rendered man better, more enlightened to his duties, more faithful in their performance? Have those marvellous systems, or those sophistical inventions by which they have been supported, carried conviction to his mind, reason into his conduct, virtue into his heart? Alas! all these things have done nothing more than plunge the human understanding into that darkness, from which it is difficult to be withdrawn; sown in man's heart the most dangerous errours, of which it is scarcely possible to divest him; given birth to those fatal passions, in which may be found the true source of those evils with which his species is afflicted.

Cease then, O mortal! to let thyself be disturbed with phantoms, which thine own imagination or imposture hath created. Renounce thy vague hopes; disengage thyself from thine overwhelming fears, follow without inquietude the necessary routine which nature has marked out for thee; strew the road with flowers if thy destiny permits; remove, if thou art able, the thorns scattered over it. Do not attempt to plunge thy views into an impenetrable futurity; its obscurity ought to be sufficient to prove to thee that it is either useless or dangerous to fathom. Only think then, of making thyself happy in that existence which is known to thee. If thou wouldst preserve thyself, be temperate, moderate, and reasonable: if thou seekest to render thy existence durable, be not prodigal of pleasure. Abstain from every thing that can be hurtful to thyself, or to others. Be truly intelligent; that is to say, learn to esteem thyself, to preserve thy being, to fulfil that end which at each moment thou proposest to thyself. Be virtuous, to the end that thou mayest render thyself solidly happy, that thou mayest enjoy the affections, secure the esteem, partake of the assistance of those beings whom nature has made necessary to thine own peculiar felicity. Even when they should be unjust, render thyself worthy of thine own love and applause, and thou shalt live content, thy serenity shall not be disturbed : the end of thy career shall not slander a life which will be exempted from remorse. Death will be to thee the door to a new existence, a new order in which thou wilt be submitted, as thou art at present, to the eternal laws of fate, which ordains, that to live happy here below, thou must make others happy.˙ Suffer thyself, then, to be drawn gently along thy journey, until thou shalt sleep peaceably on that bosom which has given thee birth.

For thou, wicked unfortunate! who art found in continual contradiction with thyself; thou whose disorderly machine can neither accord with thine own peculiar nature, nor with that of thine associates; whatever may be thy crimes, whatever may be thy fears of punishment in another life, thou art at least already cruelly punished in this? Do not thy follies, thy shameful habits, thy debaucheries, damage thine health? Dost thou not linger out life in disgust, fatigued with thine own excesses? Does not listlessness, punish thee for thy satiated passions? Has not thy vigour, thy gayety, already yielded to feebleness, to infirmities, and to regret? Do not thy vices every day dig thy grave? Every time thou hast stained thyself with crime, hast thou dared without horrour to return into thyself? Hast thou not found remorse, terrour, shame, established in thine heart? Hast thou not dreaded the scrutiny of thy fellow man? Hast thou not trembled when alone, that truth, so terrible for thee, should unveil thy dark transgressions, throw into light thine enormous iniquities? Do not then any longer fear to part with thine existence, it will at least put an end to those richly merited torments thou hast inflicted on thyself; death, in delivering the earth from an incommodious burden, will also deliver thee from thy most cruel enemy, *thyself.*

CHAPTER XVIII.

The Origin of Man's Ideas upon the Divinity.

IF man possessed the courage to recur to the source of those opinions which are most deeply engraven on his brain; if he rendered to himself a faithful account of the reasons which make him hold these opinions as sacred ; if he coolly examined the basis of his hopes, the foundation of his fears, he would find that it very frequently happens, those objects, or those ideas which move him most powerfully, either have no real existence, are words devoid of meaning, or phantoms engendered by a disordered imagination, modified by ignorance. Distracted by contending passions, which prevent him from either reasoning justly, or consulting experience in his judgment, his intellectual faculties are thrown into confusion, his ideas bewildered.

A sensible being placed in a nature where every part is in motion, has various feelings, in consequence of either the agreeable or disagreeable effects which he is obliged to experience ; in consequence he either finds himself happy or miserable ; and, according to the quality of the sensations excited in him, he will love or fear, seek after or fly from, the real or supposed causes of such marked effects operated on his machine. But if he is ignorant or destitute of experience, he will frequently deceive himself as to these causes; and he will neither have a true knowledge of their energy, nor a clear idea of their mode of acting : thus until reiterated experience shall have formed his judgment, he will be involved in trouble and incertitude.

Man is a being who brings with him nothing into the world, save an aptitude to feeling in a manner more or less lively according to his individual organization: he has no knowledge of any of the causes that act upon him: by degrees his faculty of feeling discovers to him their various qualities ; he learns to judge of them; time familiarizes him with their properties; he attaches ideas to them, according to the manner in which they have affected him ; and these ideas are correct or otherwise, in a ratio to the soundness of his organic structure, and in proportion as these organs are competent to afford him sure and reiterated experience.

The first movements of man are marked by his wants; that is to say, the first impulse he receives is to conserve his existence; this he would not be able to maintain without the concurrence of many analogous causes: these wants in a sensible being manifest themselves by a general languor, a sinking, a confusion in his machine, which gives him the consciousness of a painful sensation: this derangement subsists and is augmented, until the cause suitable to remove it re-establishes the harmony so necessary to the existence of the human frame. Want, therefore, is the first evil man experiences; nevertheless it is requisite to the maintenance of his existence.— Was it not for this derangement of his body, which obliges him to furnish its remedy, he would not be warned of the necessity of preserving the existence he has received. Without wants man would be an insensible machine, similar to a vegetable, and like it, he would be incapable of preserving himself or of using the means required to conserve his being. To his wants are to be ascribed his passions, his desires, the exercise of his corporeal and intellectual faculties: they are his wants that oblige him to think, to will, to act; it is to satisfy them, or rather to put an end to the painful sensations excited by their presence, that, according to his capacity, to the energies which are peculiar to himself, he exerts the activity of his bodily strength, or displays the extensive powers of his mind. His wants being perpetual, he is obliged to labour without relaxation to procure objects competent to satisfy them. In a word, it is owing to his multiplied wants that man's energy is kept in a state of continual activity: as soon as he ceases to have wants, he falls into inaction— becomes listless—declines into apathy —sinks into a languor that is incommodious to his feelings or prejudicial to his existence: this lethargic state of weariness lasts until new wants rouse his dormant faculties, and destroy the sluggishness to which he had become a prey.

From hence it will be obvious that *evil* is necessary to man; without it he would neither be in a condition to know that which injures him, to avoid its presence, or to seek his own welfare: he would differ in nothing from insensible, unorganized beings, if those evanescent evils which he calls *wants*, did not oblige him to call forth his faculties, to set his energies in motion, to cull experience, to compare objects, to discriminate them, to separate those which have the capabilities to injure him, from those which possess the means to benefit him. In short, without evil man would be ignorant of good; he would be continually exposed to perish. He would resemble an infant, who, destitute of experience, runs the risk of meeting his destruction at every step he takes: he would be unable to judge of any thing; he would have no preference; his will would be without volition, he would be destitute of passions, of desire: he would not revolt at the most disgusting objects; he would not strive to put them away; he would neither have stimuli to love, nor motives to fear any thing; he would be an insensible automaton—he would no longer be a man.

If no evil had existed in this world, man would never have dreamt of the divinity. If nature had permitted him easily to satisfy all his regenerating wants, if she had given him none but agreeable sensations, his days would have uninterruptedly rolled on in one perpetual uniformity, and he would never have had motives to search after the unknown causes of things. To meditate is pain: therefore man, always contented, would only have occupied himself with satisfying his wants, with enjoying the present, with feeling the influence of objects that would unceasingly warn him of his existence in a mode that he must necessarily approve; nothing would alarm his heart; every thing would be analogous to his existence: he would neither know fear, experience distrust, nor have inquietude for the future: these feelings can only be the consequence of some troublesome sensation, which must have anteriorly affected him, or which, by disturbing the harmony of his machine, has interrupted the course of his happiness.

Independent of those wants which in man renew themselves every instant, and which he frequently finds it impossible to satisfy, every individual experiences a multiplicity of evils; he suffers from the inclemency of the seasons, he pines in penury, he is infected with plague, he is scourged by war, he is the victim of famine, he is afflicted with disease, he is the sport of a thousand accidents, &c. This is the reason why all men are fearful and diffident. The knowledge he has of pain alarms him upon all unknown causes, that is to say, upon all those of which he has not yet experienced the effect; this experience made with precipitation, or if it be preferred, by instinct, places him on his guard against all those objects from the operation of which he is ignorant what consequences may result to himself. His inquietude and his fears keep pace with the extent of the disorder which these objects produce in him; they are measured by their rarity, that is to say, by the inexperience he has of them; by his natural sensibility, and by the ardour of his imagination. The more ignorant man is, the less experience he has, the more he is susceptible of fear; solitude, the obscurity of a forest, silence, and the darkness of night, the roaring of the wind, sudden, confused noises, are objects of terrour to all who are unaccustomed to these things. The uninformed man is a child whom every thing astonishes; but his alarms disappear, or diminish, in proportion as experience familiarizes him, more or less, with natural effects; his fears cease entirely, as soon as he understands, or believes he understands, the causes that act, and when he knows how to avoid their effects. But if he cannot penetrate the causes which disturb him, or by whom he suffers, if he cannot find to what account to place the confusion he experiences, his inquietude augments; his fears redouble; his imagination leads him astray; it exaggerates his evil; paints in a disorderly manner these unknown objects of his terrour; then making an analogy between them and those terrific objects with whom he is already acquainted, he suggests to himself the means he usually takes to mitigate their anger; he employs similar measures to soften the anger and to disarm the power of the concealed cause which gives birth to his inquietudes, and alarms his fears. It is thus his weak-

ness, aided by ignorance, renders him superstitious.

There are very few men, even in our own day, who have sufficiently studied nature, who are fully apprised of physical causes, or with the effects they must necessarily produce. This ignorance, without doubt, was much greater in the more remote ages of the world, when the human mind, yet in its infancy, had not collected that experience, and made those strides towards improvement, which distinguishes the present from the past. Savages dispersed, knew the course of nature either very imperfectly or not at all; society alone perfects human knowledge: it requires not only multiplied but combined efforts to unravel the secrets of nature. This granted, all natural causes were mysteries to our wandering ancestors; the entire of nature was an enigma to them; all its phenomena were marvellous, every event inspired terrour to beings who were destitute of experience; almost every thing they saw must have appeared to them strange, unusual, contrary to their idea of the order of things.

It cannot then furnish matter for surprise, if we behold men in the present day trembling at the sight of those objects which have formerly filled their fathers with dismay. Eclipses, comets, meteors, were in ancient days, subjects of alarm to all the people of the earth: these effects so natural in the eyes of the sound philosopher, who has by degrees fathomed their true causes, have yet the right to alarm the most numerous and the least instructed part of modern nations. The people of the present day, as well as their ignorant ancestors, find something marvellous and supernatural in all those objects to which their eyes are unaccustomed, or in all those unknown causes that act with a force of which their mind has no idea it is possible the known agents are capable. The ignorant see wonders, prodigies, miracles, in all those striking effects of which they are unable to render themselves a satisfactory account; all the causes which produce them they think *supernatural*; this, however, really implies nothing more than that they are not familiar to them, or that they have not hitherto witnessed natural agents whose energy was equal to the production of effects so astonishing as those with which their sight has been appalled.

Besides the ordinary phenomena to which nations were witnesses without being competent to unravel the causes, they have, in times very remote from ours, experienced calamities, whether general or local, which filled them with the most cruel inquietude, and plunged them into an abyss of consternation. The traditions and annals of all nations, recall, even at this day, melancholy events, physical disasters, dreadful catastrophes, which had the effect of spreading universal terrour among our forefathers. But when history should be silent on these stupendous revolutions, would not our own reflection on what passes under our eyes be sufficient to convince us, that all parts of our globe have been, and following the course of things, will necessarily be again, violently agitated, overturned, changed, overflowed, in a state of conflagration? Vast continents have been inundated: seas breaking their limits have usurped the dominion of the earth; at length, retiring, these waters have left striking proofs of their presence, by the marine vestiges of shells, skeletons of sea-fish, &c. which the attentive observer meets with at every step in the bowels of those fertile countries we now inhabit. Subterraneous fires have opened to themselves the most frightful volcanoes, whose craters frequently issue destruction on every side. In short, the elements unloosed, have, at various times, disputed among themselves the empire of our globe; this exhibits evidence of the fact, by those vast heaps of wreck, those stupendous ruins spread over its surface. What, then, must have been the fears of mankind, who in those countries believed he beheld the entire of nature armed against his peace, and menacing with destruction his very abode? What must have been the inquietude of a people taken thus unprovided, who fancied they saw nature cruelly labouring to their annihilation? Who beheld a world ready to be dashed into atoms, the earth suddenly rent asunder, whose yawning chasm was the grave of large cities, whole provinces, entire nations? What ideas must mortals, thus overwhelmed with

terrour, form to themselves of the irresistible cause that could produce such extended effects? Without doubt they did not attribute these wide-spreading calamities to nature; they could not suspect she was the author, the accomplice of the confusion she herself experienced; they did not see that these tremendous revolutions, these overpowering disorders, were the necessary result of her immutable laws, and that they contributed to the general order by which she subsists.*

It was under these astounding circumstances, that nations, not seeing on this mundane ball causes sufficiently powerful to operate the gigantic phenomena that filled their minds with dismay, carried their streaming and tremulous eyes towards heaven, where they supposed these unknown agents, whose unprovoked enmity destroyed their earthly felicity, could alone reside.

It was in the lap of ignorance, in the season of alarm and calamity, that mankind ever formed his first notions of the Divinity. From hence it is obvious that his ideas on this subject are to be suspected as false, and that they are always afflicting. Indeed, upon whatever part of our sphere we cast our eyes, whether it be upon the frozen climates of the north, upon the parching regions of the south, or under the more temperate zones, we every where behold the people when assailed by misfortunes, have either made to themselves national Gods, or else have adopted those which have been given them by their conquerors; before these beings, either of their own creation or adoption, they have tremblingly prostrated themselves in the hour of calamity. Thus Idea of these powerful agents, was always associated with that of terrour; their name was never pronounced without recalling to man's mind either his own particular calamities or those of his fathers: man trembles at this day, because his progenitors have trembled

* In point of fact, there is nothing more surprising in the inundation of large portions of the earth, in the swallowing up an entire nation, in a volcanic conflagration, spreading destruction over whole provinces, than there is in a stone falling to the earth, or the death of a fly: each equally has its spring in the necessity of things.

thousands of years ago. The thought of Gods always awakens in man the most afflicting ideas: if he recurred to the source of his actual fears, to the commencement of those melancholy impressions that stamp themselves in his mind when his name is pronounced, he would find it in the deluges, in the revolutions, in those extended disasters, that have at various times destroyed large portions of the human race, and overwhelmed with dismay those miserable beings who escaped the destruction of the earth; these, in transmitting to posterity the tradition of such afflicting events, have also transmitted to him their fears, and those gloomy ideas which their bewildered imaginations, coupled with their barbarous ignorance of natural causes, had formed to them of the anger of their irritated Gods, to which their alarm falsely attributed these disasters.*

If the Gods of nations had their birth in the bosom of alarm, it was again in that of despair that each individual formed the unknown power that he made exclusively for himself. Ignorant of physical causes, unpractised in their mode of action, unaccustomed to their effects, whenever he experienced any serious misfortune, or any grievous sensation, he was at a loss how to account for it. The motion which in despite of himself was excited in his machine, his diseases, his troubles, his passions, his inquietude, the painful alterations his frame underwent without his being able to fathom the true causes, at length death, of which the aspect is so formidable to a being strongly

* An English author has very correctly remarked that the universal deluge has been perhaps no less fatal to the moral than to the physical world, the human brain retaining to this day an impression of the shock it then received. See *Philemon and Hydaspis*, p. 355.

It is not at all probable that the deluge mentioned in the sacred books of the Jews and Christians, was universal; but there is every reason to believe that all parts of the earth have at different times been inundated. This is proved by the uniform tradition of every nation in the world, and also by the remains of marine bodies found in every country, imbedded to greater or less depths.' Yet it might be possible that a comet coming in contact with our globe, should have produced such a shock as to submerge at once whole continents! for this a miracle was not necessary!

attached to existence, were effects he looked upon as either supernatural, or else he conceived they were repugnant to his actual nature; he attributed them to some mighty cause, which maugre all his efforts, disposed of him at each moment. His imagination, thus rendered desperate by his endurance of evils which he found inevitable, formed to him those phantoms before whom he trembled from a consciousness of his own weakness. It was then he endeavoured by prostration, by sacrifices, by prayers, to disarm the anger of these imaginary beings to which his trepidation had given birth; whom he ignorantly imagined to be the cause of his misery, whom his fancy painted to him as endowed with the power of alleviating his sufferings: it was then, in the extremity of his grief, in the exarcerbation of his mind, weighed down with misfortune, that unhappy man fashioned the phantom God.

Man never judges of those objects of which he is ignorant but through the medium of those which come within his knowledge: thus man, taking himself for the model, ascribed will, intelligence, design, projects, passions; in a word, qualities analogous to his own, to all those unknown causes of which he experienced the action. As soon as a visible or supposed cause affects him in an agreeable manner, or in a mode favourable to his existence, he concludes it to be good, to be well intentioned towards him: on the contrary, he judges all those to be bad in their nature, and to have the intention of injuring him, which cause him many painful sensations. He attributes views, plans, a system of conduct like his own, to every thing which to his limited ideas appears of itself to produce connected effects, to act with regularity, to constantly operate in the same manner, that uniformly produces the same sensations in his own person. According to these notions, which he always borrows from himself, from his own peculiar mode of action, he either loves or fears those objects which have affected him: he in consequence approaches them with confidence or timidity; seeks after them or flies from them in proportion as the feelings they have excited are either pleasant or painful. He presently addresses them; he invokes their aid; prays to them for succour; conjures them to cease his afflictions; to forbear tormenting him; as he finds himself sensible to presents, pleased with submission, he tries to win them to his interests by humiliation, by sacrifices; he exercises towards them the hospitality he himself loves; he gives them an asylum; he builds them a dwelling; he furnishes them with all those things which he thinks will please them the most, because he himself places the highest value on them. These dispositions enable us to account for the for mation of tutelary Gods, which every man makes to himself in savage and unpolished nations. Thus we perceive that weak mortals, regard as the arbiters of their fate, as the dispensers of good and evil, animals, stones, unformed inanimate substances, which they transform into Gods, whom they invest with intelligence, whom they clothe with desires, and to whom they give volition.

Another disposition which serves to deceive the savage man, which will equally deceive those whom reason shall not enlighten on these subjects, is the fortuitous concurrence of certain effects, with causes which have not produced them, or the co-existence of these effects with certain causes which have not the slighest connexion with them. Thus the savage attributes bounty or the will to render him service, to any object whether animate or inanimate, such as a stone of a certain form, a rock, a mountain, a tree, a serpent, an owl, &c., if every time he encounters these objects in a certain position, it should so happen that he is more than ordinarily successful in hunting, that he should take an unusual quantity of fish, that he should be victorious in war, or that he should compass any enterprise whatever, that he may at that moment undertake.— The same savage will be quite as gratuitous in attaching malice or wickedness to either the same object in a different position, or any others in a given posture, which may have met his eyes on those days when he shall have suffered some grievous accident: incapable of reasoning he connects these effects with causes that are entirely

due to physical causes, to necessary circumstances, over which neither himself nor his omens have the least controul: nevertheless, he finds it much easier to attribute them to these imaginary causes, he therefore *deifies* them, endows them with passions, gives them design, intelligence, will, and invests them with supernatural powers. The savage in this is never more than an infant that is angry with the object that displeases him, just like the dog who gnaws the stone by which he has been wounded, without recurring to the hand by which it was thrown.

Such is the foundation of man's faith in either happy or unhappy omens: devoid of experience, he looks upon them as warnings given him by his ridiculous Gods, to whom he attributes the faculties of sagacity and foresight, of which he is himself deficient. Ignorance, when involved in disaster, when immersed in trouble, believes a stone, a reptile, a bird, much better instructed than himself. The slender observation of the ignorant only serves to render him more superstitious; he sees certain birds announce by their flight, by their cries, certain changes in the weather, such as cold, heat, rain, storms; he beholds at certain periods vapours arise from the bottom of some particular caverns; there needs nothing further to impress upon him the belief, that these beings possess the knowledge of future events and enjoy the gifts of prophecy.

If by degrees experience and reflection arrive at undeceiving him with respect to the power, the intelligence, the virtues, actually residing in these objects; if he at least supposes them put in activity by some secret, some hidden cause, whose instruments they are, to this concealed agent he addresses himself; pays him his vows; implores his assistance; deprecates his wrath; seeks to propitiate him to his interests; is willing to soften his anger; and for this purpose he employs the same means of which he avails himself either to appease or gain over the beings of his own species. Societies in their origin, seeing themselves frequently afflicted by nature, supposed that either the elements, or the concealed powers who regulated them, possessed a will, views, wants, desires, similar to their own. From hence, the sacrifices imagined to nourish them; the libations poured out to them; the steams, the incense to gratify their olfactery nerves. They believed these elements or their irritated movers were to be appeased like irritated man, by prayers, by humiliation, by presents. Their imagination was ransacked to discover the presents that would be most acceptable to these mute beings who did not make known their inclinations. Thus some brought the fruits of the earth, others offered sheaves of corn; some strewed flowers over their fanes; some decorated them with the most costly jewels; some served them with meats; others sacrificed lambs, heifers, bulls. As they appeared to be almost always irritated against man, they stained their altars with human gore, and made oblations of young children. At length, such was their delirium, such the wildness of their imaginations, that they believed it impossible to appease with oblations from the earth the supposed agents of nature, who therefore required the sacrifice of a God! It was presumed that an infinite being could not be reconciled to the human race but by an infinite victim.

The old men as having the most experience, were usually charged with the conduct of these peace-offerings.* These accompanied them with ceremonies, instituted rites, used precautions, adopted formalities, retraced to their fellow citizens the notions transmitted to them by their forefathers; collected the observations made by their ancestors; repeated the fables they had received. It is thus the sacerdotal order was established; thus that public worship was established;

* The Greek word Πρισβυς, from whence is derived the name *priest*, signifies an old man. Men have always felt respect for that which bore the character of antiquity, as they have always associated with it the idea of wisdom and consummate experience. It is probably in consequence of this prejudice that men, when in doubt, generally prefer the authority of antiquity and the decisions of their ancestors to those of good sense and reason. This we see every day in matters appertaining to religion, which is supposed to have been pure and undefiled in its infancy, although this idea is certainly without foundation.

No. VI.—22

by degrees each community formed a body of tenets to be observed by the citizens; these were transmitted from race to race.† Such were the unformed, the precarious elements of which rude nations every where availed themselves to compose their religions: they were always a system of conduct invented by imagination, conceived in ignorance, to render the unknown powers, to whom they believed nature was submitted, favourable to their views. Thus some irascible, at the same time placable being, was always chosen for the basis of the adopted religion; it was upon these puerile tenets, upon these absurd notions, that the priests founded their rights; established their authority: erected temples, raised altars, loaded them with wealth, rested their dogmas. In short, it was from such rude foundations that arose the structure of all religions; under which man trembled for thousands of years: and although these religions were originally invented by savages, they still have the power of regulating the fate of the most civilized nations. These systems, so ruinous in their principles, have been variously modified by the human mind, of which it is the essence to labour incessantly on unknown objects; it always commences by attaching to these a very first-rate importance, which it afterwards never dares coolly to examine.

† At length it was deemed sacrilege even to doubt these pandects in any one particular; he that ventured to reason upon them, was looked upon as an enemy to the commonwealth; as one whose impiety drew down upon them the vengeance of these adored beings, to which alone imagination had given birth. Not contented with adopting rituals, with following the ceremonies invented by themselves, one community waged war against another, to oblige it to receive their particular creeds; which the knaves who regulated them, declared would infallibly win them the favour of their tutelary Deities: thus very often to conciliate their favour, the victorious party immolated on the altars of their Gods, the bodies of their unhappy captives; and frequently they carried their savage barbarity the length of exterminating whole nations, who happened to worship Gods different from their own: thus it frequently happened, that the friends of the serpent, when victorious, covered his altars with the mangled carcasses of the worshippers of the stone whom the fortune of war had placed in their hands.

Such was the fate of man's imagination in the successive ideas which he either formed to himself, or which he received upon the divinity. The first theology of man was grounded on fear, modelled by ignorance : either afflicted or benefited by the elements, he adored these elements themselves, and extended his reverence to every material, coarse object; he afterwards rendered his homage to the agents he supposed presiding over these elements ; to powerful genii ; to inferior genii ; to heroes, or to men endowed with great qualities. By dint of reflection, he believed he simplified the thing in submitting the entire of nature to a single agent—to a sovereign intelligence—to a spirit—to a universal soul, which put this nature and its parts in motion. In recurring from cause to cause, man finished by losing sight of every thing, and in this obscurity, in this dark abyss, he placed his God, and formed new chimeras which will afflict him until a knowledge of natural causes undeceives him with regard to those phantoms he had always so stupidly adored.

If a faithful account was rendered of man's ideas upon the Divinity, he would be obliged to acknowledge, that the word *God* has only been used to express the concealed, remote, unknown causes of the effects he witnessed; he uses this term only when the spring of natural and known causes ceases to be visible: as soon as he loses the thread of these causes, or as soon as his mind can no longer follow the chain, he solves the difficulty, terminates his research, by ascribing it to God ; thus giving a vague definition to an unknown cause, at which either his idleness, or his limited knowledge, obliges him to stop. When, therefore, he ascribes to God the production of some phenomenon, of which his ignorance precludes him from unravelling the true cause, does he, in fact, do any thing more than substitute for the darkness of his own mind, a sound to which he has been accustomed to listen with reverential awe ? Ignorance may be said to be the inheritance of the generality of men ; these attribute to the Divinity not only those uncommon effects that burst upon their senses with an astounding force, but also the most simple events the causes of

which are the most easy to be known to whoever shall be willing to meditate upon them.* In short, man has always respected those unknown causes, those surprising effects which his ignorance prevented him from fathoming.

It remains, then, to inquire, if man can reasonably flatter himself with obtaining a perfect knowledge of the power of nature ;† of the properties of

* If there be a God, can it be possible we are acting rationally, eternally to make him the agent of our stupidity, of our sloth, of our want of information on natural causes ? Do we, in fact, pay any kind of adoration to this being, by thus bringing him forth on every trifling occasion, to solve the difficulties ignorance throws in our way ? Of whatever nature the *Cause of causes* may be, it is evident to the slightest reflection that it has been sedulous to conceal itself from our view; that it has rendered it impossible for us to have the least acquaintance with it, except through the medium of nature, which is unquestionably competent to every thing: this is the rich banquet spread before man ; he is invited to partake, with a welcome he has no right to dispute ;. to enjoy therefore is to obey; *to be happy himself is to make others happy ; to make others happy is to be virtuous ; to be virtuous he must revere truth : to know what truth is, he must examine with caution, scrutinize with severity, every opinion he adopts ;* this granted, is it not insulting to a God to clothe him with our wayward passions ; to ascribe to him designs similar to our narrow view of things ; to give him our filthy desires; to suppose he can be guided by our finite conceptions; to bring him on a level with frail humanity, by investing him with our qualities, however much we may exaggerate them; to indulge an opinion that he can either act or think as we do; to imagine he can in any manner resemble such a feeble plaything, as is the greatest, the most distinguished man ? No! it is to fall back into the depth of Cimmerian darkness. Let man therefore sit down cheerfully to the feast; let him contentedly partake of what he finds; but let him not worry his way to a God with his useless prayers: these supplications are, in fact, at once to say, that with our limited experience, with our slender knowledge, we better understand what is suitable to our condition, what is convenient to our welfare, than the *Cause of all causes* who has left us in the hands of nature.

† How many discoveries in the great science of natural philosophy has mankind progressively made, which the ignorant prejudices of our forefathers on their first announcement considered as impious, as displeasing to the Divinity, as heretical profanations, which could only be expiated by the sacrifice of the inquiring individuals, to whose labour their posterity owes such an infinity of gratitude. Even in modern days we have seen a Socrates de-

the beings she contains; of the effects which may result from their various combinations? Do we know why the magnet attracts iron? Are we better acquainted with the cause of polar attraction? Are we in a condition to explain the phenomena of light, electricity, elasticity? Do we understand the mechanism by which that modification of our brain, which we call volition, puts our arm or our legs into motion? Can we render to ourselves an account of the manner in which our eyes behold objects, in which our ears receive sounds, in which our mind conceives ideas? If then we are incapable of accounting for the most ordinary phenomena, which nature daily exhibits to us, by what chain of reasoning do we refuse to her the power of producing other effects equally incomprehensible to us? Shall we be more instructed, when every time we behold an effect of which we are not in a capacity to develop the cause, we may idly say, this effect is produced by the power, by the will of God?—that is to say, by an agent of which we have no knowledge whatever, and of which we are more ignorant than of natural causes. Does then, a sound, to which we can-

stroyed, a Galileo condemned, whilst multitudes of other benefactors to mankind have been held in contempt by their uninformed contemporaries, for those very researches into nature which the present generation hold in the highest veneration. *Whenever ignorant priests are permitted to guide the opinions of nations, science can make but a very slender progress:* natural discoveries will be always held inimical to the interest of bigoted religious men. It may, to the minds of infatuated mortals, to the shallow comprehension of prejudiced beings, appear very pious to reply on every occasion, our God do this, our God do that; but to the contemplative philosopher, to the man of reason, it will never be convincing that a sound, a mere word, can attach the reason of things; can have more than a fixed sense; can suffice to explain problems. The word God is used to denote the impenetrable cause of those effects which astonish mankind; which man is not competent to explain. But is not this wilful idleness? Is it not inconsistent with our nature thus to give the answer of a child to every thing we do not understand; or rather which our own sloth, or our own want of industry has prevented us from knowing? Ought we not rather to redouble our efforts to penetrate the cause of those phenomena which strike our mind? When we have given this answer, what have we said? Nothing but what every one knows.

not attach any fixed sense, suffice to explain problems? Can the word God signify any thing else but the impenetrable cause of those effects which we cannot explain?

When we shall be ingenuous with ourselves, we shall be obliged to agree that it was uniformly the ignorance in which our ancestors were involved, their want of knowledge of natural causes, their unenlightened ideas on the powers of nature, which gave birth to the Gods; that it is, again, the impossibility which the greater part of mankind find to withdraw themselves out of this ignorance, the difficulty they consequently find to form to themselves simple ideas of the formation of things, the labour that is required to discover the true sources of those events which they either admire or fear, that make them believe the idea of a God is necessary to enable them to render an account of those phenomena, the true cause of which they cannot discover. Here, without doubt, is the reason they treat all those as irrational who do not see the necessity of admitting an unknown agent, or some secret energy, which, for want of being acquainted with nature, they have placed out of herself.

The phenomena of nature necessarily breed various sentiments in man: some he thinks favourable to him, some prejudicial; some excite his love, his admiration, his gratitude; others fill him with trouble, cause aversion, drive him to despair. According to the various sensations he experiences, he either loves or fears the causes to which he attributes the effects which produce in him these different passions: these sentiments are commensurate with the effects he experiences; his admiration is enhanced, his fears are augmented, in the same ratio as the phenomena which strike his senses are more or less extensive, more or less irresistible or interesting to him. Man necessarily makes himself the centre of nature; indeed he can only judge of things, as he is himself affected by them; he can only love that which he thinks favourable to his being; he hates, he fears every thing which causes him to suffer: in short, as we have seen, he calls confusion every thing that deranges the economy of his machine, and he believes all is in order, as soon as he experiences nothing but what is suitable to his peculiar mode of existence. By a necessary consequence of these ideas, man firmly believes that the entire of nature was made for him alone; that it was only himself which she had in view in all her works; or rather that the powerful causes to which this nature was subordinate, had only for object man and his convenience, in all the effects which are produced in the universe.

If there existed on this earth other thinking beings besides man, they would fall exactly into similar prejudices with himself; it is a sentiment founded upon that predilection which each individual necessarily has for himself; a predilection that will subsist until reason, aided by experience, shall have rectified his errours.

Thus, whenever man is contented, whenever every thing is in order with respect to himself, he either admires or loves the cause to which he believes he is indebted for his welfare; when he becomes discontented with his mode of existence, he either fears or hates the cause which he supposes has produced these afflicting effects. But his welfare confounds itself with his existence; it ceases to make itself felt when it has become habitual and of long continuance; he then thinks it is inherent to his essence; he concludes from it that he is formed to be always happy; he finds it natural that every thing should concur to the maintenance of his being. It is by no means the same when he experiences a mode of existence that is displeasing to himself: the man who suffers is quite astonished at the change which has taken place in his machine; he judges it to be contrary to nature, because it is incommodious to his own particular nature; he imagines those events by which he is wounded, to be contrary to the order of things; he believes that nature is deranged every time she does not procure for him that mode of feeling which is suitable to his ideas; and he concludes from these suppositions that nature, or the agent who moves her, is irritated against him.

It is thus that man, almost insensible to good, feels evil in a very lively manner; the first he believes natural,

the other he thinks opposed to nature. He is either ignorant, or forgets, that he constitutes part of a whole, formed by the assemblage of substances, of which some are analogous, others heterogeneous; that the various beings of which nature is composed, are endowed with a variety of properties, by virtue of which they act diversely on the bodies who find themselves within the sphere of their action; he does not perceive that these beings, destitute of goodness, devoid of malice, act only according to their respective essences and the laws their properties impose upon them, without being in a capacity to act otherwise than they do. It is, therefore, for want of being acquainted with these things, that he looks upon the author of nature, as the cause of those evils to which he is submitted, that he judges him to be wicked or exasperated against him.

The fact is, man believes that his welfare is a debt due to him from nature; that when he suffers evil she does him an injustice; fully persuaded that this nature was made solely for himself, he cannot conceive she would make him suffer, if she was not moved thereto by a power who is inimical to his happiness—who has reasons for afflicting and punishing him. From hence it will be obvious, that evil, much more than good, is the true motive of those researches which man has made concerning the Divinity—of those ideas which he has formed of himself—of the conduct he has held towards him. The admiration of the works of nature, or the acknowledgment of its goodness, would never alone have determined the human species to recur painfully by thought to the source of these things; familiarized at once with all those effects which are favourable to his existence, he does not by any means give himself the same trouble to seek the causes, that he does to discover those which disquiet him, or by which he is afflicted. Thus, in reflecting upon the Divinity, it was always upon the cause of his evils that man meditated; his meditations were fruitless, because the evils he experiences, as well as the good he partakes, are equally necessary effects of natural causes, to which his mind ought rather to have bent its force, than to have invented fictitious causes of which he never could form to himself any but false ideas, seeing that he always borrowed them from his own peculiar manner of existing, and feeling. Obstinately refusing to see any thing but himself, he never became acquainted with that universal nature of which he constitutes such a very feeble part.

The slightest reflection, however, would have been sufficient to undeceive him on these erroneous ideas. Every thing tends to prove that good and evil are modes of existence that depend upon causes by which a man is moved, and that a sensible being is obliged to experience them. In a nature composed of a multitude of beings infinitely varied, the shock occasioned by the collision of discordant matter must necessarily disturb the order, derange the mode of existence of those beings who have analogy with them : these act in every thing they do after certain laws; the good or evil, therefore, which man experiences, are necessary consequences of the qualities inherent to the beings, within whose sphere of action he is found. Our birth, which we call a benefit, is an effect as necessary as our death, which we contemplate as an injustice of fate : it is of the nature of all analogous beings to unite themselves to form a whole: it is of the nature of all compound beings to be destroyed, or to dissolve themselves; some maintain their union for a longer period than others, and some disperse very quickly. Every being in dissolving itself gives birth to new beings; these are destroyed in their turn, to execute eternally the immutable laws of a nature that only exists by the continual changes that all its parts undergo. Thus nature cannot be accused of either goodness or malice, since every thing that takes place in it is necessary—is produced by an invariable system, to which every other being, as well as herself, is eternally subjected. The same igneous matter that in man is the principle of life, frequently becomes the principle of his destruction, either by the conflagration of a city, or the explosion of a volcano. The aqueous fluid that circulates through his machine, so essentially necessary to his actual existence, frequently becomes too abundant, and ter-

minates him by suffocation, is the cause of those inundations which sometimes swallow up both the earth and its inhabitants. The air, without which he is not able to respire, is the cause of those hurricanes, of those tempests, which frequently render useless the labour of mortals. These elements are obliged to burst their bonds, when they are combined in a certain manner, and their necessary consequences are those ravages, those contagions, those famines, those diseases, those various scourges, against which man, with streaming eyes and violent emotions, vainly implores the aid of those powers who are deaf to his cries : his prayers are never granted but when the same necessity which afflicted him, the same immutable laws which overwhelmed him with trouble, replaces things in the order he finds suitable to his species : a relative order of things which was, is, and always will be, the only standard of his judgment.

Man, however, made no such simple reflections; he did not perceive that every thing in nature acted by invariable laws; he continued in contemplating the good of which he was partaker as a favour, and the evil he experienced, as a sign of anger in this nature, which he supposed to be animated by the same passions as himself; or at least that it was governed by a secret agent who obliged it to execute their will, that was sometimes favourable, sometimes inimical to the human species. It was to this supposed agent, with whom in the sunshine of his prosperity he was but little occupied, in the bosom of his calamity he addressed his prayers; he thanked him, however, for his favours, fearing lest his ingratitude might further provoke his fury : thus when assailed by disaster, when afflicted with disease, he invoked him with fervour: he required him to change in his favour the mode of acting which was the very essence of beings; he was willing that to make the slightest evil that he experienced cease, that the eternal chain of things might be broken or arrested.

It was upon such ridiculous pretensions, that were founded those fervent prayers, which mortals, almost always discontented with their fate, and never in accord in their respective desires, addressed to the Divinity. They were unceasingly prostrate before the imaginary power whom they judged had the right of commanding nature;—whom they supposed to have sufficient energy to divert her course; and whom they considered to possess the means to make her subservient to his particular views; thus each hoped by presents, by humiliation, to induce him to oblige this nature to satisfy the discordant desires of their race. The sick man, expiring in his bed, asks that the humours accumulated in his body, should in an instant lose those properties which render them injurious to his existence; that, by an act of his puissance, his God should renew or recreate the springs of a machine worn out by infirmities. The cultivator of a low swampy country, makes complaint of the abundance of rain with which the fields are inundated; whilst the inhabitant of the hill, raises his thanks for the favours he receives, and solicits a continuance of that which causes the despair of his neighbour. In this, each is willing to have a God for himself, and asks according to his momentary caprices, to his fluctuating wants, that the invariable essence of things should be continually changed in his favour.

From this it must be obvious, that man every moment asks a *miracle* to be wrought in his support. It is not, therefore, at all surprising that he displayed such ready credulity, that he adopted with such facility the relation of the marvellous deeds which were universally announced to him as the acts of the power, or the effects of the benevolence of the Divinity, and as the most indubitable proof of his empire over nature, in the expectation, that if he could gain them over to his interest, this nature, which he found so sullen, so little disposed to lend herself to his views, would then be controuled in his own favour.*

By a necessary consequence of these ideas, nature was despoiled of all power; she was contemplated only as a passive instrument, who acted at the will, under the influence of the numerous, all-powerful agents to whom she was subordinate. It was thus for want of contemplating nature under her true point of view, that man has mistaken her entirely, that he believed her incapable of producing any thing by herself; that he ascribed the honour of all those productions, whether advantageous or disadvantageous to the human species, to fictitious powers, whom he always clothed with his own peculiar dispositions, only he aggrandized their force. In short it was upon the ruins of nature, that man erected the imaginary colossus of the Divinity.

If the ignorance of nature gave birth to the Gods, the knowledge of nature is calculated to destroy them. As soon as man becomes enlightened, his powers augment, his resources increase in a ratio with his knowledge; the sciences, the protecting arts, industrious application, furnish him assistance; experience encourages his progress, or procures for him the means of resisting the efforts of many causes, which cease to alarm him as soon as he obtains a correct knowledge of them. In a word, his terrours dissipate in proportion as his mind becomes enlightened. Man, when instructed, ceases to be superstitious.

CHAPTER XIX.

Of Mythology, and Theology.

THE elements of nature were, as we have shown, the first divinities of man; he has generally commenced with adoring material beings; each individual, as we have already said, and as may be still seen in savage nations, made to himself a particular God of some physical object, which he supposed to be the cause of those events in which he was himself interested; he never wandered to seek out of visible nature the source either of what happened to himself, or of those phenomena

---

* It was easy to perceive that nature was deaf, or at least that it never interrupted its march; therefore men deemed it their interest to submit the entire of nature to an intelligent agent, whom, reasoning by analogy, they supposed better disposed to listen to them than an insensible nature which they were not able to controul. Now it remains to be shown, whether the selfish interest of man is a proof sufficient of the existence of an agent

endowed with intelligence—whether, because a thing may be very convenient, it follows that it is so!

nomena to which he was a witness. As he every where saw only material effects, he attributed them to causes of the same genus; incapable in his infancy of those profound reveries, of those subtile speculations, which are the result of leisure, he did not imagine any cause distinguished from the objects that met his sight, nor of any essence totally different from every thing he beheld. [*]

The observation of nature was the first study of those who had leisure to meditate: they could not avoid being struck with the phenomena of the visible world. The rising and setting of the sun, the periodical return of the seasons, the variations of the atmosphere, the fertility and sterility of the earth, the advantages of irrigation, the damages caused by floods, the useful effects of fire, the terrible consequences of conflagration, were proper and suitable objects to occupy their thoughts. It was natural for them to believe that those beings they saw move of themselves, acted by their own peculiar energies; according to their influence over the inhabitants of the earth was either favourable or otherwise, they concluded them to have either the power to injure them, or the disposition to confer benefits. Those who first acquired the knowledge of gaining the ascendency over man, then savage, wandering, unpolished, or dispersed in woods, with but little attachment to the soil, of which he had not yet learned to reap the advantage, were always more practised observers—individuals more instructed in the ways of nature, than the people, or rather the scattered hordes, whom they found ignorant and destitute of experience. Their superior knowledge placed them in a capacity to render them services—to discover to them useful inventions, which attracted the confidence of the unhappy beings to whom they came to offer an assisting hand; savages who were naked, half famished, exposed to the injuries of the weather, and to the attacks of ferocious beasts, dispersed in caverns, scattered in forests, occupied with hunting, painfully labouring to procure themselves a very precarious subsistence, had not sufficient leisure to make discoveries calculated to facilitate their labour, or to render it less in-

cessant. These discoveries are generally the fruit of society: isolated beings, detached families, hardly ever make any discoveries—scarcely ever think of making any. The savage is a being who lives in a perpetual state of infancy, who never reaches maturity unless some one comes to draw him out of his misery. At first repulsive, unsociable, intractable, he by degrees familiarizes himself with those who render him service; once gained by their kindness, he readily lends them his confidence; in the end he goes the length of sacrificing to them his liberty.

It was commonly from the bosom of civilized nations that have issued those personages who have carried sociability, agriculture, arts, laws, Gods, religious opinions, forms of worship, to those families or hordes as yet scattered, who were not formed into nations. These softened their manners—gathered them together—taught them to reap the advantages of their own powers—to render each other reciprocal assistance—to satisfy their wants with greater facility. In thus rendering their existence more comfortable, they attracted their love, obtained their veneration, acquired the right of prescribing opinions to them, made them adopt such as they had either invented themselves, or else drawn up in the civilized countries from whence they came. History points out to us the most famous legislators as men, who, enriched with useful knowledge they had gleaned in the bosom of polished nations, carried to savages without industry and needing assistance, those arts, of which, until then, these rude people were ignorant: such were the Bacchus's, the Orpheus's, the Triptolemus's, the Moses's, the Numas, the Zamolixis's; in short, all those who first gave to nations their Gods—their worship—the rudiments of agriculture, of science, of theology, of jurisprudence, of mysteries, &c. It will perhaps be inquired, if those nations which at the present day we see assembled, were all originally dispersed? We reply, that this dispersion may have been produced at various times, by those terrible revolutions, of which it has before been remarked our globe has more than once been the theatre, in times so remote that history has not been able to transmit to us the detail. Perhaps the approach of more than one comet may have produced on our earth several universal ravages, which have at each time annihilated the greater portion of the human species. Those who were able to escape from the ruin of the world, filled with consternation, plunged in misery, were but little conditioned to preserve to their posterity a knowledge, effaced by those misfortunes of which they had been both the victims and the witnesses: overwhelmed with dismay, trembling with fear, they were not able to hand down the history of their frightful adventures, except by obscure traditions; much less to transmit to us the opinions, the systems, the arts, the sciences, anterior to these revolutions of our sphere. There have been perhaps men upon the earth from all eternity; but at different periods they may have been nearly annihilated, together with their monuments, their sciences, and their arts; those who outlived these periodical revolutions, each time formed a new race of men, who by dint of time, labour, and experience, have by degrees withdrawn from oblivion the inventions of the primitive races. It is, perhaps, to these periodical revolutions of the human species, that is to be ascribed the profound ignorance in which we see man plunged upon those objects that are the most interesting to him. This is, perhaps, the true source of the imperfection of his knowledge—of the vices of his political and religious institutions over which terrour has always presided; here, in all probability, is the cause of that puerile inexperience, of those jejune prejudices, which every where keep man in a state of infancy, and which render him so little capable of either listening to reason or of consulting truth. To judge by the slowness of his progress, by the feebleness of his advance, in a number of respects, we should be inclined to say, the human race has either just quitted its cradle, or that he was never destined to attain the age of virility or of reason.*

However it may be with these conjectures, whether the human race may always have existed upon the earth, or whether it may have been a recent proeral deluge, but even a great number since the existence of our planet; this globe itself may have been a new production in nature; it may not always have occupied the place it does at present.—See Ch. VI. Whatever idea may be adopted on this subject, it is very certain that, independent of those exterior causes which are competent to totally change its face, as the impulse of a comet may do, this globe contains within itself a cause adequate to alter it entirely, since, besides the diurnal and sensible motion of the earth, it has one extremely slow, almost imperceptible, by which every thing must eventually be changed in it: this is the motion from whence depends the precession of the equinoctial points, observed by Hipparchus and other mathematicians; by this motion, the earth must at the end of several thousand years change totally: this motion will at length cause the ocean to occupy that space which at present forms the lands or continents. From this it will be obvious that our globe, as well as all the beings in nature, has a continual disposition to change. This motion was known to the ancients, and was what gave rise to what they called their great year, which the Egyptians fixed at thirty-six thousand, five hundred and twenty-five years: the Sabines at thirty-six thousand, four hundred and twenty-five, whilst others have extended it to one hundred thousand, some to even seven hundred and fifty-three thousand years.— Again, to those general revolutions which our planet has at different times experienced, may be added those that have been partial, such as inundations of the sea, earthquakes, subterraneous conflagrations, which have sometimes had the effect of dispersing particular nations, and to make them forget all those sciences with which·they were before acquainted. It is also probable that the first volcanic fires, having had no previous vent, were more central, and greater in quantity,. before they burst the crust of earth; as the sea washed the whole, it must have rapidly sunk down into every opening, where, falling on the boiling lava, it was instantly expanded into steam, producing irresistible explosion; whence it is reasonable to conclude, that the primeval earthquakes were more widely extended, and of much greater force, than those which occur in our days. Other vapours may be produced by intense heat, possessing a much greater elasticity, from substances that evaporate, such as mercury, diamonds, &c.; the expansive force of these vapours would be much greater than the steam of water, even at redhot heat; consequently they may have had sufficient energy to raise islands, continents, or even to have detached the moon from the earth; if the moon, as has been supposed by some philosophers, were thrown out of the great cavity which now contains the South Sea; the immense quan-

---

* These hypotheses will unquestionably appear bold to those who have not sufficiently meditated on nature, but to the philosophic inquirer they are by no means inconsistent. There may have not only been one gen-

duction of nature,* it is extremely easy to recur to the origin of many existing nations: we shall find them always in the savage state; that is to say, composed of wandering hordes; these were collected together, at the voice of some missionary or legislator, from whom they received benefits, who gave them Gods, opinions, and laws. These personages, of whom the people, newly congregated, readily acknowledged the superiority, fixed the national Gods, leaving to each individual those which he had formed to himself, according to his own peculiar ideas, or else substituting others brought from those regions from whence they themselves had emigrated.

The better to imprint their lessons on the minds of their new subjects, these men became the guides, the priests, the sovereigns, the masters, of these infant societies; they spoke to

---

tity of water flowing in from the original ocean, and which then covered the earth, would much contribute to leave the continents and islands, which might be raised at the same time, above the surface of the water. In later days we have accounts of huge stones falling from the firmament, which may have been thrown by explosion from some distant earthquake, without having been impelled with a force sufficient to cause them to circulate round the earth, and thus produce numerous small moons or satellites.

---

* It may be that the larger animals we now behold were originally derived from the smallest microscopic ones, who have increased in bulk with the progression of time, or that, as the Egyptian philosophers thought, mankind were originally hermaphrodites, who, like the *aphis*, produced the sexual distinction after some generations. This was also the opinion of Plato, and seems to have been that of Moses, who was educated amongst the Egyptians, as may be gathered from the 27th and 28th verses of the first chapter of Genesis: "So God created man in his own image, in the image of God created he him; male and female created he them. And God blessed them, and God said unto them, be fruitful and multiply, and replenish the earth, and subdue it: and have dominion over the fish of the sea, and over the fowl of the air, and over every living thing that moveth upon the earth:" it is not therefore presuming too much to suppose, as the Egyptians were a nation very fond of explaining their opinions by hieroglyphics, that that part which describes Eve as taken out of Adam's rib, was an hieroglyphic emblem, showing that mankind were in the primitive state of both sexes, united, who were afterwards divided into males and females.

No. VI.—23

the imagination of their auditors.—Poetry, by its images, its fictions, its numbers, its rhyme, its harmony, conspired to please their fancy, and to render permanent the impressions it made: thus, the entire of nature, as well as all its parts, was personified; at its voice, trees, stones, rocks, earth, air, fire, water, took intelligence, held conversation with man, and with themselves; the elements were deified.—The sky, which, according to the then philosophy, was an arched concave, spreading over the earth, which was supposed to be a level plain, was itself made a God; Time, under the name of Saturn, was pictured as the son of heaven;† the igneous matter, the ethereal electric fluid, that invisible fire which vivifies nature, that penetrates all beings, that fetilizes the earth, which is the great principle of motion, the source of heat, was deified under the name of Jupiter: his combination with every being in nature was expressed by his metamorphoses—by the frequent adulteries imputed to him. He was armed with thunder, to indicate he produced meteors, to typify the electric fluid that is called lightning. He married the winds, which were designated under the name of Juno, therefore called the Goddess of the Winds; their nuptials were celebrated with

---

† Saturn was represented as an inexorable divinity—naturally artful, who devoured his own children—who revenged the anger of his mother upon his father, for which purpose she armed him with a scythe, formed of metals drawn from her own bowels, with which he struck Cœlus, in the act of uniting himself to Thea, and so mutilated him that he was ever after incapacitated to increase the number of his children: he was said to have divided the throne with Janus, king of Italy; whose reign seems to have been so mild, so beneficent, that it was called the *golden age*; human victims were sacrificed on his altars, until abolished by Hercules, who substituted small images of clay. Festivals in honour of this God, called Saturnalia, were instituted long antecedent to the foundation of Rome: they were celebrated about the middle of December, either on the 16th, 17th, or 18th; they lasted in latter times several days, originally but one. Universal liberty prevailed at the celebration, slaves were permitted to ridicule their masters—to speak freely on every subject—no criminals were executed—war never declared; the priests made their human offerings with their heads uncovered; a circumstance peculiar to the Saturnalia, not adopted at other festivals.

great solemnity.* Thus, following the same fictions, the sun, that beneficent star which has such a marked influence over the earth, became an Osiris, a Belus, a Mithras, an Adonis, an Apollo. Nature, rendered sorrowful by his periodical absence, was an Isis, an Astarte, a Venus, a Cybele.†

In short, every thing was personified: the sea was under the empire of Neptune; fire was adored by the Egyptians under the name of Serapis; by the Persians, under that of Ormus or Oromaze; and by the Romans, under that of Vesta and Vulcan.

Such was the origin of mythology: it may be said to be the daughter of natural philosophy, embellished by poetry, and only destined to describe nature and its parts. If antiquity is consulted, it* will be perceived without much trouble, that those famous sages, those legislators, those priests, those conquerors, who were the instructers

of infant nations, themselves adored active nature, or the great whole considered relatively to its different operations or qualities; that this was what caused the ignorant savages whom they had gathered together to adore.‡ It was the great whole they deified; it was its various parts which they made their inferior gods; it was from the necessity of her laws they made ᶴate. Allegory masked its mode of action: it was at length parts of this great whole that idolatry represented by statues and symbols.§

To complete the proofs of what has been said; to show distinctly that it was the great whole, the universe, the nature of things, which was the real object of the worship of Pagan antiquity, we shall here give the hymn of Orpheus addressed to the God Pan:—
"O Pan! I invoke thee, O powerful God! O universal nature! the heavens, the sea, the earth, who nourish all, and the eternal fire, because these are thy members, O all powerful Pan," &c. Nothing can be more suitable to confirm these ideas, than the ingenious explanation which is given of the fable of Pan, as well as of the figure under which he is represented. It is said.

* All the Gods, the entire brute creation, and the whole of mankind attended these nuptials, except one young woman named Chelone, who laughed at the ceremonies, for which impiety she was changed by Mercury into a tortoise, and condemned to perpetual silence. He was the most powerful of all the Gods, and considered as the king and father both of Gods and men: his worship was very extended, performed with greater solemnity, than that of any other God. Upon his altars smoked goats, sheep, and white bulls, in which he is said to have particularly delighted: the oak was rendered sacred to him, because he taught mankind to live upon acorns; he had many oracles where his precepts were delivered: the most celebrated of these were at Dodona and Ammon in Libya; He was supposed to be invisible to the inhabitants of the earth; the Lacedemonians erected his statue with four heads, thereby indicating that he listened readily to the solicitations of every quarter of the earth.— Minerva is represented as having no mother, but to have come completely armed from his brains, when his head was opened by Vulcan; by which it is meant to infer that wisdom is the result of this ethereal fluid.

† Astarte had a magnificent temple at Hieropolis, served by three hundred priests, who were always employed in offering sacrifices. The priests of Cybele, called Corybantes, also Galli, were not admitted to their sacred functions without previous mutilation. In the celebration of their festivals these priests used all kinds of indecent expressions, beat drums, cymbals, and behaved just like madmen: his worship extended all over Phrygia, and was established in Greece under the name of Eleusinian mysteries.

‡ The Greeks called nature a divinity who had a thousand names (Μηρίωνυμα). All the divinities of Paganism, were nothing more than nature considered according to its different functions, and under its different points of view. The emblems with which they decorated these divinities again prove this truth. These different modes of considering nature have given birth to Polytheism and Idolatry. See the critical remarks against Toland by M. Benoist, page 258.

§ To convince ourselves of this truth, we have only to open the ancient authors. "I believe," says Varro, "that God is the soul of the universe, which the Greeks called ΚΟΣΜΟΣ, and that the universe itself is God." Cicero says, "cos qui dii appellantur rerum natura esse." See de Natura Deorum, lib. iii. cap. 24. The same Cicero says, that in the mysteries of Samothracia, of Lemnos, of Eleusis, it was nature much more than the Gods they explained to the initiated. Rerum magis natura cognoscitur quam deorum. Join to these authorities the Book of Wisdom, chap. xiii ver. 10, and xiv. 15 and 22. Pliny says, in a very dogmatical style, "We must believe that the world, or that which is contained under the vast extent of the heavens, is the Divinity itself, eternal, immense, without beginning or end." See Plin. Hist. Nat. lib. ii. cap. 1, init.

"Pan, according to the signification of his name, is the emblem by which the ancients have designated the great assemblage of things: he represents the universe; and, in the mind of the wisest philosophers of antiquity, he passed for the greatest and most ancient of the Gods. The features under which he is delineated form the portrait of nature, and of the savage state in which she was found in the beginning. The spotted skin of the leopard, which serves him for a mantle, represented the heavens filled with stars and constellations. His person was compounded of parts, some of which were suitable to a reasonable animal, that is to say, to man; and others to the animal destitute of reason, such as the goat. It is thus," says he, "that the universe is composed of an intelligence that governs the whole, and of the prolific, fruitful elements of fire, water, earth, air. Pan, loved to drink and to follow the nymphs; this announces the occasion nature has for humidity in all her productions, and that this God, like nature, is strongly inclined to propagation. According to the Egyptians, and the most ancient Grecian philosophers, Pan had neither father nor mother; he came out of Demogorgon at the same moment with the Destinies, his fatal sisters; a fine method of expressing that the universe was the work of an unknown power, and that it was formed after the invariable relations, the eternal laws of necessity; but his most significant symbol, that most suitable to express the harmony of the universe, is his mysterious pipe, composed of seven unequal tubes, but calculated to produce the nicest and most perfect concord. The orbs which compose the seven planets of our solar system, are of different diameters; being bodies of unequal mass, they describe their revolutions round the sun in various periods; nevertheless it is from the order of their motion that results the harmony of the spheres." &c.*

Here then is the great macrocosr the mighty whole, the assemblage things, adored and deified by the ph losophers of antiquity, whilst the ui informed stopped at the emblem und which this nature was depicted, at th symbols under which its various part its numerous functions were personif ed; his narrow mind, his barbarous i norance, never permitted him to mou higher; they alone were deemed wo thy of being initiated into the myst ries, who knew the realities maske under these emblems.

Indeed, the first institutors of nation and their immediate successors in au thority, only spoke to the people by fa bles, allegories, enigmas, of which the reserved to themselves the right of giv ing an explanation. This mysteriou tone they considered necessary, wheth er it were to mask their own ignorance or whether it were to preserve their pow er over the uninformed, who for th most part only respect that which i above their comprehension. Their ex plications were always dictated eithe by interest, by a delirious imagination or by imposture; thus from age to age they did no more than render nature and its parts, which they had original ly depicted, more unknown, until they completely lost sight of the primitive ideas; these were replaced by a multitude of fictitious personages, under whose features this nature had primarily been represented to them. The people adored these personages, without penetrating into the true sense of the emblematical fables recounted to them. These ideal beings, with material figures, in whom they believed there resided a mysterious virtue, a divine power, were the objects of their worship, of their fears, of their hopes. The wonderful, the incredible actions

---

* This passage is taken from an English book entitled, *Letters concerning Mythology.* We can hardly doubt that the wisest among the Pagans adored nature, which mythology, or the Pagan theology, designated under an infinity of names and different emblems. Apuleius, although he was a Platonist and accustomed to the mysterious and unintelligible no-

tions of his master, calls nature "rerum natura parens, elementorum omnium Domina, sæculorum progenies initialis . . . . . Matrem siderum, parentem temporum, orbisque totius dominam." It is this nature that some adored under the name of the mother of the Gods, others under the names of Ceres, Venus, Minerva, &c. In short, the Pantheism of the Pagans is clearly proved by these remarkable words in the maxims of Medaura, who in speaking of nature says, "ita fit ut; dum ejus quasi membra carptim, variis supplicationibus prosequimur, totum colere profecto videamur."

ascribed to these fancied divinities, were an inexhaustible fund of admiration, which gave perpetual play to the fancy; which delighted not only the people of those days, but even the children of latter ages. Thus were transmitted from age to age those marvellous accounts, which, although necessary to the existence of the ministers of the Gods, did nothing more than confirm the blindness of the ignorant: these never supposed that it was nature, its various operations, the passions of man and his divers faculties, that lay buried under a heap of allegories;* they had no eyes but for these emblematical persons, under which nature was masked: they attributed to their influence the good, to their displeasure the evil, which they experienced: they entered into every kind of folly, into the most delirious acts of madness, to render them propitious to their views; thus, for want of being acquainted with the reality of things, their worship frequently degenerated into the most cruel extravagance, into the most ridiculous folly.

Thus it is obvious, that every thing proves nature and its various parts to have every where been the first divinities of man. Natural philosophers studied them either superficially or profoundly, explained some of their properties; detailed some of their modes of action. Poets painted them to the imagination of mortals, imbodied them, and furnished them with reasoning faculties. The statuary executed the ideas of the poets. The priests decorated these Gods with a thousand marvellous qualities—with the most terrible passions—with the most inconceivable attributes. The people adored them; prostrated themselves before these Gods, who were neither susceptible of love or hatred, goodness, or mal-

ice; and they became persecuting, malevolent, cruel, unjust, in order to render themselves acceptable to powers generally described to them under the most odious features.

By dint of reasoning upon nature thus decorated, or rather disfigured, subsequent speculators no longer recollected the source from whence their predecessors had drawn their Gods, and the fantastic ornaments with which they had embellished them. Natural philosophers and poets were transformed by leisure into metaphysicians and theologians; tired with contemplating what they could have understood, they believed they had made an important discovery by subtilly distinguishing nature from herself—from her own peculiar energies—from her faculty of action. By degrees they made an incomprehensible being of this energy, which as before they personified: this they called the mover of nature, or God. This abstract, metaphysical being, or rather, word, became the subject of their continual contemplation;* they looked upon it not only as a real being, but also as the most important of beings; and by thus dreaming, nature quite disappeared; she was despoiled of her rights; she was considered as nothing more than an unwieldy mass, destitute of power, devoid of energy, and as a heap of ignoble matter purely passive, who, incapable of acting by herself, was not competent to any of the operations they beheld, without the direct, the immediate agency of the moving power they had associated with her. Thus man ever preferred an unknown power, to that of which he was enabled to have some knowledge if he had only deigned to consult his experience, but he presently ceases to respect that which he understands, and to estimate those objects which are familiar to him: he figures to himself something marvellous in every thing he does not comprehend; his mind, above all, labours to seize upon that which appears to escape his consideration; and, in default of experience, he no longer consults any thing

---

* The passions and faculties of human nature were used as emblems, because man was ignorant of the true cause of the phenomena he beheld. As strong passions seemed to hurry man along, in despite of himself, they either attributed these passions to a God, or deified them; it was thus love became a deity; that eloquence, poetry, industry, were transformed into Gods under the names of Hermes, Mercury, Apollo; the stings of conscience were called Furies. Christians have also deified reason under the name of the eternal word.

* The Greek word ΘΕΟΣ comes from πίθηρι, pono or rather from ΘΕΑΟΜΑΙ, specto, contemplor, to take a view of hidden and secret things.

but his imagination, which feeds him with chimeras. In consequence, those speculators who have subtilly distinguished nature from her own powers, have successively laboured to clothe the powers thus separated with a thousand incomprehensible qualities: as they did not see this being, which is only a mode, they made it a spirit—an intelligence—an incorporeal being; that is to say, of a substance totally different from every thing of which we have a knowledge. They never perceived that all their inventions, that all the words which they imagined, only served to mask their real ignorance; and that all their pretended science was limited to saying, in what manner nature acted, by a thousand subterfuges which they themselves found it impossible to comprehend. Man always deceives himself for want of studying nature; he leads himself astray, every time he is disposed to go out of it; he is always quickly necessitated to return or to substitute words which he does not himself understand for things which he would much better comprehend if he was willing to look at them without prejudice.

Can a theologian ingenuously believe himself more enlightened, for having substituted the vague words, *spirit, incorporeal substance, Divinity,* &c. to the more intelligible terms nature, matter, mobility, necessity? However this may be, these obscure words once imagined, it was necessary to attach ideas to them; in doing this, he has not been able to draw them from any other source than the beings of this despised nature, which are ever the only beings of which he is enabled to have any knowledge. Man, consequently, drew them up in himself; his own mind served for the model of the universal mind of which indeed according to some it only formed a portion; his own mind was the standard of the mind that regulated nature; his own passions, his own desires, were the prototypes of those by which he actuated this being; his own intelligence was that from which he formed that of the supposed mover of nature; that which was suitable to himself, he called the order of nature; this pretended order was the scale by which he measured the wisdom of this being; in short, those qualities which he calls

perfections in himself, were the arch types, in miniature, of the Divine perfections. It was thus, that in despite of all their efforts, the theologiar were, and always will be true Anthr[c] pomorphites. Indeed, it is very diff cult, if not impossible to prevent ma from making himself the sole model [c] his divinity.* Indeed, man sees in hi God nothing but a man. Let hir subtilize as he will, let him extend hi own powers as he may, let him swell hi own perfections to the utmost, he wil have done nothing more than make [a] gigantic, exaggerated man, whom h[c] will render illusory by dint of heaping together incompatible qualities. H[e] will never see in God, but a being o[r] the human species, in whom he wil[l] strive to aggrandize the proportions, until he has formed a being totally inconceivable. It is according to these dispositions that he attributes intelligence, wisdom, goodness, justice, science, power, to the Divinity, because he is himself intelligent; because he has the idea of wisdom in some beings of his own species; because he loves to find in them ideas favourable to himself: because he esteems those who display equity; because he has a knowledge, which he holds more extensive in some individuals than himself; in short, because he enjoys certain faculties which depend on his own organization. He presently extends or exaggerates all these qualities; the sight of the phenomena, of nature, which he feels he is himself incapable of either producing or imitating, obliges him to make this difference between his God and himself; but he knows not at what point to stop; he fears lest he should deceive himself if he should see any

* Montaign says, "Man is not able to be other than he is, nor imagine but after his capacity; let him take what pains he may, he will never have a knowledge of any soul but his own." Xenophanes said, "If the ox or the elephant understood either sculpture or painting, they would not fail to represent the Divinity under their own peculiar figure; that in this, they would have as much reason as Polyclitus or Phidias, who gave him the human form." It was said to a very celebrated man that "God made man after his own image;" "Man has returned the compliment," replied the philosopher; and L'amotte le Vayer used to remark, that "*theanthropy was the foundation of every system of Christianity.*"

limits to the qualities he assigns; the word infinite, therefore, is the abstract, the vague term which he uses to characterize them. He says that his power is infinite, which signifies that when he beholds those stupendous effects which nature produces, he has no conception at what point his power can rest; that his goodness, his wisdom, his knowledge are infinite: this announces that he is ignorant how far these perfections may be carried in a being whose power so much surpasses his own. He says that his God is eternal, that is, of infinite duration, because he is not capable of conceiving he could have had a beginning or can ever cease to be, and this he considers a defect in those transitory beings of whom he beholds the dissolution, whom he sees are subjected to death. He presumes the cause of those effects to which he is a witness, is immutable, permanent, not subjected to change like all the evanescent beings whom he knows are submitted to dissolution, to destruction, to change of form. This pretended mover of nature being always invisible to man, his mode of action being impenetrable, he believes that, like the concealed principle which animates his own body, this God is the moving power of the universe. Thus when by dint of subtilizing, he has arrived at believing the principle by which his body is moved is a spiritual, immaterial substance, he makes his God spiritual or immaterial in like manner: he makes it immense, although without extent; immoveable, although capable of moving nature: immutable, although he supposes him to be the author of all the changes operated in the universe.

The idea of the unity of God, was a consequence of the opinion that this God was soul of the universe; however it was only the tardy fruit of human meditation.* The sight of those opposite, frequently contradictory effects, which man saw take place in the

world, had a tendency to persuade him there must be a number of distinct powers or causes independent of each other. He was unable to conceive that the various phenomena he beheld, sprung from a single, from a unique cause; he therefore admitted many causes or Gods, acting upon different principles; some of which he considered friendly, others as inimical to his race. Such is the origin of that doctrine, so ancient, so universal, which supposed two principles in nature, or two powers of opposite interest, who were perpetually at war with each other; by the assistance of this he explained that constant mixture of good and evil, that blending of prosperity with misfortune, in a word, those eternal vicissitudes to which in this world the human being is subjected. This is the source of those combats which all antiquity has supposed to exist between good and wicked Gods, between an Osiris and a Typhœus; between an Orosmadis and an Arimanis; between a Jupiter and the Titanes; between a Jehovah and a Satan. In these rencounters man for his own peculiar interest always gave the palm of victory to the beneficent Deity; this, according to all the traditions handed down, ever remained in possession of the field of battle; it was evidently for the benefit of mankind that the good God should prevail over the wicked.

Even when man acknowledged only one God, he always supposed the different departments of nature were confided to powers subordinate to his supreme orders, under whom the sovereign of the Gods discharged his care in the administration of the world.— These subaltern Gods were prodigiously multiplied; each man, each town, each country, had their local, their tutelary Gods; every event, whether fortunate or unfortunate, had a divine cause, and was the consequence of a sovereign decree; each natural effect, every operation of nature, each passion, depended upon a Divinity, which theological imagination, disposed to see Gods every where, and always mistaking nature, either embellished or disfigured. Poetry tuned its harmonious lays on these occasions, exaggerated the details, animated its pictures; credulous ignorance received the portraits with

---

* The idea of the unity of God cost Socrates his life. The Athenians treated as an atheist a man who believed only in one God. Plato did not dare to break entirely with the doctrine of Polytheism; he preserved Venus, an all-powerful Jupiter, and a Pallas, who was the Goddess of the country. The Christians were looked upon as Atheists by Pagans, because they adored only one God.

eagerness, and heard the doctrines with submission.

Such is the origin of Polytheism: such are the foundations, such the titles of the hierarchy, which man established between himself and the Gods, because he felt he was incapable of immediately addressing himself to the incomprehensible being whom he had acknowledged for the only sovereign of nature, without even having any distinct idea on the subject. Such is the true genealogy of those inferior Gods whom the uninformed place as a proportional means between themselves and the first of all other causes. In consequence, among the Greeks and the Romans, we see the deities divided into two classes : the one were called *great Gods,** who formed a kind of aristocratic order distinguished from the minor Gods, or from the multitude of ethnic divinities. Nevertheless, the first rank of these Pagan divinities, like the latter, were submitted to Fate, that is, to destiny, which obviously is nothing more than nature acting by immutable, rigorous, and necessary laws; this destiny was looked upon as the God of Gods; it is evident that this was nothing more than necessity personified, and that therefore it was a weakness in the heathens to fatigue with their sacrifices, to solicit with their prayers, those Divinities whom they themselves believed were submitted to the decrees of an inexorable destiny, of which it was never possible for them to alter the mandates. But man ceases always to reason whenever his theological notions are brought into question.

What has been already said, serves to show the common source of that multitude of intermediate powers, subordinate to the Gods, but superior to man, with which he filled the universe :†

* The Greeks called the great Gods Θεοι *αβιροι—Cabiri;* the Romans called them *Dii majorum gentium or Dii consentes,* because the whole world were in accord in deifying the most striking and active parts of nature, such as the sun, fire, the sea, time, &c., whilst the other Gods were entirely local, that is to say, were reverenced only in particular countries, or by individuals, as in Rome, where every citizen had Gods for himself alone, whom he adored under the names of *Penates, Lares,* &c.

† Among the Romans they were called *Dii*

they were venerated under the name of nymphs, demi-gods, angels, demons good and evil genii, spirits, heroes saints, &c. These constitute different classes of intermediate divinities, wh became either the foundation of thei hopes, the object of their fears, th means of consolation, or the source o; dread to those very mortals who onl; invented them when they found it im possible to form to themselves distinct perspicuous ideas of the incomprehen sible Being who governed the world i; chief, or when they despaired of bein; able to hold communication with hin directly.

By dint of meditation and reflectio; some, who gave the subject more con sideration than others, reduced the whole to one all-powerful Divinity whose power and wisdom sufficed to govern it. This God was looked upor as a monarch jealous of nature. They persuaded themselves that to give rivals and associates to the monarch to whom all homage was due would offend him—that he could not bear a division of empire—that infinite power and unlimited wisdom had no occasion for a division of power nor for any assistance. Thus some would-be-thoughtprofound-thinkers have admitted one God, and in doing so have flattered themselves with having achieved a most important discovery. And yet, they must at once have been most sadly perplexed by the contradictory actions of this *one* God ; so much so that they were obliged to heap on him the most incompatible and extravagant qualities to account for those contradictory effects which so palpably and clearly gave the lie to some of the attributes they assigned to him. In supposing a God, the author of every thing, man is obliged to attribute to him unlimited goodness, wisdom, and power, agreeable to the kindness, to the order he fancied he saw in the universe, and according to the wonderful effects he witnessed ; but, on the other hand, how could ha avoid attributing to this God malice, improvidence, and caprice, seeing the frequent disorders and numberless evils *mediozimi*—intermediate Gods; they were looked upon as mediators, or intercessors; as powers whom it was necessary to reverence in order either to obtain their favour, appease their anger, or divert their malignant intentions.

.to which the human race is so often liable? How can man avoid taxing him with improvidence, seeing that he is continually employed in destroying the work of his own hands? How is it possible not to suspect his impotence, seeing the perpetual non-performance of those projects of which he is supposed to be the contriver?

To solve these difficulties, man created enemies to the Divinity, who although subordinate to the supreme God, were nevertheless competent to disturb his empire, to frustrate his views; he had been made a king, and adversaries, however impotent, were found willing to dispute his diadem. Such is the origin of the fable of the Titanes, or of the *rebellious angels*, whose presumption caused them to be plunged into the abyss of misery—who were changed into *demons*, or into evil genii: these had no other functions, than to render abortive the projects of the Almighty, and to seduce, to raise to rebellion, those who were his subjects.[*]

In consequence of this ridiculous fable the monarch of nature was represented as perpetually in a scuffle with the enemies he had himself created; notwithstanding his infinite power, either he would not or could not totally subdue them; he was in a continual state of hostility, rewarding those who obeyed his laws, and punishing those who had the misfortune to enter into the conspiracies of the enemies of his glory. As a consequence of these ideas, borrowed from the conduct of earthly monarchs who are almost always in a state of war, some men claimed to be the ministers of God: they made him speak; they

unveiled his concealed intentions, and denounced the violation of his laws as the most horrible crime: the ignorant multitude received these without examination; they did not perceive that it was man and not a God who thus spoke to them; they did not reflect that it was impossible for weak creatures to act contrary to the will of a God whom they supposed to be the creator of all beings, and therefore who could have no enemies in nature but those he himself had created. It was pretended that man, spite of his natural dependance and the infinite power of his God, was able to offend him, was capable of thwarting him, of declaring war against him, of overthrowing his designs, and of disturbing the order he had established. This God, no doubt, to make a parade of his power, was supposed to have created enemies against himself, so that he might have the pleasure of fighting them, although he is not willing either to destroy them or to change their bad dispositions. In fine, it was believed that he had granted to his rebellious enemies, as well as to all mankind, the liberty of violating his commands, of annihilating his projects, of kindling his wrath, and of arresting his goodness. Hence, all the benefits of this life were considered as rewards, and its evils as merited punishments. In fact, the system of man's free will seems to have been invented only to enable him to sin against God, and to acquit this last of the evil he brings upon man for exercising the fatal liberty given him.

These ridiculous and contradictory notions served nevertheless for the basis of all the superstitions of the world, believing that they thereby accounted for the origin of evil and the cause of man's misery. And yet man could not but see that he frequently suffered or earth without having committed any crime, without any known transgression to provoke the anger of his God; he perceived that even those who complied in the most faithful manner with his pretended orders were often involved in the same ruin with the boldest violator of his laws. In the habit of bending to power, to tremble before his terrestrial sovereign, to whom he allowed the privilege of being iniquitous, never disputing his titles, nor ever criti-

---

[*] The fable of the Titanes, or *rebellious angels*, is extremely ancient and very generally diffused over the world; it serves for the foundation of the theology of the Brahmins of Hindostan, as well as for that of the European priesthood. According to the Brahmins, all living bodies are animated by fallen angels, who, under these forms, expiate their rebellion. This fable, as well as that of *demons*, makes the Divinity play a very ridiculous part; in fact it supposes that God gives existence to adversaries to keep himself employed, or in *training*, and to show his power. Yet there is no display whatever of this power, since, according to theological notions, the Devil has many more adherents than the Divinity.

cising the conduct of those who had the power in their hands, man dared still less to examine into the conduct of his God, or to accuse him of motiveless cruelty. Besides, the ministers, the celestial monarch invented means of justifying him, and of making the cause of those evils, or of those punishments which men experience fall upon themselves; in consequence of the liberty which they pretended was given to creatures, they supposed that man had sin, that his nature was perverted, that the whole human race carried with it the punishment incurred by the faults of his ancestors, which their implacable monarch still avenged upon their innocent posterity. Men found this vengeance perfectly legitimate, because according to the most disgraceful prejudices they proportioned the punishments much more to the power and dignity of the offended, than to the magnitude or reality of the offence. In consequence of this principle they thought that a God had an indubitable right to avenge, without proportion and without end, the outrages committed against his divine majesty. In a word, the theological mind tortured itself to find men culpable, and to exculpate the Divinity from the evils which nature made the former necessarily experience. Man invented a thousand fables to give a reason for the mode in which evil entered into this world; and the vengeance of heaven always appeared to have sufficient motives, because he believed that crimes committed against a being infinitely great and powerful ought to be infinitely punished.

Moreover, man saw that the earthly powers, even when they committed the most barefaced injustice, never suffered him to tax them with being unjust, to entertain a doubt of their wisdom, to murmur at their conduct. He was not going then to accuse of injustice the despot of the universe, to doubt his rights, or to complain of his rigour: he believed that God could commit every thing against the feeble work of his hands, that he owed nothing to his creatures, that he had a right to exercise over them an absolute and unlimited dominion. It is thus that the tyrants of the earth act; and their arbitrary conduct serves for the model of

that which they accord to the Divinity: it was upon their absurd and unreasonable mode of governing, that they made a peculiar jurisprudence for God.— Hence we see that the most wicked of men have served as a model for God, and that the most unjust governments were made the model of his divine administration. In despite of his cruelty and his unreasonableness, man never ceases to say, that he is most just and full of wisdom.

Men, in all countries, have paid adoration to fantastical, unjust, sanguinary, implacable Gods, whose rights they have never dared to examine.— These Gods were every where cruel, dissolute. and partial; they resembled those unbridled tyrants who riot with impunity in the misery of their subjects, who are too weak, or too much hoodwinked to resist them, or to withdraw themselves from under that yoke with which they are overwhelmed. It is a God of this hideous character which they make us adore, even at the present day; the God of the Christians, like those of the Greeks and Romans, punishes us in this world, and will punish us in another, for those faults of which the nature he has given us has rendered us susceptible. Like a monarch, inebriated with his authority, he makes a vain parade of his power, and appears only to be occupied with the puerile pleasure of showing that he is master, and that he is not subjected to any law. He punishes us for being ignorant of his inconceivable essence and his obscure will. He punishes us for the transgressions of our fathers; his despotic caprice decides upon our eternal destiny; it is according to his fatal decrees, that we become, in despite of ourselves, either his friends or his enemies: he makes us free only that he may have the barbarous pleasure of chastising us for those necessary abuses which our passions or our errours cause us to make of our liberty. In short, theology shows us, in all ages, mortals punished for inevitable and necessary faults, and as the unfortunate playthings of a tyrannical and wicked God.*

* The Pagan theology never showed the people in the persons of their Gods any thing more than men who were dissolute, adulterers, vindictive, and punishing with rigour those

It was upon these unreasonable notions that the theologians throughout the whole earth, founded the worship which man ought to render to the Divinity, who, without being attached to them, had the right of binding them to himself: his supreme power dispensed him from all duty towards his creatures; and they obstinately persisted in looking upon themselves as culpable every time they experienced calamities. Do not let us then be at all astonished if the religious man was in continual fears; the idea of God always recalled to him that of a pitiless tyrant, who sported with the miseries of his subjects and these, even without knowing it, could, at each moment, incur his displeasure; yet they never dared tax him with injustice, because they believed that justice was not made to regulate the actions of an all-powerful monarch, whose elevated rank placed him infinitely above the human species, although they had imagined, that he had formed the universe entirely for man.

It is then for want of considering good and evil as effects equally necessary; it is for want of.attributing them to their true cause, that men have created to themselves fictitious causes, and malicious divinities, respecting whom nothing is able to undeceive them.— In considering nature they however would have seen that physical evil is a necessary consequence of the particular properties of some beings; they would have acknowledged that plagues, contagions, diseases, are due to phys-

necessary crimes which were predicted by the oracles. The Judaical and Christian theology shows us a partial God who chooses or rejects, who loves or hates, according to his caprice; in short, a tyrant who plays with his creatures; who punishes in this world the whole human species for the crimes of a single man; who *predestinates* the greater number of mortals to be his enemies, to the end that he may punish them to all eternity, for having received from him the liberty of declaring against him. All the religions of the world have for basis the omnipotence of God over men; his despotism over men, and his Divine injustice. From thence, among the Christians, the doctrine of *original sin;* from thence, the theological notions upon pardon, upon the necessity of a mediator; in short, hence that ocean of absurdities with which *Christian theology* is filled. It appears, generally, that a reasonable God would not be convenient to the interests of priests.

ical causes and particular circumstances—to combinations which, although extremely natural, are fatal to their species; and they would have sought in nature herself the remedies suitable to diminish or cause those under which they suffer to cease. They would have seen in like manner that moral evil was only a necessary consequence of their bad institutions ; that it was not to the God of heaven, but to the injustice of the princes of the earth to which those wars, that poverty, those famines,.those reverses, those calamities, those vices, and those crimes under which they groan so frequently, were to be ascribed. Thus to throw aside these evils they should not have uselessly extended their trembling hands towards phantoms incapable of relieving them, and who were not the authors of their sorrows ; they should have sought in a more rational administration, in more equitable laws, in more reasonable institutions, the remedies for these misfortunes which they falsely attributed to the vengeance of a God, who is painted to them under the character of a tyrant, at the same time that they are defended from entertaining a doubt of his justice and his goodness.

Indeed priests never cease repeating that their God is infinitely good ; that he only wishes the good of his creatures; that he has made every thing only for them: and in despite of these assurances, so flattering, the idea of his wickedness will necessarily be the strongest ; it is much more likely to fix the attention of mortals than that of his goodness ; this gloomy idea is always the first that presents itself to the human mind, whenever it is occupied with the Divinity. The idea of evil necessarily makes a much more lively impression upon man than that of good ; in consequence, the beneficent God will always be eclipsed by the dreadful God. Thus, whether they admit a plurality of Gods of opposite interests, whether they acknowledge only one monarch in the universe, the sentiment of fear will necessarily prevail over love; they will only adore the good God that they may prevent him from exercising his caprice, his phantasms, his malice ; it is always inquietude and terrour that throws man at his feet; it is his rigour and his severity which they seek to disarm. In short, although they every where assure us that the Divinity is full of compassion, of clemency, and of goodness, it is always a malicious genius, a capricious master, a formidable demon, to whom every where they render servile homage, and a worship dictated by fear.

These dispositions have nothing in them that ought to surprise us; we can accord with sincerity our confidence and our love only to those in whom we find a permanent will to render us service; as soon as we have reason to suspect in them the will, the power, or the right to injure us, their idea afflicts us, we fear them, we mistrust them, and we take precautions against them; we hate them from the bottom of our hearts, even without daring to avow our sentiments. If the Divinity must be looked upon as the common source of the good and evil which happens in this world; if he has the will sometimes to render men happy, and sometimes to plunge them in misery, or punish them with rigour, men must necessarily dread his caprice or his severity, and be much more occupied with these, which they see him resolved upon so frequently, than with his goodness. Thus the idea of their celestial monarch must always make man uneasy; the severity of his judgments must cause him to tremble much oftener than his goodness is able to console or encourage him.

If we pay attention to this truth, we shall feel why all the nations of the earth have trembled before their Gods and have rendered them the most fantastical, irrational, mournful and cruel worship; they have served them as they would despots but little in accord with themselves, knowing no other rule than their fantasies, sometimes favourable, and more frequently prejudicial to their subjects; in short, like inconstant masters, who are less amiable by their kindness, than dreadful by their punishments, by their malice, and by those rigours which they still never dared to find unjust or excessive. Here is the reason why we see the adorers of a God, whom they unceasingly show to mortals as the model of goodness, of equity, and every perfection, deliver themselves up to the most cruel extravagances against themselves, with a view of punishing themselves, and of preventing the celestial vengeance, and at the same time commit the most hideous crimes against others, when they believe that by so doing they can disarm the anger, appease the justice, and recall the clemency of their God. All the religious systems of men, their sacrifices, their prayers, their customs and their ceremonies have never had for object any thing else than to avert the fury of the Divinity, prevent his caprice, and excite in him those sentiments of goodness, from which they see him deviate every instant. All the efforts, all the subtilties of theology have never had any other end than to reconcile in the sovereign of nature those discordant ideas which it has itself given birth to in the minds of mortals. We might justly define this end the art of composing chimeras, by combining together qualities which it is impossible to reconcile with each other.

THE

SYSTEM OF NATURE:

OR,

LAWS OF THE MORAL AND PHYSICAL WORLD.

BY BARON D'HOLBACH,
AUTHOR OF "GOOD SENSE," ETC.

A NEW AND IMPROVED EDITION, WITH NOTES BY DIDEROT.

TRANSLATED, FOR THE FIRST TIME,
BY H. D. ROBINSON.

TWO VOLUMES IN ONE.

VOL. II.

STEREOTYPE EDITION.

BOSTON:
PUBLISHED BY J. P. MENDUM,
84 Washington Street.
1868.

THE SYSTEM OF NATURE.

CHAPTER I.

Of the confused and contradictory ideas of Theology.

EVERY thing that has been said, proves pretty clearly, that in despite of all his efforts, man has never been able to prevent himself from drawing together from his own peculiar nature, the qualities he has assigned to the being who governs the universe. The contradictions necessarily resulting from the incompatible assemblage of these human qualities, which cannot become suitable to the same subject, seeing that the existence of one destroys the existence of the other, have been shown:—the theologians themselves have felt the insurmountable difficulties which their Divinities presented to reason: they were so substantive, that as they felt the impossibility of withdrawing themselves out of the dilemma, they endeavoured to prevent man from reasoning, by throwing his mind into confusion—by continually augmenting the perplexity of those ideas, already so discordant, which they offered him of their God. By this means they enveloped him in mystery, covered him with dense clouds, rendered him inaccessible to mankind: thus they themselves became the interpreters, the masters of explaining, according either to their fancy or their interest, the ways of that enigmatical being they made him adore. For this purpose they exaggerated him more and more—neither time nor space, nor the entire of nature could contain his immensity — every thing became an impenetrable mystery. Although man has originally borrowed from himself the traits, the colours, the primitive lineaments of which he composed his God; although he has made him a jealous powerful, vindictive

monarch, yet his theology, by force of dreaming, entirely lost sight of human nature; and in order to render his Divinities still more different from their creatures, it assigned them, over and above tne usual qualities of man, properties so marvellous, so uncommon, so far removed from every thing of which his mind could form a conception, that he lost sight of them himself. From thence he persuaded himself these qualities were divine, because he could no longer comprehend them; he believed them worthy of God, because no man could figure to himself any one distinct idea of him. Thus theology obtained the point of persuading man he must believe that which he could not conceive; that he must receive with submission improbable systems; that he must adopt, with pious deference, conjectures contrary to his reason; that this reason itself was the most agreeable sacrifice he could make on the altars of his fantastical master who was unwilling he should use the gift he had bestowed upon him. In short, it had made mortals implicitly believe that they were not formed to comprehend the thing of all others the most important to themselves.* On the other hand, man persuaded himself that the gigantic, the truly incomprehensible attributes which were assigned to his celestial monarch, placed between him and his slaves a distance so immense, that this proud master could not be by any means offended with the comparison; that these distinctions rendered him still greater; made him more powerful, ——

* It is quite evident that every religion is founded upon the absurd principle, that man is obliged to accredit finally, that which he is in the most complete impossibility of comprehending. According even to theological notions, man, by his nature, must be in an *invincible ignorance* relatively to God.

191

more marvellous, more inaccessible to observation. Man always entertains the idea, that what he is not in a condition to conceive, is much more noble, much more respectable, than that which he has the capacity to comprehend: he imagines that his God, like tyrants, does not wish to be examined too closely.

These prejudices in man for the marvellous, appear to have been the source that gave birth to those wonderful, unintelligible qualities with which theology clothed the sovereign of the world. The invincible ignorance of the human mind, whose fears reduced him to despair, engendered those obscure, vague notions, with which he decorated his God. He believed he could never displease him, provided he rendered him incommensurable, impossible to be compared with any thing of which he had a knowledge; either with that which was most sublime, or that which possessed the greatest magnitude. From hence the multitude of negative attributes with which ingenious dreamers have successively embellished their phantom God, to the end that they might more surely form a being distinguished from all others, or which possessed nothing in common with that which the human mind had the faculty of being acquainted with.

The theological metaphysical attributes, were in fact nothing but pure negations of the qualities found in man, or in those beings of which he has a knowledge; by these attributes their God was supposed exempted from every thing which they considered weakness or imperfection in him, or in the beings by whom he is surrounded. To say that God is infinite, as has been shown, is only to affirm, that unlike man, or the beings with whom he is acquainted, he is not circumscribed by the limits of space; this, however, is what he can never in any manner comprehend, because he is himself finite.*

When it is said that God is eternal, it signifies he has not had, like man or like every thing that exists, a beginning, and that he will never have an end : to say he is immutable, is to say that unlike himself or every thing which he sees, God is not subject to change : to say he is immaterial, is to advance, that their substance or essence is of a nature not conceivable by himself, but which must from that very circumstance be totally different from every thing of which he has cognizance.

It is from the confused collection of these negative qualities, that has resulted the theological God ; the metaphysical whole of which it is impossible for man to form to himself any correct idea. In this abstract being every thing is infinity—immensity—spirituality — omniscience — order — wisdom — intelligence — omnipotence. In combining these vague terms, or these modifications, the priests believed they formed something, they extended these qualities by thought, and they imagined they made a God, whilst they only composed a chimera. They imagined that these perfections or these qualities must be suitable to this God, because they were not suitable to any thing of which they had a knowledge; they believed that an incomprehensible being must have inconceivable qualities. These were the materials of which theology availed itself to compose the inexplicable phantom before which they commanded the human race to bend the knee.

Nevertheless, a being so vague, so impossible to be conceived, so incapable of definition, so far removed from every thing of which man could have any knowledge, was but little calculated to fix his restless views; his mind requires to be arrested by qualities which he is capacitated to ascertain—of which he is in a condition to form a judgment. Thus after it had subtilized this metaphysical God, after it

* Hobbes, in his *Leviathan*, says: "Whatsoever we imagine, is finite. Therefore there is no idea, or conception of any thing we call infinite. No man can have in his mind' an image of infinite magnitude, nor conceive infinite swiftness, infinite time, infinite force, or infinite power. When we say any thing is infinite, we signify only, that we are not able to conceive the ends and bound of the thing named, having no conception of the thing, but of our own inability." Sherlock says: "The word infinite is only a negation, which signifies that which has neither end, nor limits, nor extent, and, consequently, that which has no positive and determinate nature, and is therefore nothing ;" he adds, " that nothing · but custom has caused this word to be adopted, which without that, would appear devoid of sense, and a contradiction."

nad rendered him so different in idea, from every *thing that acts upon the senses. theology found itself under the necessity of again assimilating him to man, from whom it had so far'removed him: it therefore again made him human by the moral qualities which it assigned him; it felt that without this it would not be able to persuade mankind there could possibly exist any relation between him and the vague, ethereal, fugitive, incommensurable being they are called upon to adore. They perceived that this marvellous God was only calculated to exercise the imagination of some few thinkers, whose minds were accustomed to labour upon chimerical subjects, or to take words for realities; in short it found, that for the greater number of the material children of the earth it was necessary to have a God more analogous to themselves, more sensible, more known to them. In consequence the Divinity was reclothed with human qualities; theology never felt the incompatibility of these qualities with a being it had made essentially different from man, who consequently could neither have his properties, nor be modified like himself. It did not see that a God who was immaterial, destitute of corporeal organs, was neither able to think nor to act as material beings, whose peculiar organizations render them susceptible of the qualities, the feelings, the will, the virtues, that are found in them. The necessity it felt to assimilate God to their worshippers, to make an affinity between them, made it pass over without consideration these palpable contradictions, and thus theology obstinately continued to unite those incompatible qualities, that discrepance of character, which the human mind attempted in vain either to conceive or to reconcile: according to it, a pure spirit was the mover of the material world; an immense being was enabled to occupy space, without however excluding nature; an immutable deity was the cause of those continual changes operated in the world: an omnipotent being did not prevent those evils which were displeasing to him; the source of order submitted to confusion: in short. the wonderful properties of this theological being every moment contradicted themselves.

No. VII.—25

There is not less discrepance, less incompatibility, less discordance in the human· perfections, less contradiction in the moral qualities attributed to them, to the end that man might be enabled to form to himself some idea of this being. These were all said to be *eminently* possessed by God, although they every moment contradicted each other: by this means they formed a kind of patch-work character, a heterogenous being, entirely inconceivable to man, *because nature had never constructed any thing like him, whereby he was enabled to form a judgment. Man was assured that God was eminently good—that it was visible in all his actions. Now goodness is a known quality, recognisable in some beings of the human species; this is, above every other, a property he is desirous to find in all those upon whom he is in a state of dependance; but he is unable to bestow the title of good on any among his fellows, except their actions produce on him those effects which he approves—that he finds in unison with his existence—in conformity with his own peculiar modes of thinking. It was evident, according to this reasoning, that God did not impress him with this idea ; he was said to be equally the author of his pleasures, as of his pains, which were to be either secured or averted by sacrifices or prayers: but when man suffered by contagion, when he was the victim of shipwreck, when his country was desolated by war, when he saw whole nations devoured by rapacious earthquakes, when he was a prey to the keenest sorrows, how could he conceive the bounty of that being? How could he perceive the order he had introduced into the world, while he groaned under such a multitude of calamities? How was he able to discern the beneficence of a God whom he beheld sporting as it were with his species ? How could he conceive the consistency of that being who destroyed that which he was assured he had taken such pains to establish, solely for his own peculiar happiness? What becomes of those final causes, which, without any ground, they give as the most incontestable proof of the existence of an omnipotent and wise God, who, nevertheless, can preserve his work only by destroying it, and who has not been able to give it all at once that degree of perfection and consistency, of which it was susceptible. God is said to have created the universe only for man, and was willing that, under him, he should be king of nature. Feeble monarch! of whom a grain of sand, some atoms of bile, some misplaced humours, destroy at once the existence and the reign: yet thou pretendest that a good God has made every thing for thee! Thou desirest that the entire of nature should be thy domain, and thou canst not even defend thyself from the slightest of her shocks! Thou makest to thyself a God for thyself alone; thou supposest that he watcheth for thy preservation; thou supposest that he unceasingly occupieth himself only for thy peculiar happiness; thou imaginest every thing was made solely for thy pleasure; and, following up thy presumptuous ideas, thou hast the audacity to call him good! seest thou not that the kindness exhibited towards thee, in common with other beings, is contradicted? Dost thou not see that those beasts which thou supposest submitted to thine empire, frequently devour thy fellow-creatures; that fire consumeth them; that the ocean swalloweth them up; that those elements which thou admirest the order, frequently sweep them off the face of the earth? Dost thou not see that this power, which thou callest God, which thou pretendest laboureth only for thee, which thou supposest entirely occupied with thy species, flattered by thy homage, touched with thy prayers, cannot be called good, since he acts necessarily? Indeed, according to thy own ideas, dost thou not admit that thy God is the universal cause of all, who must think of maintaining the great whole, from which thou hast so foolishly distinguished him. Is he not then according to thyself, the God of nature—of the ocean—of rivers—of mountains—of the earth, in which thou occupiest so very small a space—of all those other globes that thou seest roll in the regions of space—of those orbs that revolve round the sun that enlighteneth thee?—Cease, then, obstinately to persist in beholding nothing but thyself in nature; do not flatter thyself that the human race, which reneweth itself, which disappeareth like the leaves on

the trees, can absorb all the care, can engross all the tenderness of the universal being, who,.according to thyself, ruleth the destiny of all things. What is the human race compared to the earth? What is this earth compared to the sun? What is our sun compared to those myriads of suns which at immense distances occupy the regions of space? not for the purpose of diverting thy weak eyes; not with a view to excite thy stupid admiration, as thou vainly imaginest; since multitudes of them are placed out of the range of thy visual organs, but to occupy the place which necessity hath assigned them. Mortal, feeble and vain! restore thyself to thy proper sphere; acknowledge every where the effect of necessity; recognise in thy benefits, behold in thy sorrows, the different modes of action of those various beings endowed with such a variety of properties of which nature is the assemblage; and do not any longer suppose that its pretended mover can possess such incompatible qualities as would be the result of human views, or of visionary ideas, which have no existence but in thyself.

Notwithstanding experience, which contradicts at each moment the beneficent views which man supposes in his God, theologians do not cease to call him good: when he complains of the disorders and calamities of which he is so frequently the victim, they assure him that these evils are only apparent: they tell him, that if his limited mind were capable of fathoming the depths of divine wisdom and the treasures of his goodness, he would always find the greatest benefits to result from that which he calls evil. But in spite of these frivolous answers, man will never be able to find good but in those objects which impel him in a manner favourable to his actual mode of existence; he shall always be obliged to find confusion and evil in every thing that painfully affects him, even cursorily: if God is the author of those two modes of feeling, so very opposite to each other, he must naturally conclude that this being is sometimes good and sometimes wicked; at least, if he will not allow either the one or the other, it must be admitted that he acts necessarily. A world where man experiences so much evil cannot be submitted to a God who is perfectly good; on the other hand, a world where he experiences so many benefits, cannot be governed by a wicked God. Thus he is obliged to admit of two principles equally powerful, who are in hostility with each other; or rather, he must agree that the same God is alternately kind and unkind; this after all is nothing more than avowing he cannot be otherwise than he is; in this case is it not useless to sacrifice to him, to pray, seeing it would be nothing but *destiny*—the necessity of things submitted to invariable rules.

In order to justify this God from the evils the human species experience, the deist is reduced to the necessity of calling them punishments inflicted by a *just* God for the transgressions of man. If so, man has the power to make his God suffer. To offend presupposes relations between the one who offends and another who is offended; but what relations can exist between the infinite being who has created the world and feeble mortals? To offend any one, is to diminish the sum of his happiness; it is to afflict him, to deprive him of something, to make him experience a painful sensation. How is it possible man can operate on the well-being of the omnipotent sovereign of nature, whose happiness is unalterable? How can the physical actions of a material substance have any influence over an immaterial substance, devoid of parts, having no point of contact? How can a corporeal being make an incorporeal being experience incommodious sensations? On the other hand, *justice*, according to the only ideas man can ever form of it, supposes a permanent disposition to render to each what is due to him; the theologian will not admit that God owes any thing to man; he insists that the benefits he bestows are all the gratuitous effects of his own goodness; that he has the right to dispose of the work of his hands according to his own pleasure; to plunge it if he please into the abyss of misery. But it is easy to see, that according to man's idea of justice, this does not even contain the shadow of it; that it is, in fact, the mode of action adopted by what he calls the most fright-

ful tyrants. How then can he be induced to call God just who acts after this manner? Indeed, while he sees innocence suffering, virtue in tears, crime triumphant, vice recompensed, and at the same time is told the being whom theology has invented is the author, he will never be able to acknowledge them to have *justice*.* But, says the deist, these evils are transient; they will only last for a time: very well, but then your God is unjust, at least for a time. It is for their good that he chastises his friends. But if he is good, how can he consent to let them suffer even for a time? If he knows every thing why reprove his favourites from whom he has nothing to fear? If he is really omnipotent, why not spare them these transitory pains, and procure them at once a durable and permanent felicity? If his power cannot be shaken, why make himself uneasy at the vain conspiracies they would form against him?

Where is the man filled with kindness, endowed with humanity, who does not desire with all his heart to render his fellow-creatures happy? If God really had man's qualities augmented, would he not by the same reasoning, exercise his infinite power render them all happy? Nevertheless we scarcely find any one who is perfectly satisfied with his condition on earth: for one mortal that enjoys, we behold a thousand who suffer; for one rich man who lives in the midst of abundance, there are thousands of poor who want common necessaries: whole nations groan in indigence, to satisfy the passions of some avaricious princes, of some few nobles, who are not thereby rendered more contented—who do not acknowledge themselves more fortunate on that account. In short, un-

---

* *Dies deficiet si velim numerare quibus bonis male evenerit; nec minus si commemorem quibus malis optime.*

Cicer. de Nat. Deor. lib. iii.

If a virtuous king possessed the ring of *Gyges*, that is to say, had the faculty of rendering himself invisible, would he not make use of it to remedy abuses to reward the good, to prevent the conspiracies of the wicked, to make order and happiness reign throughout his states? God is an invisible and all-powerful monarch, nevertheless his states are the theatre of crime, of confusion: he remedies nothing.

der the dominion of an omnipotent God, whose goodness is infinite, the earth is drenched with the tears of the miserable. What must be the inference from all this? That God is either negligent of, or incompetent to, his happiness. But the deist will tell you coolly, that the judgments of his God are impenetrable! How do we understand this term? Not to be taught—not to be informed—impervious—not to be pierced: in this case it would be an unreasonable question to inquire by what authority do you reason upon them? How do you become acquainted with these impenetrable mysteries? Upon what foundation do you attribute virtues which you cannot penetrate? What idea do you form to yourself of a justice that never resembles that of man?

To withdraw themselves from this, deists will affirm that the justice of their God is tempered with mercy, with compassion, with goodness: these again are human qualities: what, therefore, shall we understand by them? What idea do we attach to mercy? Is it not a derogation from the severe rules of an exact, a rigorous justice, which causes a remission of some part of a merited punishment? In a prince, clemency is either a violation of justice, or the exemption from a too severe law: but the laws of a God infinitely good, equitable, and wise, can they ever be too severe, and, if immutable, can he alter them? Nevertheless, man approves of clemency in a sovereign, when its too great facility does not become prejudicial to society; he esteems it, because it announces humanity, mildness, a compassionate, noble soul; qualities he prefers in his governors to rigour, cruelty, inflexibility: besides, human laws are defective; they are frequently too severe; they are not competent to foresee all the circumstances of every case: the punishments they decree are not always commensurate with the offence: he therefore does not always think them just; but he feels very well, he understands distinctly, that when the sovereign extends his mercy, he relaxes from his justice—that if mercy be merited, the punishment ought not to take place—that then its exercise is no longer clemency, but justice: thus he feels,

that in his fellow-creatures these two qualities cannot exist at the same moment. How then is he to form his judgment of a being who is represented to possess both in the extremest degree?

They then say, well, but in the next world this God will reward you for all the evils you suffer in this: this, indeed, is something to look to, if it had not been invented to shelter divine justice, and to exculpate him from those evils which he so frequently causes his greatest favourites to experience in this world: it is there, deists tell us, that the celestial monarch will procure for his elect that unalterable happiness, which he has refused them on earth; it is there he will indemnify those whom he loves for that transitory injustice, those afflicting trials, which he makes them suffer here below. In the meantime, is this invention calculated to give us those clear ideas suitable to justify providence? If God owes nothing to his creatures, upon what ground can they expect, in a future life, a happiness more real, more constant, than that which they at present enjoy? It will be founded, say theologians, upon his promises contained in his revealed oracles. But are they quite certain that these oracles have emanated from him? On the other hand, the system of another life does not justify this God for the most fleeting and transitory injustice; for does not injustice, even when it is transient, destroy that immutability which they attribute to the Divinity? In short, is not that omnipotent being whom they have made the author of all things, himself the first cause or accomplice of the offences which they commit against him? Is he not the true author of evil, or of the sin which he permits, whilst he is able to prevent it; and in this case can he, consistently with justice, punish those whom he himself renders culpable?

We have already seen the multitude of contradictions, the extravagant hypotheses, which the attributes theology gives to its God, must necessarily produce. A being clothed at one time with so many discordant qualities, will always be undefinable; they only present a train of ideas which will destroy each other, and he will in consequence remain a being of the imagination. This God has, say they, created the heavens, the earth, and the creatures who inhabit it, to manifest his own peculiar glory: but a monarch who is superior to all beings, who has neither rivals nor equals in nature, who cannot be compared to any of his creatures, is he susceptible of the desire of glory? Can he fear to be debased and degraded in the eyes of his fellow-creatures? Has he occasion for the esteem, the homage, or the admiration of men? The love of glory is in us only the desire of giving our fellow-creatures a high opinion of ourselves; this passion is laudable, when it stimulates us to perform great and useful actions; but more frequently it is only a weakness attached to our nature, it is only a desire in us to be distinguished from those beings with whom we compare ourselves. The God of whom they speak to us, ought to be exempt from this passion; according to theology he has no fellow-creatures, he has no competitors, he cannot be offended with those ideas which we form of him. His power cannot suffer any diminution, nothing is able to disturb his eternal felicity; must we not conclude from this that he cannot be either susceptible of desiring glory, or sensible to the praises and esteem of men? If this God is jealous of his prerogatives, of his titles, of his rank, and of his glory, wherefore does he suffer that so many men should offend him? Why does he permit so many others to have such unfavourable opinions of him? Why allows he others to have the temerity to refuse him that incense which is so. flattering to his pride?—How comes he to permit that a mortal like me, should dare attack his rights, his titles, and even his existence? It is in order to punish thee, you will say, for having made a bad use of his favours. But why does he permit me to abuse his kindness? Or why are not the favours which he confers on me sufficient to make me act agreeably to his views? It is because he has made thee free. Why has he given me liberty, of which he must have foreseen that I should be inclined to make an improper use? Is it then a present worthy of his goodness, to give me a faculty that enables me to brave his

omnipotence, to detach from him his adorers, and thus render myself eternally miserable? Would it not have been much more advantageous for me never to have been born, or at least to have been placed in the rank of brutes or stones, than to have been in despite of myself placed amongst intelligent beings, there to exercise the fatal power of losing myself without redemption, by offending or mistaking the arbiter of my fate? Had not God much better have shown his omnipotent goodness, and would he not have laboured much more efficaciously to his true glory, if he had obliged me to render him homage, and thereby to have merited an ineffable happiness?

The system of the liberty of man, which we have already destroyed, was visibly imagined to wipe from the author of nature the reproach which they must offer him in being the author, the source, the first cause of the crimes of his creatures. In consequence of this fatal present given by a beneficent God, men, according to the sinister ideas of theology, will for the most part be eternally punished for their faults in this world. Farfetched and endless torments are by the justice of a merciful and compassionate God, reserved for fragile beings, for transitory offences, for false reasonings, for involuntary errours, for necessary passions, which depend on the temperament this God has given them; circumstances in which he has has placed them, or, if they will, the abuse of this pretended liberty, which a provident God ought never to have accorded to beings capable of abusing it. Should we call that father good, rational, just, clement, or compassionate, who should arm with a dangerous and sharp knife the hands of a petulant child, with whose imprudence he was acquainted, and who should punish him all his life, for having wounded himself with it?— Should we call that prince just, merciful, and compassionate, who did not proportion the punishment to the offence, who should put no end to the torments of that subject who in a state of inebriety should have transiently wounded his vanity, without however causing him any real injustice—above all, after having himself taken pains to intoxicate him? Should we look upon that monarch as all-powerful, in whose dominions should be in such a state of anarchy, that, with the exception of a small number of faithful subjects, all the others should have the power every instant to despise his laws, insult him, and frustrate his will? O, theologians! confess that your God is nothing but a heap of qualities, which form a whole as perfectly incomprehensible to your mind as to mine; by dint of overburdening him with incompatible qualities, ye have made him truly a chimera, which all your hypotheses cannot maintain in the existence you are anxious to give him.

They will, however, reply to these difficulties, that goodness, wisdom, and justice, are, in God, qualities so eminent, or have such little similarity to ours, that they have no relation with these qualities when found in men. But I shall answer, how shall I form to myself ideas of these divine perfections, if they bear no resemblance to those of the virtues which I find in my fellow-creatures, or to the dispositions which I feel in myself? If the justice of God is not that of men; if it operates in that mode which men call injustice, if his goodness, his clemency, and his wisdom do not manifest themselves by such signs, that we are able to recognise them; if all his divine qualities are contrary to received ideas; if in theology all the human actions are obscured or overthrown, how can mortals like myself pretend to announce them, to have a knowledge of them, or to explain them to others? Can theology give to the mind the ineffable boon of conceiving that which no man is in a capacity to comprehend? Can it procure to its agents the marvellous faculty of having precise ideas of a God composed of so many contradictory qualities? In short, is the theologian himself a God?

They silence us by saying, that God himself has spoken, that he has made himself known to men. But when, where, and to whom has he spoken? Where are these divine oracles? A hundred voices raise themselves in the same moment, a hundred hands show them to me in absurd and discordant collections: I run them over, and through the whole I find that the God

of wisdom has spoken an obscure, insidious, and irrational language. I see that the *God of goodness* has been cruel and sanguinary; that the *God of justice* has been unjust and partial, has ordered iniquity; that the *God of mercies* destines the most hideous punishments to the unhappy victims of his anger. Besides, obstacles present themselves when men attempt to verify the pretended relations of a Divinity, who, in two countries, has never literally holden the same language; who has spoken in so many places, at so many times, and always so variously, that he appears every where to have shown himself only with the determined design of throwing the human mind into the strangest perplexity.

Thus, the relations which they suppose between men and their God can only be founded on the moral qualities of this being; if these are not known to men, they cannot serve them for models. It is needful that these qualities were natural in a known being in order to be imitated; how can I imitate a God of whom the goodness and the justice do not resemble mine in any thing, or rather are directly contrary to that which I call either just or good? If God partakes in nothing of that which forms us, how can we even distantly, propose to ourselves the imitating him, the resembling him, the following a conduct necessary to please him by conforming ourselves to him? What can in effect, be the motives of that worship, of that homage, and of that obedience, which we are told to render to the Supreme Being, if we do not establish them upon his goodness, upon his veracity, upon his justice, in short, upon qualities which we are able to understand? How can we have clear ideas of these qualities in God if they are no longer of the same nature as our own?

They will no doubt tell us, that there cannot be any proportion between the creator and his work; that the clay has no right to demand of the potter who has formed it, *why have you fashioned me thus?* But if there be no proportion between the workman and his work; if there be no analogy between them, what can be the relations which will subsist between them? If God is incorporated, how does he act upon bodies, or how can corporeal beings be able to act upon him, offend him, disturb his repose, excite in him emotions of anger? If man is relatively to God only an *earthen vase*, this *vase* owes neither prayers nor thanks to the potter for the form which he has been pleased to give it. If this potter irritates himself against his *vase* for having formed it badly, or for having rendered it incapable of the uses to which he had destined it, the potter, if he is not an irrational being, ought to take to himself the defects which he finds in it; he certainly has the power to break it, and the *vase* cannot prevent him; it will neither have motives nor means to soften his anger, but will be obliged to submit to its destiny; and the potter would be completely deprived of reason if he were to punish his vase, rather than, by forming it anew, give it a figure more suitable to his designs.

We see, that according to these notions, men have no more relation with God than stones. But if God owes nothing to men, if he is not bound to show them either justice or goodness, men cannot possibly owe any thing to him. We have no knowledge of any relations between beings which are not reciprocal; the duties of men amongst themselves are founded upon their mutual wants; if God has not occasion for them, they cannot owe him any thing, and men cannot possibly offend him. In the meantime, the authority of God can only be founded on the good which he does to men, and the duties of these towards God, can have no other motives than the hope of that happiness which they expect from him; if he does not owe them this happiness, all their relations are annihilated, and their duties no longer exist. Thus, in whatever manner we view the theological system, it destroys itself. Will theology never feel that the more it endeavours to exalt its God, to exaggerate his grandeur, the more incomprehensible it renders him to us? That the farther it removes him from man, or the more it debases this man, the more it weakens the relations which they have supposed between this God and him; if the sovereign of nature is an infinite being and totally different from our species, and if man is only in his eyes a worm or a speck of dirt, it is

clear there cannot be any *moral relations* between two beings so little analogous to each other; and again it is still more evident that the *vase* which he has formed is not capable of reasoning upon him.

It is, however, upon the relation subsisting between man and his God that all worship is founded, and all the religions of the world have a despotic God for their basis; but is not despotism an unjust and unreasonable power? Is it not equally to undermine his goodness, his justice, and his infinite wisdom, to attribute to the Divinity the exercise of such a power? Men in seeing the evils with which they are frequently assailed in this world, without being able to guess by what means they have deserved the divine anger, will always be tempted to believe that the master of nature is a *sultan*, who owes nothing to his subjects, who is not obliged to render them any account of his actions, who is not bound to conform himself to any law, and who is not himself subjected to those rules which he prescribes for others; who in consequence can be unjust, who has the right to carry his vengeance beyond all bounds; in short, the theologians pretend that God would have the right of destroying the universe, and replunging it into the chaos from whence his wisdom has withdrawn it; whilst the same theologians, quote to us the order and marvellous arrangement of this world, as the most convincing proof of his existence.[*]

In short, theology invests their God with the incommunicable privilege of acting contrary to all the laws of nature and of reason, whilst it is upon his reason, his justice, his wisdom and his fidelity in the fulfilling his pretended engagements, that they are willing to establish the worship which we owe him, and the duties of morality. What an ocean of contradictions! A being who can do every thing, and who owes nothing to any one, who, in his eternal decrees, can elect or reject, predestinate to happiness or to misery, who has the right of making men the playthings of his caprice, and to afflict them

[*] "We conceive, at least," says *Doctor Castrill*, "that God is able to overturn the universe, and replunge it into chaos." *See his Defence of Religion, Natural and Revealed.*

without reason, who could go so far as even to destroy and annihilate the universe, is he not a tyrant or a demon? is there any thing more frightful than the immediate consequences to be drawn from these revolting ideas given to us of their God, by those who tell us to love him, to serve him, to imitate him, and to obey his orders? Would it not be a thousand times better to depend upon blind matter, upon a nature destitute of intelligence, upon chance, or upon nothing, upon a God of stone or of wood, than upon a God who is laying snares for men, inviting them to sin, and permitting them to commit those crimes which he could prevent, to the end that he may have the barbarous pleasure of punishing them without measure, without utility to himself, without correction to them, and without their example serving to reclaim others? A gloomy terrour must necessarily result from the idea of such a being; his power will wrest from us much servile homage; we shall call him good to flatter him or to disarm his malice; but, without overturning the essence of things, such a God will never be able to make himself beloved by us, when we shall reflect that he owes us nothing, that he has the right of being unjust, that he has the power to punish his creatures for making a bad use of the liberty which he grants them, or for not having had that grace which he has been pleased to refuse them.

Thus, in supposing that God is not bound towards us by any rules, theologians visibly sap the foundation of all religion. A theology which assures us that God has been able to create men for the purpose of rendering them eternally miserable, shows us nothing but an evil and malicious genius, whose malice is inconceivable, and infinitely surpasses the cruelty of the most depraved beings of our species. Such is nevertheless the God which they have the confidence to propose for a model to the human species! Such is the Divinity which is adored even by those nations who boast of being the most enlightened in this world!

It is however upon the moral character of the Divinity, that is to say, upon his goodness, his wisdom, his equity, and his love of order, that they

pretend to establish our morals, or the science of those duties which connect us to the beings of our species. But as his perfections and his goodness are contradicted very frequently and give place to weakness, to injustice, and to cruelties, we are obliged* to pronounce him changeable, fickle, capricious, unequal in his conduct, and in contradiction with himself, according to the various modes of action which they attribute to him. Indeed, we sometimes see him favourable to, and sometimes disposed to injure the human species; sometimes a friend to reason and the happiness of society; sometimes he interdicts the use of reason, he acts as the enemy of all virtue, and he is flattered with seeing society disturbed. However, as we have seen mortals crushed by fear, hardly ever daring to avow that their God was unjust or wicked, to persuade themselves that he authorized them to be so, it was concluded simply that every thing which they did according to his pretended order or with the view of pleasing him, was always good, however prejudicial it might otherwise appear in the eyes of reason. They supposed him the master of creating the just and the unjust, of changing good into evil, and evil into good, truth into falsehood, and falsehood into truth: in short, they gave him the right of changing the eternal essence of things; they made this God superior to the laws of nature, of reason, and virtue; they believed they could never do wrong in following his precepts, although the most absurd, the most contrary to morals, the most opposite to good sense, and the most prejudicial to the repose of society. With such principles do not let us be surprised at those horrours which religion causes to be committed on the earth. The most atrocious religion was the most consistent.*

---

* The modern religion of Europe has visibly caused more ravages and troubles than any other known superstition; it was in that respect very consistent with its principles. They may well preach tolerance and mildness in the name of a despotic God, who alone has a right to the homage of the earth, who is extremely jealous, who wills that they should admit some doctrines, who punishes cruelly for erroneous opinions, who demands zeal from his adorers, such a God must make fanatical persecutors of all consistent men. The

In founding morals upon the immoral character of a God, who changes his conduct, man will never be able to ascertain what conduct he ought to pursue with regard to that which he owes to God, or to others. Nothing then was more dangerous than to persuade him there existed a being superior to nature, before whom reason must remain silent; to whom, to be happy hereafter, he must sacrifice every thing here. His pretended orders, and his example must necessarily be much stronger than the precepts of human morals; the adorers of this God, cannot then listen to nature and good sense, but when by chance they accord with the caprice of their God, in whom they suppose the power of annihilating the invariable relation of beings, of changing reason into folly, justice into injustice, and even crime into virtue. By a consequence of these ideas, the religious man never examines the will and the conduct of this celestial despot according to ordinary rules; every inspired man that comes from him, and those who shall pretend they are charged with interpreting his oracles, will always assume the right of rendering him irrational and criminal; his first duty will always be to obey his God without murmuring.

Such are the fatal and necessary consequences of the moral character which they give to the Divinity, and of the opinion which persuades mortals they ought to pay a blind obedience to the absolute sovereign whose arbitrary and fluctuating will regulates all duties. Those who first had the confidence to tell men, that in matters of religion, it was not permitted them to consult their reason, nor the interests of society, evidently proposed to themselves to make them the sport of the instruments of their own peculiar wickedness. It theology of the present day is a subtile venom, calculated to infect all by the importance which attached to it. By dint of *metaphysics*, modern theologians have become systematically absurd and wicked: by once admitting the odious ideas which they gave of the Divinity, it was impossible to make them understand that they ought to be humane, equitable, pacific, indulgent, or tolerant; they pretended **and** proved that these humane and social virtues, were not seasonable in the cause of religion, and would be treason and crimes in the eyes of the celestial Monarch, to whom every thing ought to be sacrificed.

No. VII.—26

is from this radical errour, then, that have sprung all those extravagances, which the different religions have introduced upon the earth; that sacred fury which has deluged it in blood; those inhuman persecutions which have so frequently desolated nations; in short, all those horrid tragedies, of which the name of the Most High have the cause and the pretext. Whenever they have been desirous to render men unsociable, they have cried out that it was the will of God they should be so. Thus the theologians themselves have taken pains to calumniate and to defame the phantom which they have erected upon the ruins of human reason, of a nature well known, and a thousand times preferable to a tyrannical God, whom they render odious to every honest man. These theologians are the true destroyers of their own peculiar idol, by the contradictory qualities which they accumulate on him: it is these theologians, as we shall yet prove in the sequel, who render morals uncertain and fluctuating, by founding them upon a changeable and capricious God, much more frequently unjust and cruel, than good: it is they who overturn and annihilate him, by commanding crime, carnage, and barbarity, in the name of the sovereign of the universe, and who interdict us the use of reason, which alone ought to regulate all our actions and ideas.

However, admitting for a moment that God possesses all the human virtues in an infinite degree of perfection, we shall presently be obliged to acknowledge that he cannot connect them with those metaphysical, theological, and negative attributes, of which we have already spoken. If God is a spirit, how can he act like man, who is a corporeal being? A pure spirit sees nothing; it neither hears our prayers nor our cries, it cannot be conceived to have compassion for our miseries, being destitute of those organs by which the sentiments of pity can be excited in us. He is not *immutable*, if his disposition can change: he is not *infinite*, if the totality of nature, without being him, can exist conjointly with him; he is not *omnipotent*, if he permits, or if he does not prevent disorder in the world: he is not *omnipresent*, if he is not in the man who sins, or if he leaves at the moment in which he commits the sin. Thus, in whatever manner we consider this God, the human qualities which they assign him, necessarily destroy each other; and these same qualities cannot, in any possible manner, combine themselves with the supernatural attributes given him by theology.

With respect to the pretended *revelation* of the will of God, far from being a proof of his goodness, or of his commiseration for men, it would only be a proof of his malice. Indeed, all revelation supposes the Divinity guilty of leaving the human species, during a considerable time, unacquainted with truths the most important to their happiness. This revelation, made to a small number of chosen men, would moreover show a partiality in this being, an unjust predilection but little compatible with the goodness of the common Father of the human race. This revelation destroys also the divine immutability, since, by it, God would have permitted at one time, that men should be ignorant of his will, and at another time, that they should be instructed in it. This granted, all revelation is contrary to the notions which they give us of the justice or of the goodness of a God, who they tell us is immutable, and who, without having occasion to reveal himself, or to make himself known to them by miracles, could easily instruct and convince men, and inspire them with those ideas, which he desires ; in short, dispose of their minds and of their hearts. What if we should examine in detail all those pretended revelations, which they assure us have been made to mortals ? We shall see that this God only retails fables unworthy of a wise being; as is in them, in a manner contrary to the natural notions of equity ; announces enigmas and oracles impossible to be comprehended ; paints himself under traits incompatible with his infinite perfections; exacts puerilities which degrade him in the eyes of reason ; deranges the order which he has established in nature, to convince creatures, whom he will never cause to adopt those ideas, those sentiments, and that conduct, with which he would inspire them. In short, we shall find, that God has never manifested himself, but to announce

inexplicable mysteries, unintelligible doctrines, ridiculous practices ; to throw the human mind into fear, distrust, perplexity, and above all, to furnish a never-failing source of dispute to mortals.*

We see, then, that the ideas which theology gives us of the Divinity will always be confused and incompatible, and will necessarily disturb the repose of human nature. These obscure notions, these vague speculations, would be of great indifference, if men did not regard their reveries on this unknown being, upon whom they believe they depend, as important, and if they did not draw from them conclusions pernicious to themselves. As they never will have a common and fixed standard, whereby to form a judgment on this being, to whom various and diversely modified imaginations have given birth, they will never be able either to understand each other, or to be in accord with each other upon those ideas they shall form to themselves of him. From hence, that necessary diversity of religious opinions, which, in all ages, has given rise to the most irrational disputes which they always look upon as very essential, and which has consequently always interested the tranquillity of nations. A man with a heated imagination, will not accommodate himself to the God of a phlegmatic and tranquil man ; and infirm, bilious, discontented man, will never see this God in the same point of view as he who enjoys a constitution more sound, whence commonly results gayety, contentment, and peace. An equitable, kind, compassionate, tender-hearted man, will not delineate to himself the same portrait of his God, as the man who is of a

---

* It is evident that all revelation, which is not clear, or which teaches *mysteries,* cannot be the work of a wise and intelligent being : as soon as he speaks, we ought to presume, it is for the purpose of being understood by those to whom he manifests himself. To speak so as not to be understood, only shows folly or want of good faith. It is, then, very clear, that all things which the priesthood have called *mysteries,* are inventions, made to throw a thick veil over their own peculiar contradictions, and their own peculiar ignorance of the Divinity. But they think to solve all difficulties by saying *it is a mystery ;* taking care, however, that men should know nothing of that pretended science, of which they have made themselves the depositaries.

harsh, unjust, inflexible, wicked character. Each individual will modify his God after his own peculiar manner of existing, after his own mode of thinking, according to his particular mode of feeling. A wise, honest, rational man will never figure to himself that a God can be unjust and cruel.

Nevertheless, as fear necessarily presided at the formation of those Gods man set up for the object of his worship; as the ideas of the Divinity was always associated with that of terrour; as the recollection of sufferings, which he attributed to God, often made him tremble; frequently awakened in his mind the most afflicting reminiscence; sometimes filled him with inquietude, sometimes inflamed his imagination, sometimes overwhelmed him with dismay. The experience of all ages proves, that this vague name became the most important of all considerations—was the affair which most seriously occupied the human race: that it every where spread consternation—produced the most frightful ravages, by the delirious inebriation resulting from the opinions with which it is intoxicated the mind. Indeed, it is extremely difficult to prevent habitual fear, which of all human passions is the most incommodious, from becoming a dangerous leaven, which, in the long run, will sour, exasperate, and give malignancy to the most moderate temperament.

If a misanthrope, in hatred of his race, had formed the project of throwing man into the greatest perplexity—if a tyrant, in the plenitude of his unruly desire to punish, had sought out the most efficacious means; could either the one or the other have imagined that which was so well calculated to gratify their revenge, as thus to occupy him unceasingly with being not only unknown to him, but which can never be known, which, notwithstanding, they should be obliged to contemplate as the centre of all their thoughts —as the only model of their conduct—as the end of all their actions—as the subject of all their research — as a thing of more importance to them than life itself, upon which all their present felicity, all their future happiness, must necessarily depend ? If man was subjected to an absolute monarch, to a sultan who should keep himself secluded from

his subjects; who followed no rule but his own desires; who did not feel himself bound by any duty; who could for ever punish the offences committed against him; whose fury it was easy to provoke; who was irritated even by the ideas, the thoughts of his subjects; whose displeasure might be incurred without even their own knowledge; the name of such a sovereign would assuredly be sufficient to carry trouble, to spread terrour, to diffuse consternation into the very souls of those who should hear it pronounced; his idea would haunt them every where—would unceasingly afflict them—would plunge them into despair. What tortures would not their mind endure to discover this formidable being, to ascertain the secret of pleasing him ! What labour would not their imagination bestow, to discover what mode of conduct might be able to disarm his anger ! What fears would assail them, lest they might not have justly hit upon the means of assuaging his wrath! What disputes would they not enter into upon the nature, the qualities of a ruler, equally unknown to them all ! What a variety of means would not be adopted, to find favour in his eyes; to avert his chastisement !

Such is the history of the effects the name of God has produced upon the earth. Man has always been panic-struck at it, because he never was able to form any correct opinion, any fixed ideas upon the subject ; because every thing conspired either to give his ideas a fallacious turn, or else to keep his mind in the most profound ignorance ; when he was willing to set himself right, when he was sedulous to examine the path which conducted to his felicity, when he was desirous of probing opinions so consequential to his peace, involving so much mystery, yet combining both his hopes and his fears, he was forbidden to employ the only proper method—*his reason*, guided by his experience; he was assured this would be an offence the most indelible. If he asked, wherefore his reason had been given him, since he was not to use it in matters of such high behest ? he was answered, those were mysteries of which none but the initiated could be informed ; that it sufficed for him to know that the reason which he seemed so highly to prize, which he held in so much esteem, was his most dangerous enemy—his most inveterate, most determined foe. He is told that he must believe in God, not question the mission of the priests; in short, that he had nothing to do with the laws he imposed, but to obey them: when he then required that these laws might at least be made comprehensible to him; that he might be placed in a capacity to understand them; the old answer was returned, that they were *mysteries;* he must not inquire into them. Thus he had nothing steady.; nothing permanent, whereby to guide his steps; like a blind man left to himself in the streets, he was obliged to grope his way at the peril of his existence. This will serve to show the urgent necessity there is for truth to throw its radiant lustre on systems big with so much importance; that are so calculated to corroborate the animosities, to confirm the bitterness of soul, between those whom nature intended should always act as brothers.

By the magical charms with which this God was surrounded, the human species has remained either as if it was benumbed, in a state of stupid apathy, or else it has become furious with fanaticism: sometimes, desponding with fear, man cringed like a slave who bends under the scourge of an inexorable master, always ready to strike him; he trembled under a yoke made too ponderous for his strength: he lived in continual dread of a vengeance he was unceasingly striving to appease, without ever knowing when he had succeeded: as he was always bathed in tears, continually enveloped in misery—as he was never permitted to lose sight of his fears as he was continu ally exhorted to nourish his alarm, he could neither labour for his own happiness nor contribute to that of others; nothing could exhilarate him; he became the enemy of himself, the persecutor of his fellow-creatures, because his felicity here below was interdicted; he passed his time in heaving the most bitter sighs; his reason being forbidden him, he fell into either a state of infancy or delirium, which submitted him to authority; he was destined to this servitude from the hour he quitted his mother's womb, until that in which

he was returned to his kindred dust; tyrannical opinion bound him fast in her massive fetters; a prey to the terrours with which he was inspired, he appeared to have come upon the earth for no other purpose than to dream—with no other desire than to groan—with no other motives than to sigh; his only view seemed to be to injure himself; to deprive himself of every rational pleasure; to embitter his own existence; to disturb the felicity of others. Thus, abject, slothful, irrational, he frequently became wicked, under the idea of doing honour to his God; because they instilled into his mind that it was his duty to avenge his cause, to sustain his honour, to propagate his worship.

Mortals were prostrate from race to race, before vain idols to which fear had given birth in the bosom of ignorance, during the calamities of the earth; they tremblingly adored phantoms which credulity had placed in the recesses of their own brain, where they found a sanctuary which time only served to strengthen; nothing could undeceive them; nothing was competent to make them feel, it was themselves they adored—that they bent the knee before their own work—that they terrified themselves with the extravagant pictures they had themselves delineated: they obstinately persisted in prostrating themselves, in perplexing themselves, in trembling; they even made a crime of endeavouring to dissipate their fears; they mistook the production of their own folly; their conduct resembled that of children, who having disfigured their own features, become afraid of themselves when a mirror reflects the extravagance they have committed. These notions so afflicting for themselves, so grievous to others, have their epoch in the calamitous idea of a God; they will continue, perhaps augment, until their mind, enlightened by discarded reason, illumined by truth, shall attach no more importance to this unintelligible word; until man, bursting the chains of superstition, taking a rational view of that which surrounds him, shall no longer refuse to contemplate nature under her true character; shall no longer persist in refusing to acknowledge she contains within herself the cause of that wonderful phenomena which strikes on the dazzled optics of man: until thoroughly persuaded of the weakness of their claims to the homage of mankind, he shall make one simultaneous, mighty effort, and overthrow the altars of God and his priests.

CHAPTER IV.

Examination of the Proofs of the Existence of the Divinity, as given by Clarke.

THE unanimity of man in acknowledging the Divinity, is commonly looked upon as the strongest proof of his existence. There is not, it is said, any people on the earth who have not some ideas, whether true or false, of an all-powerful agent who governs the world. The rudest savages, as well as the most polished nations, are equally obliged to recur by thought to the first cause of every thing that exists; thus it is affirmed, the cry of nature herself ought to convince us of the existence of a God, of which she has taken pains to engrave the notion in the minds of men: they therefore conclude, that the idea of God is innate. But if this existence rests upon no better foundations than the unanimity of man on this subject, it is not placed upon so solid a rock as those who make this asseveration may imagine: the fact is, man is not generally agreed upon this point; if he was, superstition could have no existence; the idea of God cannot be *innate*, because, independent of the proofs offered on every side of the almost impossibility of innate ideas, one simple fact will set such an opinion for ever at rest, except with those who are obstinately determined not to be convinced by even their own arguments: if this idea was innate, it must be every where the same; seeing that that which is antecedent to man's being, cannot have experienced the modifications of his existence, which are posterior. Even if it were waived, that the same idea should be expected from all mankind, but that only every nation should have their ideas alike on this subject, experience will not warrant the assertion, since nothing can be better established than that the idea is not uniform even in the same town; now this would be an insuperable quality in an innate idea. It not unfrequently happens, that in the endeavour to prove too much, that which stood firm before the attempt is weakened; thus a bad advocate frequently injures a good cause, although he may not be able to overturn the rights on which it is rested. It would, therefore, perhaps, come nearer to the point if it was said, that the natural curiosity of mankind has in all ages, and in all nations, led him to seek after the primary cause of the phenomena he beholds; that owing to the variations of his climate, to the difference of his organization, the greater or less calamity he has experienced, the variety of his intellectual faculties, and the circumstances under which he has been placed, man has had the most opposite, contradictory, extravagant notions of this Divinity.

If disengaged from prejudice, we analyze this proof, we shall see that the universal consent of man, so diffused over the earth, actually proves little more than that he has been in all countries exposed to frightful revolutions, experienced disasters, been sensible to sorrows of which he has mistaken the physical causes; that those events to which he has been either the victim or the witness, have called forth his admiration or excited his fear; that for want of being acquainted with the powers of nature, for want of understanding her laws, for want of comprehending her infinite resources, for want of knowing the effects she must necessarily produce under given circumstances, he has believed these phenomena were due to some secret agent of which he has had vague ideas—to beings whom he has supposed conducted themselves after his own manner; who were operated upon by similar motives with himself.

The consent then of man in acknowledging a God, proves nothing, except that in the bosom of ignorance he has either admired the phenomena of nature, or trembled under their influence; that his imagination was disturbed by what he beheld or suffered; that he has sought in vain to relieve his perplexity, upon the unknown cause of the phenomena he witnessed, which frequently obliged him to quake with terrour: the imagination of the human race has laooured variously upon these causes, which have almost always been incomprehensible to him; although every thing confessed his ignorance, his inability to define this cause, yet he maintained that he was assured of its existence; when pressed, he spoke of a spirit, (a word to which it was impossible to attach any determinate idea,) which taught nothing but the sloth, which evidenced nothing but the stupidity of those who pronounced it.

It ought, however, not to excite any surprise that man is incapable of forming any substantive ideas, save of those things which act, or which have heretofore acted upon his senses; it is very evident that the only objects competent to move his organs are material— that none but physical beings can furnish him with ideas—a truth which has been rendered sufficiently clear in the commencement of this work, not to need any further proof. It will suffice therefore to say, that the idea of a God is not an innate, but an acquired notion; that it is the very nature of this notion to vary from age to age; to differ in one country from another; to be viewed variously by individuals. What do I say? It is, in fact, an idea hardly ever constant in the same mortal. This diversity, this fluctuation, this change, stamps it with the true character of an acquired opinion. On the other hand, the strongest proof that can be adduced that these ideas are founded in errour, is, that man by degrees has arrived at perfectioning all the sciences which have any known objects for their basis, whilst the science of deism has not advanced; it is almost every where at the same point; men seem equally undecided on this subject; those who have most occupied themselves with it, have effected but little; they seem, indeed, rather to have rendered the primitive ideas man formed to himself on this head more obscure.

As soon as it is asked of man, what is the God before whom he prostrates himself, forthwith his sentiments are divided. In order that his opinions should be in accord, it would be requisite that uniform ideas, analogous sensations, unvaried perceptions, should every where have given birth to his notions upon this subject: but this

would suppose organs perfectly similar, modified by sensations which have a perfect affinity: this is what could not happen: because man, essentially different by his temperament, who is found under circumstances completely dissimilar, must necessarily have a great diversity of ideas upon objects which each individual contemplates so variously. Agreed in some general points, each made himself a God after his own manner; he feared him, he served him, after his own mode. Thus the God of one man, or of one nation, was hardly ever that of another man, or of another nation. The God of a savage, unpolished people, is commonly some material object, upon which the mind has exercised itself but little; this God appears very ridiculous in the eyes of a more polished community, whose minds have laboured more intensely upon the subject. A spiritual God, whose adorers despise the worship paid by the savage to a coarse, material object, is the subtile production of the brain of thinkers, who, lolling in the lap of polished society quite at their leisure, have deeply meditated, have long occupied themselves with the subject. The theological God, although incomprehensible, is the last effort of the human imagination; it is to the God of the savage, what an inhabitant of the city of Sybaris, where effeminacy and luxury reigned, where pomp and pageantry had reached their climax, clothed with a curiously embroidered purple habit of silk, was to a man either quite naked, or simply covered with the skin of a beast, perhaps newly slain. It is only in civilized societies, that leisure affords the opportunity of dreaming— that ease procures the facility of reasoning; in these associations, idle speculators meditate, dispute, form metaphysics: the faculty of thought is almost void in the savage, who is occupied either with hunting, with fishing, or with the means of procuring a very precarious subsistence by dint of almost incessant labour. The generality of men, even among us, have not more elevated notions of the Divinity, have not analyzed him more than the savage. A spiritual, immaterial God, is formed only to occupy the leisure of some subtile men, who have no occasion to labour for a subsistence. Theology, although a science so much vaunted, considered so important to the interests of man, is only useful to those who live at the expense of others; or of those who arrogate to themselves the privilege of thinking for all those who labour.— This futile science becomes, in some polished societies, who are not on that account more enlightened, a branch of commerce extremely advantageous to its professors, but equally unprofitable to the citizens; above all when these have the folly to take a very decided interest in their unintelligible opinions.

What an infinite distance between an unformed stone, an animal, a star, a statue, and the abstracted Deity, which theology has clothed with attributes under which it loses sight of him itself! The savage without doubt deceives himself in the object to which he addresses his vows; like a child he is smitten with the first object that strikes his sight—that operates upon him in a lively manner; like the infant, his fears are alarmed by that from which he conceives he has either received an injury or suffered disgrace; still his ideas are fixed by a substantive being, by an object which he can examine by his senses. The Laplander who adores a rock—the negro who prostrates himself before a monstrous serpent, at least see the objects they adore. The idolater falls upon his knees before a statue, in which he believes there resides some concealed virtue, some powerful quality, which he judges may be either useful or prejudicial to himself; but that subtile reasoner, called a theologian, who in consequence of his unintelligible science, believes he has a right to laugh at the savage, to deride the Laplander, to scoff at the negro, to ridicule the idolater, does not perceive that he himself is prostrate before a being of his own imagination, of which it is impossible he should form to himself any correct idea, unless, like the savage, he re-enters into visible nature, to clothe him with qualities capable of being brought within the range of his comprehension.

Thus the notions on the Divinity, which obtain credit even at the present day, are nothing more than a general

terrour diversely acquired, variously modified in the mind of nations, which do not tend to prove any thing, save that they have received them from their trembling, ignorant ancestors. These Gods have been successively altered, decorated, subtilized, by those thinkers, those legislators, those priests, who have meditated deeply upon them; who have prescribed systems of worship to the uninformed; who have availed themselves of their existing prejudices, to submit them to their yoke; who have obtained a dominion over their minds by seizing on their credulity—by making them participate in their errours—by working on their fears; these dispositions will always be a necessary consequence of man's ignorance, when steeped in the sorrows of his heart.

If it be true, as asserted, that the earth has never witnessed any nation so unsociable, so savage, to be without some form of religious worship—who did not adore some God—but little will result from it respecting its reality.— The word God, will rarely be found to designate more than the unknown cause of those effects which man has either admired or dreaded. Thus, this notion so generally diffused, upon which so much stress is laid, will prove little more than that man in all generations has been ignorant of natural causes—that he has been incompetent, from some cause or other, to account for those phenomena which either excited his surprise or roused his fears. If at the present day a people cannot be found destitute of some kind of worship, entirely without superstition, who ·do not acknowledge a God, who have not adopted a theology more or less subtile, it is because the uninformed ancestors of these people have all endured misfortunes—have been alarmed by terrifying effects, which they have attributed to unknown causes—have beheld strange sights, which they have ascribed to powerful agents, whose existence they could not fathom; the details of which, together with their own bewildered notions, they have handed down to their posterity who have not given them any kind of examination.

Besides, the universality of an opinion by no means proves its truth. Do we not see a great number of ignorant prejudices, a multitude of barbarous errours, even at the present day, receive the almost universal sanction of the human race? Are not all the inhabitants of the earth imbued with the idea of magic—in the habit of acknowledging occult powers—given to divination—believers in enchantment—the slaves to omens—supporters of witchcraft—thoroughly persuaded of the existence of ghosts? If some of the most enlightened persons are cured of these follies, they still find very zealous partisans in the greater number of mankind, who accredit them with the firmest confidence. It would not, however, be concluded by men of sound sense, that therefore these chimeras actually have existence, although sanctioned with the credence of the multitude. Before Copernicus, there was no one who did not believe that the earth was stationary, that the sun described his annual revolution round it. Was, however, this universal consent of man, which endured for so many thousand years, less an errour on that account?*

Each man has his God: but do all these Gods exist? In reply it will be said each man has his ideas of the sun; do all these suns exist? However narrow may be the pass by which superstition imagines it has thus guarded its favourite hypothesis, nothing will perhaps be more easy than the answer: the existence of the sun is a fact verifi-

ed by the daily use of the senses; all the world see the sun; no one hath ever seen God; nearly all mankind has acknowledged the sun to be both luminous and hot: however various may be the opinions of man, upon this luminary, no one has ever yet pretended there was more than one attached to our planetary system, or *that the sun is not luminous and hot;* but we find many very rational men have said, THERE IS NO GOD. Those who think this proposition hideous and irrational, and who affirm that God exists, do they not tell us at the same time that they have never seen him, and therefore know nothing of him? Theology is a science, where every thing is built upon laws inverted from those common to the globe we inhabit.

If man, therefore, had the courage to throw aside his prejudices, which every thing conspires to render as durable as himself—if divested of fear he would examine coolly—if guided by reason he would dispassionately view the nature of things, the evidence adduced in support of any given doctrine; he would, at least, be under the necessity to acknowledge, that the idea of the Divinity is not innate—that it is not anterior to his existence—that it is the production of time, acquired by communication with his own species*—that, consequently, there was a period, when it did not actually exist in him: he would see clearly, that he holds it by tradition from those who reared him: that these themselves received it from their ancestors: that thus tracing it up, it will be found to have been derived in the last resort, from ignorant savages, who were our first fathers.—The history of the world will show

---

* Yet to have doubted the truth of such a generally-diffused opinion, one that had received the sanction of so many learned men—that was clothed with the sacred vestments of so many ages of credulity—that had been adopted by Moses, acknowledged by Solomon, accredited by the Persian magi—that Elijah had not refuted—that obtained the fiat of the most respectable universities, the most enlightened legislators, the wisest kings, the most eloquent ministers: in short, a principle that embraced all the stability that could be derived from the universal consent of all ranks: to have doubted of this, would at one period been held as the highest degree of profanation, as the most presumptuous scepticism, as an impious blasphemy, that would have threatened the very existence of that unhappy country from whose unfortunate bosom such a venomous, sacrilegious mortal could have arisen. It is well known what opinion was entertained of Galileo for maintaining the existence of the antipodes. Pope Gregory excommunicated as atheists all those who gave it credit.

* When men shall be willing coolly to examine the proof of the existence of a God, drawn from general consent, they will acknowledge, that they can gather nothing from it, except that all men have guessed that there existed in nature unknown motive-powers, unknown causes; a truth of which no one has ever doubted, seeing that it is impossible to suppose effects without causes. Thus the only difference betwixt the ATHEISTS and the THEOLOGIANS, or the WORSHIPPERS OF GOD, is, that the first assign to all the phenomena *material, natural, sensible,* and *known* causes; whereas, the last assign them *spiritual, supernatural, unintelligible,* and *unknown* causes. The God of the theologians, is it in effect any other thing than an *occult power?*

that crafty legislators, ambitious tyrants, blood-stained conquerors, have availed themselves of the ignorance, the fears, the credulity of his progenitors, to turn to their own profit an idea to which they rarely attached any other substantive meaning than that of submitting them to the yoke of their own domination.

Without doubt, there have been mortals who have boasted they have seen the Divinity; but the first man who dared to say this was a liar, whose object was to take advantage of the simplicity of some, or an enthusiast, who promulgated for truths, the crazy reveries of his own distempered imagination? Nevertheless, is it not a truth, that these doctrines of crafty men are at this day the creed of millions, transmitted to them by their ancestors, rendered sacred by time, read to them in their temples, and adorned with all the ceremonies of religious worship? Indeed that man, would not experience the most gentle treatment from the infuriated Christian, who should to his face venture to dispute the divine mission of his Jesus. Thus the ancestors of the Europeans have transmitted to their posterity, those ideas of the Divinity which they manifestly received from those who deceived them; whose impositions, modified from age to age, subtilized by the priests, clothed with the reverential awe inspired by fear, have by degrees acquired that solidity, received that corroboration, attained that veteran stability, which is the natural result of public sanction, backed by theological parade.

The word God is, perhaps, among the first that vibrate on the ear of man; it is reiterated to him incessantly; he is taught to lisp it with respect; to listen to it with fear; to bend the knee when it is reverberated: by dint of re-petition, by listening to the fables of antiquity, by hearing it pronounced by all ranks and persuasions, he seriously believes all men bring the idea with them into the world. He thus confounds a mechanical habit with instinct; whilst it is for want of being able to recall to himself the first circumstances under which his imagination was awakened by this name; for want of recollecting all the recitals made to him during the course of his

infancy; for want of accurately defining what was instilled into him by his education; in short, because his memory does not furnish him with the succession of causes that have engraven it on his brain, that he believes this idea is really inherent to his being; innate in all his species.*

It is, however, uniformly by habit, that man admires, that he fears a being, whose name he has attended to from his earliest infancy. As soon as he hears it uttered, he, without reflection, mechanically associates it with those ideas with which his imagination has been filled by the recitals of others; with those sensations which he has been instructed to accompany it. Thus, if for a season man would be ingenuous with himself, he would concede that the idea of a God, and of those attributes with which he is clothed, have their foundation, take their rise in, and are the fruit of the opinions of his fathers, traditionally infused into · him by education—confirmed by habit —corroborated by example—enforced by authority. That it very rarely happens he examines these ideas; that they are for the most part adopted by inexperience, propagated by tuition, rendered sacred by time, inviolable from respect to his progenitors, reverenced as forming a part of those institutions he has most learned to value. He thinks he has always had them, because he has had them from his infancy; he considers them indubitable, because he is never permitted to question them—because he never has the intrepidity to examine their basis.

If it had been the destiny of a Brahmin, or a Mussulman, to have drawn his first breath on the shores of Africa, he would adore, with as much simplicity, with as much fervour, the serpent reverenced by the negroes, as he does

* Iamblicus, who was a Pythagorean philosopher not in the highest repute with the learned world, although one of those visionary priests in some estimation with theologians, (at least, if we may venture to judge by the unlimited draughts they have made on the bank of his doctrines) who was unquestionably a favourite with the emperor Julian, says, that "anteriorly to all use of reason, the notion of the Gods is inspired by nature, and that we have even a sort of feeling of the Divinity, preferable to the knowledge of him."

the God his own metaphysicians have offered to his reverence. He would be equally indignant if any one should presumptuously dispute the divinity of this reptile, which he would have learned to venerate from the moment he quitted the womb of his mother, as the most zealous, enthusiastic fakir, when the marvellous wonders of his prophet should be brought into question; or as the most subtile theologian when the inquiry turned upon the incongruous qualities with which he has decorated his Gods. Nevertheless, if this serpent God of the negro should be contested, they could not at least dispute his existence. Simple as may be the mind of this dark son of nature, uncommon as may be the qualities with which he has clothed his reptile, he still may be evidenced by all who choose to exercise their organs of sight. It is by no means the same with the immaterial, incorporeal, contradictory God, or with the deified man, which our modern thinkers have so subtilly composed. By dint of dreaming, of reasoning, and of subtilizing, they have rendered his existence impossible to whoever shall dare to examine it coolly. We shall never be able to figure to ourselves a being, who is only composed of abstractions and of negative qualities; that is to say, who has no one of those qualities, which the human mind is susceptible of judging. Our theologians do not know that which they adore; they have not one real idea of the being with which they unceasingly occupy themselves; this being would have been long since annihilated, if those to whom they announced him had dared to examine into his existence.

Indeed, at the very first step we find ourselves arrested; even the existence of this most important and most revered being, is yet a problem for whoever shall coolly weigh the proofs which theology gives of it; and although, before reasoning or disputing upon the nature and the qualities of a being, it was necessary to verify his existence, that of the Divinity is very far from being demonstrable to any man who shall be willing to consult good sense.— What do I say? The theologians themselves have scarcely ever been in unison upon the proofs of which they

have availed themselves to establish the divine existence. Since the human mind has occupied itself with its God (and when has it not been occupied with it?) it has not hitherto arrived at demonstrating the existence of this interesting object, in a manner satisfactory to those themselves who are anxious that we should be convinced of it. From age to age, new champions of the Divinity, profound philosophers, and subtile theologians, have sought new proofs of the existence of God, because they were, without doubt, but little satisfied with those of their predecessors. Those thinkers who flattered themselves with having demonstrated this great problem, were frequently accused of ATHEISM, and of having betrayed the cause of God, by the weakness of those arguments with which they had supported it.* Men of very great genius, have indeed successively miscarried in their demonstrations, or in the solutions which they have given of it; in believing they had surmounted a difficulty, they have continually given birth to a hundred others. It is to no purpose that the greatest metaphysicians have exhausted all their efforts to prove that God existed, to reconcile his incompatible attributes, or to reply to the most simple objections; they have not yet been able to succeed. The difficulties which are opposed to them, are sufficiently clear to be understood by an infant; whilst, in the most learned nations, they will be troubled to find twelve men capable of understanding the demonstrations, the solutions, and the replies of a *Des-cartes*, of a *Leibnitz*, and of a *Clarke*, when they endeavour to prove to us the existence of the Divinity. Do not let us be at all astonished; men never understand themselves when they speak to us of God, how then should they be

* Descartes, Paschal, and Dr. Clarke himself, have been accused of ATHEISM by the theologians of their time; this has not prevented subsequent theologians from making use of their proofs, and giving them as extremely valid. *See further on, the tenth chapter.* Not long since, a celebrated author, under the name of Doctor BOWMAN, published a work, in which he pretends, that all the proofs of the existence of God hitherto offered, are crazy and futile: he substitutes his own in their place, full as little convincing as the others.

able to understand each other, or agree amongst themselves, when they reason upon the nature and the qualities of a being, created by various imaginations, which each man is obliged to see diversely, and upon the account of whom men will always be in an equal state of ignorance, for want of having a common standard upon which to form their judgments of him.

To convince ourselves of the little solidity of those proofs which they give us of the existence of the theological God, and of the inutility of those efforts which they have made to reconcile his discordant attributes, let us hear what the celebrated *Doctor Samuel Clarke* has said, who, in his treatise *concerning the being and attributes of God*, is supposed to have spoken in the most convincing manner.*

* Although many people look upon the work of Doctor Clarke, as the most solid and the most convincing, it is well to observe, that many theologians of his time, and of his country, have by no means judged of it in the same manner, and have looked upon his proofs as insufficient, and his method as dangerous to his cause. Indeed, Doctor Clarke has pretended to prove the existence of God *a priori*, this is what others deem impossible, and look upon it, with reason, as *begging the question*. This manner of proving it has been rejected by the school-men, such as *Albert the Great, Thomas d'Aquinus, John Scot,* and by the greater part of the moderns, with the exception of *Suarez*. They have pretended that the existence of God was impossible to be demonstrated *a priori*, seeing that there is nothing anterior to the first of causes; but that this existence could only be proved *a posteriori;* that is to say, by its effects. In consequence, the work of Doctor Clarke was attacked rudely by a great number of theologians, who accused him of innovation, and of deserting their cause, by employing a method unusual, rejected, and but little suitable to prove any thing. Those who may wish to know the reasons which have been used against the demonstrations of *Clarke,* will find them in an English work, entitled, *An Inquiry into the ideas of Space, Time, Immensity, &c. by Edmund Law,* printed at *Cambridge,* 1734. If the author proves in it with success, that the demonstrations a *priori,* of *Doctor Clarke,* are false; it will be easy to convince ourselves by every thing which is said in our work, that all the demonstrations a *posteriori,* are not better founded. For the rest, the great esteem in which they hold the book of *Clarke* at the present day, proves that the theologians are not in accord amongst themselves, frequently changing their opinions, and are not difficult upon the demonstrations which they give of the existence of a being which hitherto is by no means demon-

Those who have followed him, indeed, have done no more than repeat his ideas, or present his proofs under new forms. After the examination which we are going to make, we dare say it will be found that his proofs are but little conclusive, that his principles are unfounded, and that his pretended solutions are not suitable to resolve any thing. In short, in the God of *Doctor Clarke,* as well as in that of the greatest theologians, they will only see a chimera established upon gratuitous suppositions, and formed by the confused assemblage of extravagant qualities, which render his existence totally impossible; in a word, in this God will only be found a vain phantom, substituted to the energy of nature, which has always been obstinately mistaken. We are going to follow, step by step, different propositions in which this learned theologian develops the received opinions upon the Divinity. Dr. Clarke sets out with saying:—

1st. "*Something existed from all eternity.*"

This proposition is evident—has no occasion for proofs. Matter has existed from all eternity, its forms alone are evanescent; matter is the great engine used by nature to produce all her phenomena, or rather it is nature herself. We have some idea of matter, sufficient to warrant the conclusion that this has always existed. First, that which exists, supposes existence essential to its being. That which cannot annihilate itself, exists necessarily; it is impossible to conceive that that which cannot cease to exist, or that which cannot annihilate itself, could ever have had a beginning. If matter cannot be annihilated, it could not commence to be. Thus we say to Dr. Clarke, that it is matter, that is nature, acting by her own peculiar energy, of which no particle is ever in an absolute state of rest, which has always existed. The various material bodies which this nature contains often change their form, their combination, their properties, their mode of action; but their principles or elements

strated. However, it is certain that the work of *Clarke,* in despite of the contradictions which he has experienced, enjoys the greatest reputation.

are indestructible—have never been able to commence. What the doctor actually understands, when he makes the assertion, that "an eternal duration is now actually past," is not quite so clear; yet he affirms, that "not to believe it would be a real and express contradiction."

2d. "*There has existed from eternity some one "unchangeable and independent being.*"

We may fairly inquire what is this being? Is it independent of its own peculiar essence, or of those properties which constitute it such as it is? We shall further inquire, if this being, whatever it may be, can make the other beings which it produces, or which it moves, act otherwise than they do, according to the properties which it has given them? And in this case we shall ask, if this being, such as it may be supposed to be, does not act necessarily; if it is not obliged to employ indispensable means to fulfil its designs, to arrive at the end which it either has, or may be supposed to have in view? Then we shall say, that nature is obliged to act after her essence; that every thing which takes place in her is necessary; and that if they suppose it governed by a Deity, this God cannot act otherwise than he does, and consequently is himself subjected to necessity. A man is said to be independent, when he is determined in his actions only by the general causes which are accustomed to move him; he is equally said to be dependant on another, when he cannot act but in consequence of the determination which this last gives him. A body is dependant on another body when it owes to it its existence, and its mode of action. A being existing from eternity cannot owe his existence to any other being; he cannot then be dependant upon him, except he owes his action to him; but it is evident that an eternal or self-existent being contains in his own nature every thing that is necessary for him to act: then. matter being eternal, is necessarily independent in the sense we have explained; of course, it has no occasion for a mover upon which it ought to depend.

This eternal being is also immutable, if by this attribute be understood that he cannot change his nature; but if it be intended to infer by it that he cannot change his mode of action or existence, it is without doubt deceiving themselves, since even in supposing an immaterial being, they would be obliged to acknowledge in him different modes of being, different volitions, different ways of acting; particularly if he was not supposed totally deprived of action, in which case he would be perfectly useless. Indeed, it follows of course that to change his mode of action he must necessarily change his manner of being. From hence it will be obvious, that the theologians, in making their God immutable, render him immoveable; consequently he cannot act. An immutable being, could evidently neither have successive volition, nor produce successive action ; if this being has created matter, or given birth to the universe, there must have been a time in which he was willing that this matter, this universe, should exist ; and this time must have been preceded by another time, in which he was willing that it might not yet exist. If God be the author of all things, as well as of the motion and of the com-. binations of matter, he is unceasingly occupied in producing and destroying; in consequence, he cannot be called immutable, touching his mode of existing. The material world always maintains itself by motion, and the continual change of its parts; the sum of the beings who compose it, or of the elements which act in it, is invariably the same ; in this sense the immutability of the universe is much more easy of comprehension, much more demonstrable than that of any other being to whom they would attribute all the effects, all the mutations which take place. Nature is not more to be accused of mutability, on account of the succession of its forms, than the eternal being is by the theologians, by the diversity of his decrees.*

* Here we shall be able to perceive that, supposing the laws by which nature acts to be immutable, it does not require any of these logical distinctions to account for the changes that take place: the mutation which results, is, on the contrary, a striking proof of the immutability of the system which produces them ; and completely brings nature under the range of this second proposition as stated by Dr. Clarke.

3d. " That unchangeable and independent being, which has existed from eternity without any eternal cause of its existence, must be self-existent, that is, necessarily existing."

This proposition is merely a repetition of the first; we reply to it by inquiring, Why matter, which is indestructible, should not be self-existent? It is evident that a being who had no beginning, must be self-existent; if he had existed by another, he would have commenced to be; consequently he would not be eternal. Those who make matter coeternal with God, do no more than multiply beings without necessity.

4th. " What the substance or essence of that being which is self-existent, or necessarily existing, is, we have no idea; neither is it at all possible for us to comprehend it."

Dr. Clarke would have spoken more correctly if he had said his essence is impossible. Nevertheless, we shall readily concede that the essence of matter is incomprehensible, or at least, that we conceive it very feebly by the manner in which we are affected by it; but we must also concede that we are much less able to conceive the Divinity, who is impervious on any side. Thus it must necessarily be concluded, that it is folly to argue upon it, since it is by matter alone we could have any knowledge of him; that is to say, by which we could assure ourselves of his existence—by which we could at all guess at his qualities. In short, we must conclude, that every thing related of the Divinity, either proves him material, or else proves the impossibility in which the human mind will always find itself, of conceiving any being different from matter; without extent, yet omnipresent; immaterial, yet acting upon matter; spiritual, yet producing matter; immutable, yet putting every thing in activity, &c.

Indeed it must be allowed that the incomprehensibility of the Divinity does not distinguish him from matter; this will not be more easy of comprehension when we shall associate it with a being much less comprehensible than itself; and of this last we have some slender knowledge through some of its parts. We do not certainly know the essence of any being, if by that word

we are to understand that which constitutes its peculiar nature. We only know matter by the sensations, the perceptions, the ideas which it furnishes; it is according to these that we judge it to be either favourable or unfavourable, following the particular disposition of our organs. But when a being does not act upon any part of our organic structure, it does not exist for us, we cannot, without exhibiting folly, without betraying our ignorance, without falling into obscurity, either speak of its nature, or assign its qualities; our senses are the only channel by which we could have formed the slightest idea of it. The incomprehensibility of the Divinity ought to convince man that it is folly to seek after it; but this, however, would not suit with those priests who are willing to reason upon him continually, to show the depth of their learning—to persuade the uninformed they understand that which is incomprehensible to all men; by which they expect to be able to submit him to their own views. Nevertheless, if the Divinity be incomprehensible, we must conclude that a priest, does not comprehend him better than other men: and the wisest or the surest way is, not to be guided by the imagination of a theologian.

5th. " Though the substance, or essence of the self-existent being, is in itself absolutely incomprehensible to us, yet many of the essential attributes of his nature are strictly demonstrable, as well as his existence. Thus, in the first place, the self-existent being must of necessity be eternal."

This proposition differs in nothing from the first, except Dr. Clarke does here understand that as the self-existent being had no beginning, he can have no end. However this may be, we must ever inquire, Why should not this be matter? We shall further observe, that matter not being capable of annihilation, exists necessarily, consequently will never cease to exist; that the human mind has no means of conceiving how matter should originate from that which is not itself matter: is it not obvious, that matter is necessary; that there is nothing, except its powers, its arrangement, its combinations, which are contingent or evanes-

cent? The general motion is necessary, but the, given motion is not so;
only during the season that the particular combinations subsist, of which
this motion is the consequence, or the
effect: we may be competent to change
the direction, to either accelerate or retard, to suspend or arrest, a particular
motion, but the general motion can
never possibly be annihilated. Man,
in dying, ceases to live; that is to say,
he no longer either walks, thinks or
acts in the mode which is peculiar to
human organization: but the matter
which composed his body, the matter
which formed his mind, does not cease
to move on that account: it simply becomes susceptible of another species
of motion.

6th. " The self-existent being must
of necessity be infinite and omnipresent."

The word infinite presents only a
negative idea which excludes all
bounds: it is evident that a being who
exists necessarily, who is independent,
cannot be limited by any thing which
is out of himself; he must consequently be his own limits: in this sense we
may say he is infinite.

Touching what is said of his omnipresence, it is equally evident that if
there be nothing exterior to this being,
either there is no place in which he
must not be present, or that there will
be only himself and the vacuum. This
granted, I shall inquire if matter exists; if it does not at least occupy a
portion of space? In this case, matter, or the universe, must exclude every
other being who is not matter, from that
place which the material beings occupy in space. In asking whether the
God of the theologians be by chance
the abstract being which they call the
vacuum or space, they will reply, no!
They will further insist, that their God,
who is not matter, penetrates that which
is matter. But it must be obvious, that
to penetrate matter, it is necessary to
have some correspondence with matter,
consequently to have extent; now to
have extent, is to have one of the properties of matter. If the Divinity penetrates matter, then he is material; by a
necessary deduction he is inseparable
from matter; then if he is omnipresent,
he will be in every thing. This the theologian will not allow: he will say it

is a mystery; by which I shall understand that he is himself ignorant how
to account for the existence of his God;
this will not be the case with making
nature act after immutable laws; she
will of necessity be every where, in my
body, in my arm, in every other material
being, because matter composes them all.

7th. " The self-existent being must
of necessity be but one."

If there be nothing exterior to a being
who exists necessarily, it must follow
that he is unique. It will be obvious
that this proposition is the same with
the preceding one; at least, if they are
not willing to deny the existence of
the material world or to say with Spinosa, that there is not, and that we cannot conceive any other substance than
God. Præter Deum neque dari neque concipi potest substantia, says this
celebrated athiest, in his fourteenth
proposition.

8th. " The self-existent and original cause of all things, must be an intelligent being."

Here Dr. Clarke most unquestionably assigns a human quality: intelligence is a faculty appertaining to organized or animated beings, of which
we have no knowledge out of these beings. To have intelligence, it is necessary to think; to think, it is requisite
to have ideas; to have ideas, supposes
senses; when senses exist they are material; when they are material, they
cannot be a pure spirit, in the language
of the theologian.

The necessary being who comprehends, who contains, who produces animated beings, contains, includes, and
produces intelligence. But has the
great whole a peculiar intelligence,
which moves it, which makes it act,
which determines it in the mode that
intelligence moves and determines animated bodies; or rather, is not this intelligence the consequence of immutable laws, a certain modification resulting from certain combinations of
matter, which exists under one form of
these combinations, but is wanting under another form? This is assuredly
what nothing is competent to prove.
Man having placed himself in the first
rank in the universe, has been desirous to judge of every thing after what
he saw within himself, because he has
pretended that in order to be perfect it

was necessary to be like himself. Here is the source of all his erroneous reasoning upon nature and his Gods. He has therefore concluded that it would be injurious to the Divinity not to invest him with a quality which is found estimable in man—which he prizes highly--to which he attaches the idea of perfection—which he considers as a manifest proof of superiority. He sees his fellow-creature is offended when he is thought to lack intelligence; he therefore judges it to be the same with the Divinity. He denies this quality to nature, because he considers her a mass of ignoble matter, incapable of self-action, although she contains and produces intelligent beings. But this is rather a personification of an abstract quality, than an attribute of the Deity, with whose perfections, with whose mode of existence, he cannot by any possible means become acquainted according to the fifth proposition of Dr. Clarke himself. It is in the earth that is engendered those living animals called worms; yet we do not say the earth is a living creature. The bread which man eats, the wine that he drinks, are not themselves thinking substances; yet they nourish, sustain, and cause those beings to think, who are susceptible of this modification of their existence. It is likewise in nature, that is formed intelligent, feeling, thinking beings; yet it cannot be rationally said, that nature feels, thinks, and is intelligent after the manner of these beings, who nevertheless spring out of her bosom.

How, they will say to us, refuse to the Creator, these qualities which we discover in his creatures! The work would then be more perfect than the workman! *God who hath made the eye, shall he not see? God, who hath formed the ear, shall he not hear?* But if we adopt this mode of reasoning, ought we not to attribute to God all the other qualities that we shall meet with in his creatures? Should we not say, with equal foundation, that the God who has made matter, is himself matter; that the God who has fashioned the body, must possess a body; that the God who has made so many irrational beings, is irrational himself; that the God who has created man who sins, is liable himself to sin? If, because the works of God possess certain qualities, and are susceptible of certain modifications, we conclude that God possesses them also, we shall be obliged by parity of reasoning to conclude that God is material, has extent, has gravity, is wicked, &c.

To attribute wisdom, or an infinite intelligence to God, that is to say, to the universal mover of nature, there should be neither folly, nor evil, nor wickedness, nor confusion on the earth. They will perhaps tell us, that, even according to our own principles, evil and disorder are necessary; but our principles do not admit of a wise and intelligent God, who should have the power of preventing them. If, in admitting such a God, evil is not less necessary, what end can this God, so wise, so powerful, and so intelligent, be able to serve, seeing that he is himself subjected to necessity? From thence he is no longer independent, his power vanishes, he is obliged to admit a free course to the essence of things; he cannot prevent causes from producing their effects; he cannot oppose himself to evil; he cannot render man more happy than he is; he cannot, consequently, be good; he is perfectly useless; he is no more than the unconcerned witness of that which must necessarily happen; he cannot do otherwise than will every thing which takes place in the world. Nevertheless, they tell us, in the succeeding proposition, that—

9th. " *The self-existent and original cause of all things, is not a necessary agent, but a being endowed with liberty and choice.*"

Man is called *free*, when he finds within himself motives which determine him to action, or when his will finds no obstacle to the performance of that to which his motives have determined him. God, or the necessary being, of which question is here made, does he not find obstacles to the execution of his projects? Is he willing that evil should be committed, or can he not prevent it? In this case, he is not free, and the will meets with continual obstacles; or else, we must say, he consents to the commission of sin; that he is willing we should offend him; that he suffers men to restrain his liberty, and derange his projects. How will the theologians draw themselves out of this perplexing intricacy?

On the other hand, the God whom they suppose cannot act, but in conse- quence of the laws of his peculiar ex- istence, we should be enabled then to call a *being endowed with liberty*, as far as his actions should not be determined by any thing which should be exterior to himself, but this would visibly be an abuse of terms: indeed, we cannot say, that a being who is not capable of act- ing otherwise than he does, and who can never cease to act, but in virtue of the laws of his peculiar existence, is a being possessed of liberty—there is evidently necessity in all his actions. Ask a theologian, if God has power to reward crime, and punish virtue? Ask him again, if God can love him, or if he is a free agent when the action of a man necessarily produces in him a new will? A man is a being exterior to God, and nevertheless they pretend, that the conduct of this man has an influence on this being endowed with liberty, and necessarily determines his will. In short, we demand if God can avoid to will that which he willeth, and not do that which he doeth? Is not his will necessitated by intelligence, wisdom and views which they suppose him to have? If God be thus connect- ed, he is no more a free agent than man: if every thing which he does be necessary, he is nothing more than destiny, fatality, the *fatum* of the an- cients, and the moderns have not chang- ed the Divinity, although they have changed his name.

They will, perhaps, tell us, that God is free, insomuch that he is not bound by the laws of nature, or by those which he imposes on all beings. Never- theless, if it be true that he has made these laws, if they are the effect of his infinite wisdom, of his supreme intel- ligence, he is by his essence obliged to follow them, or else it must be acknow- ledged that it would be possible for God to act irrationally. Theologians fearing, without doubt, to restrain the liberty of God, have supposed that he was not subjected to any laws, as we have before proved; in consequence, they have made him a despotic, fantas- tical, and strange being, whose power gives him the right to violate all the laws which he has himself established. The pretended miracles which they have attributed to him, he derogates from the

laws of nature; by the conduct which they have supposed him to hold, he acts very frequently in a mode contra- ry to his divine wisdom, and to the rea- son which he has given to men, to re- gulate their judgments. If God is a free agent in this sense, all religion is useless; it can only found itself upon those immutable rules which this God has prescribed to himself, and upon those engagements which he has enter- ed into with the human species? As soon as religion does not suppose him bound by his covenants, it destroys it- self, it commits suicide.

10th. "*The self-existent being, the supreme cause of all things, must of necessity have infinite power.*"

There is no power but in him, this power then has no limits; but if it is God who enjoys this power, man ought not to have the power of doing evil; without which he would be in a state to act contrary to the divine power; there would be exteriorly to God a power capable of counterbalancing his, or of preventing it from producing those effects which he proposes to himself; the Divinity would be obliged to suffer that evil which he could prevent.

On the other hand, if man is free to sin, God is not himself a free agent, his conduct is necessarily determined by the actions of man. An equitable monarch is not a free agent when he believes himself obliged to act confor- mably to the laws which he has sworn to observe, or which he cannot violate without wounding his justice. A mon- arch is not powerful when the least of his subjects has the power of insulting him, of openly resisting him, or secret- ly making all his projects miscarry. Nevertheless, all the religions of the world, show us God under the charac- ter of an absolute sovereign, of whom nothing is capable to constrain the will, nor limit the power; whilst on the other hand, they assure us that his sub- jects have at each instant the power and the liberty to disobey him and an- nihilate his designs: from whence it is evident that all the religions of the world destroy with one hand what they establish with the other: so that, ac- cording to the ideas with which they furnish us, their God is neither free, powerful, nor happy.

11th. "*The supreme cause and*

author of things, must of necessity be infinitely wise."

Wisdom and folly are qualities founded on our peculiar judgment; now in this world, which God is supposed to have created, to preserve, to move, and to penetrate, there happens a thousand things, which appear to us as follies, and even the creatures for whom we imagine the universe to have been made, are frequently much more foolish and irrational than prudent and wise. The author of every thing which exists, must be equally the author of that which we call irrational, and of that which we judge to be extremely wise. On the other hand, to judge of the intelligence and the wisdom of a being, it were necessary, at least, to foresee the end which he proposes to himself. What is the aim of God? It is, they tell us, his own peculiar glory; but does this God attain this end, and do not sinners refuse to glorify him? Besides, suppose God is sensible to glory, is not this supposing him to have our follies and our weaknesses? Is not this saying he is haughty? If they tell us that the aim of the divine wisdom is to render men happy, I shall always ask, wherefore these men, in despite of his views, so frequently render themselves miserable? If they tell me, the views of God are impenetrable to us, I shall reply, in the first place, that in this case it is at random that they tell me the Divinity proposes to himself the happiness of his creatures, an object, which, in fact, is never attained; I shall, in the second place, reply, that, ignorant of his real aim, it is impossible for us to judge of his wisdom, and that to be willing to reason upon it shows madness.

12th. " *The supreme cause and author of all things, must of necessity be a being of infinite goodness, justice, and truth, and all other moral perfections; such as become the supreme governor and judge of the world.*"

The idea of *perfection* is an abstract, metaphysical, negative idea, which has no archetype, or model, exterior to ourselves. A perfect being would be a being similar to ourselves, whom, by thought, we should divest of all those qualities which we find prejudicial to us, and which, for that reason, we call *imperfections;* it is always relatively

to ourselves, and to our mode of feeling and of thinking, and not in itself, that a thing is perfect or imperfect; it is according to this, that a thing is more or less useful or prejudicial, agreeable or disagreeable. In this sense, how can we attribute perfection to the self-existent being? Is God perfectly good relatively to men? But men are frequently wounded by his works, and are obliged to complain of the evils, which they suffer in this world. Is God perfect, relative to his works? But do we not frequently see the most complete disorder, range itself on the side of order? These works of the Divinity so perfect, are they not changed, are they not destroyed unceasingly; do they not oblige us to experience, in despite of ourselves, those sorrows and troubles which balance the pleasures and the benefits which we receive from nature? Do not all the religions of the world suppose a God continually occupied in remaking, repairing, undoing, and rectifying his marvellous works? They will not fail telling us, that God cannot possibly communicate to his works that perfection, which he himself possesses. In this case, we shall say, that the imperfections of this world, being necessary for God himself, he never will be able to remedy them, even in another world; and we shall conclude, that this ·God, cannot be to us of any utility whatever. .

The metaphysical or theological attributes of the Divinity, make him an abstract and inconceivable being as soon as they distinguish him from nature and from all the beings which she contains: the moral qualities make him a being of the human species although by negative attributes it is endeavoured to remove him to a distance from man. The theological God is an insulated being, who in truth cannot have any relation with any of the beings of which we have a knowledge. The moral God is never more than a man who is believed to be rendered perfect, in removing from him by thought, the imperfections of human nature. The moral qualities of men are founded upon the relations between them, and upon their mutual wants. The theological God cannot certainly have moral qualities, or human perfections; he has no occasion for men, he has no re-

.ation with them, seeing that no relations can exist which are not reciprocal. A pure spirit cannot assuredly
have relations with material beings, at
least in parts; an infinite being cannot
be susceptible of any relation with an
finite beings; an eternal being cannot
have relations with perishable and
transitory beings. The one being who
has neither species nor cause, who has
no fellow creatures, who does not live
in society, who has nothing in common with his creatures, if he really
existed, could not possess any of those
qualities, which we call perfections;
he would be of an order so different
from man, that we should not be able
to assign him either vices or virtues.
It is unceasingly repeated to us, that
God owes us nothing; that no being is
comparable to him; that our limited
understanding cannot conceive his perfections; that the human mind is not
formed to comprehend his essence: but
do they not, even by this, destroy our
relation with this being, so dissimilar,
so disproportionate, so incomprehensible to us? All relation supposes a
certain analogy; all duties suppose a
resemblance, and reciprocal wants, to
render to any one the obligations we
owe him, it is necessary to have a
knowledge of him.

They will, without doubt, tell us, that
God has made himself known by revelation. But does not this revelation
suppose the existence of the God we dispute? Does not this revelation itself
destroy the moral perfections, which
they attribute to him? Does not all
revelation suppose in men, ignorance,
imperfection, and perversity, which a
beneficent, wise, omnipotent, and provident God, ought to have prevented?
Does not all particular revelation suppose in this God a preference, a predilection, and an unjust partiality for
some of his creatures; dispositions that
visibly contradict his infinite goodness
and justice? Does not this revelation
announce in him aversion, hatred, or at
least indifference for the greater number of the inhabitants of the earth, or
even a fixed design of blinding them,
in order that they may lose themselves?
In short, in all the known revelations,
is not the Divinity, instead of being represented as wise, equitable, and filled
with tenderness for man, continually

depicted to us as a fantastical, iniquitous, and a cruel being; as one who is
willing to seduce his children; as one
who is laying snares for them, or making them lay snares for themselves:
and as one who punishes them for having fallen into them? The truth is,
the God of *Doctor Clarke*, and of the
Christians, cannot be looked upon as
a perfect being, at least, if in theology
they do not call those qualities *perfections*, which reason and good sense
call striking imperfections or odious dispositions. Nay more, there are not in
the human race individuals so wicked,
so vindictive, so unjust, so cruel, as the
tyrant on whom the Christians prodigally lavish their servile homage, and
on whom their theologians heap those
perfections which the conduct they ascribe to him contradicts every moment.

The more we consider the theological God, the more impossible and contradictory will he appear; theology
seems only to have formed him, immediately to destroy him. What is this,
in fact, but a being of whom they can
affirm nothing that is not instantly contradicted? What is this but a good
God who is unceasingly irritating himself; an omnipotent God who never
arrives at the end of his designs; a
God infinitely happy, whose felicity is
perpetually disturbed; a God who loves
order, and who never maintains it; a
just God who permits his most innocent
subjects to be exposed to continual injustice? What is this but a pure spirit
who creates and who moves matter?
What is this but an immutable being
who is the cause of the motion and
those changes which are each moment
operating in nature? What is this but
an infinite being who is, however, coexistent with the universe? What is
this but an omniscient being who believes himself obliged to make trial of
his creatures? What is this but an
omnipotent being who never can communicate to his works that perfection
which he would find in them? What
is this but a being clothed with every
divine quality, and of whom the conduct is always human? What is this but
a being who is able to do every thing,
and who succeeds in nothing, who
never acts in a manner worthy of himself? Like man, he is wicked, unjust,
cruel, jealous, irascible, and vindictive;

like man, he miscarries in all his projects; and this with all the attributes capable of guarantying him from the defects of our species. If we would but be ingenuous, we should confess, that this being is nothing; and we shall find the phantom imagined to explain nature, is perpetually in contradiction with this very nature, and that instead of explaining any thing, it only serves to throw every thing into perplexity and confusion.

According to Clarke himself, "*nothing is that of which every thing can truly be denied, and nothing can truly be affirmed. So that the idea of nothing, if I may so speak, is absolutely the negation of all ideas. The idea, therefore, either of a finite or infinite nothing is a contradiction in terms.*" Let them apply this principle to what our author has said of the Divinity, and they will find that he is by his own confession, an *infinite nothing*, since the idea of this Divinity is the *absolute negation of all ideas* which men are capable of forming to themselves. Spirituality is indeed a mere negation of corporeity; to say God is spiritual, is it not affirming to us that they do not know what he is? They tell us there are substances which we can neither see nor touch, but which do not exist the less on that account. Very well, but then we can neither reason upon them nor assign them qualities. Can we have a better conception of infinity which is a mere negation of those limits which we find in all beings? Can the human mind comprehend what is infinite, and in order to form to itself a kind of a confused idea is it not obliged to join limited quantities to other quantities which again it only conceives as limited? Omnipotence, eternity, omniscience, and perfection, are they any thing else but abstractions or mere negations of the limitation of power, of duration, and of science? If it is pretended that God is nothing of which man can have a knowledge, can see, can feel; if nothing can be said positively, it is at least permitted us to doubt his existence; if it is pretended that God is what our theologians describe him, we cannot help denying the existence or the possibility of a being who is made the subject of those qualities which the

human mind will never be able to conceive or reconcile.

According to Clarke, "*the self-existent being must be a most simple, unchangeable, incorruptible being; without parts, figure, motion, divisibility, or any other such properties as we find in matter. For all these things do plainly and necessarily imply finiteness in their very notion, and are utterly inconsistent with complete infinity.*" Indeed! and is it possible to form any true notion of such a being? The theologians themselves agree, that men cannot have a complete notion of God; but that which they have here presented us, is not only incomplete, but it also destroys in God all those qualities upon which our mind is capable of fixing any judgment. Doctor Clarke is obliged to avow, that, "*as to the particular manner of his being infinite, or everywhere present, in opposition to the manner of created things being present in such or such infinite places; this is as impossible for our finite understandings to comprehend or explain, as it is for us to form an adequate idea of infinity.*" But what is this but a being which no man can either explain or comprehend? It is a chimera, which, if it existed, could not possibly interest man.

Plato, the great creator of chimeras, says that "*those who admit nothing but what they can see and feel, are stupid and ignorant beings, who refuse to admit the reality of the existence of invisible things.*" Our theologians hold the same language to us: our European religions, have visibly been infected with the reveries of the Platonists, which evidently are no more than the result of obscure notions, and of the unintelligible metaphysics of the Egyptian Chaldean, and Assyrian priests, among whom Plato drew up his pretended philosophy. Indeed, if philosophy consists in the knowledge of nature, we shall be obliged to agree, that the Platonic doctrines in nowise merit this name, seeing that he has only drawn the human mind from the contemplation of visible nature, to throw it into an intellectual world, where it finds nothing but chimeras. Nevertheless, it is this fantastical philosophy, which regulates all our opinions at present. Our theologians, still guided by the enthusiasm of Plato, discourse with their followers only of *spirits ; intelligent, incorporeal substances ; invisible powers ; angels ; demons of mysterious virtues ; supernatural effects ; divine inspiration ; innate ideas,* &c., &c.* To believe in them, our senses are entirely useless; experience is good for nothing, imagination, enthusiasm, fanaticism, and the workings of fear, which our religions prejudices give birth to in us, are *celestial inspirations,* divine warnings, natural sentiments, which we ought to prefer to reason, to judgment, and to good sense. After having imbued us from our infancy with these maxims, so proper to hoodwink us, and to lead us astray, it is very easy for them to make us admit the greatest absurdities under the imposing name of *mysteries,* and to prevent us from examining that which they tell us to believe. Be this as it may, we shall reply to Plato, and to all those doctors, who, like him, impose upon us the necessity of believing that which we cannot comprehend, that to believe a thing exists, it is at least necessary to have some idea of it ; that this idea can only come to us by the medium of our senses; that every thing which our senses do not give us a knowledge of, is nothing to us ; that if there is an absurdity in denying the existence of that which we

* Whoever will take the trouble to read the works of *Plato* and his disciples, such as *Proclus, Jamblicus, Plotinus,* &c. will find in them almost all the doctrines and metaphysical subtilties of the *Christian Theology.* Moreover, they will find the origin of the *symbols,* the *rites,* the *sacraments,* in short, of the *theurgy,* employed in Christian worship, who, as well in their religious ceremonies as in their doctrines, have done no more than follow, more or less faithfully, the road which had been traced out for them by the *priests* of *paganism.* Religious follies are not so various as they are imagined.

With respect to the ancient philosophy, with the exception of that of Democritus and Epicurus, it was, for the most part, a true *Theosophy,* imagined by the Egyptian and Assyrian priests : Pythagoras and Plato have been no more than theologians, filled with enthusiasm, and perhaps with knavery. At least, we find in them a *sacerdotal* and mysterious mind, which will always indicate, that they seek to deceive, or that they are not willing men should be enlightened. It is in *nature,* and not in *theology,* that we must draw up an intelligible and true philosophy.

do not know, there is extravagance in assigning to it unknown qualities, and that there is stupidity in trembling before true phantoms, or in respecting vain idols, clothed with incompatible qualities, which our imaginations have combined, without ever being able to consult experience and reason.

This will serve as a reply to Doctor Clarke, who says : "*How weak then, and foolish is it to raise objections against the being of God from the incomprehensibleness of his essence!—and to represent it as a strange and incredible thing, that there should exist any incorporeal substance, the essence of which we are not able to comprehend!*" He had said, a little higher : "*There is not so mean and contemptible a plant or animal that does not confound the most enlarged understanding upon earth : nay, even the simplest and plainest of all inanimate beings have their essence or substance hidden from us in the deepest and most impenetrable obscurity. How weak then, and foolish it is to raise objections against the being of God from the incomprehensibleness of his essence!*"

We shall reply to him, first, that the idea of an immaterial substance or being, without extent, is only an absence of ideas, a negation of extent, and that when they tell us a being is not matter, they speak to us of that which is not, and do not teach us that which is ; and that in saying a being cannot act upon our senses, they teach us that we have no means of assuring ourselves whether he exists or not.

Secondly, we shall confess, without hesitation, that men of the greatest genius, are not acquainted with the essence of stones, plants, animals, nor the secret springs which constitute some and which make others vegetate or act ; but that at least we see them, that our senses at least have a knowledge of them in some respects ; that we can perceive some of their effects, according to which we judge them well or ill; whilst our senses cannot compass, on any side, an immaterial being, and, consequently, cannot furnish us with any one idea of it ; such a being is to us an *occult quality,* or rather *a being of the imagination :* if we are ignorant of the essence or of the intimate combination of the most material beings, we shall at least discover, with the help of experience, some of their relations with ourselves: we know their surface, their extent, their form, their colour, their softness, and their hardness, by the impressions which they make on us: we are capable of distinguishing them, of comparing them, of judging of them, of seeing them, and of flying from them, according to the different modes in which we are affected by them: we cannot have the same knowledge of the immaterial God, nor of those spirits, of whom men, who cannot have more ideas of them than other mortals, are unceasingly talking to us.

Thirdly, we have a knowledge of modifications in ourselves which we call sentiments, thoughts, will, and passions; for want of being acquainted with our own peculiar essence, and the energy of our particular organization, we attribute these effects to a concealed cause, and one distinguished from ourselves which we call a *spiritual* being, because it appeared they acted differently from our body: nevertheless, reflection proves to us that material effects can only emanate from a material cause. We only see even in the universe, physical and material effects, which can only be produced by an analogous cause, and which we shall attribute, not to a spiritual cause of which we are ignorant, but to nature itself, which we may know in some respects if we will deign to meditate with attention.

If the incomprehensibility of God is not a reason for denying his existence, it is not one to establish that he is immaterial, and we shall yet less comprehend him as spiritual than as material, since materiality is a known quality, and spirituality is an occult or unknown quality, or rather a mode of speaking of which we avail ourselves only to throw a veil over our ignorance. It would be bad reasoning in a man born blind, if he denied the existence of colours, although these colours can have no relation with the senses in the absence of sight, but merely with those who have it in their power to see and know them ; this blind man, however, would appear perfectly ridiculous, if he undertook to define them. If there

were beings who had real ideas of God and of a pure spirit, and our theologians should thence undertake to define them, they would be just as ridiculous as the blind man.

We are repeatedly told that our senses only show us the external things, that our limited senses are not able to conceive a God ; we agree therein, but these same senses do not even show us the external of this Divinity that our theologians would define to us, to whom they ascribe attributes upon which they unceasingly dispute, though even to this time they are not come to the proof of his existence. " *I greatly esteem,*" says Locke, " *all those who faithfully defend their opinion; but there are so few persons who after the manner they do defend them, appear fully convinced of the opinions they profess, that I am tempted to believe there are more sceptics in the world than are generally imagined.*"[*] Abbadie tells us, that " *the question is, whether there be a God, and not what this God is.*" But how assure ourselves of the existence of a being concerning which we shall never be able to have a knowledge ? If they do not tell us what this being is, how shall we feel ourselves capacitated to judge whether or not his existence be possible ? We have seen the ruinous foundation upon which men have hitherto erected the phantom created by their imagination ; we have examined the proofs of which they avail themselves to establish his existence ; we have pointed out the numberless contradictions which result from those irreconcilable qualities with which they pretend to decorate him. What conclusion must we draw from all this, except that he does not exist? It is true, they assure us, *there are no contradictions between the divine attributes, but there is a disproportion between our understanding and the nature of the Supreme being.* This granted, what standard is it necessary man should possess to enable him to judge of his God ? Are not they men, who have imagined this being, and who have clothed him with attributes ascribed him by themselves ? If it

* See his *Familiar Letters.* Hornes says, that if men found their interest in it, they would doubt the truth of *Euclid's Elements.*

needs an infinite mind to comprehend the Divinity, can the theologians boast of being themselves in a capacity to conceive him? To what purpose then do they speak of him to others? Man who will never be an infinite being, will he be more capable of conceiving his God in a future world, than he is in the one which he at this day inhabits? If hitherto we have no knowledge of God, we can never flatter ourselves with obtaining it hereafter, seeing that we shall never be Gods.

Nevertheless, it is pretended that it is necessary to know this God; but how prove the necessity of having a knowledge of that which is impossible to be known? We are then told, that good sense and reason are sufficient to convince us of the existence of a God. But, on the other hand, am I not told that reason is a treacherous guide in religious matters? Let them at least show us the precise time when we must abandon this reason, which shall have conducted us to the knowledge of God. Shall we consult it again, when there shall be a question to examine whether what they relate of this God is probable, if he can unite the discordant qualities which they ascribe to him, if he has spoken the language which they have attributed to him? Our priests never will permit us to consult reason upon these things; they will still pretend that we ought blindly to believe that which they tell us, and that the most certain way is to submit ourselves to that which they have thought proper to decide on the nature of a being, concerning whom they avow they are ignorant, and who is in nowise within the reach of mortals. Besides, our reason cannot conceive infinity, therefore it cannot convince us of the existence of a God; and if our priests have a more sublime reason than that which is found in us, it will be then on the words of our priests that we shall believe in God; we shall never be ourselves perfectly convinced: intimate conviction can only be the effect of evidence and demonstration. A thing is demonstrated to be impossible, not only as soon as we are incapable of having true ideas of it, but also whenever the ideas we can form of it contradict themselves, destroy themselves, and are repugnant to

one another. We can have no true ideas of a spirit; the ideas we are able to form of it are contradictory, when we say that a being, destitute of organs and of extent, can feel, can think, can have will or desires. The theological God cannot act: it is repugnant to his divine essence to have human qualities; and if we suppose these qualities infinite, they will only be more unintelligible, and more difficult or impossible to be reconciled.

If God is to the human species what colours are to the man born blind, this God has no existence with relation to us; if it is said that he unites the qualities which are assigned to him, this God is impossible. If we are blind, let us not reason either upon God or upon his colours; let us not ascribe to him attributes; let us not occupy ourselves with him. The theologians are blind men, who would explain to others, who are also blind, the shades and the colours of a portrait representing an original which they have not even stumbled upon in the dark.* Let us not be told then that the original, the portrait, and his colours do not exist the less, because the blind man cannot explain them to us nor form to himself an idea of them, by the evidence of those men who enjoy the faculty of sight: but where are those quicksighted mortals who have seen the Divinity, who have a better knowledge of him than ourselves, and who have the right to convince us of his existence?

Doctor Clarke tells us, *it is sufficient that the attributes of God may be possible, and such as there is no demon-*

* I find, in the work of Doctor Clarke, a passage of Melchior Canus, bishop of the Canaries, which could be opposed to all the theologians in the world, and all their arguments: Puderet me dicere non me intelligere, si ipsi intelligerent qui trnctarunt. Heraclitus said, if it were demanded of a blind man what a sight was, he would reply that it was blindness. St. Paul announced his God to the Athenians as being precisely the *unknown* God to whom they had raised an altar. St. Denis, the areopagite, says, it is when they acknowledge they do not know God, that they know him the best. Tunc deum maxime cognoscimus, cum ignorare eum cognoscimus. It is upon this *unknown* God that all theology is founded! It is upon this *unknown* God that they reason unceasingly!! It is for the honour of this *unknown* God, that they cut the throats of men!!!

stration to the contrary. Strange method of reasoning! Would theology then be the only science in which it was permitted to conclude, that a thing is, as soon as it is possible to be? After having brought forward reveries without foundation, and propositions which nothing support, has he quitted them to say that they are truths, because the contrary cannot be demonstrated?—Nevertheless it is extremely possible to demonstrate that the theological God is impossible; to prove it, it is sufficient to make it seen, as we have not ceased to do, that a being formed by the monstrous combination of contrasts, the most offensive to reason, cannot exist.

Nevertheless, it is always insisted upon, and we are told that it is not possible to conceive that intelligence or thought can be properties and modifications of matter, of which, however, Doctor Clarke avows we ignore the energy and the essence, or of which he has said that men of the greatest genius have had but superficial or incomplete ideas. But could it not be asked of him if it is easier to conceive that intelligence and thought may be properties of spirit, of which we have certainly far less ideas than we have of matter? If we have only obscure and imperfect ideas of the most sensible and gross bodies, should we be able to have a more distinct knowledge of an immaterial substance, or of a spiritual God, who does not act upon any one of our senses, and who if he did act upon them, would cease from thence to be immaterial?

Doctor Clarke has no foundation for telling us that *" immaterial substances are not impossible;"* or that *" a substance immaterial is not a contradictory notion. Now whoever asserts that it is contradictory, must affirm that whatever is not matter is nothing."* Every thing that acts upon our senses, is matter; a substance destitute of extent or of the properties of matter cannot make itself felt by us, nor consequently give us perceptions or ideas: constituted as we are, that of which we have no ideas has no existence with relation to us. Thus, there is no absurdity in maintaining that all which is not matter is nothing; on the contrary, this is a truth so stri-

king, that there is nothing short of the most inveterate prejudice or knavery that can doubt or deny it.

Our learned adversary does not remove the difficulty in asking: *"Are our five senses, by an absolute necessity in the nature of the thing, all and the only possible ways of perception? And is it impossible and contradictory there should be any being in the universe endued with ways of perception different from those which are the result of our present composition? Or, are these things, on the contrary, purely arbitrary; and the same power that gave us these may have given others to other beings, and might, if he had pleased, have given to us others in this present state?"* I reply, first, that before we presume what God can or cannot do, it were necessary to have proved his existence. I reply also, that we have in fact but five senses;* that by their aid it is impossible for man to conceive such a being as they suppose the theological God to be; that we are absolutely ignorant what would be the extent of our conception if we had more senses. Thus to demand what God could have done in such a case, is also to suppose the thing in question, seeing that we cannot have a knowledge how far can go the power of a being of which we have no idea. We have no more knowledge of that which angels, beings different from ourselves, intelligences superior to us, can feel and know. We are ignorant of the mode in which plants vegetate; how should we know any thing of beings of an order entirely distinguished from our own? At least we can rest assured that if God is infinite, as it is said he is, neither the angels nor any subordinate intelligence can conceive him. If man is an enigma to himself, how should he be able to comprehend that which is not himself? It is necessary then that we confine ourselves

* The theologians frequently speak to us of an *intimate sense,* of a *natural instinct,* by the aid of which we discover or feel the divinity and the pretended truths of religion. But if we only examine these things, we shall find that this *intimate sense* and this *instinct* are no more than the effects of habit, of enthusiasm, of inquietude, and of prejudice, which, frequently in despite of all reason, lead us back to prejudices which our mind, when tranquil, cannot but reject.

to judge with the five senses we have. A blind man has only the use of four senses; he has not the right of denying that there does exist an extra sense for others; but he can say, with truth and reason, that he has no idea of the effects which would be produced with the sense which he lacks. It is with these five senses that we are reduced to judge of the Divinity which no one amongst the theologians can show us, or see better than ourselves. Would not a blind man, surrounded with other men devoid of sight, be authorized to demand of them by what right they spoke to him of a sense which they themselves did not possess, or of a being upon which their own peculiar experience taught them nothing ?*

In short, we can again reply to Doctor Clarke, that, according to his system, the supposition is impossible, and ought not to be made, seeing that God having, according to himself, made man, was willing, without doubt, that he should have no more than five senses, or that he was what he actually is, because it was necessary that he should be thus to conform to the wise views and to the immutable designs which theology gives him.

Doctor Clarke, as well as all other theologians, found the existence of their God upon the necessity of a power that may have the ability to begin motion. But if matter has always existed, it has always had motion, which as we have proved, is as essential to it as its extent, and flows from its primitive properties. There is, then, motion only in matter, and mobility is a consequence of its existence; not that the great whole can itself occupy other parts of space than those which it actually occupies, but its parts can change, and do change continually their respective situations; it is from thence results the conservation and the life of nature, which is always immutable in its whole. But in supposing, as is done every day, that matter is inert,

that is to say, incapable of producing any thing by itself without the assistance of a moving power which gives it motion, can we ever conceive that material nature receives its motion from a power that has nothing material? Can man really figure to himself that a substance, which has no one of the properties of matter, can create matter, draw it from its own peculiar source, arrange it, penetrate it, direct its motion, and guide its course?

Motion, then, is coeternal with matter. From all eternity the particles of the universe have acted one upon the other in virtue of their energies, of their peculiar essences, of their primitive elements, and of their various combinations. These particles must have combined in consequence of their analogy or relations, attracted and repelled each other, have acted and reacted, gravitated one upon the other, been united and dissolved, received their forms, and been changed by their continual collisions. In a material world, the acting-power must be material; in a whole, of which the parts are essentially in motion, there is no occasion for an acting power distinguished from itself; the whole must be in perpetual motion by its own peculiar energy. The general motion, as we have elsewhere proved, has its birth from the particular motions which beings communicate to each other without interruption.

We see, then, that theology, in supposing a God who gives motion to nature, and who was distinguished from it, has done no more than multiply beings, or rather has only personified the principle of mobility inherent in matter: in giving to this principle human qualities, it has only lent its intelligence, thought—perfections which can in nowise be suitable to it. Every thing which Doctor Clarke, and all the modern theologians, tell us of their God, becomes, in some respects, sufficiently intelligible as soon as we apply it to nature and to matter; it is eternal, that is to say, it cannot have had a commencement, and it will never have an end; it is infinite, that is to say, we have no conception of its limits, &c. But human qualities, always borrowed from ourselves, cannot be suitable to it, seeing that these qualities are modes of being, or modes of which

---

* In supposing, as the theologians do, that God imposes upon men the necessity of knowing him, their pretension appears as irrational as would be the idea of a landholder to whom they should ascribe the whim that the ants of his garden could know him and might reason pertinently upon him.

only belong to particular beings, and not to the whole which contains them.

Thus, to resume the answers which have been given to Doctor Clarke, we shall say, first, we can conceive that matter has existed from all eternity, seeing that we cannot conceive it to have had a beginning. Secondly, that matter is independent, seeing there is nothing exterior to it; that it is immutable, seeing it cannot change its nature, although it is unceasingly changing its form or combination. Thirdly, that matter is self-existent; since, not being able to conceive that it can be annihilated, we cannot conceive it can possibly have commenced to exist.—Fourthly, that we do not know the essence or true nature of matter, although we have a knowledge of some of its properties and qualities according to the mode in which it acts upon us; this is what we cannot say of God. Fifthly, that matter, not having had a beginning, will never have an end, although its combinations and its forms have a commencement and an end. Sixthly, that if all which exists, or every thing that our mind can conceive, is matter, this matter is infinite; that is to say, cannot be limited by any thing; that it is omnipresent, if there is no place exterior to itself; indeed, if there was a place exterior to it, this would be a vacuum, and then God would be the vacuum. Seventhly, that nature is only one, although its elements or its parts may be varied to infinity, and indued with properties extremely different. Eighthly, that matter, arranged, modified, and combined, in a certain mode, produces in some beings, that which we call intelligence; it is one of its modes of being, but it is not one of its essential properties. Ninthly, that matter is not a free agent, since it cannot act otherwise than it does in virtue of the laws of its nature, or of its existence, and consequently, that heavy bodies must necessarily fall, light bodies must rise, fire must burn; man must feel good and evil, according to the 'nature of the beings of which he experiences the action. Tenthly, that the power or the energy of matter has no other bounds than those which are prescribed by its own nature. Eleventhly, that wisdom, justice, goodness, &c. are qualities peculiar to matter combined and modified as it is found in some beings of the human species, and that the idea of perfection is an abstract, negative, metaphysical idea, or a mode of considering objects which supposes nothing real to be exterior to ourselves. In fine, twelfthly, that matter is the principle of motion, which it contains within itself, since matter only is capable of giving and receiving motion : this is what cannot be conceived of an immaterial and simple being, destitute of parts; who, devoid of extent, of mass, of weight, cannot either move himself, or move other bodies—much less, create, produce, and preserve them.

CHAPTER III.

Examination of the Proofs of the Existence of God given by Descartes, Malebranche, Newton, &c.

God is incessantly spoken of, and yet no one has hitherto arrived at demonstrating his existence ; the most sublime geniuses have been obliged to run aground against this rock ; the most enlightened men have done no more than stammer upon a matter which every one concurred in considering the most important ; as if it could be necessary to occupy ourselves with objects inaccessible to our senses, and of which our mind cannot take any hold !

To the end that we may convince ourselves of the little solidity which the greatest men have given to those proofs by which they have successively imagined to establish the existence of God, let us briefly examine what the most celebrated philosophers have said ; and let us begin with Descartes, the restorer of modern philosophy. This great man himself tells us : " All the strength of argument which I have hitherto used to prove the existence of God, consists in this, that I acknowledge it would not be possible my nature was such as it is, that is to say, that I should have in me the idea of a God, if God did not truly exist; this same God, I say, of whom the idea is in me, that is to say, who possesses all those *high perfections* of which our mind can have some slight idea, without, how-

ever, being able to comprehend them."
See *Meditation III.*, *upon the Existence of God*, *p.* 71-2.

He had said, a little before, page 69:
"We must necessarily conclude from
this alone, that because I exist and
have the idea of a most perfect being,
that is to say, of God, the existence of
God is most evidently demonstrated."

First, we reply to Descartes, that we
have no right to conclude that the
thing exists because we have an idea
of it; our imagination presents to us
the idea of a sphynx or of a hippogriff,
without having the right from that circumstance to conclude that these things
really exist.

Secondly, we say to Descartes, that it
is not possible he should have a positive and true idea of God, of whom, as
well as the theologians, he would
prove the existence. It is impossible
for men, for material beings, to form
to themselves a real and true idea of a
spirit; of a substance destitute of extent; of an incorporeal being, acting
upon nature, which is corporeal and
material; a truth which we have already sufficiently proved.

Thirdly, we shall say to him, that it
is impossible man should have any
positive and real idea of perfection, of
infinity, of immensity, and of the other
attributes which theology assigns to
the Divinity. We shall then make the
same reply to Descartes, which we
have already made in the preceding
chapter to the twelfth proposition of
Doctor Clarke.

Thus nothing is less conclusive than
the proofs upon which Descartes rests
the existence of God. He makes of
this God thought and intelligence; but
how conceive intelligence or thought,
without a subject to which these qualities may adhere? Descartes pretends
that we cannot conceive God, but "*as
a power which applies itself successively to the parts of the universe.*"
He again says, that "*God cannot be
said to have extent but as we say of
fire contained in a piece of iron,
which has not, properly speaking,
any other extension than that of the
iron itself.*" But, according to these
notions, we have the right to tax him
with announcing in a very clear manner, that there is no other God than
nature; this is a pure *Spinosism.* In

fact, we know that it is from the principles of Descartes that Spinosa drew
up his system, which flows from them
necessarily.

We might, then, with great reason,
accuse Descartes of atheism, seeing
that he destroys in a very effectual
manner the feeble proofs which he
gives of the existence of a God. We
have then foundation for saying to him
that his system overturns the idea of
the creation. Indeed, before God had
created matter, he could not co-exist
nor be co-extended with it; and in this
case according to Descartes, there was
no God; seeing that by taking from
the modifications their subject, these
modifications must themselves disappear. If God, according to the Carte-
sians, is nothing but nature, they are
quite *Spinosians;* if God is the motive-power of this nature, if God no
longer exists by himself, he exists no
longer than the subject to which he
is inherent subsists; that is to say,
nature, of which he is the motive-
power. Thus, God no longer exists
by himself, he will only exist as long
as the nature which he moves; without matter, or without a subject to
move, to conserve, to produce, what
will become of the motive-power of
the universe? If God is this motive-
power, what will become of him without a world, in which he can make use
of his action?*

We see, then, that Descartes, far
from establishing on a solid foundation
the existence of a God, totally destroys
him. The same thing will happen
necessarily to all those who shall reason upon him; they finish always by
confuting him, and by contradicting
themselves. We shall find the same
want of just inference, and the same
contradictions, in the principles of the
celebrated father Malebranche, which,
if considered with the slightest attention, appear to conduct us directly to
Spinosism; indeed, what can be more
conformable to the language of Spinosa, than to say, that "*the universe is
only an emanation from God; that
we see every thing in God; that every
thing which we see is only God; that*

_____
* See *The Impious Man Convinced, or a
Dissertation against Spinosa*, page 115, and
sequel. *Amsterdam*, 1685.

God alone does every thing that is done; that all the action, and every operation which takes place in all nature is himself; in a word, that God is every being, and the only being?" Is not this formally saying that nature is God? Besides, at the same time that Malebranche assures us we see every thing in God, he pretends, that *"it is not yet clearly demonstrated that matter and bodies have existence, and that faith alone teaches us these mysteries, of which, without it, we should not have any knowledge whatever."* In reply, it may be reasonably asked of him, how the existence of God, who has created matter, can be demonstrated, if the existence of this matter itself is yet a problem? Malebranche himself acknowledges that we can have no precise demonstration of the existence of any other being than of that which is necessary; he adds, that *"if it be closely examined, it will be seen that it is not even possible to know, with certitude, if God be or be not truly the creator of a material and sensible world."* After these notions, it is evident, that according to *Father* Malebranche, men have only their faith to guaranty the existence of God; but faith itself supposes this existence; if it be not certain that God exists, how shall we be persuaded that we must believe that which it is reported he says?

On the other hand, these notions of Malebranche evidently overturn all theological doctrines. How can the liberty of man be reconciled with the idea of a God who is the motive-power of all nature; who immediately moves matter and bodies, without whose consent nothing is done in the universe; who predetermines the creatures to every thing which they do? How can they, with this belief, pretend that human souls have the faculty of forming thoughts, wills; of moving and of modifying themselves? If it be supposed, with the theologians, that the conservation of his creatures is a continued creation, is it not God who, in preserving them, enables them to commit evil? It is evident, that, according to the system of Malebranche, God does every thing, and that his creatures are no more than passive instruments in

his hands; their sins, as well as their virtues, appertain to him; men can neither have merit nor demerit; this is what annihilates all religion. It is thus that theology is perpetually occupied with destroying itself.*

Let us now see if the immortal Newton will give us ideas more true, and proofs more certain, of the existence of God. This man, whose extensive genius, has unravelled nature and its laws has bewildered himself as soon as he lost sight of them; a slave to the prejudices of his infancy, he has not had the courage to hold the flambeau of his enlightened understanding up to the chimera which they have gratuitously associated with nature; he has not allowed that its own peculiar powers were sufficient for it to produce all those phenomena which he has himself so happily explained. In short, the sublime Newton is no more than an infant, when he quits physics and demonstration, to lose himself in the imaginary regions of theology. Here is the manner in which he speaks of the Divinity :†

"This God," says he, "governs all, not as the soul of the world, but as the lord and sovereign of all things. It is in consequence of his sovereignty that he is called the Lord God, Παντοκράτωρ, the universal emperor. Indeed, the word God is relative and relates to slaves; the deity is the dominion or the sovereignty of God, not over his own body, as those who look upon God as the soul of the world think, but over slaves."

We see from thence that Newton, as well as all the theologians, makes of his God a pure spirit, who presides over the universe; a monarch, a lord paramount, a despot, that is to say, a powerful man; a prince, whose government takes for a model that which the kings of the earth sometimes exercise over their subjects, transformed into slaves, whom ordinarily they make to feel, in a very grievous manner, the weight of their authority. Thus the God of Newton is a despot, that is to say, a man, who has the privilege of

---

* See *The Impious Man Convinced*, p. 143 and 214.

† See *Principia Mathematica*, page 528, and sequel. London edition. 1726.

being good when it pleases him, unjust and perverse when his fancy so determines him. But, according to the ideas of Newton, the world has not existed from all eternity, the *slaves* of God have been formed in the course of time, therefore, we must conclude from it that before the creation of the world, the God of Newton was a sovereign without subjects and without estates. Let us see if this great philosopher is more in accord with himself, in the subsequent ideas which he gives us of his deified despot.

"The supreme God," he says, "is an eternal, infinite, and absolutely perfect being, but however perfect a being he may be, if he has no sovereignty, he is not the supreme God : the word God signifies lord, but every lord is not God ; it is the sovereignty of the spiritual being which constitutes God ; it is the true sovereignty which constitutes the true God ; it is the supreme sovereignty which constitutes the supreme God ; it is a false sovereignty which constitutes a false God. From true sovereignty, it follows, that the true God is living, intelligent and powerful ; and from his other perfections, it follows, that he is supremely or sovereignly perfect. He is eternal, infinite, omniscient ; that is to say, that he exists from all eternity, and will never have an end : *durat ab æterno, ab infinito in infinitum ;* he governs all and he knows every thing that is done, or that can be done. He is neither eternity nor infinity, but he is eternal and infinite ; he is not space nor duration, but he exists and is present (*adest*)."*

In all this unintelligible rigmarole, we see nothing but incredible efforts to reconcile the theological attributes or the abstract qualities with the human qualities·given to the deified monarch ; we see in it negative qualities, which are no longer suitable to man, given, however, to the sovereign of nature, whom they have supposed a king.— However it may be, here is always the supreme God who has occasion for subjects to establish his sovereignty ; thus God needs men for the exercise

of his empire, without which he would not be a king. When there was nothing, of what was God lord ? However this may be, this lord, this spiritual king, does he not exercise his spiritual empire in vain upon beings who frequently do not that whIch he wills they should do, who are continually struggling against him, who spread disorder in his states ? This spiritual monarch is the master of the minds, of· the souls, of the wills, and of the passions, of his subjects, to whom he has left the freedom of revolting against him. This infinite monarch, who fills every thing with his immensity, and who governs all, does he govern the man who sins, does he direct his actions, is he in him when he offends his God ? The Devil, the false God, the evil principle, has he not a more extensive empire than the true God, whose projects, according to the theologians, he is unceasingly overturning ? The true sovereign, is it not he whose power in a state influences the greater number of his subjects ? If God is omnipresent, is he not the sad witness and the accomplice of those outrages which are every where offered to his divine majesty ? If he fills all,· has he not extent, does he not correspond with various points of space, and from thence does he not cease to be spiritual ?

"God is one," continues Newton "and he is the same for ever and every where, not only by his virtue alone, or his energy, but also by his substance."

But how can a being who acts, who produces all those changes which beings undergo, always be the same ? What is understood by the virtue or energy of God ? These vague words, do they present any clear idea to our mind ? What is understood by the divine substance? If this substance is spiritual and devoid of extent, how can there exist in it any parts ? How can it put matter in motion ? How can it be conceived ?

Nevertheless, Newton tells us, that "all things are contained in him, and are moved in him, but without reciprocal action (*sed sine mutua passione*). God experiences nothing by the motion of bodies ; these experience no opposition whatever by his omnipresence." It here appears, that Newton gives

---

* The word *adest*, which Newton makes use of in the text, appears to be placed there to avoid saying that God is contained in space.

to the Divinity characters which are suitable only to vacuum and to nothing. Without that, we cannot conceive that it is possible not to have a reciprocal action or relation between those substances which are penetrated, which are encompassed on all sides. It appears evident that here the author does not understand himself.

"It is an incontestable truth that God exists necessarily, and the same necessity obliges him to exist always and every where: from whence it follows that he is in every thing similar to itself; he is all eyes, all ears, all brain, all arms, all feeling, all intelligence, and all action; but in a mode by no means human, by no means corporeal, and which is totally unknown to us. In the same manner as a blind man has no idea of colours, it is thus we have no idea of the mode in which God feels and understands."

The necessary existence of the Divinity, is precisely the thing in question; it is this existence which it is necessary to have verified by proofs as clear, and demonstration as strong, as gravitation and attraction. If the thing had been possible, the genius of Newton would, without doubt, have compassed it. But, oh man! so great and so powerful, when you were a geometrician; so little and so weak, when you became a theologian; that is to say, when you reasoned upon that which can neither be calculated nor submitted to experience; how could you think of speaking to us of a being who is, by your own confession, to you just what a picture is to a blind man? Wherefore quit nature, to seek in imaginary spaces, those causes, those powers, and that energy, which nature would have shown you in itself, if you had been willing to consult her with your ordinary sagacity? But the great Newton has no longer any courage; he voluntarily blinds himself, when the question is a prejudice which habit has made him look upon as sacred. Let us continue, however, to examine how far the genius of man is capable of leading itself astray, when once he abandons experience and reason, and suffers himself to be guided by his imagination.

"God," continues the father of modern philosophy, "is totally destitute of

body and of corporeal figure; here is the reason why he cannot be either seen or touched, or understood; and ought not to be adored under any corporeal form."

But what ideas can be formed of a being who is nothing of that of which we have a knowledge? What are the relations which can be supposed to exist between us and him? To what end adore him? Indeed if you do adore him, you will be obliged, in despite of yourself, to make him a being similar to man; sensible, like him, to homage, to presents, and to flattery; in short, you will make him a king, who, like those of the earth, exacts the respect of all who are subjected to them. Indeed, he adds :—

"We have ideas of his attributes, but we do not know that it is any one substance; we only see the figures and the colours of bodies, we only hear sounds, we only touch the exterior surfaces, we only smell odours, we only taste flavours; no one of our senses, no one of our reflections, can show us the intimate nature of substances; we have still less ideas of God."

If we have an idea of the attributes of God, it is only because we give him those belonging to ourselves, which we never do more than aggrandize or exaggerate to that height as to make them mistaken for those qualities we knew at first. If, in all those substances which strike our senses, we only know the effects which they produce on us, and after which we assign them qualities, at least these qualities are something, and give birth to distinct and clear ideas in us. That superficial knowledge, or whatever it may be, which our senses furnish us, is the only one we can possibly have; constituted as we are, we find ourselves obliged to be contented with it, and we see that it is sufficient for our wants: but we have not even the most superficial idea of a God distinguished from matter, or from all known substances; nevertheless, we are reasoning upon him unceasingly!

"We only have a knowledge of God by his attributes, by his properties, and by the excellent and wise arrangement which he has given to all things, and their *final causes;* and we admire him in consequence of his perfections."

I repeat, that we have no knowledge of God, but by those of his attributes which we borrow from ourselves; but it is evident they cannot become suitable to the universal being, who can have neither the same nature nor the same properties as· particular beings, such as ourselves. It is after ourselves, that we assign to God, intelligence, wisdom and perfection, in abstracting from him that which we call defects in ourselves. As to the order or the arrangement of the universe, of which we make a God the author, we find it excellent and wise, when it is favourable to us, or when the causes which are coexistent with ourselves do not disturb our own peculiar existence; otherwise, we complain of the confusion, and the *final causes* vanish. We attribute to an immutable God, motives, equally borrowed from our own peculiar mode of action, for deranging the beautiful order which we admire in the universe. Thus it is always in ourselves, that is in our peculiar mode of feeling, that we draw up the ideas of order, the attributes of wisdom, of excellence, and of perfection, which we give to God; whilst all the good and all the evil which happen in the world, are the necessary consequences of the essences of things, and of the general laws of matter, in short, of the gravity, of the attraction, and of the repulsion, and of the laws of motion, which Newton himself has so well developed, but which he dared not apply, as there was a question concerning the phantom to which prejudice ascribes the honour of all those effects, of which nature is itself the true cause.

"We revere and we adore God on account of his sovereignty: we worship him like his slaves; a God destitute of sovereignty, of providence, and of final causes, would be no more than nature and destiny."

It is true, we adore God like ignorant slaves, who tremble under a master whom they know not; we foolishly pray to him, although he is represented to us as immutable; although, in truth, this God is nothing more than nature acting by necessary laws necessarily personified, or destiny, to which the name of God is given.

Nevertheless, Newton tells us, "from

a physical and blind necessity, which should preside every where, and he always the same, there could not emanate any variety in beings; the diversity which we see, could only have their origin in the ideas and the will of a being which exists necessarily."

Wherefore should this diversity not happen from natural causes, from matter acting by itself, and of which the motion attracts and combines various, and yet analogous elements : or separates beings, by the aid of those substances which are not found suitable to unite ? Is not *bread* the result of the combination of flower, yest, and water. As for the blind necessity, as is elsewhere said, it is that of which we ignore the energy, or of which, being blind ourselves, we have no knowledge of the mode of action. Philosophers explain all phenomena by the properties of matter; and though they feel the want of being acquainted with the natural causes, they do not less believe them deducible from their properties or their causes. The philosophers, then, in this, are atheists! otherwise, they would reply, that it is God who is the author of all these phenomena.

"It is allegorically said, that God sees, hears, speaks, smiles, lives, hates, desires, gives, receives, rejoices, or becomes angry, fights, makes and fashions, &c. For all that is said of God, is borrowed from the conduct of men, by a kind of imperfect analogy."

Men have not been able to do otherwise, for want of being acquainted with nature and her ways ; they have imagined a peculiar energy, to which they have given the name of God, and they have made him act according to the same principles, as they are them selves made to act upon, or according to which they would act, if they were the masters: it is from this *theanthropy* that have flowed all those absurd and frequently dangerous ideas, upon which are founded all the religions of the world, who all adore in their God a powerful and wicked man. We shall see by the sequel, the fatal effects which have resulted to the human species, from those ideas which they have formed to themselves of the Divinity, whom they have never considered but as an absolute sovereign, a despot, and a tyrant. As for the present, let us continue to examine the proofs which are given to us by the deists of the existence of their God, whom they imagine they see every where.

Indeed, it is unceasingly repeated to us, that the regulated motion, the invariable order, which we see reign in the universe, those benefits which are heaped upon men, announce a wisdom, an intelligence, a goodness, which we cannot refuse acknowledging in the cause which produces such marvellous effects. We shall reply, that the regulated motion which we witness in the universe, is the necessary consequence of the laws of matter; it cannot cease to act in the manner it does, so long as the same causes act in it; these motions cease to be regulated, order gives place to disorder, as soon as new causes disturb or suspend the action of the first. Order, as we have elsewhere shown, is only the effect which results to us from a series of motion; there cannot be any real disorder relative to the great whole, where every thing that takes place is necessary and determined by laws that nothing can change. The order of nature may be contradicted or destroyed, relatively to us, but never is it contradicted relatively to itself, since it cannot act otherwise than it does. If, after the regulated and well-ordered motion which we see, we attribute intelligence, wisdom, and goodness, to the unknown or supposed cause of these effects, we are obliged in a similar manner to attribute to him extravagance and malice, every time that these motions become confused, that is to say, cease to be regulated relatively to us, or that we are ourselves disturbed by them, in our mode of existence.

It is pretended that animals furnish us with a convincing proof of a powerful cause of their existence; it is said, that the admirable harmony of their parts, which we see lend each other mutual assistance, to the end of fulfilling their functions and maintaining them together, announce to us a workman, who unites wisdom to power.*

. We cannot doubt the power of nature; she produces all the animals we see, by the aid of the combination of matter which is a continual action; the harmony that subsists between parts of these same animals, is a consequence of the necessary laws of their nature and of their combination; as soon as this accord ceases, the animal is necessarily destroyed. What becomes then of the wisdom, of the intelligence, or the goodness of that pretended cause to whom they ascribe the honour of this so much boasted harmony? These animals, so marvellous, which are said to be the work of an immutable God, are they not continually changing; and do they not always finish by decaying? Where is the wisdom, the goodness, the foresight, and the immutability, of a workman, who appears only to be occupied with deranging and breaking the springs of those machines, which are announced to us as the *chefs d'œuvres* of his power and of his ability. If this God cannot do otherwise, he is neither free nor omnipotent. If he changes his will, he is not immutable. If he permits

which prove nothing, except that there exists in nature elements suitable to unite, to arrange themselves, to co-order themselves, in a mode to form wholes, or combinations susceptible of producing particular effects. Thus these writings, loaded with erudition, only make known that there exists in nature beings diversely organized, formed in a certain manner, suitable to certain uses, who would no longer exist under the form they at present have, if their particles ceased to act as they do, that is to say, to be disposed in such a manner, as to lend each other mutual succours. To be surprised that the *brain*, the *heart*, the *eyes*, the *arteries*, and *veins*, of an animal act as we see them, that the roots of a plant attract juices, or that a tree produces fruit, is to be surprised that an animal, a plant, or a tree exists. These beings would not exist, or would no longer be that which we know they are, if they ceased to act as they do; this is what happens when they die. If their formation, their combination, their modes of action and of conserving themselves some time in life, was a proof that these beings are the effects of an intelligent cause; their destruction, their dissolution, the total cessation of their mode of acting, their death, ought to prove, in the same manner, that these beings are the effects of a cause destitute of intelligence, and of permanent views. If we are told that his views are unknown to us; we shall ask, by what right then they can ascribe them to this cause, or how it can be reasoned upon?

---

* We have already remarked, elsewhere, that many authors, with a view of proving the existence of a divine intelligence, have copied whole tracts of *anatomy* and *botany*,

those machines, which he has rendered sensible, to experience pain, he
wants goodness. If he has not been
able to render his works more solid, it
is that he wants the ability. In seeing
that animals, as well as all the other
works of the Divinity, decay, we cannot prevent ourselves from concluding
therefrom, either that every thing nature does is necessary, and is only a
consequence of its laws, or that the
workman who made it is destitute of
plan, of power, of stability, of ability,
of goodness.

Man, who looks upon himself as the
chef-d'œuvre of the Divinity, furnishes
more than every other production, a
proof of the incapacity or of the malice
of his pretended author: in this sensible, intelligent, and thinking being,
who believes himself the constant object of the divine predilection, and
who forms his God after his own peculiar model, we only see a more inconstant, more brittle machine, which is
more subject to derange itself, by its
great complication, than the grosser
beings. Beasts, destitute of our knowledge, plants, which vegetate, stones,
devoid of feeling, are, in many respects, beings more favoured than man;
they are, at least, exempted from the
sorrows of the mind; from the torments
of thought; from that devouring chagrin, of which he is so frequently the
prey. Who is he that would not be
an animal or a stone, every time he
recalls to his imagination the irrepar-
able loss of a beloved object? Would
it not be better to be an inanimate
mass, than a restless, superstitious being, who does nothing but tremble
here below under the yoke of his God,
and who again foresees infinite torments in a future life? Beings, destitute of feeling, of life, of memory, and
of thought, are not afflicted by the idea
of the past, of the present, or of the
future; they do not believe themselves
in danger of becoming eternally unhappy from having reasoned badly,
like many of those favoured beings
who pretend it is for them alone that
the architect of the world has constructed the universe.*

* Cicero says: "Inter hominem et belluam
hoc maxime interest, quod hæc ad id solum
quod adest, quod que præsens est, se accom-
modat, paululum admodum sentiens præteri-

Let us not be told that we cannot
have the idea of a work without having also that of a workman distinguished from his work. Nature is not a
work ; she has always been self-existent; it is in her bosom that every thing
is operated ; she is an immense elaboratory, provided with materials, and
who makes the instruments of which
she avails herself to act: all her works
are the effect of her own energy, and
of those agents or causes which she
makes, which she contains, which she
puts in action. Eternal, uncreated,
indestructible elements, always in motion, in combining themselves variously, give birth to all the beings and to
all the phenomena which our eyes behold ; to all the effects, good or bad,
which we feel ; to the order or the confusion which we never distinguish but
by the different modes in which we
are affected; in short, to all those
wonderful phenomena upon which we
meditate and reason. For that purpose, these elements have occasion
only for their properties, whether particular or united, and the motion which
is essential to them, without its being
necessary to recur to an unknown
workman to arrange, fashion, combine,
conserve, and dissolve them.

But, supposing, for an instant, that it
were impossible to conceive the universe without a workman, who has
formed it, and who watches over his
work, where shall we place this workman? shall it be within or without the
universe? is he matter or motion; or
rather, is he only space, nothing, or
the vacuum? In all these cases, either
he would be nothing, or he would be
contained in nature, and submitted to
her laws. If he be in nature, I can
only see matter in motion, and I must
conclude from it that the agent who

tum et futurum." Thus, what it has been
wished to make pass as a prerogative of man,
is only a real disadvantage. Seneca has said :
" Nos et venturo torquemur et præterito, timo-
ris enim tormentum memoria reducit, provi-
dentia anticipat ; nemo tantum præsentibus
miser est." Could we not demand of every
honest man, who tells us that a good God
created the universe for the happiness of our
sensible species, would you yourself have created a world which contains so many wretches ?
would it not have been better to have abstain-
ed from creating so great a number of sensible beings, than to have called them into life
for the purpose of making them suffer ?

moves it is corporeal and material, and that, consequently, he is subject to dissolution. If this agent be exterior to nature, I have then no longer any idea of the place which he occupies, neither can I conceive an immaterial being, nor the mode in which a spirit without extent, can act upon the matter from which it is separated. Those unknown spaces which the imagination has placed beyond the visible world, have no existence relatively to a being who sees with difficulty down to his feet; the ideal power which inhabits them cannot be painted to my mind, but when my imagination shall combine at random the fantastical colours which it is always obliged to draw from the world where I am; in this case, I shall do no more than reproduce in idea that which my senses shall have really perceived, and this God, which I strive to distinguish from nature, or to place out of its bosom, will always return into it necessarily and in despite of me.[*]

It will be insisted that if a statue or a watch were shown to a savage, who had never before seen either, he would not be able to prevent himself from acknowledging that these things were the works of some intelligent agent, of more ability, and more industrious than himself: it will be concluded from thence, that we are in like manner obliged to acknowledge that the machine of the universe, that man, that the various phenomena of nature, are the works of an agent, whose intelligence and power far surpasses our own.

I reply, in the first place, that we cannot doubt that nature is extremely powerful and very industrious; we admire her activity, every time that we are surprised by those extensive, various, and complicated effects which we find in those of her work, which we take the trouble to meditate upon; nevertheless, she is neither more nor less industrious in one of her works than in another. We no more understand how she has been capable of producing a stone or a metal, than a head organized like that of Newton. We call that man industrious, who can do things, which we ourselves cannot do; nature can do every thing, and as soon as a thing exists, it is a proof that she has been capable of making it. Thus it is never more than relatively to ourselves that we judge nature to be industrious; we compare her then to ourselves; and as we enjoy a quality which we call *intelligence*, by the assistance of which we produce works, or by which we show our industry, we conclude from it, that those works of nature, which astonish us the most, do not belong to her, but are to be ascribed to an intelligent workman like ourselves, but in whom we proportion intelligence to the astonishment which his works produce in us, that is to say, to our own peculiar weakness and ignorance.

In the second place, the savage to whom a statue or a watch shall be brought, will or will not have ideas of human industry: if he has ideas of it, he will feel that this watch or this statue, may be the work of a being of his own species, enjoying those faculties which he himself lacks. If the savage has no idea of human industry and of the resources of art, in seeing the spontaneous motion of a watch, he will believe that it is an animal, which cannot be the work of man. Multiplied experience, confirms the mode of thinking which I ascribe to this savage.[*] Thus in the same manner as a great many men, who believe themselves much more acute than he, this savage will attribute the strange effects he sees, to a Genius, to a Spirit, to a God; that is to say, to an *unknown power*, to whom he will assign capabilities of which he believes the beings of his own species to be absolutely destitute; by this he will prove no-

---

[*] Hobbes says: "The world is corporeal; it has the dimensions of size, that is to say, length, breadth, and depth. Each portion of a body, is a body, and has these same dimensions: consequently, each part of the universe is a body, and that which is not a body, is no part of the universe; but as the universe is every thing, that which does not make a part of it, is nothing, and can be no part." See *Hobbes' Leviathan*, chap. 46.

---

[*] The Americans took the Spaniards for Gods, because they made use of gunpowder, rode on horseback, and had vessels which sailed quite alone. The inhabitants of the island of Tenian, having no knowledge of fire before the arrival of the Europeans, took it, the first time they saw it, for an animal which devoured wood.

thing, except that he is ignorant of what man is capable of producing. It is thus that a raw, unpolished people raise their eyes to heaven, every time they witness some unusual phenomenon. It is thus that the people call *miraculous, supernatural, divine*, all those strange effects of the natural causes of which they are ignorant; and as for the greater part, they do not know the cause of any thing; every thing is a miracle to them, or at least they imagine that God is the cause of all the good and of all the evil which they experience. In short, it is thus that theologians solve all difficulties in attributing to God every thing of which they are ignorant, or of which they are not willing men should understand the true causes. •   '

In the third place, the savage, in opening the watch and examining its parts, will feel, perhaps, that these parts announce a work which can only be the result of human labour. He will see that they differ from the immediate productions of nature, whom he has not seen produce wheels made of a polished metal. He will again see that these parts, separated from each other, no longer act as they did when they were together; after these observations, the savage will attribute the watch to the ingenuity of man, that is to say, to a being like himself, of whom he has ideas, but whom he judges capable of doing things which he does not himself know how to do; in short, he will ascribe the honour of this work to a being known in some respects, provided with some faculties superior to his own, but he will be far from thinking that a material work can be the effect of an immaterial cause, or of an agent destitute of organs and of extent, of whom it in impossible to conceive the action upon material beings : whilst, for want of being acquainted with the power of nature, we ascribe the honour of her work to a being of whom we have much less knowledge than of her, and to which, without knowing it, we attribute those amongst her labours which we comprehend the least. In seeing the world, we acknowledge a material cause of those phenomena which take place in it; and this cause is nature, of whom the energy is shown to those who study her.

Let us not be told, that, according to this hypothesis, we attribute every thing to a blind cause, to the fortuitous concurrence of atoms ; to *chance*. We only call those *blind causes*, of which we know not the combination, the power, and the laws. We call *fortuitous*, those effects of which we are ignorant of the causes, and which our ignorance and inexperience prevent us from foreseeing. We attribute to *chance*, all those effects of which we do not see the necessary connexion with their causes. Nature is not a blind cause ; she does not act by chance ; nothing that she does would ever be fortuitous to him who should know her mode of acting, her resources, and her course. Every thing which she produces is necessary, and is never more than a consequence of her fixed and constant laws; every thing in her is connected by invisible bands, and all those effects which we see flow necessarily from their causes, whether we know them or not. It is very possible there should be ignorance on our part, but the words *God, Spirit, Intelligence*, will not remedy this ignorance: they will do no more than redouble it by preventing us from seeking the natural causes of those effects which our visual faculties make us acquainted with.

This may serve for an answer to the eternal objection which is made to the partisans of nature, who are unceasingly accused of *attributing every thing to chance*. Chance is a word devoid of sense, or at least it indicates only the ignorance of those who employ it. Nevertheless we are told, and it is reiterated continually, that a regular work cannot be ascribed to the combinations of chance. Never, we are informed, will it be possible to arrive at the formation of a poem, such as the Iliad, by means of letters thrown or combined together at random. We agree to it without hesitation ; but, ingeniously, are those letters thrown with the hand like dice, which compose a poem ? It would avail as much to say that we could pronounce a discourse with the feet. It is nature who combines, after certain and necessary laws, a head organized in a manner to make a poem, it is nature who gives man a brain suitable to give birth to such a work ; it is nature who, by the temperament, the imagination, the passions, which she gives to man, capacitates him to produce a *chef-d'œuvre:* it is his brain, modified in a certain manner, decorated with ideas or images, made fruitful by circumstances, which can become the only matrix in which a poem can be conceived and developed. A head organized like that of Homer, furnished with the same vigour and the same imagination, enriched with the same knowledge, placed in the same circumstances, will produce necessarily, and not by chance, the poem of the Iliad ; at least if it be not denied that causes similar in every thing, must produce effects perfectly identical.*

It is, then, puerility, or knavery, to talk of composing, by a throw of the hand, or by mingling letters together by chance, that which can only be done with the assistance of a brain organized and modified in a certain manner. The human seed does not develop itself by chance, it cannot be conceived or formed but in the womb of a woman. A confused heap of characters or of figures, is only an assemblage of signs, destined to paint ideas; but in order that these ideas may be painted, it is previously necessary that they may have been received, combined, nourished, developed, and connected, in the head of a poet, where circumstances make them fructify and ripen, on account of the fecundity, of the heat, and

* Should we not be astonished if there were in a dice-box a hundred thousand dice, to see a hundred thousand *sixes follow in succession?* Yes, without doubt, it will be said ; but if these dice were all *cogged* or *loaded*, we should cease to be surprised. Well then, the particles of matter may be compared to *cogged* dice, that is to say, always producing certain determined effects ; these particles being essentially varied in themselves, and in their combination, they are *cogged* in an infinity of different modes. The head of Homer, or the head of Virgil, was no more than the assemblage of particles, or if they choose, of dice, *cogged* by nature; that is to say, of beings combined and wrought in a manner to produce the Iliad or the Æneid. As much may be said of all the productions, whether they be those of intelligence, or of the handiwork of men. Indeed, what are men, except dice *cogged*, or machines which nature has rendered capable of producing works of a certain kind ? A man of genius produces a good work, in the same manner as a tree of good species, placed in good ground, and cultivated with care, produces excellent fruit.

of the energy of the soil where these *intellectual seeds* have been thrown. Ideas, in combining, extending, connecting, and associating themselves, form a whole, like all the bodies of nature : this whole pleases us, when it gives birth to agreeable ideas in our mind ; when they offer us pictures which move us in a lively manner ; it is thus that the poem of Homer, engendered in his head, has the power of pleasing heads analogous and capable of feeling its beauties.

We see, then, that nothing is made by chance. All the works of nature grow out of certain uniform and invariable laws, whether our mind can with facility follow the chain of the successive causes which she puts in action, or whether, in her more complicated works, we may find ourselves in the impossibility of distinguishing the different springs which she causes to act.†
It is not more difficult for nature to produce a great poet, capable of composing

† It is not often that the most sedulous attention, the most patient investigation, afford us the information we are seeking after ; sometimes, however, the unwearied industry of the philosopher is rewarded by throwing into light the most mysterious operations of Nature. Thus the keen penetration of a Newton, aided by uncommon diligence, developed the starry system, which, for so many thousand years, had eluded the research of all the astronomers by whom he was preceded. Thus the sagacity of a Harvey giving vigour to his application, brought out of the obscurity in which for almost countless centuries it had been buried, the true course pursued by the sanguinary fluid, when circulating through the veins and arteries of man, giving activity to his machine, diffusing life through his system, and enabling him to perform those actions which so frequently strike an astonished world with wonder and regret. Thus Galileo, by a quickness of perception, a depth of reasoning peculiar to himself, held up to an admiring world, the actual form and situation of the planet we inhabit, which, until then, had escaped the observation of the most profound geniuses—the most subtile metaphysicians—and which, when first promulgated, was considered so contradictory to all the then received opinions, (besides giving the lie to the story of Joshua stopping the sun, as recorded in the Holy Bible!) that he was ranked as an impious blasphemer, to hold communion with whom would infallibly secure to the communers a place in the regions of everlasting torment : indeed, Pope Gregory, who then filled the papal chair, excommunicated all those who had the temerity to accredit so abominable a doctrine !

an admirable work, than to produce a glittering metal or a stone, which gravitates towards a centre. The mode which she takes to produce these different beings, is equally unknown to us, when we have not meditated upon it. Man is born by the necessary concurrence of some elements; he increases and is strengthened in the same manner as a plant or a stone, which is, as well as he, increased and augmented by those substances which come and join themselves thereto: this man, feels, thinks, acts, and receives ideas, that is to say, is, by his peculiar organization, susceptible of modifications, of which the plant and the stone are totally incapable: in consequence, the man of genius produces works, and the plant fruits, which please and surprise us, by reason of those sensations which they operate in us; or on account of the rarity, the magnitude, and the variety of the effects which they occasion us to experience. That which we find most admirable in the productions of nature, and in those of animals or men, is never more than a natural effect of the parts of matter, diversely arranged and combined; from whence result in them organs, brains, temperaments, tastes, properties, and different talents.

Nature, then, makes nothing but what is necessary; it is not by fortuitous combinations, and by cnance throws, that she produces the beings we see; all her throws are sure, all the causes which she employs have, infallibly, their effects. When she produces extraordinary, marvellous and rare beings, it is, that, in the order of things, the necessary circumstances, or the concurrence of the productive causes of these beings, happen but seldom. As soon as these beings exist, they are to be ascribed to nature, to whom every thing is equally easy, and to whom every thing is equally possible, when she assembles the instruments or the causes necessary to act. Thus, let us never limit the powers of nature. The throws and the combinations which she makes during eternity, can easily produce all beings; her eternal course must necessarily bring and bring again the most astonishing circumstances, and the most rare, for those beings who are only for a moment enabled to consider them,

without ever having either the time or the means of searching into the bottom of causes. Infinite throws during eternity, with the elements, and combinations infinitely varied, suffice to produce every thing of which we have a knowledge, and many other things which we shall never know.

Thus, we cannot too often repeat to the Deicolists, or supporters of the being of a God, who commonly ascribe to their adversaries ridiculous opinions, in order to obtain an easy and transitory triumph in the prejudiced eyes of those who dare examine nothing deeply, that *chance is nothing* but a word, as well as the word God, imagined to cover the ignorance in which men are of the causes acting in a nature whose course is frequently inexplicable. It is not chance that has produced the universe, it is of itself that which it is; it exists necessarily and from all eternity. However concealed may be the ways of nature, her existence is indubitable; and her mode of acting is, at least, much more known to us than that of the inconceivable being, which, it has been pretended, is associated with her; which has been distinguished from her; which has been supposed necessary and self-existent, although, hitherto, it has neither been possible to demonstrate his existence, to define nim, to say any thing reasonable of him, nor to form upon his account any thing more than conjectures, which reflection has destroyed as soon as they have been brought forth.

————

CHAPTER IV.

Of Pantheism, or of the Natural Ideas of the Divinity.

WE see, by that which has preceded, that all the proofs upon which theology pretends to found the existence of its God, have their origin in the false principle that matter is not self-existent, and is, by its nature, in an impossibility of moving itself; and, consequently, is incapable of producing those phenomena which attract our wondering eyes in the wide expanse of the universe. After these suppositions, so gratuitous and so false, as we have already shown elsewhere,* it has been believed that matter did not always exist, but that it was indebted for its existence and for its motion to a cause distinguished from itself; to an unknown agent, to whom it was subordinate. As men find in themselves a quality which they call *intelligence*, which presides over all their actions, and by the aid of which they arrive at the end they propose to themselves, they have attributed intelligence to this invisible agent; but they have extended, magnified, and exaggerated this quality in him, because they have made him the author of effects of which they believed themselves incapable, or which they did not suppose natural causes had sufficient energy to produce.

As this agent could never be perceived, nor his mode of action conceived, he was made a *spirit*, a word which designates that we are ignorant what he is, or that he acts like the breath of which we cannot trace the action. Thus, in assigning him *spirituality*, we did no more than give to God an occult quality, which was judged suitable to a being always concealed, and always acting in a made imperceptible to the senses. It appears, however, that, originally, by the word *spirit* it was meant to designate a matter more subtile than that which coarsely struck the organs; capable of penetrating this matter, of communicating to it motion and life, of producing in it those combinations and those modifications which our visual organs discover. Such was, as we have seen, that Jupiter, who was originally designed to represent in the theology of the ancients the ethereal matter which penetrates, gives activity, and vivifies all the bodies of which nature is the assemblage.

Indeed it would be deceiving ourselves to believe that the idea of God's spirituality, such as we find it at the present day, presented itself in the early stages of the human mind. This *immateriality*, which excludes all analogy and all resemblance with any thing we are in a capacity to have a knowledge of, was, as we have already observed, the slow and tardy fruit of men's imagination, who, obliged to meditate, without any assistance drawn from experience, upon the concealed mover of nature, have, by degrees, arrived at forming this ideal phantom; this being, so fugitive, that we have been made to adore it without being able to designate its nature, otherwise than by a word to which it is impossible we should attach any true idea.* Thus by dint of reasoning and subtilizing, the word God no longer presents any one image; when they spoke of it, it was impossible to understand them, seeing that each painted it in his own manner, and in the portrait which he made of it, consulted only his own peculiar temperament, his own peculiar imagination, and his own peculiar reveries; if they were in unison in some points, it was to assign him inconceivable qualities, which they believed were suitable to the incomprehensible being to which they had given birth; and from the incompatible heap of these qualities resulted only a whole, perfectly impossible to have existence. In short, the master of the universe, the omnipotent mover of nature, that being which is announced as of the most importance to be

*See what has been said upon this in the *seventh chapter* of the *first part*. Although the first doctors of the *Christian Church* may, for the greater part, have drawn from the *Platonic* philosophy their obscure notions of *spirituality*, of *incorporeal, and immaterial substances*, of *intellectual powers*, &c. we have only to open their works, to convince ourselves that they had not that idea of God which the theologians of the present day give us. *Tertullian*, as we have elsewhere said, considered God as corporeal. *Seraphis* said, crying, *that they had deprived him of his God*, in making him adopt the opinion of *spirituality*, which was not, however, so much subtilized then as it has been since. Many *fathers of the Church* have given a human form to God, and have treated as heretics those who made a God spirit. The *Jupiter* of the pagan theology is looked upon as the youngest child of *Saturn* or of *Time:* the *spiritual God* of the *Christians* is a much more recent production of time; it is only by dint of subtilizing that this God, the conqueror of all those *Gods* who preceded him, has been formed by degrees. *Spirituality* is become the last refuge of theology, which has arrived at making a God more than aerial in the hope, no doubt, that such a God would be inaccessible; indeed, he is so for to attack him is to combat a mere chimera.

*See *first part, chapter second*, where we have shown that motion is essential to matter. This chapter is only a summary of the first five chapters of the first part, which it is intended to recall to the reader; he will pass to the next if these ideas are remembered.

known, was by theological reveries, reduced to be no more than a vague word destitute of sense; or, rather a vain sound, to which each attaches his own peculiar ideas. Such is the God who has been substituted to matter, to nature; such is the idol to which men are not permitted to refuse paying their homage.

There have been, however, men of sufficient courage to resist this torrent of opinion and delirium. They believed that the object which was announced as the most important for mortals, as the only centre of their actions and their thoughts, demanded an attentive examination. They apprehended that if experience, judgment, or reason, could be of any utility, it must be, without doubt, to consider the sublime monarch who governed nature, and who regulated the destiny of all those beings which it contains. They quickly saw they could not subscribe to the general opinion of the uninformed, who examine nothing; and much less with their guides, who, deceivers or deceived, forbade others to examine it, or perhaps, were themselves incapable of making such an examination. Thus, some thinkers had the temerity to shake off the yoke which had been imposed upon them in their infancy; disgusted with the obscure, contradictory, and nonsensical notions which they had been made, by habit, to attach mechanically to the vague name of a God, impossible to be defined; supported by reason against the terrours with which this formidable chimera was environed; revolting at the hideous paintings under which it was pretended to represent him, they had the intrepidity to tear the veil of delusion and imposture, they considered, with a calm eye, this pretended power, become the continual object of the hopes, the fears, the reveries, and the quarrels of blind mortals. The spectre quickly disappeared before them; the tranquillity of their mind permitted them to see every where, only a nature acting after invariable laws, of whom the world is the theatre; of whom men, as well as all other beings, are the works and the instruments, obliged to accomplish the eternal decrees of *necessity*.

Whatever efforts we make to penetrate into the secrets of nature, we never find in them, as we have many times repeated, more than matter, various in itself, and diversely modified by the assistance of motion. Its whole, as well as all its parts, show us only necessary causes and effects, which flow the one from the other, and of which, by the aid of experience, our mind is more or less capable of discovering the connexion. In virtue of these specific properties, all the beings we see gravitate, attract and repel each other; are born and dissolved, receive and communicate motion, qualities, modifications, which maintain them, for a time, in a given existence, or which make them pass into a new mode of existence. It is to these continual vicissitudes that are to be ascribed all the phenomena, great or small, ordinary or extraordinary, known or unknown, simple or complicated, which we see operated in the world. It is by these changes that we have a knowledge of nature; she is only mysterious to those who consider her through the veil of prejudice, her course being always simple to those who look at her without prepossession.

To attribute the effects our eyes witness to nature, to matter variously combined, to the motion which is inherent in it, is to give them a general and known cause; to penetrate deeper, is to plunge ourselves in imaginary regions, where we only find an abyss of incertitudes and obscurities. Let us not seek, then, a moving principle out of nature of which the essence always was to exist and to move itself; which cannot be conceived to be without properties, consequently, without motion; of which all the parts are in action, reaction, and continual efforts; where a single molecule cannot be found that in in absolute repose, and which does not necessarily occupy the place assigned to it by necessary laws. What occasion is there to seek out of matter a motive-power to give it play, since its motion flows necessarily from its existence, its extent, its forms, its gravity, &c., and since nature in inaction would no longer be nature?

If it be demanded how we can figure to ourselves, that matter, by its own peculiar energy, can produce all the effects we witness? I shall reply, that if by matter it is obstinately determined to understand nothing but a dead and inert mass, destitute of every property, without action, and incapable of moving itself, we shall no longer have a single idea of matter. As soon as it exists, it must have properties and qualities; as soon as it has properties, without which it could not exist, it must act by virtue of those properties, since it is only by its action that we can have a knowledge of its existence and its properties. It is evident, that if by matter be understood that which it is not, or if its existence be denied, those phenomena which strike our visual organs, cannot be attributed to it. But if by *nature* be understood that which she truly is, a heap of existing matter, furnished with properties, we shall be obliged to acknowledge that nature must move herself, and by her diversified motion be capable, without foreign aid, of producing the effects which we behold; we shall find that nothing can be made from nothing; that nothing is made by chance; that the mode of acting of every particle of matter is necessarily determined by its own peculiar essence, or by its individual properties.

We have elsewhere said, that that which cannot be annihilated or destroyed cannot have commenced to have existence. That which cannot have had a beginning, exists necessarily, or contains within itself the sufficient cause of its own peculiar existence. It is, then, useless to seek out of nature or of a self-existent cause, which is known to us at least in some respects, another cause whose existence is totally unknown. We know some general properties in matter, we discover some of its qualities; wherefore seek for its existence in an unintelligible cause, which we cannot know by any one property? Wherefore recur to the inconceivable and chimerical operation which has been designated by the word *creation*?*   Can we conceive that an

immaterial being has been able to draw matter from his own peculiar source? If creation is *an eduction from nothing* must we not conclude from it that God, who has drawn it from his own peculiar source, has drawn it from nothing, and is himself nothing? Do those who are continually talking to us of this act of the divine omnipotence, by which an infinite mass of matter has been, all at once, substituted to nothing, well understand what they tell us? Is here a man on earth, who conceives that a being devoid of extent, can exist, become the cause of the existence of beings, who have extent; act upon matter, draw it from his own peculiar essence, and set it in motion? In truth, the more we consider theology, and its ridiculous romances, the more we must be convinced that it has done no more than invent words, devoid of sense, and substituted sounds to intelligible realities.

For want of consulting experience, of studying nature and the material world, we have thrown ourselves into an intellectual world, which we have peopled with chimeras. We have not stooped to consider matter, nor to follow it through its different periods and changes. We have either ridiculously or knavishly confounded dissolution, decomposition, the separation of the elementary particles of which bodies are composed, with their radical destruction; we have been unwilling to see that the elements were indestructible, although their forms were fleeting and depended upon transitory combinations. We have not distinguished the change of figure, of position, of texture, to which matter is liable from its annihilation, which is totally impossible; we have falsely concluded that matter was not a necessary being, that it had commenced to exist, that it owed its existence to an unknown being, more necessary than itself; and this ideal being has become the creator, the motive-power, the preserver of the whole of nature. Thus a vain name only has been substituted for matter, which furnishes us with true ideas of nature, of which, at each moment we experience the action and the power,

---

* Some theologians have frankly confessed that the theory of the creation was founded on an hypothesis supported by very little probability, and which had been invented some centuries after Jesus Christ. An author, who endeavoured to refute Spinoza, assumes that Tertullian was the first who advanced this opinion against another Christian philosopher who maintained the eternity of matter. *See* " *The Impious Man Convinced*," end of the advertisement. Even the author of this work

admits that it is impossible to combat Spinosa without admitting the eternal coexistence of matter with God.

and of which we should have a much better knowledge if our abstract opinions did not continually place a bandage before our eyes.

Indeed, the most simple notions of philosophy show us, that although bodies change and disappear, nothing is, however, lost in nature; the various produce of the decomposition of a body serves for elements, for materials, and for basis to the formation, to the accretion, to the maintenance of other bodies. The whole of nature subsists and is conserved only by the circulation, the transmigration, the exchange, and the perpetual displacing of insensible particles and atoms, or of the sensible combinations of matter. It is by this *palingenesia*, or regeneration, that the great whole subsists, who, like the Saturn of the ancients, is perpetually occupied with devouring his own children. But it may be said, in some respects, that the metaphysical God, who has usurped his throne, has deprived him of the faculty of procreating and of acting, ever since he has been put in his place.

Let us acknowledge then, that matter is self-existent, that it acts by its own peculiar energy, and that it will never be annihilated. Let us say, that matter is eternal, and that nature has been, is, and ever will be occupied with producing, with destroying, with doing, and undoing ; with following laws resulting from its necessary existence. For every thing that she does, she needs only to combine elements and matter, essentially diverse, which attract and repel each other, dash against each other or unite themselves, remove from or approximate each other, hold themselves together or separate themselves. It is thus, that she brings forth plants, animals, men ; organized, sensible and thinking beings, as well as those destitute of feeling and of thought. All these beings act only for the term of their respective duration, according to invariable laws, determined by their properties, by their configuration, their masses, their weight, &c. Here is the true origin of every thing which presents itself to our view, showing the mode in which nature, by its own peculiar power, is in a state to produce all those effects, of which our eyes witness, as well as all the bodies which

act diversely upon the organs with which we are furnished, and of which we judge only according to the manner in which these organs are affected. We say they are good, when they are congenial to us, or contribute to maintain harmony in ourselves ; we say they are bad, when they disturb this harmony and we ascribe, in consequence, an aim, ideas, designs, to the being, whom we make the motive-power of a nature which we see destitute of projects and intelligence.

Nature is effectually destitute of them ; she has no intelligence or end ; she acts necessarily, because she exists necessarily. Her laws are immutable and founded upon the essence of things. It is the essence of the seed of the male, composed of the primitive elements, which serve for the basis of an organized being, to unite itself with that of the female, to fructify it, to produce, by its combination with it, a new organized being, who, feeble in his origin, for want of a sufficient quantity of particles of matter, suitable to give him consistence, strengthens himself by degrees, by the daily and continual addition of particles, analogous and appropriate to his being; thus he lives, he thinks, he is nourished, and he engenders, in his turn, organized beings similar to himself. By a consequence of permanent and physical laws, generation does not take place, except when the circumstances necessary to produce it find themselves united. Thus, this procreation is not operated by chance ; the animal does not produce but with an animal of his own species, because this is the only one analogous to himself, or who unites the qualities suitable to the producing a being similar to himself ; without this, he would not produce any thing, he would only produce a being, denominated *monstrous*, because it would be dissimilar to himself. It is of the essence of the grain of plants, to be fructified by the seed of the stamina of the flower, to develop themselves in consequence in the bowels of the earth, to grow with the assistance of water, to attract for that purpose analogous particles, to form by degrees a plant, a shrub, a tree susceptible of the life, the action, the motion, suitable to vegetables. It is of the essence of particles of earth, attenuated, divided.

elaborated by water and by heat, to unite themselves, in the bosom of mountains, with those which are analogous to them, and to form by their aggregation, according as they are more or less similar or analogous, bodies, more or less solid and pure, which we denominate crystals, stones, metals, minerals. It is the essence of the exhalations, raised by the heat of the atmosphere, to combine, to collect themselves, to dash against each other, and, by their combination or their collision, to produce meteors and thunder. It is the essence of some inflammable matter to collect itself, to ferment, to heat itself in the caverns of the earth, to produce those terrible explosions and those earthquakes which destroy mountains, plains, and the habitations of alarmed nations; these complain to an unknown being, of the evils which nature makes them experience as necessarily as those benefits which fill them with joy. In short, it is the essence of certain climates, to produce men so organized and modified, that they become either extremely useful, or very prejudicial to their species, in the same manner as it is the property of certain portions of soil to bring forth agreeable fruits, or dangerous poisons.

In all this, nature has no end; she exists necessarily, her modes of acting are fixed by certain laws, which flow themselves from the constituent properties of the various beings which she contains, and those circumstances which the continual motion she is in must necessarily bring about. It is ourselves who have a necessary aim, which is our own conservation; it is by this that we regulate all the ideas we form to ourselves of the causes which act upon us, and by which we judge of them. Animated and living ourselves, we, like the savages, ascribe a soul and life to every thing that acts upon us: thinking and intelligent ourselves, we ascribe to every thing intelligence and thought; but as we see matter incapable of so modifying itself, we suppose it to be moved by another agent or cause, which we always make similar to ourselves. Necessarily attracted by that which is advantageous to us, and repelled by that which is prejudicial, we cease to reflect that our modes of feeling are due to

No. VIII.—31

our peculiar organization, modified by physical causes, which, in our ignorance we mistake for instruments employed by a being to whom we ascribe our ideas, our views, our passions, our mode of thinking and of acting.

If it be asked of us, after this, what is the end of nature? we shall reply, that it is to act, to exist, to conserve her whole. If it be asked of us, wherefore she exists? we shall reply, that she exists necessarily, and that all her operations, her motions, and her works, are necessary consequences of her necessary existence. There exists something that is necessary, this is nature or the universe, and this nature acts necessarily as she does. If it be wished to substitute the word God to that of *nature*, it may be demanded with equal reason, wherefore this God exists, as well as it can be asked, what is the end of the existence of nature. Thus, the word God will not instruct us as to the end of his existence. But in speaking of nature or of the material universe, we shall have fixed and determinate ideas of the cause of which we speak; whilst in speaking of a theological God, we shall never know what he can be, or whether he exists, nor the qualities which we can with justice assign him. If we give him attributes, it will always be ourselves who must conjecture them, and it will be for ourselves alone that the universe will be formed: ideas which we have already sufficiently refuted. To undeceive ourselves, it is sufficient to open our eyes, and see that we undergo, in our mode, a destiny, of which we partake in common with all the beings of which nature is the assemblage; like us, they are subjected to necessity, which is no more than the sum total of those laws which nature is obliged to follow.

Thus every thing proves to us, that nature or matter exists necessarily, and cannot swerve from those laws which its existence imposes on it. If it cannot be annihilated, it cannot have commenced to be. The theologians themselves agree that it were necessary to have an act of the divine omnipotence or that which they call a miracle, to annihilate a being; but a necessary being, cannot perform a miracle; he cannot derogate from the necessary laws of his existence; we must conclude, then, that if God is the necessary being, every thing that he does, is a consequence of the necessity of his existence, and that he can never derogate from its laws. On the other hand, we are told, that the creation is a miracle, but this creation would be impossible to a necessary being, who could not act freely in any one of his actions. Besides, a miracle is to us only a rare effect, the natural cause of which we ignore; thus, when we are told that *God works a miracle*, we are taught nothing, save that an unknown cause has produced, in an unknown manner, an effect that we did not expect, or which appears strange to us. This granted, the intervention of a God, far from removing the ignorance in which we find ourselves respecting the power and the effects of nature, serves only to augment it. The creation of matter, and the cause to whom is ascribed the honour of this creation, are to us, things as incomprehensible, or as impossible, as is its annihilation.

Let us then conclude, that the word *God*, as well as the word *create*, not presenting to the mind any true idea, ought to be banished the language of all those who are desirous to speak so as to be understood. These are abstract words, invented by ignorance; they are only calculated to satisfy men destitute of experience, too idle, or too timid to study nature and its ways; to content those enthusiasts, whose curious imagination pleases itself with springing beyond the visible world, to run after chimeras. In short, these words are useful to those only, whose sole profession is to feed the ears of the uninformed with pompous words, which are not understood by themselves, and upon the sense and meaning of which they are never in harmony with each other.

Man is a material being; he cannot have any ideas whatever but of that which is material like himself; that is to say, of that which can act upon his organs, or of that which, at least, has qualities analogous to his own. In despite of himself, he always assigns material properties to his God, which the impossibility of compassing has made him suppose to be spiritual, and

distinguished from nature or the material world. Indeed, either he must be content not to understand himself, or he must have material ideas of a God, who is supposed to be the creator, the mover, the conserver of matter; the human mind may torture itself as long as it will, it will never comprehend that material effects can emanate from an immaterial cause, or that this cause can have any relation with material beings. Here is, as we have seen, the reason why men believe themselves obliged to give to God those moral qualities, which they have themselves; they forget that this being, who is purely spiritual, cannot, from thence, have either their organization, or their ideas, or their modes of thinking and acting, and that, consequently, he cannot possess that which they call intelligence, wisdom, goodness, anger, justice, &c. Thus, in truth, the moral qualities which have been attributed to the Divinity, suppose him material, and the most abstract theological notions are founded upon a true and undeniable *anthropomorphism*.*

The theologians, in despite of all their subtilties, cannot do otherwise; like all the beings of the human species, they have a knowledge of matter alone, and have no real idea of a pure spirit. When they speak of intelligence, of wisdom, and of design in the Divinity, they are always those of men which they ascribe to him, and which they obstinately persist in giving to a being, of whom the essence they ascribe to him, does not render him susceptible. How shall we suppose a being, who has occasion for nothing, who is sufficient for himself, whose projects must be executed as soon as they are formed, to have wills, passions, and desires? How shall we attribute anger to a being who has neither blood nor bile? How an omnipotent being, whose wisdom and the beautiful order which he has himself established in the universe we admire, can permit that this beautiful order should be continually disturbed, either by the elements in discord, or by the crimes of human creatures? In short, a God, such as

* *Anthropomorphism* is supposing God to have a bodily shape: a sect of this persuasion appeared in Egypt in 359 of the Christian era.

he has been depicted to us, cannot have any of the human qualities, which always depend on our peculiar organization, on our wants, on our institutions, and which are always relative to the society in which we live. The theologians vainly strive to aggrandize, to exaggerate in idea, to carry to perfection, by dint of abstractions, the moral qualities which they assign to their God; in vain they tell us that they are in him of a different nature from what they are in his creatures; that they are *perfect, infinite, supreme, eminent*; in holding this language, they no longer understand themselves; they have no one idea of the qualities of which they are speaking to us, seeing that a man cannot conceive them but inasmuch as they bear an analogy to the same qualities in himself.

It is thus, that by subtilizing, mortals have not one fixed idea of the God to whom they have given birth. But little contented with a physical God, with an active nature, with matter capable of producing every thing, they must despoil it of the energy which it possesses in virtue of its essence, in order to invest it in a pure spirit, which they are obliged to remake a material being, as soon as they are inclined to form an idea of it themselves, or make it understood by others. In assembling the parts of man, which they do no more than enlarge and spin out to infinity, they believe they form a God. It is upon the model of the human soul, that they form the soul of nature, or the secret agent from which she receives impulse. After having made man double, they make nature double, and they suppose that this nature is vivified by an intelligence. In the impossibility of knowing this pretended agent, as well as that which they have gratuitously distinguished from their own body, they have called it spiritual, that is to say, of an unknown substance; from this, of which they have no ideas whatever, they have concluded that the spiritual substance was much more noble than matter, and that its prodigious subtility, which they have called *simplicity*, and which is only an effect of their metaphysical abstractions, secured it from decomposition, from dissolution, and from all

the revolutions to which material bodies are evidently exposed.

It is thus, that men always prefer the marvellous to the simple; that which they do not understand, to that which they can understand; they despise those objects which are familiar to them, and estimate those alone which they are not capable of appreciating: from that of which they have only had vague ideas, they have concluded that it contains something important, supernatural, and divine. In short, they need mystery to move their imagination, to exercise their mind, to feed their curiosity, which is never more in labour than when it is occupied with enigmas impossible to be guessed at, and which they judge, from thence, extremely worthy of their researches.* This, without doubt, is the reason why they look upon matter, which they have under their eyes, which they see act and change its forms, as a con temptible thing, as a contingent being, which does not exist necessarily and by itself. This is the reason why they imagine a spirit, which they will never be able to conceive, and which, for this reason, they declare to be superior to matter, existing necessarily by himself, anterior to nature; its creator, its mover, its preserver, and its master. The human mind found food in this mystical being; it was occupied by it unceasingly; the imagination embel-

* A great many nations have adored the sun; the sensible effects of this star, which appears to infuse life into all nature, must naturally have induced men to worship it.— Yet, whole people have abandoned this God so visible, to adopt an abstract and metaphysical God. If the reason of this phenomenon should be asked, we shall reply, that the God who is most concealed, most mysterious, and most unknown, must always, for that very reason, be more pleasing to the imagination of the uninformed, than the God whom they see daily. An unintelligible and mysterious tone is essentially necessary to the ministers of all religions: a clear, intelligible religion, without mystery, would appear less divine to the generality of men, and would be less useful to the sacerdotal order, whose interest it is that the people should comprehend nothing of that which they believe to be the most important to them. This, without doubt, is the secret of the clergy. The priest must have an unintelligible God, whom he makes to speak and act in an unintelligible manner, reserving to himself the right of explaining his orders after his own manner.

lished it in its own manner; ignorance fed itself with the fables which had been recounted of it; habit identified this phantom with the existence of man, it became necessary to him; man believed he fell into a vacuum when it was tried to detach him from it, to lead him back to a nature which he had long ago learnt to despise, or to consider only as an impotent mass of matter, inert, dead, and without energy; or as a contemptible assemblage of combinations and of forms subject to perish.

In distinguishing nature from its mover, men have fallen into the same absurdity as when they have distinguished their soul from their body, life from the living being, the faculty of thought from the thinking being. Deceived on their own peculiar nature, and upon the energy of their organs, men have in like manner been deceived upon the organization of the universe; they have distinguished nature from herself; the life of nature from living nature; the action of this nature from acting nature. It was this soul of the world, this energy of nature, this active principle which men personified, then separated by abstraction, sometimes decorated with imaginary attributes, sometimes with qualities borrowed from their own peculiar essences. Such were the aerial materials of which men availed themselves to compose their God; their own soul was the model; deceived upon the nature of this, they never had any true ideas of the Divinity, who was only a copy exaggerated or disfigured to that degree, as to mistake the prototype upon which it had been originally formed.

If, because man has been distinguished from himself, it has been impossible ever to form any true ideas of him, it is also for having distinguished nature from herself that nature and her ways were always mistaken. Men have ceased to study nature, to recur by thought to her pretended cause, to her concealed motive-power, to the sovereign which has been given her. This motive-power has been made an inconceivable being, to whom every thing that takes place in the universe has been attributed; his conduct has appeared mysterious and marvellous because he was a continual contradic-

tion; it has been supposed that his wisdom and his intelligence were the sources of order; that his goodness was the spring of every benefit; that his rigid justice or his arbitrary power, was the supernatural cause of the confusion and the evils with which we are afflicted. In consequence, instead of applying himself to nature, to discover the means of obtaining her favours, or of throwing aside his misfortunes; in the room of consulting experience; in lieu of labouring usefully to his happiness, man was only occupied with addressing himself to the fictitious cause which he had gratuitously associated with nature; he rendered his homage to the sovereign which he had given to this nature; he expected every thing from him and no longer relied either upon himself or upon the assistance of a nature become impotent and contemptible in his eyes.

Nothing could be more prejudicial to the human species than this extravagant theory, which, as we shall presently prove, has become the source of all his evils. Solely occupied with the imaginary monarch which they had elevated to the throne of nature, mortals no longer consulted any thing; they neglected experience, they despised themselves, they mistook their own powers, they laboured not to their own well-being; they became slaves, trembling under the caprices of an ideal tyrant, from whom they expected every good, or from whom they feared those evils which afflicted them.— Their life was employed in rendering servile homage to an idol, of whom they believed themselves eternally interested in meriting the goodness, in disarming the justice, in calming the wrath; they were only happy when consulting reason, taking experience for their guide, and making an abstraction of their romantic ideas, they took courage, gave play to their industry, and applied themselves to nature, who alone can furnish the means of satisfying their wants and their desires, and of throwing aside or diminishing those evils which they are obliged to experience.

Let us, then, reconduct bewildered mortals to the altar of nature; let us destroy those chimeras which their ignorant and disordered imagination believed it was bound to elevate to her throne. Let us say to them, that there is nothing either above or beyond nature; let us teach them that nature is capable of producing, without any foreign aid, all those phenomena which they admire, all the benefits which they desire, as well as all the evils which they apprehend. Let us inform them, that experience will conduct them to a knowledge of this nature; that she takes a pleasure in unveiling herself to those who study her; that she discovers her secrets to those who by their labour dare wrest them from her, and that she always rewards elevation of soul, courage, and industry. Let us tell them, that reason alone can render them happy; that that reason is nothing more than the science of nature applied to the conduct of men in society; let us instruct them that those phantoms with which their minds have been so long and so vainly occupied, can neither procure them the happiness which they demand with loud cries, nor avert from their heads those inevitable evils to which nature has subjected them, and which reason ought to teach them to support, when they cannot avoid them by natural means. Let us teach them that every thing is necessary; that their benefits and their sorrows are the effects of a nature, who in all her works follows laws which nothing can make her revoke. In short, let us unceasingly repeat to them, that it is in rendering their fellow-creatures happy, that they will themselves arrive at a felicity which they will expect in vain from heaven, when the earth refuses it to them.

Nature is the cause of every thing; she is self-existent; she will always exist; she is her own cause; her motion is a necessary consequence of her necessary existence; without motion, we could have no conception of nature; under this collective name we designate the assemblage of matter acting in virtue of its own peculiar energies. This granted, for what purpose should we interpose a being more incomprehensible than herself to explain her modes of action, marvellous, no doubt, to every one, but much more so to those who have not studied her? Will men be more advanced or more

instructed when they shall be told that a being which they are not formed to comprehend, is the author of those visible effects, the natural causes of which they cannot unravel? In short, will the unaccountable being which they call *God*, enable them to have a better knowledge of nature which is acting perpetually upon them?*

Indeed, if we were desirous to attach some sense to the word *God*, of which mortals form such false and such obscure ideas, we should find that it can designate only active nature, or the sum total of the unknown powers which animate the universe, and which oblige beings to act in virtue of their own peculiar energies, and, consequently, according to necessary and immutable laws. But, in this case, the word *God* will only be synonymous to *destiny, fatality, necessity;* it is however to this abstract idea, personified and deified, that they attribute *spirituality,* another abstract idea, of which we cannot form any conception. It is to this abstraction that is assigned intelligence, wisdom, goodness, and justice, of which such a being cannot be the subject. It is with this metaphysical idea, as it is pretended, that the beings of the human species have direct relation. It is to this idea, personified, deified, humanized, spiritualized, decorated with the most incompatible qualities, to which are attributed will, passions, desires, &c. It is this personified idea which is made to speak in the different revelations which are announced in every country as emanating from heaven!

Every thing proves to us, then, that it is not out of nature we ought to seek the Divinity. When we shall be disposed to have an idea of him, let us say that nature is God; let us say that nature contains every thing we can have a knowledge of, since it is the assemblage of all the beings capable of acting upon us, and which can, consequently, be interesting to us. Let us say, that it is this nature which does every thing; that that which she does not do, is impossible to be done; that that which is said to exist out of her

---

* Let us say, with Cicero: *Magna stultitia est earum rerum deos facere effectores, causas rerum non quærere.* Cic. de divinitat. lib. ii

does not exist, and cannot have existence; seeing that there can be nothing beyond the great whole. In short, let us say, that those invisible powers, which the imagination has made the movers of the universe, either are no more than the powers of acting nature, or are nothing.

If we have only an incomplete knowledge of nature and her ways, if we have only superficial and imperfect ideas of matter, why should we flatter ourselves with knowing or having any certain ideas of a being much more fugitive and more difficult to compass in thought than the elements, than the constituent principles of bodies, than their primitive properties, than their modes of acting and existing? If we cannot recur to the first cause, let us content ourselves with second causes, and those effects which experience shows us; let us gather true and known facts; they will suffice to make us judge of that which we know not; let us confine ourselves to the feeble glimmerings of truth with which our senses furnish us, since we have no means whereby to acquire greater.

Do not let us take for real sciences those which have no other basis than our imagination; they can only be visionary. Let us cling to nature which we see, which we feel, which acts upon us, of which, at least, we know the general laws. If we are ignorant of the secret principles which she employs in her complicated works, we are at least certain that she acts in a permanent, uniform, analogous and necessary manner. Let us then observe this nature; let us never quit the routine which prescribes for us; if we do, we shall infallibly be punished with numberless errours, which will darken our mind, estrange us from reason—the necessary consequence of which will be countless sorrows, which we may otherwise avoid. Let us not adore, let us not flatter after the manner of men, a nature who is deaf, and who acts necessarily, and of which nothing can derange the course. Do not let us implore a whole which can only maintain itself by the discord of elements, from whence the universal harmony and the stability of the whole has its birth. Let us consider that we are sensible parts of a whole destitute of

feeling, in which all the forms and the combinations are destroyed after they are born, and have subsisted for a longer or shorter time. Let us look upon nature as an immense laboratory which contains every thing necessary for her to act and to produce all those works which are displayed to our sight. Let us acknowledge her power to be inherent in her essence. Do not let us attribute her works to an imaginary cause which has no other existence than in our brain. Let us rather for ever banish from our mind a phantom calculated to disturb it, and to prevent our pursuing the simple, natural, and certain means which can conduct us to happiness. Let us, then, restore this nature, so long mistaken, to her legitimate rights: let us listen to her voice, of which reason is the faithful interpreter; let us silence that enthusiasm and imposture which, unfortunately, have drawn us aside from the only worship suitable to intelligent beings.

* * *

CHAPTER V.

Of Theism or Deism; of the System of Optimism; and of Final Causes.

VERY few men have the courage to examine the God which every one is in agreement to acknowledge; there is scarcely any one who dares to doubt his existence although it has never been proved; each receives in infancy, without any examination, the vague name of God, which his fathers transmit him, which they consign to his brain with those obscure ideas which they themselves have attached to it, and which every thing conspires to render habitual in him: nevertheless, each modifies it In his own manner, indeed, as we have frequently observed, the unsteady notions of an imaginary being cannot be the same in all the individuals of the human species; each man has his mode of considering him: each man makes to himself a God in particular, after his own peculiar temperament, his natural dispositions, his imagination, more or less exalted, his individual circumstances, the prejudices he has received, and the mode in which he is affected at different times. The contented and healthy man does not see his God with the same eyes as the man who is chagrined and sick; the man who has a heated blood, an ardent imagination, or is subject to bile, does not see him under the same traits as he who enjoys a more peaceable soul, who has a cooler imagination, who is of a more phlegmatic character. This is not all: even the same man does not see him in the same manner in the different periods of his life; his God undergoes all the variations of his own machine, all the revolutions of his own temperament, those continual vicissitudes which his being experiences. The idea of the Divinity, whose existence is looked upon as so demonstrable, this idea which is pretended to be innate or infused in all men, this idea, of which we are assured that the whole of nature is earnest in furnishing us with proofs, is perpetually fluctuating in the mind of each individual, and varies, at each moment, in all the beings of the human species; there are not two who admit precisely the same God, there is not a single one who, in different circumstances, does not see him variously.

Do not, then, let us be surprised at the weakness of those proofs which are furnished of the existence of a being which men will never see but within themselves. Do not let us be astonished at seeing them so little in harmony with each other upon the various systems which they set up relatively to him, upon the worship which they render to him; their disputes on his account, the want of just inference in their opinions, the little consistency and connexion in their systems, the contradictions in which they are unceasingly falling when they would speak of him, the incertitude in which their mind finds itself every time it is occupied with this being so arbitrary, ought not to appear strange to us; they must necessarily dispute when they reason upon an object seen diversely under various circumstances, and upon which there is not a single man that can be constantly in accord with himself.

All men are agreed upon those objects which they are enabled to submit to the test of experience; we do not hear any disputes upon the principles of geometry; those truths which are evident and demonstrated, never vary in our mind; we never doubt that the part is less than the whole, that two and two make four, that benevolence is an amiable quality, that equity is necessary to man in society. But we find nothing but disputes, but incertitude, but variations. upon all those systems which have the Divinity for their object; we see no harmony in the principles of theology; the existence of God, which is announced to us every where as an evident and demonstrated truth, is only so for those who have not examined the proofs upon which it is founded. These proofs frequently appear false or feeble to those themselves who, otherwise, do not by any means doubt of his existence; the inductions or the corollaries which are drawn from this pretended truth, said to be so clear, are not the same in two nations or even in two individuals; the thinkers of all ages and of all countries unceasingly quarrel amongst themselves upon religion, upon their theological hypothesis, upon the fundamental truths which serve for the basis of them, upon the attributes and the qualities of a God with whom they vainly occupy themselves, and the idea of whom varies continually in their own brain.

These disputes and these perpetual variations ought at least to convince us, that the ideas of the Divinity have neither the evidence nor the certitude which are attributed to them, and that it may be permitted to doubt the reality of a being which men see so diversely, and upon which they are never in accord, and of which the image so often varies with themselves. In despite of all the efforts and the subtilties of its most ardent defenders, the existence of a God is not even probable, and if it should be, can all the probabilities of the world acquire the force of a demonstration? Is it not astonishing, that the existence of the being the most important to believe and to know has not even probability in its favour, whilst truths much less important, are clearly demonstrated to us? Should it not be concluded from this that no man is fully assured of the existence of a being which he sees so subject to vary within himself, and which for two days in succession does not present itself under the same traits to his mind? There is nothing short of evidence that can fully convince us. A truth is not evident to

, us, but when constant experience and reiterated reflections always show it to us under the same point of view. From the constant relation which is made by well-constituted senses, results that evidence and that certitude which can alone produce full conviction. What becomes, then, of the certitude of the existence of the Divinity? Can his discordant qualities have existence in the same subject? And has a being, who is nothing but a heap of contradictions, probability in his favour? Can those who admit it, be convinced of it themselves? And, in this case, ought they not to permit that we should doubt those pretended truths, which they announce as demonstrated and as evident, whilst they themselves feel that they are wavering in their own heads? The existence of this God and of the divine attributes cannot be things demonstrated for any man on earth; his non-existence and the impossibility of the incompatible qualities which theology assigns him will be evidently demonstrated to whoever shall be disposed to feel that it is impossible the same subject can unite those qualities which reciprocally destroy each other, and which all the efforts of the human mind will never be able to conciliate.*

---

* Cicero has said, Plura discrepantia vera esse non possunt. From whence we see, that no reasoning, no revelation, no miracle can render that false which experience has demonstrated to us as evident; that there is nothing short of a confusion, an overturning of the brains, that can cause contradictions to be admitted. According to the celebrated Wolfe, in his Ontology, § 99: Possibile est quod nullum in se repugnantium habet, quod contradictione caret. After this definition the existence of God must appear impossible, seeing that there is a contradiction in saying that a spirit without extent can exist in extension, or move matter which has extent.— Saint Thomas, says that ens est quod non repugnat esse. This granted, a God, such as he is defined to be, is only a being of the imagination, since he can have existence no where. According to Bilfinger, de deo. anima et mundo, § 5, Essentia est primus rerum conceptus constitutivus vel quidditativus, cujus ope cætera, quæ de re aliqua dicentur, demonstrari possunt. In this case, could it not be demanded of him, if any one has an idea of the divine essence? Which is the understanding that constitutes God that which he is, and whence flows the demonstration of every thing which is said of him? Ask a the-

However it may be with these qualities, whether irreconcilable, or totally incomprehensible, which the theologians assign to a being already inconceivable by himself, whom they make the author or the architect of the world, what can result to the human species in supposing him to have intelligence and views? Can a universal intelligence, whose care ought to be extended to every thing that exists, have relations more direct and more intimate with man, who only forms an insensible portion of the great whole? Is it, then, to make joyful the insects and the ants of his garden, that the Monarch of the universe has constructed and embellished his habitation? Should we be better capacitated to have a knowledge of his projects, to divine his plan, to measure his wisdom with our feeble eyes, and could we better judge of his works from our own narrow views? The effects, good or bad, favourable or prejudicial to ourselves, which we may imagine to emanate from his omnipotence and from his providence, will they be less the necessary effects of his wisdom, of his justice, of his eternal decrees? In this case, can we suppose that a God so wise, so just, so intelligent, will change his plan for us? Overcome by our prayers and our servile homage, will he, to please us, reform his immutable decrees? Will he take away from beings their essences and their properties? Will he abrogate, by miracles, the eternal laws of a nature, in which his wisdom and his goodness are admired? Will he cause that in our favour, fire shall cease to burn when we shall approach it too nearly? Will he so order it, that fever, that the gout shall cease to torment us when we shall have amassed those humours of which these infirmities are the necessary consequence? Will he prevent an edifice that tumbles in ruins from crushing us by its fall, when we shall pass beside it? Will our vain cries and the most fervent supplications, prevent our country from being unhappy,

ologian if God can commit crime? He will tell you no, seeing that crime is repugnant to justice, which is his essence. But this theologian does not see that, in supposing God a spirit, it is full as repugnant to his essence to have created or to move matter, as to commit a crime repugnant to his justice.

when it shall be devastated by an ambitious conqueror, or governed by tyrants who oppress it?

If this infinite intelligence is always obliged to give a free course to those events which his wisdom has prepared; if nothing happens in this world but after his impenetrable designs, we have nothing to ask of him; we should be madmen to oppose ourselves to them; we should offer an insult to his prudence if we were desirous to regulate them. Man must not flatter himself with being wiser than his God, with being capable of engaging him to change his will; with having the power to determine him to take other means than those which he has chosen to accomplish his decrees; an intelligent God can only have taken those measures which are the most just, and those means which are the most certain, to arrive at his end; if he were capable of changing them, he neither could be called wise, immutable, nor provident. If God did suspend, for an instant, those laws which he himself fixed, if he could change any thing in his plan, it is because he could not have foreseen the motives of this suspension, or of this change; if he had not made these motives enter into his plan, it is that he had not foreseen them; if he has foreseen them, without making them enter into his plan, it is that he has not been able. Thus, in whatever manner these things are contemplated, the prayers which men address to the Divinity, and the different worships which they render him, always suppose they believe they have to deal with a being whose wisdom and providence are small, and capable of change, or who, in despite of his omnipotence, cannot do that which he is willing, or that which would be expedient for men, for whom, nevertheless, it is pretended that he has created the world.

It is, however, upon these notions, so badly directed, that are founded all the religions of the earth. We every where see man on his knees before a wise God, of whom he strives to regulate the conduct, to avert the decrees, to reform the plans; every where man is occupied with gaining him to his interests; by meannesses and presents; in overcoming his justice by dint of prayers, by practices, by ceremonies,

No. VIII.—32

and by expiations, which he believes will make him change his resolutions, every where man supposes that he can offend his creator, and disturb his eternal felicity; every where man is prostrate before an omnipotent God, who finds himself in the impossibility of rendering his creatures such as they ought to be, to accomplish his divine views, and fulfil his wisdom!

We see, then, that all the religions of the world are only founded upon those manifest contradictions into which men will fall every time they mistake nature, and attribute the good or the evil which they experience at her hands, to an intelligent cause, distinguished from herself, of which they will never be able to form to themselves any certain ideas. Man will always be reduced, as we have so frequently repeated, to the necessity of making a man of his God; but man is a changeable being, whose intelligence is limited, whose passions vary, who, placed in diverse circumstances, appears to be frequently in contradiction with himself: thus, although man believes he does honour to his God, in giving him his own peculiar qualities, he does no more than lend him his inconstancy, his weakness, and his vices. The theologians, or the fabricators of the divinity, may distinguish, subtilize, and exaggerate his pretended perfections; render them as unintelligible as they please, it will ever be, that a being who is irritated and is appeased by prayers, is not immutable; that a being who is offended, is neither omnipotent nor perfectly happy; that a being who does not prevent the evil he can restrain, consents to evil; that a being who gives liberty to sin, has resolved, in his eternal decrees, that sin should be committed; that a being who punishes those faults which he has permitted to be done, is sovereignly unjust and irrational; that an infinite being who contains qualities infinitely contradictory, is an impossible being, and is only a chimera.

Let us then be no longer told, that the existence of God is at least a problem. A God, such as the theologians depict him, is totally impossible; all the qualities which can be assigned to him, all the perfections with which they shall embellish him, will still be found every moment in contradiction. As for the abstract and negative qualities with which they may invest him, they will always be unintelligible, and will only prove the inutility of the efforts of the human mind, when it wishes to define beings which have no existence. As soon as men believe themselves greatly interested in knowing a thing, they labour to form to themselves an idea of it; if they find great obstacles, or even an impossibility of enlightening their ignorance, then the small success that attends their researches, disposes them to credulity; hence crafty knaves or enthusiasts profit of this credulity. to make their inventions or their reveries (which they deliver out as permanent truths, of which it is not permitted to doubt) pass current. It is thus that ignorance, despair, idleness, the want of reflecting habits, place the human species in a state of dependence on those who are charged with the care of. building up those systems upon those objects of which he has no one idea. As soon as there is a question of the Divinity and of religion, that is to say, of objects of which it is impossible to comprehend any thing, men reason in a very strange mode, or are the dupes of very deceitful reasonings. As soon as men see themselves in a total impossibility of understanding what is said, they imagine that those who speak to them are better acquainted with the things of which they discourse than themselves; these do not fail to repeat to them that *the most certain way is to agree with that which they tell them,* to allow themselves to be guided by them, and to shut their eyes: they menace them with the anger of the irritated phantom, if they refuse to believe what they tell them; and this argument, although it only supposes the thing in question, closes the mouth of poor mortals, who, convinced by this victorious reasoning, fear to perceive the palpable contradictions of the doctrines announced to them—blindly agree with their guides, not doubting that they have much clearer ideas of those marvellous objects with which they unceasingly entertain them, and on which their profession obliges them to meditate. The uninformed believes his priests have more senses than himself, he

takes them for divine men, or demigods. He sees in that which he adores only what the priests tell him, and from every thing which they say of him, it results, to every man who thinks, that God is only a being of the imagination, a phantom clothed with those qualities which the priests have judged suitable to give him, to redouble the ignorance, the incertitude, and the fear of mortals. It is thus the authority of the priests decide, without appeal, on the thing, which is useful only to the priesthood.

When we shall be disposed to recur to the origin of things, we shall always find that it is ignorance and fear which have created Gods; that it is imagination, enthusiasm, and imposture, which have embellished or disfigured them; that it is weakness that adores them, that it is credulity which nourishes them, that it is habit which respects them, that it is tyranny who sustains them, to the end that tyrants may profit by the blindness of men. We are unceasingly told of the advantages that result to men, from the belief of a God. We shall presently examine if these advantages be as real as they are said to be; in the mean time, let us ascertain whether the opinion of the existence of a God be an errour or a truth? If it is an errour, it cannot be useful to the human species; if it is a truth, it ought to be susceptible of proofs so clear as to be compassed by all men, to whom this truth is supposed to be necessary and advantageous. On the other hand, the utility of an opinion does not render it more certain on that account. This suffices to reply to Doctor Clarke, who asks, *if it is not a thing very desirable, and which any wise man would wish to be true, for the great benefit and happiness of men, that there were a God, an intelligent and wise, a just and good being, to govern the world?* We shall say to him, first, that the supposed author of a nature in which we are obliged to see, at each instant, confusion by the side of order, wickedness by the side of goodness, folly by the side of wisdom, justice by the side of injustice, can no more be qualified to be good, wise, intelligent, and just, than to be wicked, irrational, and perverse; at least, as far as the two principles in nature are equal in power, of which the one unceasingly destroys the works of the other. We shall say, secondly, that the benefit which can result from a supposition, neither renders it either more certain, or more probable. Indeed, where should we be, if, because a thing would be useful, we went so far as to conclude from it that it really existed? We shall say, thirdly, that every thing which has been related until the present moment, proves, that it is repugnant to all common notions, and impossible to be believed, that there should be a being associated with nature. We shall further say, that it is impossible to believe, very sincerely the existence of a being of which we have not any real idea, and to which we cannot attach any that does not instantly destroy it. Can we believe the existence of a being of which we can affirm nothing, who is only a heap of negations of every thing of which we have a knowledge? In short, is it possible firmly to believe the existence of a being upon which the human mind cannot fix any judgment which is not found to be instantly contradicted?

But the happy enthusiast, when the soul is sensible of its enjoyments, and when the softened imagination has occasion to paint to itself a seducing object to which it can render thanks for its pretended kindness, will ask, "Wherefore deprive me of a God whom I see under the character of a sovereign, filled with wisdom and goodness? What comfort do I not find in figuring to myself a powerful, intelligent, and good monarch, of whom I am the favourite, who occupies himself with my well-being, who unceasingly watches over my safety, who administers to my wants, who consents that under him I command the whole of nature? I believe I behold him unceasingly diffusing his benefits on man; I see his Providence labouring for him without relaxation; he covers the earth with verdure, and loads the trees with delicious fruits to gratify his palate; he fills the forest with animals suitable to nourish him; he suspends over his head planets and stars, to enlighten him during the day, to guide his uncertain steps during the night; he extends around him the azure firmament;

to rejoice his eyes, he decorates the meadows with flowers; he washes his residence with fountains, with rivulets, and with rivers. Ah! suffer me to thank the author of so many benefits. Do not deprive me of my charming phantom; I shall not find my illusions so sweet in a severe and rigid necessity, in a blind and inanimate matter, in a nature destitute of intelligence and feeling."

"Wherefore," the unfortunate will say, from whom his destiny has rigorously withheld those benefits which have been lavished on so many others, "Wherefore ravish from me an errour that is dear to me? Wherefore annihilate to me a God, whose consoling idea dries up the source of my tears, and serves to calm my sorrows?—Wherefore deprive me of an object, which I represent to myself as a compassionate and tender father, who reproves me in this world, but into whose arms I throw myself with confidence, when the whole of nature appears to have abandoned me? Supposing even that this God is no more than a chimera, the unhappy have occasion for him, to guaranty them from a frightful despair: is it not inhuman and cruel to be desirous to plunge them into a vacuum, by seeking to undeceive them? Is it not a useful errour, preferable to those truths which deprive the mind of every consolation, and which do not hold forth any relief from its sorrows?"

No! I shall reply to these enthusiasts, truth can never render you unhappy; it is this which really consoles us; it is a concealed treasure, which, much superior to those phantoms invented by fear, can cheer the heart and give us courage to support the burdens of life; it elevates the mind, it renders it active, it furnishes it with means to resist the attacks of fate, and to combat, with success, bad fortune. I shall then ask them upon what they found this goodness, which they foolishly attribute to their God? But this God I shall say to them, is he then benevolent to all men? For one mortal who enjoys abundance and the favours of fortune, are there not millions who languish in want and misery? Those who take for model the order on which they suppose this God the author, are they then the most happy in this world? The goodness of this being, to some favourite individual, does it never contradict itself? Even those consolations which the imagination seeks in his bosom, do they not announce misfortunes brought on by his decrees, and of which he is the author? Is not the earth covered with unfortunates, who appear to come upon it only to suffer, to groan, and to die? Does this divine providence give itself up to sleep during those contagions, those plagues, those wars, those disorders, those moral and physical revolutions, of which the human race is continually victim? This earth, of which the fecundity is looked upon as a benefit from heaven, is it not in a thousand places dry, barren, and inexorable?— Does it not produce poisons, by the side of the most delicious fruits?— Those rivers and those seas, which are believed to be made to water our abode, and to facilitate our commerce, do they not frequently inundate our fields, overturn our dwellings, and carry away men and their flocks?* In short, this God, who presides over the universe, and who watches unceasingly for the preservation of his creatures, does he not almost always deliver them up to the chains of many inhuman sovereigns, who make sport of the misery of their unhappy subjects, whilst these unfortunates vainly address themselves to heaven, that their multiplied calamities may cease, which are visibly due to an irrational administration and not to the wrath of Heaven?

The unhappy man, who seeks con solation in the arms of his God, ought at least to remember that it is this same God, who being the master of all, distributes the good and the evil: if nature is believed to be subjected to his supreme orders, this God is as frequently unjust, filled with malice, with imprudence, with irrationality, as with goodness, wisdom, and equity. If the devotee, less prejudiced and more consistent, would reason a little, he would suspect that his God was a capricious God, who frequently made him suffer; he would not seek to console himself in the arms of his executioner, whom he has the folly to mistake for a friend or for his father.

Do we not, indeed, see in nature a constant mixture of good and evil? To obstinately see only the good, is as irrational as only to perceive the evil. We see the calm succeed to the storm, sickness to health, and peace to war. The earth produces in every country plants necessary to the nourishment of man, and plants suitable to his destruction. Each individual of the human species is a necessary compound of good and bad qualities; all nations present us with the variegated spectacle of vices and virtue; that which rejoices one individual, plunges many others into mourning and sadness; there happens no event that has not advantages for some, and disadvantages for others. Nature, considered in its whole, shows us beings alternately subjected to pleasure and grief, born to die, and exposed to those continual vicissitudes from which no one of them is exempt. The most superficial glance of the eye will suffice, then to unde-

*Nevertheless, on the whole, there is no such a thing as real evil. Insects find a safe retreat in the ruins of the palace which crushes man in its fall; man by his death furnishes food for myriads of contemptible insects, whilst animals may destroy by thousands that he may increase his bulk, and linger out for a season a feverish existence. The halcyon, delighted with the tempest, voluntarily mingles with the storm--rides contentedly upon the surge; rejoiced by the fearful howlings of the northern blast, plays with happy buoyancy upon the foaming billows, that have ruthlessly dashed in pieces the vessel of the unfortunate mariner, who, plunged into an abyss of misery, with tremulous emotion clings to the wreck—views with horrific despair the premature destruction of his indulged hopes--sighs deeply at the thoughts of home,—with aching heart thinks of the cherished friends his streaming eyes will never more behold—in agony dwells upon the faithful affection of an adored companion, who

will never again repose her drooping head upon his manly bosom—grows wild with the appalling remembrance of beloved children his wearied arms will never more encircle with parental fondness; then sinks for ever the unhappy victim of circumstances that fill with glee the fluttering bird, who sees him yield to the overwhelming force of the infuriate waves. The conqueror displays his military skill, fights a sanguinary battle, puts his enemy to the rout, lays waste his country, slaughters thousands of his fellows, plunges whole districts into tears, fills the land with the moans of the fatherless, the wailings of the widow, in order that the crows may have a banquet--that ferocious beasts may glutton ously gorge themselves with human gore— that worms may riot in luxury!

ceive us as to the idea that man is the *final cause* of the creation, the constant object of the labours of nature, or of its author, to whom they can attribute, according to the visible state of things, and the continual revolutions of the human race, neither goodness nor malice; neither justice nor injustice; neither intelligence nor irrationality. In short, in considering nature without prejudice, we shall find that all beings in the universe are equally favoured, and that every thing which exists, undergoes the necessary laws from which no being can be exempted.

Thus, when there is a question concerning an agent we see act so variously as nature, or as its pretended mover, it is impossible to assign him qualities according to his works, which are sometimes advantageous and sometimes prejudicial to the human species; or at least, each man will be obliged to judge of him after the peculiar mode in which he is himself affected; there will be no fixed point or standard in the judgments which men shall form of him; our mode of judging will always be founded upon our mode of seeing and of feeling, and our mode of feeling will depend on our temperament, on our organization, on our particular circumstances, which cannot be the same in all the individuals of our species. These different modes of being affected, then, will always furnish the colours of the portraits which men may paint to themselves of the Divinity; consequently these ideas cannot be either fixed or certain; the inductions which they may draw from them, can never be either constant or uniform; each will always judge after himself, and will never see any thing but himself, or his own peculiar situation, in his God.

This granted, men who are contented, who have a sensible soul, a lively imagination, will paint the Divinity under the most charming traits; they will believe they see in the whole of nature, which will unceasingly cause them agreeable sensations, nothing but proofs of benevolence and goodness; in their poetical ecstasy, they will imagine they perceive every where the impression of a perfect intelligence, of an infinite wisdom, of a providence tenderly occupied with the well-being of man; self-love joining itself to these exalted

qualities, will put the finishing hand to their persuasion that the universe is made solely for the human race; they will strive, in imagination, to kiss with transport the hand from whom they believe they receive so many benefits; touched with these favours, gratified with the perfume of these roses, of which they do not see the thorns, or which their ecstatic delirium prevents them from feeling, they will think they can never sufficiently acknowledge the necessary effects, which they look·up-. on as indubitable proofs of the divine predilection for man. Inebriated with these prejudices, enthusiasts will not perceive those sorrows and that confusion of which the universe is the theatre; or, if they cannot prevent themselves from seeing them, they will be persuaded that, in the views of a benevolent providence, these calamities are necessary to conduct man to a higher state of felicity; the reliance which they have placed in the Divinity, upon whom they imagine they depend, induces them to believe that man only suffers for his good, and that this being, who is fruitful in resources, will know how to make him reap advantage from the evils he experiences in this world. Their mind, thus pre-occupied, from thence sees nothing that does not excite their admiration, their gratitude, and their confidence; even those effects which are the most natural and the most necessary, appear to them to be miracles of benevolence and goodness; obstinately persisting in seeing wisdom and intelligence every where, they shut their eyes to the disorders which could contradict those amiable qualities they attribute to the being with whom• their hearts are engrossed; the most cruel calamities, the most afflicting events to the human race, cease to appear to them disorders, and only furnish them with new proofs of the divine perfections: they persuade themselves that what appears. to them defective or imperfect, is only so in appearance; and they admire the wisdom and the bounty of their God, even in those effects which are the most terrible, and the most suitable to discourage them.

It is, without doubt, to this stupid intoxication, to this strange infatuation, to which is to be ascribed the system of *optimism*, by which enthusiasts, furnished with a romantic imagination, appear to have renounced the evidence of their senses, and thus they find, that, even for man, *every thing is good* in a nature where the good is found constantly accompanied with evil, and where minds less prejudiced, and imaginations less poetical, would judge that every thing is only that which it can be ; that the good and the evil are equally necessary ; that they emanate from the nature of things, and not from a fictitious hand, which, if it really existed, or did every thing that we see, could be called wicked with as much reason as he is inaptly said to be filled with goodness. Besides, to be enabled to justify providence for the evils, the vices, and the disorders which we see in the whole which is supposed to be the work of his hands, we should know the aim of the whole. Now, the whole cannot have an aim, because, if it had an aim, a tendency, an end, it would no longer be the whole.

We shall be told, that the disorders and the evils which we see in this world, are only relative and apparent, and prove nothing against divine wisdom and goodness. But can it not be replied, that the so much boasted benefits, and he marvellous order, upon which the wisdom and goodness of God are founded, are, in a like manner, only relative and apparent ? It is uniformly our mode of feeling, and of co-existing with those causes by which we are encompassed, which constitutes the order of nature with relation to ourselves, and which authorizes us to ascribe wisdom or goodness to its author ; ought not our modes of feeling and of existing authorize us to call that disorder which injures us, and to ascribe imprudence or malice to the being whom we shall suppose to put nature in motion ? In short, that which we see in the world conspires to prove that every thing is necessary ; that nothing is done by chance ; that all the events, good or bad, whether for us, whether for beings of a different order, are brought about by causes, acting after certain and determinate laws ; and that nothing can authorize us to ascribe any one of our human qualities either to nature, or to the motive-power that has been given to her.

With respect to those who pretend

that the supreme wisdom will know how to draw the greatest benefits for us, even out of the bosom of those evils which he permits us to experience in this world, we shall ask them if they are themselves the confidants of the Divinity ; or upon what they found their flattering hopes ? They will tell us, without doubt, that they judge of the conduct of God by analogy ; and that, by the actual proofs of his wisdom and goodness, they have a just right to conclude in favour of his bounty and wisdom. We shall reply to them, that they admit, according to these gratuitous suppositions, that the goodness and the wisdom of their God contradict themselves so frequently in this world, that nothing can assure them that his conduct will ever cease to be the same with respect to those men who experience here below sometimes his kindness and sometimes his disfavour. If, in despite of his omnipotent goodness, God has not been either able or willing to render his beloved creatures completely happy in this world, what reason is there to believe that he either will be able or willing to do it in another?

Thus, this language only founds itself upon ruinous hypotheses which have for basis only a prejudiced imagination ; it only shows that men, once persuaded, without motives and without cause, of the goodness of their God, cannot figure to themselves that he will consent to render his creatures constantly unhappy. But on the other hand, what real and known good do we see result to the human species from those sterilities, from those famines, from those contagions, from those sanguinary combats, which cause so many millions of men to perish, and which unceasingly depopulate and desolate the world which we inhabit ? Is there any one capable to ascertain the advantages which result from all those evils which besiege us on all sides ? Do we not see daily, beings consecrated to misfortune from the moment they quit the womb of their mother, until that in which they descend into the silent grave, who, with great difficulty, found time to respire, and lived the constant sport of affliction, of grief, and of reverses of fortune ? How, or when, will this God, so bountiful, draw good from the evils which he causes mankind to suffer?

The most enthusiastic *optimists*, the *theists*, themselves, the partisans of *natural religion*, (which is any thing but *natural* or founded upon reason,) are, as well as the most credulous and superstitious, obliged to recur to the system of another life, to exculpate the Divinity from those evils which he decrees to be suffered in this by those themselves whom they suppose to be most agreeable in his eyes. Thus, in setting forth the idea that God is good and filled with equity, we cannot dispense with admitting a long series of hypotheses which, as well as the existence of this God, have only imagination for a basis, and of which we have already shown the futility. It is necessary to recur to the doctrine, so little probable. of a future life, and of the immortality of the soul, to justify the Divinity; we are obliged to say, that for want of having been able or willing to render man happy in this world, he will procure him an unalterable happiness when he shall no longer exist, or when he shall no longer have those organs by the aid of which he is enabled to enjoy it at present.

And after all, these marvellous hypotheses are insufficient to justify the Divinity for his wickedness or for his transitory injustice. If God has been unjust and cruel for an instant, God has derogated, at least for that moment, from his divine perfections; then he is not immutable; his goodness and justice are then subject to contradict themselves for a time; and, in this case, who can guaranty that the qualities which we confide in, will not contradict themselves even in a future life invented to exculpate God for those digressions which he permits in this world? What is this but a God who is perpetually obliged to depart from his principles, and who finds himself unable to render those whom he loves happy, without unjustly doing them evil, at least during their abode here below? Thus, to justify the Divinity it will be necessary to recur to other hypotheses; we must suppose that man can offend his God, disturb the order of the universe, be injurious to the felicity of a being sovereignly happy, and derange the designs of the om-

nipotent being. To reconcile many things, we must recur to the system of the liberty of man.* At length, we shall find ourselves obliged to admit, one after another, the most improbable, the most contradictory, and the most false ideas, as soon as we admit that the universe is governed by an intelligence filled with wisdom, with justice, and with goodness; this principle alone, if we are consistent, is sufficient to lead us insensibly into the grossest absurdities.

This granted, all those who speak of the divine goodness, wisdom, and intelligence, which are shown in the works of nature; who offer these same works as incontestable proofs of the existence of a God, or of a perfect agent, are men prejudiced or blinded by their own imagination, who see only a corner of the picture of the universe, without embracing the whole. Intoxicated with the phantom which their mind has formed to itself, they resemble those lovers who do not perceive any defect in the objects of their affection; they conceal, dissimulate, and justify their vices and deformities, and frequently end with mistaking them for perfections.

We see, then, that the proofs of the existence of a sovereign intelligence, drawn from the order, from the beauty, from the harmony of the universe, are quite ideal, and have no reality but for those who are organized and modified in a certain mode, or whose cheerful imagination is constructed to give birth to agreeable chimeras which they embellish according to their fancy. These illusions, however, must be frequently dissipated even in themselves whenever their machine becomes deranged; the spectacle of nature, which, under certain circumstances, has appeared to them so delightful and so seducing, must then give place to disorder and confusion. A man of a melancholy temperament, soured by misfortunes or infirmities, cannot view nature and its author under the same perspective as the healthy man of a sprightly humour,

---

* Is there any thing more inconclusive than the ideas of some *Theists* who deny the liberty of man, and who, notwithstanding, obstinately persist in speaking of an avenging and remunerating God? How can a just God punish necessary actions?

and contented with every thing. Deprived of happiness, the peevish man can only find disorder, deformity, and subjects to afflict himself with ; he only contemplates the universe as the theatre of the malice or the vengeance of an angry tyrant; he cannot sincerely love this malicious being, he hates him at the bottom of his heart, even when rendering him the most servile homage : trembling, he adores a hateful monarch, of whom the idea produces only sentiments of mistrust, of fear, of pusillanimity ; in short, he becomes superstitious, credulous, and very often cruel after the example of the master whom he believes himself obliged to serve and to imitate.

In consequence of these ideas, which have their birth in an unhappy temperament and a peevish humour, the superstitious are continually infected with terrours, with mistrusts, with alarms. Nature cannot have charms for them ; they do not participate in her cheerful scenes, they only look upon this world, so marvellous and so good to the contented enthusiast, as a *valley of tears*, in which a vindictive and jealous God has placed them only to expiate crimes committed either by themselves or their fathers ; they consider themselves to be here the victims and the sport of his despotism, to undergo continual trials, to the end that they may arrive for ever at a new existence, in which they shall be happy or miserable, according to the conduct which they shall have held towards the fantastical God who holds their destiny in his hands.

These are the dismal ideas which have given birth to all the worships, to all the most foolish and the most cruel superstitions, to all the irrational practices, all the absurd systems, all the extravagant notions and opinions, all the doctrines, the ceremonies, the rites, in short, to all the religions on the earth ;*

* History abounds with details of the most atrocious cruelties under the imposing name of " God's will," " God's judgments:" nothing has been considered either too fantastical or too flagitious by the votaries of superstition. Parents have immolated their children ; lovers have sacrificed the objects of their affection ; friends have destroyed each other ; the most bloody disputes have been fomented ; the most interminable animosities have been engendered. to gratify the whim of implacable

they have been, and always will be, an eternal source of alarm, of discord, and of delirium, for those dreamers who are nourished with bile, or intoxicated with divine fury, whose atrabilious humour disposes to wickedness, whose wandering imagination disposes to fanaticism, whose ignorance prepares them for credulity, and who blindly submit to their priests : these, for their own interests, avail themselves frequently of their fierce and austere God to excite them to crimes, and to induce them to ravish from others that repose of which they are themselves deprived.

It is, then, in the diversity of temperaments and passions that we must seek the difference we find between the God of the theist, the optimist, the happy enthusiast, and that of the devotee, the superstitious, the zealot, whose intoxication so frequently renders him unsociable and cruel. They are all equally irrational ; they are the dupes of their imagination ; the one, in the transport of their love, see God only on the favourable side; the others never see him but on the unfavourable side. Every time we set forth a false supposition, all the reasonings we make on it are only a long series of errours; every time we renounce the evidence of our senses, of experience, of nature, and of reason, it is impossible to calculate the bounds at which the imagination will stop. It is true the ideas of the happy enthusiast will be less dangerous to himself and to others, than those of the superstitious atrabilious man, whose temperament shall render him both cowardly and cruel; nevertheless the Gods of the one and the other are not the less chimerical ; that of the first is the produce of agreeable dreams, that of the second is the fruit of a peevish transport of the brain.

There will never be more than a step between theism and superstition. The smallest revolution in the machine, a slight infirmity, an unforseen affliction, suffices to change the course of the humours, to vitiate the temperament, to overturn the system of opinions of the theist, or of the happy devotee ; as soon as the portrait of his God is found disfigured, the beautiful order

priests, who, by crafty inventions, have obtained an influence over the people.

of nature will be overthrown relatively to him, and melancholy will, by degrees, plunge him into superstition, into pusillanimity, and into all those irregularities which produce fanaticism and credulity.

The Divinity, existing but in the imagination of men, must necessarily take its complexion from their character, he will have their passions; he will constantly follow the revolutions of their machine, he will be lively or sad, favourable or prejudicial, friendly or inimical, sociable or fierce, humane or cruel, according as he who carries him in his brain shall be himself disposed. A mortal, plunged from a state of happiness into misery, from health into sickness, from joy into affliction, cannot, in these vicissitudes, preserve the same God. What is this but a God who depends at each instant upon the variations which natural causes make the organs of man undergo? A strange God, indeed! of whom the floating idea depends on the greater or less portion of heat and fluidity of our blood!

No doubt that a God constantly good, filled with wisdom, embellished with qualities amiable and favourable to man, would be a more seductive chimera than the God of the fanatic and of the superstitious; but he is not less for that reason a chimera, that will become dangerous when the speculators, who shall be occupied with it, shall change their circumstances or their temperament; these, looking upon him as the author of all things, will see their God change, or will at least be obliged to consider him as a being full of contradictions, upon which there is no depending with certainty; from thence incertitude and fear will possess their mind; and this God, whom at first they fancied so charming, will become a subject of terrour to them, likely to plunge them in the most gloomy superstition, from which, at first sight, they appeared to be at an infinite distance.

Thus theism, or the pretended *natural religion*, cannot have certain principles, and those who profess it are necessarily subject to vary in their opinions of the Divinity, and in their conduct which flows from them. Their system, originally founded upon a wise and intelligent God, whose goodness

No. IX.—33

can never contradict itself, as soon as circumstances change, must presently be converted into fanaticism, and into superstition. This system, successively meditated by enthusiasts of different characters, must experience continual variations, and very quickly depart from its pretended primitive simplicity. The greater part of those philosophers have been disposed to substitute theism to superstition, but they have not felt that theism was formed to corrupt itself and to degenerate. Indeed, striking examples prove this fatal truth; theism is every where corrupted; it has by degrees formed those superstitions, those extravagant and prejudicial sects, with which the human species is infected. As soon as man consents to acknowledge invisible powers out of nature, upon which his restless mind will never be able invariably to fix its ideas, and which his imagination alone will be capable of painting to him; whenever he shall not dare to consult his reason relatively to these imaginary powers, it must necessarily be, that this first false step leads him astray, and that his conduct, as well as his opinions, becomes, in the long run, perfectly absurd.*

We call *theists* or *deists*, among

* The religion of Abraham appears to have originally been a theism imagined to reform the superstition of the Chaldeans; the theism of Abraham was corrupted by Moses, who availed himself of it to form the Judaical superstition. Socrates was a theist, who, like Abraham, believed in divine inspirations; his disciple, Plato, embellished the theism of his master with the mystical colours which he borrowed from the Egyptian and Chaldean priests, and which he modified himself in his poetical brain. The disciples of Plato such as Proclus, Jamblichus, Plotinus, Porphyrus, &c. were true fanatics, plunged in the grossest superstition. In short, the first doctors of Christianity were Platonists, who combined the Judaical superstition, reformed by the Apostles or by Jesus, with Platonism. Many people have looked upon Jesus as a true theist, whose religion has been by degrees corrupted. Indeed, in the books which contain the law which is attributed to him, there is no mention either of worship, or of priests, or of sacrifices, or of offerings, or of the greater part of the doctrines of actual Christianity, which has become the most prejudicial of all the superstitions of the earth. Mahomet, in combating the polytheism of his country, was only desirous of bringing back the Arabs to the primitive theism of Abraham and of his son Ishmael, and yet Mahometism is divided

ourselves, those who, undeceived in a great number of grosser errours with which the uninformed and superstitious are successively filled, simply hold to the vague notion of the Divinity which they consider as an unknown agent, endued with intelligence, wisdom, power, and goodness; in short, full of infinite perfections. According to them, this being is distinguished from nature; they found his existence upon the order and the beauty which reigns in the universe. Prepossessed in favour of his benevolent providence, they obstinately persist in not seeing the evils of which this universal agent · must be the reputed cause whenever he does not avail himself of his power to prevent them. Infatuated by these ideas, of which we have shown the slender foundation, it is not surprising there should be but little harmony in their systems, and in the consequences which they draw from them. Indeed, some suppose,. that this imaginary being, retired into the profundity of his essence, after having brought matter out of nothing, abandoned it for ever to the motion which he had once given to it. They have occasion for God only to give birth to nature; this done, every thing that takes place in it is only a necessary consequence of the impulse which was given to it in the origin of things; he was willing that the world should exist; but too great to enter into the detail of its administration, he delivered all the events to second or natural causes; he lives in a state of perfect indifference as to his creatures, who have no relation whatever with him, and who can in no wise disturb his unalterable happiness. From whence we see the least superstitious of the deists make of their God a being useless to men; but they have occasion for a word to designate the first cause or the unknown power to which, for want of being acquainted with the energy of nature, they believe they ought to attribute its primitive formation, or, if they will, the arrangement of matter which is coeternal with God.

Other theists, furnished with a more lively imagination, suppose more particular relations between the universal agent and the human species; each of them, according to the fecundity of his genius, extends or diminishes these relations, supposes duties from man towards his Creator, believes that, to please him, he must imitate his pretended goodness, and, like him, do good to his creatures. Some imagine to themselves that this God, being just, reserves rewards for those who do good, and chastisements for those who commit evil to their fellow-creatures. From whence we see that these humanize their Divinity a little more than the others, in making him like unto a sovereign, who punishes or recompenses his subjects, according to their fidelity in fulfilling their duties, and the laws which he imposes on them: they cannot, like the pure deists, content themselves with an immoveable and indifferent God; they need one who approaches nearer to themselves; or who, at least, can serve them to explain some of those enigmas which this world presents. As each of these speculators, which we denominate theists to distinguish them from the first, makes a separate system of religion for himself, they are in nowise in accord in their worship, nor in their opinions; there are found between them shades frequently imperceptible, which, from simple deism, conducts some among them to superstition; in short, but little in harmony with themselves, they do not know upon what to fix.*

into seventy-two sects. All this proves that theism is always more or less mingled with fanaticism, which sooner or later finishes by producing ravages and misery.

* It is easy to perceive that the writings of the theists and of the deists are commonly as much filled with paralogisms, or fallacious syllogisms, and with contradictions, as those of the theologians; their systems are frequently in the last degree inconsequent. One says that every thing is necessary, denies the spirituality and the immortality of the soul, refusing to believe the liberty of man. Could we not ask them, in this case, of what service can be their God? They have occasion for a word, which custom has rendered necessary to them. There are very few men in the world who dare be consistent: but let us invite all the deicolists, or supporters of the existence of a God, under whatever denomination they may be designated, to inquire of themselves, if it be possible for them to attach any fixed, permanent, and invariable idea, always compatible with the nature of things, to the being whom they designate under the name of God, and they will see, that, as soon as they distinguish him from nature, they will no longer understand any thing about him. The repugnance which the greater part of men show

We must not be astonished; if the God of the *deist* is useless, that of the *theist* is necessarily full of contradictions: both of them admit a being, who is nothing but a mere fiction. Do they make him material? he returns from thence into nature. Do they make him spiritual? they have no longer any real ideas of him. Do they give him moral attributes? they immediately make a man of him, of whom they only extend the perfections, but of whom the qualities are in contradiction every moment, as soon as they suppose him the author of all things. Thus, whenever one of the human species experiences misfortunes, you will see him deny providence, laugh at final causes, obliged to acknowledge either that God is impotent, or that he acts in a mode contradictory to his goodness. Yet, those who suppose a just God, are they not obliged to suppose duties and regulations, emanating from this being, whom they cannot offend if they do not know his will? Thus the *theist*, one after another, to explain the conduct of his God, finds himself in continual embarrassment, from which he knows not how to withdraw himself; but, in admitting all the theological reveries, without even excepting those absurd fables which were imagined to render an account of the strange economy of this being, so good, so wise, so full of equity; it will be necessary, from supposition to supposition, to recur to the sin of *Adam*, or to the *fall of the rebel angels*, or to the crime of *Prometheus*, and the box of *Pandora*, to find in what manner evil has crept into a world subjected to a benevolent intelligence. It will be necessary to suppose the free agency of man; it will be necessary to acknowledge, that the creature can offend his God, provoke his anger, move his passions, and calm them afterwards by superstitious ceremonies and expiations. If they suppose nature to be subject to a concealed agent, endued with occult qualities, acting in a mys-

terious manner, wherefore should it not be supposed that ceremonies, motions of the body, words, rites, temples, and statues can equally contain secret virtues, suitable to reconcile them to the mysterious being whom they adore? Wherefore should they not give faith to the concealed powers of magic, of theurgy, of enchantments, of charms, and of talismans? Wherefore not believe in inspirations, in dreams, in visions, in omens, and in soothsayers? Who knows if the motive power of the universe, to manifest itself to men, has not been able to employ impenetrable ways, and has not had recourse to metamorphoses, to incarnations, and to transubstantiations? Do not all these reveries flow from the absurd notions which men have formed to themselves of the Divinity? All these things, and the virtues which are attached to them, are they more incredible and less possible than the ideas of *theism*, which suppose that an inconceivable, invisible, and immaterial God has been able to create and can move matter; that a God destitute of organs, can have intelligence, think like men, and have moral qualities; that a wise and intelligent God can consent to disorder; that an immutable and just God can permit that innocence should be oppressed for a time? When a God so contradictory, or so much opposed to the dictates of good sense, is admitted, there is no longer any thing to make reason revolt at. As soon as they suppose such a God, they can believe any thing; it is impossible to point out where they ought to arrest the progress of their imagination. If they presume relations between man and this incredible being, they must rear him altars, make him sacrifices, address him with continual prayers, and offer him presents. If nothing can be conceived of this being, is it not the most certain way to refer to his ministers, who by situation must have meditated upon him, to make him known to others? In short, there is no revelation, no mystery, no practice that it may not be necessary to admit upon the word of the priests, who, in each country, are in the habit of teaching to men that which they ought to think of the Gods, and of suggesting to them the means of pleasing them.

for *atheism*, perfectly resembles the *horrour of a vacuum:* they have occasion to believe something, the mind cannot remain in suspense; above all, when they persuade themselves that the thing interests them in a very lively manner; and then, rather than believe nothing, they will believe every thing that shall be desired, and will imagine that the most certain mode is to take a part.

We see, then, that the *deists* or *theists*, have no real ground to separate themselves from the superstitious, and that it is impossible to fix the line of demarcation, which separates them from the most credulous men, or from those who reason the least upon religion. Indeed, it is difficult to decide with precision the true dose of folly which may be permitted them. If the deists refuse to follow the superstitious in every step their credulity leads them, they are more inconsistent than these last, who, after having admitted upon hearsay, an absurd, contradictory, and fantastical Divinity, also adopt upon report, the ridiculous and strange means which are furnished them to render him favourable to them. The first set forth a false supposition, of which they reject the necessary consequences ; the others admit both the principle and the conclusion.* The God, who exists

only in imagination, demands an imaginary worship; all theology is a mere fiction ; there are no degrees in falsehood, no more than in truth. If God exists, every thing which his ministers say of him must be believed; all the reveries of superstition hare in them nothing more incredible than the incompatible Divinity, which serves for their foundation ; these reveries themselves, are only corollaries drawn with more or less subtlety, inductions which enthusiasts or dreamers have deduced from his impenetrable essence, from his unintelligible nature, and from his contradictory qualities. Wherefore, then, stop on the road ? Is there, in any one religion in the world, a miracle more impossible to be believed than that of the *creation*, or the eduction from nothing ?. Is there a mystery more difficult to be comprehended than a God impossible to be conceived, and whom, however, it is necessary to admit ? Is there any thing more contradictory, than an intelligent and omnipotent workman, who only produces to destroy ? Is there any thing of greater

favour of an object which we believe of the utmost importance. God is to be feared only because his interests disturb society. In the meantime, it cannot be denied that pure *theism*, or that which is called *natural religion*, is preferable to superstition, the same as the *reform* has banished many abuses from those countries which have embraced it. There is nothing short of an unlimited and inviolable liberty of thought, that can permanently assure peace to the mind. The opinions of men are only dangerous when they are restrained, or when it is imagined necessary to make others think in the same manner as we ourselves think. No opinions, not even those of superstition, would be dangerous, if the superstitious did not think themselves obliged to persecute them, and had not the power to do so; it is this prejudice, which, for the benefit of mankind, it is essential to annihilate, and if the thing be impossible, the object which philosophy may reasonably propose to itself, will be to make the depositaries of power feel that they never ought to permit their subjects to commit evil for their religious opinions. In this case, wars would be almost unheard of amongst men, and instead of beholding the melancholy spectacle of man cutting the throat of his fellow man, because he will not see his God with his own peculiar eyes, we shall see him labouring essentially to his own happiness, by promoting that of his neighbour; cultivating the fields and bringing forth the productions of nature, instead of puzzling his brain with theological disputes, which can never be of the smallest advantage to any one except the *priests*.

* A very profound philosopher has remarked, and with reason, that *deism* must be subject to as many heresies and schisms as religion. The *deists* have principles in common with the superstitious, and these have frequently the advantage in their disputes against them. If there exists a God, that is to say, a being of whom we have no idea, and who, nevertheless, has relations with us, wherefore should we not worship him ? But what rule shall we follow in the worship we ought to render him ? The most certain way will be to adopt the worship of our fathers and of our priests. It will not depend upon us to seek another; this worship, is it absurd ? It will not be permitted us to examine it. Thus, however absurd it may prove, the most certain way will be to conform to it: and we may plead as an excuse, that an unknown cause can act in a mode inconceivable to us : that the views of God *are an impenetrable abyss ;* that it is very expedient blindly to leave them to our guides : that we shall act wisely in looking upon them as *infallible*, &c. Whence we see that a consequent theism can conduct us, step by step, to the most abject credulity, to superstition, and even to the most dangerous fanaticism. . Is fanaticism, then, any other thing than an irrational passion for a being, who has no existence but in the imagination ? *Theism* is, with relation to superstition, that which *reform* or *Protestantism* has been to the *Roman Catholic religion.* The reformers, shocked at some absurd mysteries, have not contested others which were no less revolting. As soon as the theological God is admitted, there is nothing more in religion which may not be adopted. On the other hand, if, notwithstanding the *reform*, the *Protestants* have frequently been intolerant, it is to be feared that the *theists* may be the same; it is difficult not to be angry in

inutility than to associate with nature an agent, who cannot explain any one of the phenomena of nature?

Let us conclude, then, that the man who is the most credulously superstitious, reasons in a manner more conclusive, or, at least, more consistent, than those, who, after having admitted a God, of whom they have no idea, stop all at once, and refuse to admit those systems of conduct which are the immediate and necessary result of a radical and primitive errour. As soon as they subscribe to a principle opposed to reason, by what right do they dispute its consequences, however absurd they may find them?

The human mind, we cannot too often repeat for the happiness of men, may torment itself as much as it will; whenever it quits visible nature, it leads itself astray, and is presently obliged to return. If a man mistakes nature and her energy, he has occasion for a God to move her: he will no longer have any ideas of her, and he is instantly obliged to form a God, of whom he is himself the model; he believes he makes a God, in giving him his own qualities, which he believes he renders more worthy the sovereign of the world, by exaggerating them, whilst, by dint of abstractions, of negations, of exaggerations, he annihilates them, or renders them totally unintelligible. When he does no longer understand himself, and loses himself in his own fictions, he imagines he has made a God, whilst he has only made an imaginary being.] A God clothed with mortal qualities has always man for a model; a God clothed with the attributes of theology, has a model no where, and does not exist relatively to us: from the ridiculous and extravagant combination of two beings so diverse, there can only result a pure chimera, with which our mind can have no relation, and with which it is of the greatest inutility to occupy ourselves.

Indeed, what could we expect from a God such as he is supposed to be? What could we ask of him? If he is spiritual, how can he move matter, and arm it against us? If it be he who establishes the laws of nature; if it be he who gives to beings their essence and their properties; if every thing that takes place is a proof and the work of his infinite providence, of his profound wisdom, to what end address prayers to him? Shall we pray to him to alter, in our favour, the invariable course of things? Could he, even if he would, annihilate his immutable decrees, or retrace his steps? Shall we demand, that, to please us, he shall make the beings act in a mode opposite to the essence which he has given them? Can he prevent that a body, hard by its nature, such as a stone, shall not wound, in falling, a brittle body, such as the human frame, whose essence is to feel? Thus, let us not demand miracles of this God, whatever he may be; in despite of the omnipotence which he is supposed to have, his immutability would oppose itself to the exercise of his power; his goodness would oppose itself to the exercise of his rigid justice; his intelligence would oppose itself to those changes that he might be disposed to make in his plan. Whence we see, that theology itself, by dint of discordant attributes, makes of its God an immoveable being, useless to man, to whom miracles are totally impossible.

We shall perhaps be told, that the infinite science of the Creator of all things knows in the beings which he has formed resources concealed to imbecile mortals; and that without changing any thing, either in the laws of nature or in the essence of things, he is able to produce effects which surpass our feeble understanding, without, however, these effects being contrary to the order which he has himself established. I reply, that every thing which is conformable to the nature of beings, can neither be called *supernatural* nor *miraculous.* Many things are, without doubt, above our conception, but every thing that takes place in the world, is natural, and can be much more simply attributed to nature, than to an agent of whom we have no idea. In the second place, that by the word *miracle,* an effect is meant, of which, for want of knowing nature, she is believed to be incapable. In the third place, that by *miracle,* the theologians of all countries pretend to indicate, not an extraordinary operation of nature, but an effect directly opposite to her laws of this nature; to which, however, we are assured that God has prescribed *his* laws.* On the other hand, if God, in those of his works which surprise us, or which we do not comprehend, does no more than give play to springs unknown to men, there is nothing in nature that, in this sense, may not be looked upon as a miracle, seeing that the cause which makes a stone fall, is as unknown to us, as that which makes our globe turn. In short, if God, when he performs a miracle, only avails himself of the knowledge which he has of nature, to surprise us, he simply acts like some men more cunning than others, or more instructed than the uninformed, who astonish them with their tricks and their marvellous secrets, by taking advantage of their ignorance, or of their incapacity. To explain the phenomena of nature by miracles, is to say, that we are ignorant of the true causes of these phenomena : to attribute them to a God, is to confess that we do not know the resources of nature, and that we need a word to designate them ; it is to believe in magic. To attribute to an intelligent, immutable, provident, and wise being, those miracles by which he derogates from his laws, is to annihilate in him those qualities.† An omnipotent God would

* A miracle, says BUDDÆUS, is an operation by which the, laws of nature, upon which depend the order and the preservation of the universe, are suspended.—See *Treatise on Atheism,* p. 140.

† The last refuge of the deist and theologian, when driven off all other ground, is the possibility of every thing he asserts, couched in the dogma, "that nothing is impossible with God." They mark this asseveration with a degree of self-complacency, with an air of triumph, that would almost persuade one they could not be mistaken; most assuredly with him who dips no farther than the surface, they carry complete conviction. But if we examine a little the nature of this proposition, we will find that it is untenable. In the first place, the possibility of a thing, by no means proves its absolute existence : a thing may be extremely possible, and yet not be. Secondly, if this was once an admitted argument, there would be, in fact, an end of all morality. The Bishop of Chester, Dr. John Wilkins, says : "Would not such men be generally accounted out of their wits, who could please themselves by entertaining actual hopes of any thing, merely upon account of the possibility of it, or torment themselves with actual fears of all such evils as are possible? Is there any thing imaginable more wild and extravagant than this would be?" Thirdly, tho impossibility

not have occasion for miracles to govern the world, nor to convince his creatures, whose minds and hearts would be in his own hands. All the miracles announced by all the religions of the world, as proofs of the interest which the Most High takes in them, prove nothing but the inconstancy of this being, and the impossibility in which he finds himself to persuade men of that which he would inculcate.

In short, and as a last resource, it will be demanded, whether it would not be better to depend on a good, wise, intelligent being, than on a blind nature, in which we do not find any quality that is consoling to us, or on a fatal necessity always inexorable to our cries ? I reply, first, that our interest does not decide the reality of things, and that if even it should be more advantageous to us to have to do with a being as favourable as God is pointed out to us, this would not prove the existence of this being. Secondly. that this being, so good and so wise, is, on the other hand, represented to us as an irrational tyrant, and that it would be more advantageous for man to depend on a blind nature, than on a being whose good qualities are contradicted every instant by the same theology which has invented them. Thirdly, that nature, duly studied, furnishes us with every thing necessary to render us as happy as our essence admits. When, by the assistance of experience, we shall consult nature, or cultivate our reason, she will discover to us our duties, that is to say, the indispensable means to which her eternal and necessary laws have attached our preservation, our own happiness, and that of society. It is in nature that we shall find wherewith to satisfy our physical wants ; it is in nature we shall find those duties defined, without which we cannot live happy in our sphere. Out of nature, we only find

would reasonably appear to be on the other side ; so far from nothing being impossible, every thing that is erroneous, would seem to be so ; for, if a God existed, he could not possibly either love vice, cherish crime, be pleased with depravity, or commit wrong. This decidedly turns the argument against them, and leaves them no other alternative but to retire from behind the shield with which they have imagined they rendered themselves invulnerable.

prejudicial chimeras which render us doubtful as to what we owe to ourselves and to the other beings with whom we are associated.

Nature is not, then, a stepmother to us; we do not depend upon an inexorable destiny. Let us rely on nature alone; she will procure us a multitude of benefits, when we shall pay her the attention she deserves: she will furnish us wherewithal to alleviate our physical and moral evils, when we shall be disposed to consult her: she does not punish us or show us rigour, except when we despise her to prostitute our incense to the idols which our imagination has elevated to the throne that belongs to her. It is by incertitude, discord, blindness, and delirium, that she visibly chastises all those who put a monster-God in the place which she ought to occupy.

In supposing, even for an instant, this nature to be inert, inanimate, blind, or, if they will, in making chance the God of the universe, would it not be better to depend absolutely upon nothing than upon a God necessary to be known, and of whom we cannot form any one idea, or if we shall form one, to whom we are obliged to attach notions the most contradictory, the most disagreeable, the most revolting, and most prejudicial to the repose of human beings? Were it not better to depend on destiny or on fatality, than on an intelligence so irrational as to punish his creatures for the little intelligence and understanding which he has been pleased to give them? Were it not better to throw ourselves into the arms of a blind nature, destitute of wisdom and of views, than to tremble all our life under the scourge of an omnipotent intelligence, who has combined his sublime plans in such a manner that feeble mortals should have the liberty of counteracting and destroying them, and thus becoming the constant victims of his implacable wrath.*

* Lord Shaftesbury, although a very zealous *theist*, says, with reason, that "many honest people would have a more tranquil mind if they were assured that they had only a blind destiny for their guide: they tremble more in thinking that there is a God, than if they believed that he did not exist." *See his Letter on Enthusiasm;* see also *Chapter* xiii.

CHAPTER VI.

Examination of the Advantages which result to men from their Notions on the Divinity, or of their Influence upon Morals, upon Politics, upon the Sciences, upon the Happiness of Nations and Individuals.

WE have hitherto seen the slender foundation of those ideas which men form to themselves of the Divinity; the little solidity there is in the proofs by which they suppose his existence: the want of harmony in the opinions they have formed of this being, equally impossible to be known to the inhabitants of the earth: we have shown the incompatibility of those attributes which theology assigns to him: we have proved that this being, whose name alone has the power of inspiring fear, is nothing but the shapeless fruit of ignorance, of an alarmed imagination, of enthusiasm, of melancholy: we have shown that the notions which men have formed of him, only date their origin from the prejudices of their infancy, transmitted by education, strengthened by habit, nourished by fear, maintained and perpetuated by authority. In short, every thing must have convinced us, that the idea of God, so generally diffused over the earth, is no more than a universal errour of the human species. It remains now to examine if this errour be useful.

No errour can be advantageous to the human species; it is ever founded upon his ignorance, or the blindness of his mind. The more importance men shall attach to their prejudices, the more is the fatal consequences of their errours. Thus, Bacon had great reason for saying that *the worst of all things, is deified errour.* Indeed, the inconveniences which result from our religious errours have been, and always will be, the most terrible and the most extensive. The more we respect these errours, the more play they give to our passions, the more they disturb our mind, the more irrational they render us, the greater influence they have on the whole conduct of our lives. There is but little likelihood that he who renounces his reason in the thing which he considers as the most essential to his happiness, will listen to it on any other occasion.

If we reflect a little, we shall find the most convincing proof of this sad truth; we shall see in those fatal notions which men have cherished of the Divinity, the true source of those prejudices and of those sorrows of every kind to which they are the victims. Nevertheless, as we have elsewhere said, utility ought to be the only rule and the uniform standard of those judgments which are formed on the opinions, the institutions, the systems, and the actions of intelligent beings; it is according to the happiness which these things procure for us, that we ought to attach to them our esteem; whenever they are useless, we ought to despise them; as soon as they become pernicious, we ought to reject them: and reason prescribes that we should detest them in proportion to the magnitude of the evils they cause.

From these principles, founded on our nature, and which will appear incontestable to every reasonable being, let us coolly examine the effects which the notions of the Divinity have produced on the earth. We have already shown, in more than one part of this work, that morals, which have only for object that man should preserve himself and live in society, had nothing in common with those imaginary systems which he can form to himself upon a power distinguished from nature; we have proved, that it sufficed to meditate on the essence of a sensible, intelligent, and rational being, to find motives to moderate his passions, to resist his vicious propensities, to make him fly criminal habits, to render himself useful and dear to those beings for whom he has a continual occasion. These motives are, without doubt, more true, more real, more powerful, than those which it is believed ought to be borrowed from an imaginary being, calculated to be seen diversely by all those who shall meditate upon him. We have demonstrated, that education, in making us, at an early period, contract good habits, favourable dispositions, strengthened by the laws, by a respect for public opinion, by ideas of decency, by the desire of meriting the esteem of others, by the fear of losing our own esteem, would be sufficient to accustom us to a laudable conduct, and to divert us even from those secret crimes for which we are obliged to punish ourselves by fear, shame, and remorse.

Experience proves that the success of a first secret crime disposes us to commit a second, and this a third; that the first action is the commencement of a habit; that there is much less distance from the first crime to the hundredth, than from innocence to criminality; that a man who permits himself to commit a series of bad actions, in the assurance of impunity, deceives himself, seeing that he is always obliged to punish himself, and that, moreover, he cannot know where he shall stop. We have shown, that those punishments which, for its own preservation, society has the right to inflict on all those who disturb it, are, for those men who are insensible to the charms of virtue, or the advantages which result from the practice of it, more real, more efficacious, and more immediate obstacles, than the pretended wrath or the distant punishments of an invisible power, of whom the idea is effaced every time that impunity in this world is believed to be certain. In short, it is easy to feel that politics, founded upon the nature of man and of society, armed with equitable laws, vigilant with regard to the morals of men, faithful in rewarding virtue and punishing crime, would be more suitable to render morality respectable and sacred than the chimerical authority of that God who is adored by all the world, and who never restrains any but those who are already sufficiently restrained by a moderate temperament, and by virtuous principles.

On the other hand, we have proved that nothing was more absurd and more dangerous than attributing human qualities to the Divinity, which, in fact, are found in continual contradiction with themselves; a goodness, a wisdom, and an equity, which we see every instant counterbalanced or denied by wickedness, by confusion, by an unjust despotism, which all the theologians of the world have·at all times attributed to this same Divinity. It is then very easy to conclude from it that God, who is shown to us under such different aspects, cannot be the model of man's conduct, and that his moral character cannot serve for an example to beings living together in society, who are only reputed virtuous when their conduct does not deviate from that benevolence and justice which they owe to their fellow-creatures. A God superiour to every thing, who owes nothing to his subjects, who has occasion for no one, cannot be the model of creatures who are full of wants, and consequently must have duties.

Plato has said, that *virtue consisted in resembling God*. But where shall we find this God whom man ought to resemble? Is it in nature? Alas! he who is supposed to be the mover of it, diffuses indifferently over the human race great evils and great benefits; he is frequently unjust to the purest souls; he accords the greatest favours to the most perverse mortals; and if, as we are assured, he must show himself one day more equitable, we shall be obliged to wait for that time to regulate our conduct upon his own.

Shall it be in the revealed religions, that we shall draw up our ideas of virtue? Alas! do they not all appear to be in accord in announcing a despotic, jealous, vindictive, and selfish God, who knows no law, who follows his caprice in every thing, who loves or who hates, who chooses or reproves, according to his whim; who acts irrationally, who delights in carnage, rapine, and crime; who plays with his feeble subjects, who overloads them with puerile laws, who lays continual snares for them, who rigorously prohibits them from consulting their reason? What would become of morality, if men proposed to themselves such Gods for models.

It is, however, some Divinity of this temper which all nations adore. Thus, we see it is in consequence of these principles, that religion, in all countries, far from being favourable to morality, shakes it and annihilates it. It divides men in the room of uniting them; in the place of loving each other, and lending mutual succours one to the other, they dispute with each other, they despise each other, they hate each other, they persecute each other, and they frequently cut each others' throats for opinions equally irrational: the slightest difference in their religious notions, renders them from that moment enemies, separates their interests, and sets them into continual quarrels. For theological conjectures, nations be-

No. IX.—34.

come opposed to other nations; sovereign arms himself against his subjects; citizens wage war against their fellow-citizens; fathers detest their children, these plunge the sword into the bosom of their parents; husbands and wives are disunited; relations forget each other; all the social bonds are broken; society rends itself in pieces by its own hands, whilst, in the midst of this horrid confusion, each pretends that he conforms to the views of the God whom he serves, and does not reproach himself with any one of those crimes which he commits in the support of his cause.

We again find the same spirit of whim and madness in the rites, the ceremonies, and the practices, which all the worships in the world appear to have placed so much above the social or natural virtues. Here mothers deliver up their children to feed their God; there subjects assemble themselves in the ceremony of consoling their God for those pretended outrages which they have committed against him, by immolating to him human victims. In another country, to appease the wrath of his God, a frantic madman tears, himself and condemns himself for life to rigorous tortures. The Jehovah of the Jews is a suspicious tyrant, who breathes nothing but blood, murder, and carnage, and who demands that they should nourish him with the vapours of animals. The Jupiter of the Pagans is a lascivious monster. The Moloch of the Phœnicians is a cannibal; the pure mind of the Christians resolved, in order to appease his fury, to crucify his own son; the savage God of the Mexicans cannot be satisfied without thousands of mortals which are immolated to his sanguinary appetite.

Such are the models which the Divinity presents to men in all the superstitions of the world. Is it then surprising that his name has become the signal of terrour, madness, cruelty, and inhumanity for all nations, and serves as a continual pretext for the most shameful and impudent violation of the duties of morality? It is the frightful character that men every where give of their God, that banishes goodness for ever from their hearts, morality from their conduct, felicity and reason from their habitations; it is every where a God who is disturbed by the mode in which unhappy mortals think, that arms them with poniards one against the other, that makes them stifle the cries of nature, that renders them barbarous to themselves and atrocious to their fellow creatures; in short, they become irrational and furious every time that they wish to imitate the God whom they adore, to merit his love, and to serve him with zeal.

It is not, then, in heaven that we ought to seek either for models of virtue, or the rules of conduct necessary to live in society. Man needs human morality, founded upon his own nature, upon invariable experience, upon reason: the morality of the Gods will always be prejudicial to the earth; cruel Gods cannot be well served but by subjects who resemble them. What becomes, then, of those great advantages which have been imagined resulted from the notions which are unceasingly given us of the Divinity? We see that all nations acknowledge a God who is sovereignly wicked; and to conform themselves to his views, they trample under feet the most evident duties of humanity; they appear to act as if it were only by crimes and madness that they hoped to draw down upon themselves the favours of the sovereign intelligence, of whose goodness they boast so much. As soon as there is a question of religion, that is to say, of a chimera, whose obscurity has made them place him above either reason or virtue, men make it a duty with themselves to give loose to all their passions; they mistake the clearest precepts of morality, as soon as their priests give them to understand that the Divinity commands them to commit crimes, or that it is by transgressions that they will be able to obtain pardon for their faults.

Indeed, it is not in those revered men, diffused over the whole earth, to announce to men the oracles of Heaven, that we shall find real virtues. Those enlightened men, who call themselves the ministers of the Most High, frequently preach nothing but hatred, discord, and fury, in his name: the Divinity, far from having a useful influence over their own morals, commonly does no more than render them more

ambitious, more covetous, more hardened, more obstinate, and more proud. We see them unceasingly occupied in giving birth to animosities, by their unintelligible quarrels. We see them wrestling against the sovereign authority, which they pretend is subject to their's. We see them arm the chiefs of a nation against their legitimate magistrates. We see them distribute to the credulous people weapons to massacre each other with in those futile disputes, which the sacerdotal vanity makes to pass for matters of importance. Those men, so persuaded of the existence of a God, and who menace the people with his eternal vengeance, do they avail themselves of these marvellous notions to moderate their pride, their cupidity, their vindictive and turbulent humour? In those countries, where their empire is established in the most solid manner, and where they enjoy impunity, are they the enemies of that debauchery, that intemperance, and those excesses, which a severe God interdicts to his adorers? On the contrary, do we not see them from thence emboldened in crime, intrepid in iniquity, giving a free scope to their irregularities, to their vengeance, to their hatred, and suspicious cruelties? In short, it may be advanced without fear, that those who, in every part of the earth, announce a terrible God, and make men tremble under his yoke; that those men, who unceasingly meditate upon him, and who undertake to prove his existence to others, who decorate him with pompous attributes, who declare themselves his interpreters, who make all the duties of morality to depend upon him, are those whom this God contributes the least to render virtuous, humane, indulgent, and sociable. In considering their conduct, we should be tempted to believe that they are perfectly undeceived with respect to the idol whom they serve, and that no one is less the dupe of those menaces which they pronounce in his name, than themselves. In the hands of the priests of all countries, the Divinity resembles the head of *Medusa*, which, without injuring him who showed it, petrified all the others. The priests are generally the most crafty of men, the best among them are truly wicked. Does the idea of an avenging and remunerating God impose more upon those princes, on those Gods of the earth, who found their power and the titles of their grandeur upon the Divinity himself; who avail themselves of his terrific name to intimidate, and make those people hold them in reverence who are so frequently rendered unhappy by their caprice? Alas! the theological and supernatural ideas, adopted by the pride of sovereigns, have done no more than corrupt politics, and have changed them into tyranny. The ministers of the Most High, always tyrants themselves, or the cherishers of tyrants, are they not unceasingly crying to monarchs, that they are the images of the Deity? Do they not tell the credulous people, that Heaven wills that they should groan under the most cruel and the most multifarious injustice; that to suffer is their inheritance; that their princes, like the Supreme Being, have the indubitable right to dispose of the goods, the persons, the liberty, and the lives of their subjects? Do not those chiefs of nations, thus poisoned in the name of the Divinity, imagine that every thing is permitted them? Competitors, representatives, and rivals of the celestial power, do they not exercise, after his example, the most arbitrary despotism? Do they not think, in the intoxication into which sacerdotal flattery has plunged them, that, like God, they are not accountable to men for their actions, that they owe nothing to the rest of mortals, that they are bound by no bonds to their miserable subjects?

Then it is evident, that it is to theological notions, and to the loose flattery of the ministers of the Divinity, that are to be ascribed the despotism, the tyranny, the corruption, and the licentiousness of princes, and the blindness of the people, to whom, in the name of Heaven, they interdict the love of liberty, to labour to their own happiness, to oppose themselves to violence, to exercise their natural rights. These intoxicated princes, even in adoring an avenging God, and in obliging others to adore him, never cease a moment to outrage him by their irregularities and their crimes. Indeed, what morality is this, but that of men who offer themselves as living images and representatives of the Divinity?

Are they, then, atheists, those monarchs who, habitually unjust, wrest, without remorse, the bread from the hands of a famished people, to administer to the luxury of their insatiable courtiers, and the vile instruments of their iniquities? Are they atheists, those ambitious conquerors, who, but little contented with oppressing their own subjects, carry desolation, misfortune, and death, among the subjects of others? What do we see in those potentates, who reign by *divine right* over nations, except ambitious mortals, whom nothing can arrest, with hearts perfectly insensible to the sorrows of the human species? souls without energy, and without virtue, who neglect the most evident duties, with which they do not even deign to become acquainted? powerful men, who insolently place themselves above the rules of natural equity?* knaves who make sport of honesty? In the alliances which those deified sovereigns form between themselves, do we find even the shadow of sincerity? In those princes, when even they are subjected, in the most abject manner, to superstition, do we meet with the smallest real virtue? We only see in them robbers, too haughty to be humane, too great to be just, who make for themselves alone a code of perfidies, violence, and treason; we only see in them wicked beings, ready to overreach, surprise, and injure each other; we only find in them furies, always at war, for the most futile interests, empoverishing their people, and wresting from each other the bloody remnants of nations; it might be said, that they dispute who shall make the greatest number of miserable beings

* The Emperor Charles the Fifth used to say, that, *being a warrior, it was impossible for him to have either conscience or religion :* his general, the Marquis de Pescaire, said, that *nothing was more difficult, than to serve at one and the same time the God Mars and Jesus Christ.* Generally speaking, nothing is more contrary to the spirit of Christianity than the profession of arms; and yet the Christian princes have most numerous armies, and are perpetually at war. Moreover, the clergy would be extremely sorry that the maxims of the evangelists, or the Christian meekness should be rigidly followed, which in nowise accords with their interests. This clergy have occasion for soldiers to give solidity to their doctrines and their rights. This proves to what a degree religion is calculated to impose on the passions of men.

on the earth! At length, wearied with their own fury, or forced by the hand of necessity to make peace, they attest the most insidious treaties in the name of God, ready to violate their most solemn oaths, as soon as the smallest interest shall require it.*

This is the manner in which the idea of God imposes on those who call themselves his images, who pretend they have no account to render but to him alone! Amongst these representatives of the Divinity, it is with difficulty we find, during thousands of years, one who has equity, sensibility, or the most ordinary talents and virtues. The people, brutalized by superstition, suffer infants who are made giddy with flattery, to govern them with an iron sceptre; these madmen, transformed into Gods, are the masters of the law; they decide for society, whose tongue is tied; they have the power to create both the just and the unjust; they exempt themselves from those rules which their caprice imposes on others; they neither know relations nor duties; they have never learned to fear, to blush, or to feel remorse: their licentiousness has no limits, because it is assured of remaining unpunished; in consequence, they disdain public opinion, decency, and the judgments of men whom they are enabled to overwhelm by the weight of their enormous power. We see them commonly given up to vice and debauchery, because the listlessness and the disgust which follow the surfeit of satiated passions, oblige them to recur to strange pleasures and costly follies, to awaken activity in their benumbed souls. In short, accustomed only to fear God, they always conduct themselves as if they had nothing to fear.

History, in all countries, shows us only a multitude of vicious and mischievous potentates; nevertheless, it shows us but few who have been atheists. The annals of nations, on the contrary, offer to our view a great number of superstitious princes, who passed their lives plunged in luxury and effeminacy, strangers to every virtue, uniformly good to their hungry courtiers,

and insensible of the sorrows of their subjects; governed by mistresses and unworthy favourites; leagued with priests against the public happiness; in short, persecutors, who, to please their God, or expiate their shameful irregularities, joined to all their other crimes, that of tyrannising over the thought, and of murdering citizens for their opinions. Superstition in princes is allied with the most horrid crimes; almost all of them have religion, very few of them have a knowledge of true morality, or practise any useful virtue. Religious notions only serve to render them more blind and more wicked; they believe themselves assured of the favour of Heaven; they think that their Gods are appeased, if, for a little, they show themselves attached to futile customs, and to the ridiculous duties which superstition imposes on them. Nero, the cruel Nero, his hands yet stained with the blood of his own mother, was desirous to be initiated into the mysteries of Eleusis. The odious Constantine found, in the Christian priests, accomplices disposed to expiate his crimes. That infamous Philip, whose cruel ambition caused him to be called the *Demon of the South*, whilst he assassinated his wife and his son, piously caused the throats of the Batavians to be cut for religious opinions. It is thus that superstitious blindness persuades sovereigns that they can expiate crimes by crimes of still greater magnitude.

Let us conclude, then, from the conduct of so many princes, so very religious, but so little imbued with virtue, that the notions of the Divinity, far from being useful to them, only served to corrupt them, and to render them more wicked than nature had made them. Let us conclude, that the idea of an avenging God can never impose restraint on a deified tyrant, sufficiently powerful or sufficiently insensible not to fear the reproaches or the hatred of men; sufficiently hardened not to have compassion for the sorrow of the human species, from whom they believe themselves distinguished: neither heaven nor earth has any remedy for a being perverted to this degree; there is no curb capable of restraining his passions to which religion itself continually gives loose, and whom it ren-

* Nihil est quod credere de se
Non possit, cum laudatur dei æqua potestas.- Juvenal Sat., 4. v. 79.

ders more rash and inconsiderate. Every time that they flatter themselves with easily expiating their crimes, they deliver themselves up with greater facility to crime. The most dissolute men are frequently extremely attached to religion; it furnishes them with means of compensating by forms that of which they are deficient in morals: it is much easier to believe or to adopt doctrines, and to conform themselves to ceremonies, than to renounce their habits or to resist their passions.

Under chiefs, depraved even by religion, nations continued necessarily to be corrupted. The great conformed themselves to the vices of their masters; the example of these distinguished men, whom the uninformed believe to be happy, was followed by the people; courts became sinks, whence issued continually the contagion of vice. The law, capricious and arbitrary, alone delineated honesty; jurisprudence was iniquitous and partial; justice had her bandage over her eyes only to the poor; the true ideas of equity were effaced from all minds; education, neglected, served only to produce ignorant and irrational beings; devotees, always ready to injure themselves; religion, sustained by tyranny, took place of every thing; it rendered those people blind and tractable whom the government proposed to despoil.*

Thus nations, destitute of a rational administration of equitable laws, of useful instruction, of a reasonable education, and always continued by the monarch and the priest in ignorance and in chains, have become religious and corrupted. The nature of man, the true interests of society, the real advantage of the sovereign and of the people once mistaken, the morality of nature, founded upon the essence of man living in society, was equally unknown. It was forgotten that man has wants, that society was only formed to facilitate the means of satisfying them, that government ought to have for ob-

ject the happiness and maintenance of this society; that it ought, consequently, to make use of motives suitable to have a favourable influence over sensible beings. It was not seen that recompenses and punishments form the powerful springs of which public authority could efficaciously avail itself to determine the citizens to blend their interests, and to labour to their own felicity, by labouring to that of the body of which they are members. The social virtues were unknown; the love of country became a chimera; men associated, had only an interest in injuring each other, and had no other care than that of meriting the favour of the sovereign, who believed himself interested in injuring the whole.

This is the mode in which the human heart has become perverted; here is the true source of moral evil, and of that hereditary, epidemical, and inveterate depravity, which we see reign over the whole earth. It is for the purpose of remedying so many evils, that recourse has been had to religion, which has itself produced them; it has been imagined that the menaces of Heaven would restrain those passions which every thing conspired to rouse in all hearts; men foolishly persuaded themselves that an ideal and metaphysical barrier, that terrible fables, that distant phantoms, would suffice to restrain their natural desires and impetuous propensities; they believed that invisible powers would be more efficacious than all the visible powers, which evidently invite mortals to commit evil. They believed they had gained every thing in occupying their minds with dark and gloomy chimeras, with vague terrours, and with an avenging Divinity; and politics foolishly persuaded itself that it was for its own interests the people should blindly submit to the ministers of the Divinity.

What resulted from this? Nations had only a sacerdotal and theological morality, accommodated to the views and to the variable interests of priests, who substituted opinions and reveries to truth, customs to virtue, a pious blindness to reason, fanaticism to sociability. By a necessary consequence of that confidence which the people gave to ministers of the Divinity, two distinct authorities were established in

---

* Machiavelli, in Chap. 11–13 of his *Political Discourses upon Titus Livius,* endeavours to show the utility of superstition to the Roman republic; but, unfortunately, the examples by which he supports it, proves, that none but the senate profited by the blindness of the people, and availed themselves of it to keep them under their yoke.

each state, who were continually at variance and at war with each other; the priest fought the sovereign with the formidable weapon of opinion ; it generally proved sufficiently powerful to shake thrones.* The sovereign was never at rest, but when abjectly devoted to his priests, and tractably received their lessons, and lent his assistance to their phrensy. These priests, always restless, ambitious, and intolerant, excited the sovereign to ravage his own states, they encouraged him to tyranny, they reconciled him with Heaven when he feared to have outraged it. Thus, when two rival powers united themselves, morality gained nothing by the junction ; the people were neither more happy, nor more virtuous; their morals, their well-being, their liberty were overwhelmed by the united forces of the God of heaven, and the God of the earth. Princes, always interested in the maintenance of theological opinions, so flattering to their vanity, and so favourable to their usurped power, for the most part made a common cause with their priests ; they believed that that religious system which they themselves adopted must be the most convenient and useful to the interests of their subjects ; and, consequently, those who refused to adopt it, were treated by them as enemies. The most religious sovereign became, either politically, or through piety, the executioner of one part of his subjects : he believed it to be a sacred duty to tyrannise over thought, to overwhelm and to crush the enemies of his priests, whom he always believed to be the enemies of his own authority. In cutting their throats, he imagined he did that which at once discharged his duty to Heaven, and what he owed to his own security. He did not perceive, that by immolating victims to

* It is well to observe, that the priests, who are perpetually crying out to the people to submit themselves to their sovereigns, because their authority is derived from Heaven, because they are the images of the Divinity, change their language whenever the sovereign does not blindly submit to them. The clergy upholds despotism only that it may direct its blows against its enemies, but it overthrows it whenever it finds it contrary to its interests. The ministers of the invisible powers only preach up obedience to the visible powers when these are humbly devoted to them.

his priests, he strengthened the enemies of his power, the rivals of his greatness, the least subjected of his subjects.

Indeed, owing to the false notions with which the minds of sovereigns and the superstitious people have been so long prepossessed, we find that every thing in society concurs to gratify the pride, the avidity, and the vengeance of the sacerdotal order. Every where we see that the most restless, the most dangerous, and the most useless men, are those who are recompensed the most amply. We see those who are born enemies to the sovereign power, honoured and cherished by it ; the most rebellious subjects looked upon as the pillars of the throne, the corrupters of youth rendered the exclusive masters of education ; the least laborious of the citizens, richly paid for their idleness, for their futile speculations, for their fatal discord, for their inefficacious prayers, for their expiations, so dangerous to morals, and so suitable to encourage crime.

For thousands of years, nations and sovereigns have been emulously despoiling themselves to enrich the ministers of the Gods, to enable them to wallow in abundance, loading them with honours, decorating them with titles, privileges, and immunities, thus making them bad citizens. What fruits did the people and kings gather from their imprudent kindness, and from their prodigality ? Have princes become more powerful ; have nations become more happy, more flourishing, and more rational? No! without doubt; the sovereign lost the greater portion of his authority ; he was the slave of his priests, or he was obliged to be continually wrestling against them ; and the greater part of the riches of society was employed to support in idleness, luxury, and splendour, the most useless and the most dangerous of its members.

Did the morals of the people improve under these guides who were so liberally paid? Alas! the superstitious never knew them ; religion had taken place of every thing else in them ; its ministers, satisfied with maintaining the doctrines and the customs useful to their own interests, only invented fictitious crimes, multiplied painful or ridiculous customs, to the end that they might turn even the transgressions of their slaves to their own profit. Every where they exercised a monopoly of expiations; they made a traffic of the pretended pardons from above, they established a book of rates for crimes; the most serious were always those which the sacerdotal order judged the most injurious to his views. *Impiety, heresy, sacrilege, blasphemy,* &c., vague words, and devoid of sense, which have evidently no other object than chimeras, interesting only the priests, alarmed their minds much more than real crimes, and truly interesting to society. Thus, the ideas of the people were totally overturned; imaginary crimes frightened them much more than real crimes. A man whose opinions and abstract systems did not harmonize with those of the priests, was much more abhorred than an assassin, than a tyrant, than an oppressor, than a robber, than a seducer, or than a corrupter. The greatest of all wickedness, was the despising of that which the priests were desirous should be looked upon as sacred.* The civil laws concurred also to this confusion of ideas; they punished in the most atrocious manner those unknown crimes which the imagination had exaggerated; heretics, blasphemers, and infidels, were burnt; no punishment was decreed against the corrupters of innocence, adulterers, knaves, and calumniators.

Under such instructers, what could become of youth? It was shamefully sacrificed to superstition. Man from his infancy was poisoned by them with unintelligible notions; they fed him with mysteries and fables; they drenched him with a doctrine in which he was obliged to acquiesce, without being able to comprehend it; they disturbed his mind with vain phantoms; they cramped his genius with sacred trifles, with puerile duties, and with mechanical devotions.† They made him lose

his most precious time in customs and ceremonies: they filled his head with sophisms and with errours; they intoxicated him with fanaticism; they prepossessed him for ever against reason and truth; the energy of his mind was placed under continual shackles; he could never soar, he could never render himself useful to his associates; the importance which they attached to the divine science, or rather the systematic ignorance which served for the basis of religion, rendered it impossible for the most fertile soil to produce any thing but thorns.

Does a religious and sacerdotal education form citizens, fathers of families, husbands, just masters, faithful servants, humble subjects, pacific associates? No! it either makes peevish and morose devotees, incommodious to themselves and to others, or men without principles, who quickly sink in oblivion the terrours with which they have been imbued, and who never knew the laws of morality. Religion was placed above every thing; the fanatic was told, *that it were better to obey God than man;* in consequence, he believed that he must revolt against his prince, detach himself from his wife, detest his child, estrange himself from his friend, cut the throats of his fellow-citizens, every time that they questioned the interests of Heaven. In short, religious education, when it had its effect, only served to corrupt juvenile hearts, to fascinate youthful minds, to degrade young minds, to make man mistake that which he owed to himself, to society, and to the beings which surrounded him.

What advantages might not nations have reaped, if they would have employed, on useful objects, those riches which ignorance has so shamefully lavished on the ministers of imposture! What progress might not genius have made, if it had enjoyed those recompenses, granted during so many ages, to those who are at all times opposed to its elevation! To what a de-

* The celebrated Gordon says, that the most abominable of heresies is, to believe there is any other God than the clergy.

† Superstition has fascinated the human mind to such a degree, and made such mere machines of men, that there are a great many countries, in which the people do not understand the language of which they make use to speak to their God. We see *women* who

have no other occupation all their lives, than singing Latin, without understanding a word of the language. The people who comprehend no part of their worship, assist at it very punctually, under an idea that it is sufficient to show themselves to their God, who takes it kind of them that they should come and weary themselves in his temples.

gree might not the useful sciences, the arts, morality, politics, and truth, have been perfected, if they had had the same succours as falsehood, delirium, enthusiasm, and inutility!

It is, then, evident, that the theological notions were and will be perpetually contrary to sound politics and to sound morality; they change sovereigns into mischievous, restless, and jealous Divinities; they make of subjects envious and wicked slaves, who, by the assistance of some futile ceremonies, or by their exterior acquiescence to some unintelligible opinions, imagine themselves amply compensated for the evil which they commit against each other. Those who have never dared to examine into the existence of a God, who rewards and punishes; those who persuade themselves that their duties are founded upon the divine will; those who pretend that this God desires that men should live in peace, cherishing each other, lending each mutual assistance, and abstaining from evil, and that they should do good to each other, presently lost sight of these steril speculations, as soon as present interests, passions, habits, or importunate whims, hurry them away. Where shall we find the equity, the union, the peace and concord, which these sublime notions, supported by superstition and divine authority, promise to those societies under whose eyes they are unceasingly placing them? Under the influence of corrupt courts and priests, who are either impostors or fanatics, who are never in harmony with each other, I only see vicious men, degraded by ignorance, enslaved by criminal habits, swayed by transient interests, or by shameful pleasures, who do not even think of their God. In despite of his theological ideas, the courtier continues to weave his dark plots: he labours to gratify his ambition, his avidity, his hatred, his vengeance, and all those passions inherent to the perversity of his being: maugre that hell, of which the idea alone makes her tremble, the corrupt woman persists in her intrigues, her impostures, and her adulteries. The greater part of men, dissipated, dissolute, and without morals, who fill cities and courts, would recoil with horrour, if the smallest doubt was ex-

hibited to them of the existence of that God whom they outrage. What good results from the practice of this opinion, so universal and so barren, which never has any other kind of influence on the conduct, than to serve as a pretext to the most dangerous passions? On quitting that temple, in which they have been sacrificing, delivering out the divine oracles, and terrifying crime in the name of Heaven, does not the religious despot, who would scruple to omit the pretended duties which superstition imposes on him, return to his vices, his injustice, his political crimes, his transgressions against society? Does not the minister return to his vexations, the courtier to his intrigues, the woman of gallantry to her prostitution, the publican to his extortions, the merchant to his frauds and tricks?

Will it be pretended that those assassins, those robbers, those unfortunates, whom the injustice or the negligence of government multiply, and from whom laws, frequently cruel, barbarously wrest their life; will they pretend, I say, that these malefactors, who every day fill our gibbets and our scaffolds, are incredulous or atheists? No! unquestionably these miserable beings, these outcasts of society, believe in God; his name has been repeated to them in their infancy; they have been told of the punishments destined for crimes; they have been habituated in early life to tremble at his judgments; nevertheless they have outraged society; their passions, stronger than their fears, not having been capable of restraint by the visible motives, have not for much stronger reasons been restrained by invisible motives; a concealed God, and his distant punishments, never will be able to hinder those excesses, which present and assured torments are incapable of preventing.

In short, do we not, every moment, see men persuaded that their God views them, hears them, encompasses them, and yet who do not stop on that account when they have the desire of gratifying their passions, and of committing the most dishonest actions? The same man who would fear the inspection of another man, whose presence would prevent him from committing a bad action, delivering himself up to some scandalous vice, permits himself to do every thing, when he believes he is seen only by his God. What purpose, then, does the conviction of the existence of this God, of his omniscience, of his ubiquity or his presence in all parts, answer, since it imposes much less on the conduct of man, than the idea of being seen by the least of his fellow-men? He, who would not dare to commit a fault, even in the presence of an infant, will make no scruple of boldly committing it, when he shall have only his God for witness. These indubitable facts may serve for a reply to those who shall tell us, that the fear of God is more suitable to restrain the actions of men, than the idea of having nothing to fear from him. When men believe they have only their God to fear, they commonly stop at nothing.

Those persons, who do not suspect the most trivial of religious notions, and of their efficacy, very rarely employ them when they are disposed to influence the conduct of those who are subordinate to them, and to reconduct them into the paths of reason. In the advice which a father gives to his vicious or criminal son, he rather represents to him the present and temporal inconveniences to which his conduct exposes him, than the danger he encounters in offending an avenging God: he makes him foresee the natural consequences of his irregularities, his health deranged by his debaucheries, the loss of his reputation, the ruin of his fortune by play, the punishments of society, &c. Thus the deicolist himself, in the most important occasions of life, reckons much more upon the force of natural motives, than upon the supernatural motives furnished by religion: the same man who vilifies the motives which an atheist can have to do good, and abstain from evil, makes use of them on this occasion, because he feels the full force of them.

Almost all men believe in an avenging and remunerating God; nevertheless, in all countries, we find that the number of the wicked exceed by much that of honest men. If we trace the true cause of so general a corruption, we shall find it in the theological notions themselves and not in those imaginary sources which the different religions of the world have invented, in

order to account for human depravity. Men are corrupt, because they are almost every where badly governed; they are unworthily governed, because religion has deified the sovereigns; these perverted, and assured of impunity, have necessarily rendered their people miserable and wicked. Submitted to irrational masters, the people have never been guided by reason. Blinded by priests, who are impostors, their reason became useless; tyrants and priests have combined their efforts with success, to prevent nations from becoming enlightened, from seeking after truth, from meliorating their condition, ● from rendering their morals more honest, and from obtaining liberty.

It is only by enlightening men, by demonstrating truth to them, that we can promise ourselves to render them better and happier. It is by making known to sovereigns and to subjects their true relations, and their true interests, that politics will be perfected, and that it will be felt that the art of governing mortals is not the art of blinding them, of deceiving them, or of tyrannising over them. Let us, then, consult reason; let us call in experience to our aid; let us interrogate nature, and we shall find what is necessary to be done in order to labour efficaciously to the happiness of the human species. We shall see that errour is the true source of the évils of our species; that it is in cheering our hearts, in dissipating those vain phantoms, of which the idea makes us tremble, in laying the axe to the root of superstition, that we can peaceably seek after truth, and find in nature the torch that can guide us to felicity. Let us, then, study nature; let us observe its immutable laws; let us search into the essence of man; let us cure him of his prejudices, and by these means we shall conduct him, by an easy and gentle declivity, to virtue, without which he will feel that he cannot be permanently happy in the world which he inhabits.

Let us, then, undeceive mortals with regard to those Gods who every where make nothing but unfortunates. Let us substitute visible nature to those unknown powers who have in all times only been worshipped by trembling slaves, or by delirious enthusiasts. Let us tell them that, in order to be happy, they must cease to fear. The ideas of the Divinity, which, as we have seen, are of such inutility, and so contrary to sound morality, do not procure more striking advantages to individuals than to society. In every country, the Divinity was, as we have seen, represented under the most revolting traits, and the superstitious man, when consistent in his principles, was always an unhappy being; superstition is a domestic enemy which man always carries within himself. Those who shall seriously occupy themselves with this formidable phantom, will live in continual agonies and inquietude; they will neglect those objects which are the most worthy of their attention to run after chimeras; they will commonly pass their melancholy days in groaning, in praying, in sacrificing, and in expiating the faults, real or imaginary, which they believe likely to offend their rigid God. Frequently in their fury, they will torment themselves, they will make a duty of inflicting upon themselves the most barbarous punishments to prevent the blows of a God ready to strike ; they will arm themselves against themselves, in the hopes of disarming the vengeance and the cruelty of an atrocious master, whom they think they have irritated ; they will believe they appease an angry God in becoming the executioners of themselves, and doing themselves all the harm their imagination shall be capable of inventing. Society reaps no benefit from the mournful notions of these pious irrationals; their mind finds itself continually absorbed by their sad reveries, and their time is dissipated in irrational ceremonies. The most religious men are commonly misanthropists, extremely useless to the world, and injurious to themselves. If they show energy, it is only to imagine means to afflict themselves, to put themselves to torture, to deprive themselves of those objects which their nature desires. We find, in all the countries of the earth, *penitents* intimately persuaded that by dint of barbarities exercised upon themselves, and lingering suicide, they shall merit the favour of a ferocious God, of whom, however, they every where publish the goodness. We see madmen of this

species in all parts of the world ; the idea of a terrible God has in all times and in all places, given birth to the most cruel extravagances !

If these irrational devotees only injure themselves, and deprive society of that assistance which they owe it, they without doubt, do less harm than those turbulent and zealous fanatics who, filled with their religious ideas, believe themselves obliged to disturb the world, and to commit actual crimes to sustain the cause of their celestial phantom. It very frequently happens, that in outraging morality, the fanatic supposes he renders himself agreeable to his God. He makes perfection consist either in tormenting himself, or breaking, in favour of his fantastical notions, the most sacred ties which nature has made for mortals.

Let us, then, acknowledge, that the ideas of the Divinity are not more suitable to procure the well-being, the content, and peace of individuals than of the society of which they are members. If some peaceable, honest, inconclusive enthusiasts find consolation and comfort in their religious ideas, there are millions who, more conclusive to their principles, are unhappy during their whole life, perpetually assailed by the melancholy ideas of a fatal God their disordered imagination shows them every instant. Under such a formidable God, a tranquil and peaceable devotee is a man who has not reasoned upon him.

In short, every thing proves that religious ideas have the strongest influence over men to torment, divide, and render them unhappy; they inflame the mind, envenom the passions, without ever restraining them, except when the temperament proves too feeble to propel them forward.

CHAPTER VII.

Theological Notions cannot be the Basis of Morality. Comparison between Theological Morality and Natural Morality. Theology Prejudicial to the Progress of the Human Mind.

A SUPPOSITION to be useful to men, ought to render them happy. What right have we to flatter ourselves that an hypothesis which here makes only unhappy beings, may one day conduct us to permanent felicity? If God has only made mortals to tremble and to groan in this world, of which they have a knowledge, upon what foundation can they expect that he will, in the end, treat them with more gentleness in an unknown world. If we see a man commit crying injustice, even transiently, ought it not to render him extremely suspected by us, and make him forever forfeit our confidence? .

On the other hand, a supposition which should throw light on every thing, or which should give an easy solution to all the questions to which it could be applied, when even it should not be able to demonstrate the certitude, would probably be true: but a system which should only obscure the clearest notions, and render more insoluble all the problems desired to be resolved by its means, would most certainly be looked upon as false, as useless, as dangerous. To convince ourselves of this principle, let us examine, without prejudice, if the existence of the theological God has ever given the solution of any one difficulty. Has the human understanding progressed a single step by the assistance of theology? This science, so important and so sublime, has it not totally obscured morality? Has it not rendered the most essential duties of our nature doubtful and problematical? Has it not shamefully confounded all notions of justice and injustice, of vice and of virtue? Indeed, what is virtue in the ideas of our theologians? It is, they will tell us, that which is conformable to the will of the incomprehensible being who governs nature. But what is this being, of whom they are unceasingly speaking without being able to comprehend it; and how can we have a knowledge of his will? They will forthwith tell you what this being is not, without ever being capable of telling you what he is; if they do undertake to give you an idea of him, they will heap upon this hypothetical being a multitude of contradictory and incompatible attributes, which will form a chimera impossible to be conceived; or else they will refer you to those supernatural revelations, by which this phantom has made known his divine intentions to men. But how will they prove

the authenticity of these revelations? It will be by miracles! How can we believe miracles, which, as we have seen, are contrary even to those notions which theology gives us of its intelligent, immutable, and omnipotent Divinity? As a last resource, then, it will be necessary to give credit to the honesty and good faith of the priests, who are charged with announcing the divine oracles. But who will assure us of their mission? Are they not these priests themselves who announce to us, that they are the infallible interpreters of a God whom they acknowledge they do not know. This granted, the priests, that is to say, men extremely suspicious, and but little in harmony among themselves, will be the arbiters of morality; they will decide, according to their uncertain knowledge, or their passions, those laws which ought to be followed; enthusiasm or interest are the only standard of their decisions; their morality is as variable as their whims and their caprice; those who listen to them will never know to what line of conduct they should adhere; in their inspired books, we shall always find a Divinity of little morality, who will sometimes command crime and absurdity; who will sometimes be the friend and sometimes the enemy of the human race; who will sometimes be benevolent, reasonable, and just; and who will sometimes be irrational, capricious, unjust, and despotic. What will result from all this to a rational man? It will be, that neither inconstant Gods nor their priests, whose interests vary every moment, can be the models or the arbiters of a morality which ought to be as regular and as certain as the invariable laws of nature, from which we never see her derogate.

No! arbitrary and inconclusive opinions, contradictory notions, abstract and unintelligible speculations, can never serve for the basis of the science of morals. They must be evident principles, deduced from the nature of man, founded upon his wants, inspired by education, rendered familiar by habit, made sacred by laws: these will carry conviction to our minds, will render virtue useful and dear to us, and will people nations with honest men and good citizens. A God, necessarily incomprehensible, presents nothing but a vague idea to our imagination; a terrible God leads it astray; a changeable God, and who is frequently in contradiction with himself, will always prevent us from ascertaining the road we ought to pursue. The menaces made to us, on the part of a fantastical being,. who is unceasingly in contradiction with our nature, of which he is the author, will never do more than render virtue disagreeable; fear alone will make us practise that which reason and our own immediate interest ought to make us execute with pleasure. A terrible or wicked God, which is one and the same thing, will only serve to disturb honest people, without arresting the progress of the profligate and flagitious; the greater part of men, when they shall be disposed to sin, or deliver themselves up to vicious propensities, will cease to contemplate the terrible God, and will only see the merciful God, who is filled with goodness; men never view things but on the side which is most conformable to their desires.

The goodness of God cheers the wicked, his rigour disturbs the honest man. Thus, the qualities which theology attributes to its God, themselves turn out disadvantageous to sound morality. It is upon this infinite goodness that the most corrupt men will have the audacity to reckon when they are hurried along by crime, or given up to habitual vice. If, then, we speak to them of their God, they tell us that *God is good*, that his clemency and his mercy are infinite. Does not superstition, the accomplice of the iniquities of mortals, unceasingly repeat to them, that by the assistance of certain ceremonies, of certain prayers, of certain acts of piety, they can appease the anger of their God, and cause themselves to be received with open arms by this softened and relenting God? Do not the priests of all nations possess infallible secrets for reconciling the most perverse men to the Divinity?

It must be concluded from this, that under whatever point of view the Divinity is considered, he cannot serve for the basis of morality, formed to be always invariably the same. An irascible God is only useful to those who have an interest in terrifying men, that they may take advantage of their ig-

norance, of their fears, and of their expiations; the nobles of the earth, who are commonly mortals the most destitute of virtue and of morals, will not see this formidable God, when they shall be inclined to yield to their passions; they will, however, make use of him to frighten others, to the end that they may enslave them, and keep them under their guardianship, whilst they will themselves only contemplate this God under the traits of his goodness; they will always see him indulgent to those outrages which they commit against his creatures, provided they have a respect for him themselves; besides, religion will furnish them with easy means of appeasing his wrath. This religion appears to have been invented only to furnish to the ministers of the Divinity an opportunity to expiate the crimes of human nature.

Morality is not made to follow the caprices of the imagination, the passions, and the interests of men: it ought to possess stability; it ought to be the same for all the individuals of the human race; it ought not to vary in one country, or in one time, from another; religion has no right to make its immutable rules bend to the changeable laws of its Gods. There is only one method to give morality this firm solidity; we have more than once, in the course of this work, pointed it out;* there is no other way than to found it upon our duties, upon the nature of man, upon the relations subsisting between intelligent beings, who are, each of them, in love with their happiness, and occupied with conserving themselves; who live together in society, that they may more surely attain these ends. In short, we must take for the basis of morality the necessity of things.

In weighing these principles, drawn from nature, which are self-evident, confirmed by constant experience, and approved by reason, we shall have a certain morality, and a system of conduct, which will never be in contradiction with itself. Man will have no occasion to recur to theological chimeras to regulate his conduct in the visible world. We shall then be capacitated to reply to those who pretend

* See vol. i. chap. viii. of this work; also what is said in chap. xii., and at the conclusion of chap. xiv. of the same volume.

that without a God, there cannot be any morality ; and that this God, by virtue of his power and the sovereign empire which belongs to him over his creatures, has alone the right to impose laws, and to subject them to those duties to which they are compelled. If we reflect on the long train of errours and of wanderings which flow from the obscure notions we have of the Divinity, and on the sinister ideas which all religions in every country give, it would be more conformable to truth to say, that all sound morality, all morality useful to the human species, all morality advantageous to society, is totally incompatible with a being who is never presented to men but under the form of an absolute monarch, whose good qualities are continually eclipsed by dangerous caprices: consequently, we shall be obliged to acknowledge that, to establish morality upon a sure foundation, we must necessarily commence by overturning the chimerical systems upon which they have hitherto founded the ruinous edifice of supernatural morality, which, during so many ages, has been uselessly preached up to the inhabitants of the earth.

Whatever may have been the cause that placed man in the abode which he inhabits, and that gave him his faculties ; whether we consider the human species as the work of nature, or whether we suppose that he owes his existence to an intelligent being, distinguished from nature ; the existence of man, such as he is, is a fact; we see in him a being who feels, who thinks, who has intelligence, who loves himself, who tends to his own conservation ; who, in every moment of his life, strives to render his existence agreeable ; who, the more easily to satisfy his wants, and to procure himself pleasure, lives in society with beings similar to himself, whom his conduct can render favourable or disaffected to him. It is, then, upon these general sentiments, inherent in our nature, and which will subsist as long as the race of mortals, that we ought to found morality, which is only the science of the duties of men living in society.

Here, then, are the true foundations of our duties ; these duties are neces-

sary, seeing that they flow from our peculiar nature, and that we cannot arrive at the happiness we propose to ourselves, if we do not take the means without which we shall never obtain it. Then, to be permanently happy, we are obliged to merit the affection and the assistance of those beings with whom we are associated ; these will not take upon themselves to love us, to esteem us, to assist us in our projects, to labour to our peculiar felicity, but in proportion as we are disposed to labour to their happiness. It is this necessity which is called moral obligation. It is founded upon reflection, on the motives capable of determining sensible and intelligent beings, who tend towards an end, to follow the conduct necessary to arrive at it. These motives can be in us only the desire, always regenerating, of procuring ourselves good, and of avoiding evil. Pleasure and pain, the hope of happiness or the fear of misery, are the only motives capable of having an effiacious influence on the will of sensible beings; to compel them, then, it is sufficient that these motives exist, and may be understood ; to know them, it is sufficient to consider our constitution, according to which we can love or approve in ourselves only those actions from whence result our real and reciprocal utility, which constitutes virtue. In consequence, to conserve ourselves, to enjoy security, we are compelled to follow the conduct necessary to this end ; to interest others in our own conservation, we are obliged to interest ourselves in their's, or to do nothing that may interrupt in them the will of co-operating with us to our own felicity. Such are the true foundations of moral obligation.

We shall always deceive ourselves, when we shall give any other basis to morality than the nature of man ; we cannot have any that is more solid and more certain. Some authors, even of integrity, have thought, that, to render more respectable and more sacred, in the eyes of men, those duties which nature imposes on them, it was necessary to clothe them with the authority of a being, which they made superior to nature, and stronger than necessity. Theology has, in consequence, invaded morality, or has strove to connect it with the religious system; it has been thought that this union would render virtue more sacred; that the fear of the invisible power who governs nature, would give more weight and efficacy to its laws; in short, it has been imagined, that men, persuaded of the necessity of morality, in seeing it united with religion, would look upon this religion itself as necessary to their happiness. Indeed, it is the supposition that a God is necessary to support morality, that sustains the theological ideas, and the greater part of the religious systems of the earth; it is imagined that, without a God, man would neither have a knowledge of, nor practise that which he owes to others. This prejudice once established, it is always believed that the vague ideas of a metaphysical God are in such a manner connected with morality and the welfare of society, that the Divinity cannot be attacked without overturning at the same time the duties of nature. It is thought, that want, the desire of happiness, the evident interest of society, and of individuals, would be impotent motives, if they did not borrow all their force and their *sanction* from an imaginary being who has been made the arbiter of all things.

But it is always dangerous to connect fiction with truth, the unknown with the known, the delirium of enthusiasm with the tranquillity of reason. Indeed, what has resulted from the confused alliance which theology has made of its marvellous chimeras with realities? The imagination bewildered, truth is mistaken; religion, by the aid of its phantom, would command nature, make reason bend under its yoke, subject man to its own peculiar caprice, and frequently, in the name of the Divinity, it obliges him to stifle his nature, and to piously violate most evident duties of morality. When this same religion was desirous of restraining mortals whom it had taken care to render blind and irrational, it gave them only ideal curbs and motives; it could substitute only imaginary causes to true causes; marvellous and supernatural motive-powers to those which were natural and known; romances and fables, to realities. By this inversion of principles, morality no longer had any fixed basis; nature, reason,

virtue, demonstrations, depended upon an undefinable God, who never spoke distinctly, who silenced reason, who only explained himself by inspired beings, by impostors, by fanatics, whose delirium or the desire of profiting by the wanderings of men, interested them in preaching up only an abject submission, factitious virtues, frivolous ceremonies; in short, an arbitrary morality, conformable to their own peculiar passions, and frequently very prejudicial to the rest of the human species.

Thus, in making morality flow from God, they in reality subjected it to the passions of men. In being disposed to found it upon a chimera, they founded it upon nothing; in deriving it from an imaginary being, of whom every one forms to himself a different notion, of whom the obscure oracles were interpreted either by men in a delirium, or by knaves; in establishing it upon his pretended will, goodness, or malignity; in short, in proposing to man, for his model, a being who is supposed to be changeable, the theologians, far from giving to morality a steady basis, have weakened, or even annihilated that which is given by nature, and have substituted in its place nothing but incertitude. This God, by the qualities which are given him, is an inexplicable enigma, which each expounds after his own manner, which each religion explains in its own mode, in which all the theologians of the world discover every thing that suits their purpose, and according to which each man separately forms his morals, conformable to his peculiar character. If God tells the gentle, indulgent, equitable man to be good, compassionate, and benevolent, he tells the furious man, who is destitute of compassion, to be intolerant, inhuman, and without pity. The morality of this God varies in each man, from one country to another: some people shiver with horrour at the sight of those actions which other people look upon as sacred and meritorious. Some see God filled with gentleness and mercy; others judge him to be cruel, and imagine that it is by cruelties they can acquire the advantage of pleasing him.

The morality of nature is clear: it is evident even to those who outrage it. It is not so with religious morality, this is as obscure as the Divinity who prescribes it, or rather as changeable as the passions and the temperaments of those who make him speak, or who adore him. If it were left to the theologians, morality ought to be considered as a science the most problematical, the most uncertain, and the most difficult to fix. It would require the most subtile or the most profound genius, the most penetrating and active mind, to discover the principles of the duties of man towards himself and others. Are not, then, the true sources of morality calculated to be known only to a small number of thinkers or of metaphysicians? To derive it from a God, whom nobody sees but within himself, and which each modifies after his own peculiar ideas, is to submit it to the caprice of each man; to derive it from a being which no man upon the earth can boast of knowing, is to say they do not know whence it could come to us. Whatever may be the agent upon whom they make nature, and all the beings which it contains, depend, whatever power they may suppose him to have, it is very possible that man should or should not exist; but as soon as he shall have made him what he is, when he shall have rendered him sensible, in love with his own being, and living in society, he cannot, without annihilating or new-moulding him, cause him to exist otherwise than he does. According to his actual essence, qualities, and modifications, which constitute him a being of the human species, morality is necessary to him, and the desire of conserving himself will make him prefer virtue to vice, by the same necessity that it makes him prefer pleasure to pain.*

To say that man cannot possess any moral sentiments without the idea of

God, is to say that he cannot distinguish vice from virtue; it is to pretend that, without the idea of God, man would not feel the necessity of eating to live, would not make any distinction or choice in his food: it is to pretend that, without being acquainted with the name, the character, and the qualities of him who prepares a mess for us, we are not in a state to judge whether this mess be agreeable or disagreeable, good or bad. He who does not know what opinion to hold upon the existence and the moral attributes of a God, or who formally denies them, cannot at least doubt his own existence, his own qualities, his own mode of feeling and of judging: neither can he doubt the existence of other organized beings like himself, in whom every thing discovers to him qualities analogous with his own, and of whom he can, by certain actions, attract the love or the hatred, the assistance or the ill-will, the esteem or the contempt: this knowledge is sufficient to enable him to distinguish moral good and evil. In short, every man enjoying a well-ordered organization, or the faculty of making true experience, will only have to contemplate himself, in order to discover what he owes to others: his own nature will enlighten him much better upon his duties than those Gods, in which he can only consult his own passions, or those of some enthusiasts or impostors. He will allow, that to conserve himself, and secure his own permanent wellbeing, he is obliged to resist the impulse, frequently blind, of his own desires; and that to conciliate the benevolence of others, he must act in a mode conformable to their advantage; in reasoning thus, he will find out what virtue is;* if he put this

* According to theology, man has occasion for supernatural grace to do good: this doctrine was, without doubt, very hurtful to sound morality. Men always waited for the *call from above* to do good, and those who governed them never employed the *calls from below*, that is to say, the natural motives to excite them to virtue. Nevertheless, *Tertullian* says to us: "Wherefore will ye trouble yourselves, seeking after the law of God, whilst ye have that which is common to all the world, and which is written on the tablets of nature?"— *Tertull. De Corona Militis.*

* Hitherto theology has not known how to give a true definition of virtue. According to it, it is an effect of grace, that disposes us to do that which is agreeable to the Divinity. But what is the Divinity? What is grace? How does it act upon man? What is that which is agreeable to God? Wherefore does not this God give to all men the grace to do that which is agreeable in his eyes? *Adhuc sub judice lis est.* Men are unceasingly told to do good, because God requires it; never have they been informed what it was to do good, and priests have never been able to tell them what God was, nor that which he was desirous they should do.

theory into practice, he will be virtuous; he will be rewarded for his conduct, by the happy harmony of his machine, by the legitimate esteem of himself, confirmed by the kindness of others: if he act in a contrary mode, the trouble and the disorder of his machine will quickly warn him that nature, whom he thwarts, disapproves his conduct, which is injurious to himself, and he will be obliged to add the condemnation of others, who will hate him and blame his actions. If the wanderings of his mind prevent him from seeing the most immediate consequences of his irregularities, neither will he perceive the distant rewards and punishments of the invisible monarch, whom they have so vainly placed in the empyreum ; this God will never speak to him in so distinct a manner as his conscience, which will either reward him or punish him on the spot.

Every thing that has been advanced, evidently proves, that religious morality is an infinite loser, when compared with the morality of nature, with which it is found in perpetual contradiction. Nature invites man to love himself, to preserve himself, to incessantly augment the sum of his happiness: religion orders him to love only a formidable God, that deserves to be hated ; to detest himself, to sacrifice to his frightful idol the most pleasing and legitimate pleasures of his heart. Nature tells man to consult reason, and to take it for his guide : religion teaches him that his reason is corrupted, that it is only a treacherous guide, given by a deceitful God to lead his creatures astray. Nature tells man to enlighten himself, to search after truth, to instruct himself in his duties: religion enjoins him to examine nothing, to remain in ignorance, to fear truth ; it persuades him, that there are no relations more important than those which subsist between him and a being of whom he will never have any knowledge. Nature tells the being who is in love with his welfare, to moderate his passions, to resist them when they are destructive to himself, to counterbalance them by real motives borrowed from experience: religion tells the sensible being to have no passions, to be an insensible mass, or to combat his propensities by motives borrowed

from the imagination, and variable as itself. Nature tells man to be sociable, to love his fellow-creatures, to be just, peaceable, indulgent, and benevolent, to cause or suffer his associates to enjoy their opinions : religion counsels him to fly society, to detach himself from his fellow-creatures, to hate them, when their imagination does not procure them dreams conformable to his own, to break the most sacred bonds to please his God, to torment, to afflict, to persecute, and to massacre those who will not be mad after his own manner. Nature tells man in society to cherish glory, to labour to render himself estimable, to be active, courageous, and industrious : religion tells him to be humble, abject, pusillanimous, to live in obscurity, to occupy himself with prayers, with meditations, and with ceremonies ; it says to him, be useful to thyself, and do nothing for others.* Nature proposes to the citizen for a model, men endued with honest, noble, energetic souls, who have usefully served their fellow-citizens ; religion commends to them abject souls, extols pious enthusiasts, frantic penitents, fanatics, who, for the most ridiculous opinions, have disturbed empires. Nature tells the husband to be tender, to attach himself to the company of his mate, and to cherish her in his bosom : religion makes a crime of his tenderness, and frequently obliges him to look upon the conjugal bonds as a state of pollution and imperfection. Nature tells the father to cherish his children, and to make them useful members of society: religion tells him to rear them in the fear of God, and to make them blind and superstitious, incapable of serving society, but extremely well calculated to disturb its repose. Nature tells children to honour, to love, to listen to their parents, to be the support of their old age : religion tells them to prefer the oracles of their God, and to trample father and mother under feet, in support of the divine interests. Nature says to the philosopher, occupy

---

* It is very easy to perceive that religious worship does a real injury to political societies, by the loss of time, by the laziness and inaction which it causes, and of which it makes a duty. Indeed, religion suspends the most useful labours during a considerable portion of the year.

thyself with useful objects, consecrate thy cares to thy country, make for it advantageous discoveries, calculated to perfectionate its condition: religion says to him, occupy thyself with useless reveries, with endless disputes, with researches suitable to sow the seeds of discord and carnage, and obstinately maintain opinions, which thou wilt never understand thyself. Nature tells the perverse man to blush for his vices, for his shameful propensities, for his crimes; it shows him, that his most secret irregularities will necessarily have an influence on his own felicity: religion says to the most corrupted and wicked man, "Do not irritate a God, whom thou knowest not; but if, against his laws, thou deliverest thyself up to crime, remember that he will be easily appeased; go into his temple, humiliate thyself at the feet of his ministers, expiate thy transgressions by sacrifices, by offerings, by ceremonies, and by prayers: these important ceremonies will pacify thy conscience, and cleanse thee in the eyes of the Eternal."

The citizen, or the man in society, is not less depraved by religion, which is always in contradiction with sound politics. Nature says to man, *thou art free, no power on earth can legitimately deprive thee of thy rights:* religion cries out to him, that he is a slave, condemned by his God to groan all his life under the iron rod of his representatives. Nature tells man to *love the country which gave him birth,* to serve it faithfully, to blend his interests with it against all those who shall attempt to injure it: religion orders him to obey, without murmuring, the tyrants who oppress his country, to serve them against it, to merit their favours, by enslaving their fellow-citizens, under their unruly caprices. Nevertheless, if the sovereign be not sufficiently devoted to his priests, religion quickly changes its language; it calls upon subjects to become rebels, it makes it a duty in them to resist their master, it cries out to them, that it is better to obey God than man. Nature tells princes they are men; that it is not their whim that can decide what is just, and what is unjust, *that the public will maketh the law:* religion, sometimes says to them, that they are Gods, to whom nothing in this world

ought to offer resistance; sometimes it transforms them into tyrants whom enraged Heaven is desirous should be immolated to its wrath.

Religion corrupts princes; these princes corrupt the law, which, like themselves, becomes unjust; all the institutions are perverted; education forms only men who are base, blinded with prejudices, smitten with vain objects, with riches, with pleasures which they can obtain only by iniquitous means: nature is mistaken, reason is disdained, virtue, is only a chimera, quickly sacrificed to the slightest interest; and religion, far from remedying these evils, to which it has given birth, does no more than aggravate them still farther; or else only causes steril regret, which it quickly effaces; and thus man is obliged to yield to the torrent of habit, of example, of propensities, and of dissipation, which conspire to hurry all his species to commit crimes, who will not renounce their own wellbeing.

Here is the mode in which religion and politics unite their efforts to pervert, abuse, and poison the heart of man; all the human institutions appear to have only for their object to render man base or wicked. Do not, then, let us be at all astonished, if morality is every where only a barren speculation, from which every one is obliged to deviate in practice, if he will not risk the rendering himself unhappy. Men can be moral only when renouncing their prejudices, they consult their nature; but the continual impulses, which their minds are receiving every moment, on the part of more powerful motives, quickly oblige them to forget those rules which nature points out to them. They are continually floating between vice and virtue; we see them unceasingly in contradiction with themselves; if sometimes they feel the value of an honest conduct, experience very soon shows them that this conduct cannot lead them to any thing good, and can even become an invincible obstacle to that happiness which their heart never ceases to search after. In corrupt societies it is necessary to become corrupt, in order to become happy.

Citizens, led astray at the same time both by their spiritual and temporal guides, neither knew reason nor virtue. The slaves of both Gods and men, they had all the vices attached to slavery; kept in a perpetual state of infancy, they had neither knowledge nor principles; those who preached up virtue to them, knew nothing of it themselves, and could not undeceive them with respect to those playthings in which they had learned to make their happiness consist. In vain they cried out to them to stifle those passions which every thing conspired to unloose: in vain they made the thunder of the Gods roll to intimidate men, whom tumultuous passions rendered deaf. It was quickly perceived, that the Gods of heaven were much less feared than those of the earth; that the favours of these procured a much more certain wellbeing than the promises of the others; that the riches of this world were preferable to the treasures which heaven reserved for its favourites; that it was much more advantageous for men to conform themselves to the views of visible powers than to those of powers whom they never saw.

In short, society, corrupted by its chiefs, and guided by their caprices, could only bring forth corrupt children. It gave birth only to avaricious, ambitious, jealous, and dissolute citizens, who never saw any thing happy but crime, who beheld meanness rewarded, incapacity honoured, fortune adored, rapine favoured, and debauchery esteemed; who every where found talents discouraged, virtue neglected, truth proscribed, elevation of soul crushed, justice trodden under feet, moderation languishing in misery, and obliged to groan under the weight of haughty injustice.

In the midst of this disorder, of this confusion of ideas, the precepts of morality could only be vague declamations, incapable of convincing any one. What barrier can religion, with its imaginary motive-powers, oppose to the general corruption? When it spake reason, it was not heard; its Gods were not sufficiently strong to resist the torrent; its menaces could not arrest those hearts which every thing hurried on to evil; its distant promises could not counterbalance present advantages; its expiations, always

ready to cleanse mortals from their iniquities, emboldened them to persevere in crime, its frivolous ceremonies calmed their consciences; in short, its zeal, its disputes, and its whims, only multiplied and exasperated the evils with which society found itself afflicted; in the most vitiated nations, there were a multitude of devotees, and very few honest men. Great and small listened to religion when it appeared favourable to their passions; they listened to it no longer when it counteracted them. Whenever this religion was conformable to morality, it appeared incommodious it was only followed when it combated morality, or totally destroyed it. The despot found it marvellous when it assured him he was a God upon earth; that his subjects were born to adore him alone, and to administer to his phantasms. He neglected religion when it told him to be just: from hence he saw that it was in contradiction with itself, and that it was useless to preach equity to a deified mortal. Besides, he was assured that his God would pardon every thing as soon as he should consent to recur to his priests, always ready to reconcile him. The most wicked subjects reckoned, in the same manner, upon their divine assistance: thus religion, far from restraining them, assured them of impunity; its menaces could not destroy the effects which its unworthy flattery had produced in princes; these same menaces could not annihilate the hopes which its expiations furnished to all. Sovereigns, puffed up with pride, or always certain of expiating their crimes, no longer feared the Gods; become Gods themselves, they believed they were permitted to do any thing against poor pitiful mortals, whom they no longer considered in any other light than as playthings, destined to amuse them on this earth.

If the nature of man were consulted in politics, which supernatural ideas have so shamefully depraved, it would completely rectify the false notions which are entertained equally by sovereigns and subjects: it would contribute, more amply than all the religions in the world. to render society happy, powerful, and flourishing under rational authority. Nature would teach

them, that it is for the purpose of enjoying a greater quantum of happiness
that mortals live together in society;
that it is its own conservation, and its
felicity that every society should have
for its constant and invariable end;
that without equity, a nation only resembles a congregation of enemies;
that the most cruel enemy to man is
he who deceives, in order to enslave
him; that the scourge most to be
feared by him is those priests who
corrupt his chiefs, and who assure
them of impunity for their crimes, in
the name of the Gods. It would prove
to them, that association is a misfortune under unjust, and negligent, and
destructive governments.

This nature, interrogated by princes,
would teach them, that they are men,
and not Gods; that their power is
only derived from the consent of other
men; that they are citizens, charged
by other citizens with the care of
watching over the safety of the whole;
that the law ought to be only the expression of the public will, and that it
is never permitted them to counteract
nature, or to thwart the invariable end
of society. This nature would make
these monarchs feel that, in order to be
truly great and powerful, they ought
to command elevated and virtuous
minds, and not minds equally degraded
by despotism and superstition. This
nature would teach sovereigns that, in
order to be cherished by their subjects,
they ought to afford them succours,
and cause them to enjoy those benefits
which the wants of their nature demand; that they ought to maintain
them inviolably in the possession of
their rights, of which they are the defenders and the guardians. This nature would prove to all those princes
who should deign to consult her, that
it is only by good works and kindness
that they can merit the love and attachment of the people; that oppression
only raises up enemies against them;
that violence procures them only an
unsteady power; that force cannot confer any legitimate right on them; and
that beings essentially in love with
happiness, must sooner or later finish
by revolting against an authority that
only makes itself felt by violence.
This, then, is the manner in which nature the sovereign of all beings, and

to whom all are equal, would speak to
one of those superb monarchs whom
flattery has deified: "Untoward, head-
strong child! Pigmy, so proud of
commanding pigmies! Have they,
then, assured thee that thou wert a
God? Have they told thee that thou
wert something supernatural? But
know, that there is nothing superior to
me. Contemplate thine own insignifi-
cance, acknowledge thine impotence
against the slightest of my blows. I
can break thy sceptre, I can take away
thy life, I can reduce thy throne to
powder, I can dissolve thy people, I
can even destroy the earth, which thou
inhabitest: and thou believest thyself
a God! Be, then, again thyself; honestly avow that thou art a man,
made to submit to my laws, like the
least of thy subjects. Learn, then,
and never let it escape thy memory,
that thou art the man of thy people;
the minister of thy nation; the interpreter and the executor of its will; the
fellow-citizen of those whom thou hast
the right of commanding only because
they consent to obey thee, in view of
the wellbeing which thou promisest
to procure for them. Reign, then, on
these conditions; fulfil thy sacred engagements. Be benevolent, and above
all, equitable. If thou art willing to
have thy power assured to thee, never
abuse it; let it be circumscribed by
the immoveable limits of eternal justice. Be the father of thy people, and
they will cherish thee as thy children.
But if thou neglectest them; if thou
separatest thine interests from those of
thy great family; if thou refusest to
thy subjects the happiness which thou
owest them; if thou armest thyself
against them, thou shalt be like all
tyrants, the slave of gloomy care, of
alarm, and of cruel suspicion. Thou
wilt become the victim of thine own
folly. Thy people, in despair, will no
longer acknowledge thy *divine rights.*
In vain, then, thou wouldst sue for
aid to that religion which has deified
thee; it can avail nothing with those
people whom misery has rendered deaf;
Heaven will abandon thee to the fury
of those enemies which thy phrensy
shall have made thee. The Gods can
effect nothing against my irrevocable decrees, which will, that man shall be irritated against the cause of his sorrows "

In short, every thing would make known to rational princes, that they have no occasion for Heaven to be faithfully obeyed on earth ; that all the powers of Heaven will not sustain them when they shall act the tyrant, that their true friends are those who undeceive the people of their delusion ; that their real enemies are those who intoxicate them with flattery, who harden them in crime, who make the road to heaven too easy for them ; who feed them with chimeras, calculated to draw them aside from those cares and those sentiments which they owe to their nations.*

It is, then, I repeat it, only by reconducting men to nature that we can procure them evident notions, and certain knowledge ; it is only ·by showing them their true relations with each other that we can place them on the road to happiness. The human mind, blinded by theology, has scarcely advanced a single step. Man's religious systems have rendered him dubious of the most demonstrable truths. Superstition influenced every thing, and served to corrupt all. Philosophy, guided by it, was no longer any thing more than imaginary science : it quitted the real world to plunge into the ideal world of metaphysics ; it neglected nature to occupy itself with Gods, with spirits, and with invisible powers, which only served to render all questions more obscure and more complicated. In all difficulties, they brought in the Divinity, and from thence things only became more and more perplexed, until nothing could be explained. Theological notions appear to have been invented only to put man's reasons to flight, to confound his judgment, to deceive his mind, to overturn his clearest ideas of every science. In the hands of the theologians, logic, or the art of reasoning, was nothing more than an unintelligible jargon, calculated to support sophism and falsehood, and to prove the most palpable contradictions. Morality became, as we have seen, uncertain and wavering, because it was founded on an ideal being, who was never in accord with himself;

his goodness, his justice, his moral qualities, and his useful precepts, were continually contradicted by an iniquitous conduct, and the most barbarous commands. Politics, as we have said, were perverted by the false ideas which were given to sovereigns of their rights. Jurisprudence and the laws were subjected to the caprice of religion, who put shackles on the labour, the commerce, the industry, and the activity of nations. Every thing was sacrificed to the interests of the theologians ; for every science, they only taught obscure and quarrelsome metaphysics, which, hundreds of times, caused the blood of those people to flow who were incapable of understanding it.

Born an enemy to experience, theology, that *supernatural* science, was an invincible obstacle to the progress of the natural sciences, as it almost always threw itself in their way. It was not permitted for natural philosophy, for natural history, or for anatomy, to see any thing but through the medium of the jaundiced eye of superstition. The most evident facts were rejected with disdain, and proscribed with horrour, whenever they could not be made to square with the hypotheses of religion.† In short, theology unceasingly opposed itself to the happiness of nations, to the progress of the human mind, to useful researches, and to the liberty of thought : it [1] kept man in ignorance, all his steps guided by it were no more than errours. Is it resolving a question in natural philosophy, to say that an effect which surprises us, that an unusual phenomenon, that a volcano, a deluge, a comet, &c., are signs of divine wrath, or works contrary to the laws of nature ? In persuading nations, as it has done, that the calamities, whether physical

* Ad generum cereris, sine cœde et vulnere pauci.
Descendunt reges et sicca morte tyranni.
           Juvenal, sat. xv. 110.

† Virgil, the bishop of Saltzburg, was condemned by the church, for having dared to maintain the existence of the antipodes. All the world are acquainted with the persecutions which Galileo suffered for pretending that the sun did not make its revolution round the earth. Descartes was put to death in a foreign land. Priests have a right to be enemies to the sciences; the progress of reason will annihilate, sooner or later, superstitious ideas. Nothing that is founded on *nature* and on *truth* can ever be lost; the works of imagination and of imposture must be overturned first or last.

or moral, which they experience, are the effects of the will of God, or chastisements, which his power inflicts on them, is it not preventing them from seeking after remedies for these evils?* Would it not have been more useful to have studied the nature of things, and to seek in nature herself, in human industry, for succours against those sorrows with which mortals are afflicted, than to attribute the evil which man experiences to an unknown power,

* In the year 1725, the city of Paris was afflicted with a scarcity, which it was thought would cause an insurrection of the people; they brought down the shrine of St. Genevieve, the patroness or tutelary goddess of the Parisians, and it was carried in procession to cause this calamity to cease, which was brought on by monopolies in which the mistress of the then prime minister was interested.

In the year 1795, England was afflicted with a scarcity, brought on by an ill judged war against the French people, for having thrown off the tyranny of their monarchy, in which contest immense quantities of grain and other provisions were destroyed, to prevent them falling into the hands of the French republicans, and also by the dismemberment of Poland (the granary of Europe) by the king of Prussia and the emperess of Russia, whose troops laid waste every thing they came near, because a general named Kosciusko, of the most exemplary courage, had, with a chosen body of brave Poles, endeavoured, though vainly, to prevent the cruel injustice, by opposing force to force. This alarming scarcity induced a meeting, at the London Tavern, in London, to consider of the means to alleviate the distresses of the English people, which proved as fruitless as the opposition of the Poles to these crowned robbers. At this meeting, a Doctor Vincent, a Christian priest, and the then master of Westminster school, made a grave and solemn speech, in which he attributed the whole calamity to the chastisement of God for the sins of the people.

The name of this God is always made use of by wicked and abandoned men to cover their own iniquities, and screen themselves from the resentment of the people; the priests, those pests to society, who are immediately interested in their peculations and oppressions, always maintain the doctrine of these designing knaves, and the ignorance of the citizens suffers these fables to pass for incontestable truths: it is thus that *kingcraft* and *priestcraft*, in uniting their forces, always keep men in a state of degrading slavery, never suffering the bandeau of delusion to be removed from before their eyes, by decreeing, in the name of God, the most cruel punishments against those who attempt to throw the light of day on the secret caverns of imposition and despotism.

against whose will it cannot be supposed there is any relief? The study of nature, the search after truth, elevates the mind, expands the genius, and is calculated to render man active and courageous; theological notions appear to have been made to debase him, to contract his mind, to plunge him into despondency.* In the place of attributing to the divine vengeance those wars, those famines, those sterilities, those contagions, and that multitude of calamities which desolate the people, would it not have been more useful, and more consistent with truth, to have shown that these evils were to be ascribed to their own folly, or rather to the passions, to the want of energy, and to the tyranny of their princes, who sacrifice nations to their frightful delirium? These irrational people, instead of amusing themselves with expiations for their pretended crimes, and seeking to render themselves acceptable to imaginary powers, should they not have sought in a more rational administration the true means of avoiding those scourges to which they were the victims? Natural evils demand natural remedies: ought not experience long since to have convinced mortals of the inefficacy of supernatural remedies, of expiations, of prayers, of sacrifices, of fasting, of processions, &c., which all the people of the earth have vainly opposed to the disasters which they experienced?

Let us then conclude, that theology and its notions, far from being useful to the human species, are the true sources of all those sorrows which afflict the earth, of all those errours by which men are blinded, of those prejudices which benumb them, of that ignorance which renders them credulous, of those vices which torment them, of those governments which oppress them. Let us then conclude, that those divine and supernatural ideas with which we are inspired from our infancy, are the true causes of our habitual folly, of our religious quarrels, of our sacred dissensions, of our inhuman persecutions. Let us at length acknowledge, that they are the fatal ideas which have obscured morality, corrupt-

* Non enim aliunde venit animo robur, quam a bonis artibus, quam a contemplatione naturæ. *Senec. quæst. Natur.* lib. vi. chap. xxxii.

ed politics, retarded the progress of the sciences, and even annihilated happiness and peace in the heart of man. Let it then be no longer dissimulated, that all those calamities, for which man turns his eyes towards heaven, bathed in tears, are to be ascribed to those vain phantoms which his imagination has placed there; let him cease to implore them; let him seek in nature, and in his own energy, those resources which the Gods, who are deaf to his cries, will never procure for him. Let him consult the desires of his heart, and he will find that which he owes to himself, and that which he owes to others; let him examine the essence and the aim of society, and he will no longer be a slave; let him consult experience, he will find truth, and he will acknowledge that errour can never possibly render him happy.*

———

CHAPTER VIII.

Men can form no Conclusion from the Ideas which are given them of the Divinity: of the want of Just Interference in, and of the Inutility of, their Conduct on his Account.

If, as we have proved, the false ideas which men have in all times formed to themselves of the Divinity, far from being of utility, are prejudicial to morality, to politics, to the happiness of society, and the members who compose it; in short, to the progress of the human understanding; reason and our interest ought to make us feel the necessity of banishing from our mind these vain and futile opinions, which will never do more than confound it, and disturb the tranquillity of our hearts. In vain should we flatter ourselves with arriving at the rectification

* The author of the book of wisdom, has said, and with reason, infandorum enim idolorum cultura, omnis mali est causa et initium et finis. SEE CHAP. XXIV. VER. 27. He did not see that his God was an idol more prejudicial than all the others. At all events, it appears that the dangers of superstition have been felt by all those who have sincerely taken to heart the interest of the human species. This, without doubt, is the reason why philosophy, which is the fruit of reflection, was almost always at open war with religion, which, as we have shown, is itself the fruit of ignorance, of imposture, of enthusiasm, and of imagination.

of theological notions; false in their principles, they are not susceptible of reform. Under whatever shape an errour presents itself, as soon as men shall attach a great importance to it, it will end, sooner or later, by producing consequences as extensive as dangerous. Besides, the inutility of the researches which in all ages have been made after the Divinity, of whom the notions have never had any other effect than to obscure him more and more, even for those themselves who have most meditated upon him; this inutility, I say, ought it not to convince us, that these notions are not within the reach of our capacity, and that this imaginary being will not be better known by us, or by our descendants, than it has been by our ancestors, either the most savage or the most ignorant? The object which men in all ages have the most considered, reasoned upon the most, and written upon the most, remains, nevertheless, the least known; nay, time has only rendered it more impossible to be conceived. If God be such as modern theology depicts him, he must be himself a God who is capable of forming an idea of him.* We know little of man, we hardly know ourselves and our own faculties, and we are disposed to reason upon a being inaccessible to all our senses! Let us, then, travel in peace over the line described for us by nature, without diverging from it, to run after chimeras; let us occupy ourselves with our true happiness; let us profit by the benefits which are spread before us; let us labour to multiply them, by diminishing the number of our errours; let us submit to those evils which we cannot avoid; and do not let us augment them by filling our mind with prejudices calculated to lead it astray. When we shall reflect on it, every thing will clearly prove that the pretended science of God, is, in truth, nothing but a presumptuous ignorance, masked under pompous and unintelligible words. In short, let us terminate unfruitful researches; let us, at least, acknowledge

* A modern poet has composed a piece of poetry, that received the sanction of the French academy, upon the *attributes of God*, in which the following line was particularly applauded:—

"*To say what he is, 'twere need to be himself.*"

our invincible ignorance; it will be more advantageous to us than an arrogant science, which hitherto has done nothing more than sow discord on the earth and affliction in our hearts.

In supposing a sovereign intelligence, who governs the world; in supposing a God, who exacts from his creatures that they should know him, that they should be convinced of his existence, of his wisdom, of his power, and who is desirous they should render him homage, it must be allowed, that no man on earth completely fulfils in this respect the views of Providence. Indeed, nothing is more demonstrable than the impossibility in which the theologians find themselves to form to their mind any idea whatever of their Divinity.* The weakness and the obscurity of the proofs which they give of his existence; the contradictions into which they fall; the sophisms and the begging of the question which they employ, evidently prove that they are very frequently in the greatest incertitude upon the nature of the being with whom it is their profession to occupy themselves. But, granting that they have a knowledge of him, that his existence, his essence, and his attributes were so fully demonstrated to them as not to leave one doubt in their mind, do the rest of human beings enjoy the same advantage? Ingenuously, how many persons will be found in the world who have the leisure, the capacity, and the penetration necessary to understand what is meant to be designated under the name of an immaterial being, of a pure spirit, who moves matter, without being matter himself; who is the motive-power of nature, without being contained in nature, and without being able to touch it? Are there, in the most religious societies, many persons who are in a state to follow their spiritual guides in those subtile proofs which they give them of the existence of the God which they make them adore?

* Procopius, the first bishop of the Goths, says, in a very solemn manner: "I esteem it a very foolish temerity to be disposed to penetrate into the knowledge of the nature of God." And farther on he acknowledges, that he "has nothing more to say of him, except that he is perfectly good. He who knoweth more, whether he be ecclesiastic or layman, has only to tell it."

Very few men, without doubt, are capable of a profound and connected meditation; the exercise of thought is, for the greater part, a labour as painful as it is unusual. The people, obliged to toil hard in order to subsist, are commonly incapable of reflection. Nobles, men of the world, women, and young people, occupied with their own affairs, with the care of gratifying their passions, of procuring themselves pleasure, think as rarely as the uninformed. There are not, perhaps, two men in a hundred thousand, who have seriously asked themselves the question, what it is they understand by the word *God?* whilst it is extremely rare to find persons to whom the existence of God is a problem: nevertheless, as we have said, conviction supposes that evidence which can alone procure certitude to the mind. Where, then, are the men who are convinced of the existence of their God? Who are those in whom we shall find the complete certitude of this pretended truth, so important to all? Who are the persons who have given themselves an accurate account of the ideas which they have formed to themselves upon the Divinity, upon his attributes, and upon his essence? Alas! I see in the whole world only some speculators, who, by dint of occupying themselves with him, have foolishly believed they have discovered something in the confused and unconnected wanderings of their imagination; they have endeavoured to form a whole, which, chimerical as it is, they have accustomed themselves to consider as really existing: by dint of musing upon it, they have sometimes persuaded themselves they saw it distinctly, and they have succeeded in making others believe it, who have not mused upon it quite so much as themselves.

It is only upon hearsay that the mass of the people adore the God of their fathers and their priests: authority, confidence, submission, and habit, take place of conviction and proofs; they prostrate themselves, and pray, because their fathers have taught them to fall down and worship; but wherefore have these fallen upon their knees? It is because, in times far-distant, their legislators and their guides have imposed it on them us a duty. "Adore ·

and believe," have they been told, "those Gods, whom ye cannot comprehend; yield yourselves in this respect to our profound wisdom; we know more than you about the Divinity." But wherefore should I take this matter on your authority? It is because God wills it thus; it is because God will punish you, if you dare resist. But is not this God the thing in question? And yet, men have always satisfied themselves with this circle of errours; the idleness of their mind made them find it more easy to yield themselves to the judgment of others. All religious notions are uniformly founded on authority; all the religions of the world forbid examination, and are not disposed that men should reason upon them; it is authority that wills they should believe in God; this God is himself founded solely upon the authority of some men, who pretend to have a knowledge of him, and to be sent to announce him to the earth. A God made by men, has, without doubt, occasion for men to make him known to men.*

Is it not, then, for the priests, the

---

* Men are always as credulous as children upon those objects which relate to religion; as they comprehend nothing about it, and are nevertheless told that they must believe it, they imagine they run no risk in joining sentiments with their priests, whom they suppose to have been able to discover that which they do not themselves understand. The most rational people say to themselves, *What shall I do? what interest can so many people have to deceive?* I say to them, they do deceive you, either because they are themselves deceived, or because they have a great interest in deceiving you.

By the confession of the theologians themselves, men are without *religion:* they have only *superstition.* Superstition, according to them, is a *worship of the Divinity, badly understood and irrational,* or else, *a worship rendered to a false Divinity.* But where are the people or the clergy, who will allow that their Divinity is false, and their worship irrational? How shall it be decided, who is right or who is wrong? It is evident, that in this affair, all men are equally wrong. Indeed, Buddæus, in his *Treatise on Atheism,* tells us: "In order that a religion may be true, not only the object of the worship must be true, but we must also have a just idea of it. He, then, who adores God, without knowing him, adores him in a perverse and corrupt manner, and is guilty of superstition." This granted, could it not be demanded of all the theologians in the world, if they can boast of having a *just idea,* or a real knowledge of the Divinity?

inspired, and the metaphysicians, that the conviction of the existence of a God would be reserved, which is nevertheless said to be so necessary for the whole human species? But shall we find any harmony among the theological notions of the different inspired men, or those thinkers who are scattered over the earth? Those themselves, who make a profession of adoring the same God, are they in accord with respect to him? Are they contented with the proofs which their colleagues bring of his existence? Do they unanimously subscribe to the ideas which they present upon his nature, upon his conduct, upon the manner of understanding his various oracles? Is there one country on earth where the science of God is really perfectionated? Has this science obtained any degree of that consistency and uniformity which we see attached to human knowledge, in the most futile arts, or in those trades which are most despised? The words *spirit, immateriality, creation, predestination, grace;* this multitude of subtile distinctions with which theology is throughout filled in some countries; these inventions, so ingeniously imagined by those thinkers who have succeeded each other during so many ages, have done no more, alas! than perplex things; and hitherto the science the most necessary to man, has never been able to acquire the least degree of stability. For thousands of years past, these idle dreamers have been relieving each other to meditate on the Divinity, to divine his concealed ways, to invent hypotheses suitable to develop this important enigma. Their slender success has not at all discouraged theological vanity; they have always spoken of God; they have disputed, they have cut each others' throats for him; and this sublime being nevertheless remains the most unknown and the most examined.*

---

* If things were coolly examined, it would be acknowledged that religion is by no means formed for the greater part of mankind, who are utterly incapable of comprehending any of those aerial subtilties upon which it rests. Who is the man that understands any thing of the fundamental principles of his religion; of the spirituality of God; of the imateriality of the soul; of the mysteries of which he is told every day? Are there many people who

Men would have been too happy, if, confining themselves to those visible objects which interest them, they had employed, in perfectionating the real sciences, the laws, the morals, and their education, half those efforts which they have wasted in their researches after the Divinity. They had been also much wiser, and more fortunate, if they had agreed to let their idle and unemployed guides quarrel between themselves, and fathom those depths calculated to stun and amaze them without intermeddling with their irrational disputes. (But it is the essence of ignorance to attach importance to every thing it does not understand.) Human vanity makes the mind bear up against difficulties. The more an object eludes our inquiry, the more efforts we make to compass it, because, from thence, our pride is spurred on, our curiosity is irritated, and it appears interesting to us. On the other hand, the longer and more laborious our researches have been, the more importance we attach to our real or pretended discoveries, the more we are desirous not to have lost our time; besides, we are always ready to defend warmly the soundness of our judgment. Do not let us, then, be surprised at the interest which ignorant people have at all times taken in the discoveries of their priests; nor at the obstinacy which these have always manifested in their disputes. Indeed, in combating for his God, each fought only for the interests of his own vanity, which, of all human passions, is the most quickly alarmed, and the most suitable to produce very great follies.

If, throwing aside for a moment the fatal ideas which theology gives us of a capricious God, whose partial and despotic decrees decide the condition of human beings, we would only fix our eyes upon his pretended goodness, which all men, even when trembling before this God, agree to give him : if we suppose him to have in view what

they have ascribed to him; to have laboured only to his own glory; to exact the homage of intelligent beings; to seek in all his works only the well-being of the human species: how can we reconcile all this with the ignorance, truly invincible, in which this God, so glorious and so good, leaves the greater part of mankind with respect to him ? (If God is desirous to be known, cherished, and thanked, wherefore does he not show himself, under favourable traits, to all those intelligent beings, by whom he would be loved and adored ? Wherefore does he not manifest himself to all the earth in an unequivocal manner, much more likely to convince us than those particular revelations which appear to accuse the Divinity of a fatal partiality for some of his creatures ? Has the omnipotent no better means of showing himself to men than those ridiculous metamorphoses, those pretended incarnations, which are attested by writers so little in harmony with each other ? Instead of such a number of miracles, invented to prove the divine mission of so many legislators held in reverence by the different people of the world, could not the sovereign of minds have convinced at once the human mind of those things with which he was desirous it should be acquainted ? In the room of suspending a sun in the vaulted firmament; in lieu of diffusing without order the stars and constellations, which fill up the regions of space, would it not have been more conformable to the views of a God so jealous of his glory, and so well-intentioned towards man, to have written, in a manner not liable to dispute, his name, his attributes, his everlasting will, in indelible characters, and equally legible to all the inhabitants of the earth ?* No one, then, could have doubted the existence of a God, of his manifest will, of his visible intentions; no mortal would have dared

* I foresee that the theologians will oppose to this passage, their cœli enarrant gloriam Dei. But we shall reply to them, that the heavens prove nothing, except the power of nature, the immutability of its laws, the power of attraction, of repulsion, of gravitation, the energy of matter; and that the heavens in no way announce the existence of an immaterial cause, of a God who is in contradiction with himself, and who can never do that which he wishes to do.

can boast of perfectly understanding the state of the question in those theological speculations, which have frequently the power of disturbing the repose of mankind ? Nevertheless, even women believe themselves obliged to take a part in the quarrels excited by idle speculators, who are of less utility to society than the meanest artisan.

to place himself in a situation to attract his wrath; in short, no man would have had the audacity to have imposed on men in his name, or to have interpreted his will, according to his own whim and caprice.

Theology is truly the vessel of the Danaides. By dint of contradictory qualities and bold assertions, it has so shackled its God, as to make it impossible for him to act. Indeed, when even we should suppose the existence of the theological God, and the reality of those attributes, so discordant, which are given him, we can conclude nothing from them to authorize the conduct or sanction the worship which they prescribed. If God be infinitely good, what reason have we to fear him? If he be infinitely wise, wherefore disturb ourselves with our condition? If he be omniscient, wherefore inform him of our wants, and fatigue him with our prayers? If he be omnipresent, wherefore erect temples to him? If he be Lord of all, wherefore make sacrifices and offerings to him? If he be just, wherefore believe that he punishes those creatures whom he has filled with imbecility? If his grace works every thing in man, what reason has he to reward him? If he be omnipotent, how can he be offended; and how can we resist him? If he be rational, how can he be enraged against those blind mortals to whom he has left the liberty of acting irrationally? If he be immutable, by what right shall we pretend to make him change his decrees? If he be inconceivable, wherefore should we occupy ourselves with him? If he has spoken, wherefore is the universe not convinced? If the knowledge of a God be the most necessary thing, wherefore is it not more evident and more manifest?

But, on the other hand, the theological God has two faces. Nevertheless, if he be wrathful, jealous, vindictive, and wicked, as theology supposes him, to be, without being disposed to allow it, we shall no longer be justified in addressing our prayers to him, nor in sorrowfully occupying ourselves with his idea. On the contrary, for our present happiness, and for our quiet, we ought to make a point of banishing him from our thought; we ought to place him in the rank of those neces-

sary evils, which are only aggravated by a consideration of them. Indeed, if God be a tyrant, how is it possible to love him? Are not affection and tenderness sentiments incompatible with habitual fear? How could we experience love for a master who gives to his slaves the liberty of offending him, to the end that he may take them on their weak side, and punish them with the utmost barbarity? If to this odious character, God has joined omnipotence; if he hold in his hands the unhappy playthings of this fantastic cruelty, what can we conclude from it? Nothing; save that, whatever efforts we may make to escape our destiny, we shall always be incapacitated to withdraw ourselves from it. If a God, cruel or wicked by his nature, be armed with infinite power, and take pleasure in rendering us eternally miserable, nothing will divert him from it; his wickedness will always pursue its course; his malice would, without doubt, prevent him from paying any attention to our cries; nothing would be able to soften his obdurate heart.

Thus, under whatever point of view we contemplate the theological God, we have no worship to render him, no prayers to offer up to him. If he be perfectly good, intelligent, equitable, and wise, what have we to ask of him? If he be supremely wicked, if he be gratuitously cruel, as all men believe, without daring to avow it, our evils are without remedy; such a God would deride our prayers, and, sooner or later, we should be obliged to submit to the rigour of the lot which he has destined for us.

This granted, he who can undeceive himself with regard to the afflicting notions of the Divinity, has this advantage over the credulous and trembling superstitious mortal, that he establishes in his heart a momentary tranquillity, which, at least, renders him happy in this life. If the study of nature has banished from him those chimeras with which the superstitious man is infested, he enjoys a security of which this one is himself deprived. In consulting nature, his fears are dissipated; his opinions, true or false, become steady; and a calm succeeds the storm which panic terrours and wavering notions excite in the hearts of all men who occupy themselves with the Divinity. If the human soul, cheered by philosophy, had the boldness to consider things coolly, it would no longer behold the universe governed by an implacable tyrant, always ready to strike. If he were rational, he would see that, in committing evil, he did not disturb nature ; that he did not outrage his author ; he injures himself alone, or he injures other beings, capable of feeling the effects of his conduct ; from thence, he knows the line of his duties ; he prefers virtue to vice, and for his own permanent repose, satisfaction, and felicity in this world, he feels himself interested in the practice of virtue, in rendering it habitual to his heart, in avoiding vice, in detesting crime, during the whole time of his abode amongst intelligent and sensible beings, from whom he expects his happiness. By attaching himself to these rules, he will live contented with himself, and be cherished by those who shall be capable of experiencing the influence of his actions ; he will expect, without inquietude, the term when his existence shall have a period ; he will have no reason to dread the existence which shall follow the one he at present enjoys ; he will not fear to be deceived in his reasonings ; guided by demonstration and honesty, he will perceive, that, if contrary to his expectation, there did exist a good God, he would not punish him for his involuntary errours, depending upon the organization he should have received.

Indeed, if there did exist a God ; if God were a being full of reason, equity, and goodness, and not a ferocious, irrational, and malicious genius, such as religion is pleased so frequently to depict him ; what could a virtuous atheist have to apprehend, who, believing at the moment of his death he falls asleep for ever, should find himself in the presence of a God whom he should have mistaken and neglected during his life ?

"O, God!" would he say, "father, who hast rendered thyself invisible to thy child ! inconceivable and hidden author, whom I could not discover ! pardon me, if my limited understanding has not been able to know thee in a nature where every thing has appeared to me to be necessary ! Excuse

me, if my sensible heart has not discerned thine august traits under those of the austere tyrant whom superstitious mortals tremblingly adore. I could only see a phantom in that assemblage of irreconcilable qualities, with which the imagination has clothed thee. How should my coarse eyes perceive thee in a nature in which all my senses have never been able to know but material beings and perishable forms ? Could I, by the aid of these senses, discover thy spiritual essence, of which they could not furnish any proof ? How should I find the invariable demonstration of thy goodness in thy works, which I saw as frequently prejudicial as favourable to the beings of my species ? My feeble brain, obliged to form its judgments after its own capacity, could it judge of thy plan, of thy wisdom, of thine intelligence, whilst the universe presented to me only a continued mixture of order and confusion, of good and of evil, of formation and destruction ? Have I been able to render homage to thy justice, whilst I so frequently saw crime triumphant and virtue in tears ? Could I acknowledge the voice of a being filled with wisdom, in those ambiguous, contradictory, and puerile oracles which impostors published in thy name, in the different countries of the earth which I have quitted ? If I have refused to believe thine existence, it is because I have not known, either what thou couldst be, or where thou couldst be placed, or the qualities which could be assigned to thee. My ignorance is excusable, because it was invincible : my mind could not bend itself under the authority of some men, who acknowledged themselves as little enlightened upon thine essence as myself, and who, for ever disputing amongst themselves, were in harmony only in imperiously crying out to me to sacrifice to them that reason which thou hast given men. But, O God ! if thou cherishest thy creatures, I also have cherished them like thee ; I have endeavoured to render them happy in the sphere in which I have lived. If thou art the author of reason, I have always listened to it, and followed it ; if virtue please thee, my heart has always honoured it ; I have never outraged it ; and, when my powers have permitted me, I have myself practised it; I was an affectionate husband, a tender father, a sincere friend, a faithful and zealous citizen. I have held out consolation to the afflicted: if the foibles of my nature have been injurious to myself, or incommodious to others, I have not, at least, made the unfortunate groan under the weight of my injustice; I have not devoured the substance of the poor; I have not seen without pity the widow's tears; I have not heard without commiseration the cries of the orphan. If thou didst render man sociable, if thou wast disposed that society should subsist and be happy, I have been the enemy of all those who oppressed him, or deceived him, in order that they might take advantage of his misfortunes.

"If I have thought amiss of thee, it is because my understanding could not conceive thee; if I have spoken ill of thee, it is because my heart, partaking too much of human nature, revolted against the odious portrait which was painted of thee. My wanderings have been the effect of a temperament which thou hast given me; of the circumstances in which, without my consent, thou hast placed me; of those ideas which, in despite of me, have entered into my mind. If thou art good and just, as we are assured thou art, thou canst not punish me for the wanderings of my imagination, for faults caused by my passions, which are the necessary consequence of the organization which I have received from thee. Thus, I cannot fear thee, I cannot dread the condition which thou preparest for me. Thy goodness cannot have permitted that I should incur punishments for inevitable errours. Wherefore didst thou not rather prevent my being born, than have called me into the rank of intelligent beings, there to enjoy the fatal liberty of rendering myself unhappy? If thou punishest me with severity, and eternally, for having listened to the reason which thou gavest me; if thou correctest me for my illusions; if thou art wroth, because my feebleness has made me fall into those snares which thou hast every where spread for me; thou wilt be the most cruel and the most unjust of tyrants; thou wilt not be a God, but a malicious demon, to whom I shall be obliged to yield, and

satiate the barbarity; but of whom I shall at least congratulate myself to have for some time shook off the insupportable yoke."

It is thus that a disciple of nature would speak, who, transported all at once into the imaginary regions, should there find a God, of whom all the ideas were in direct contradiction to those which wisdom, goodness, and justice furnish us here. Indeed, theology appears to have been invented only to overturn in our mind all natural ideas. This illusory science seems to be bent on making its God a being the most contradictory to human reason. It is, nevertheless, according to this reason that we are obliged to judge in this world; if in the other, nothing is conformable to this, nothing is of more inutility than to think of it, or reason upon it. Besides, wherefore leave it to the judgment of men, who are themselves only enabled to judge like us?

However, in supposing God the author of all, nothing is more ridiculous than the idea of pleasing him, or irritating him by our actions, our thoughts, or our words; nothing is more inconclusive than to imagine that man, the work of his hands, can have merits or demerits with respect to him. It is evident that he cannot injure an omnipotent being, supremely happy by his essence. It is evident that he cannot displease him, who has made him what he is: his passions, his desires, and his propensities, are the necessary consequence of the organization which he has received; the motives which determine his will towards good or evil, are evidently due to qualities inherent to the beings which God places around him. If it be an intelligent being who has placed us in the circumstances in which we are, who has given the properties to those causes which, in acting upon us, modify our will, how can we offend him? If I have a tender, sensible, and compassionate soul, it is because I have received from God organs easily moved, from whence results a lively imagination, which education has cultivated. If I am insensible and cruel, it is because he has given me only refractory organs, from whence results an imagination of little feeling, and a heart difficult to be touched. If I profess a religion, it is because I have received it from parents, from whom it did not depend upon me to receive my birth, who professed it before me, whose authority, example, and instructions, have obliged my mind to conform itself to theirs. If I am incredulous, it is because, little susceptible of fear or enthusiasm for unknown objects, my circumstances have so ordered it, that I should undeceive myself of the chimeras with which I had occupied myself in my infancy.

It is, then, for want of reflecting on his principles, that the theologian tells us that man can please or displease the powerful God who has formed him. Those who believe they have merited well, or deserved punishment of their God, imagine that this being will be obliged to them for the organization which he has himself given them, and will punish them for that which he has refused them. In consequence of this idea, so extravagant, the affectionate and tender devotee flatters himself he shall be recompensed for the warmth of his imagination. The zealous devotee doubts not that his God will some day reward him for the acrimony of his bile or the heat of his blood. Penitent, frantic, and atrabilious beings, imagine that God will keep a register of those follies which their vicious organization or their fanaticism make them commit; and, above all, will be extremely contented with their melancholy humour, the gravity of their countenance, and their antipathy to pleasure. Devotees, zealous, obstinate, and quarrelsome beings, cannot persuade themselves that their God, which they always form after their own model, can be favourable to those who are more phlegmatic, who have less bile in their composition, or have a cooler blood circulating through their veins. Each mortal believes his own organization is the best, and the most conformable to that of his God.

What strange ideas must these blind mortals have of their Divinity, who imagine that the absolute master of all can be offended with the motions which take place in their body or in their mind! What contradiction, to think that his unalterable happiness can be disturbed, or his plan deranged, by the transitory shocks which the impercep-

tible fibres of the brain of one of his creatures experience. Theology gives us very feeble ideas of a God, of whom, however, it is unceasingly exalting the power, the greatness, and the goodness.

Without a very marked derangement of our organs, our sentiments hardly ever vary upon those objects which our senses, experience, and reason have clearly demonstrated to us. In whatever circumstances we are found, we have no doubt either upon the whiteness of snow, the light of day, or the utility of virtue. It is not so with those objects which depend solely on our imagination, and which are not proved to us by the constant evidence of our senses; we judge of them variously, according to the disposition in which we find ourselves. These dispositions vary by reason of the involuntary impressions which our organs receive at each instant on the part of an infinity of causes, either exterior to us, or contained within our own machine. These organs are, without our knowledge, perpetually modified, relaxed, or bent, by the greater or less weight or elasticity in the air; by heat or cold, by dryness or humidity, by health or sickness, by the heat of the blood, by the abundance of the bile, by the state of the nervous system, &c. These different causes necessarily have an influence on the momentary ideas, thoughts, and opinions of man. He is, consequently, obliged to see variously those objects which his imagination presents to him, without being able to be corrected by experience and memory. Here is the reason why man is obliged continually to see his God and his religious chimeras under different aspects. In a moment when his fibres find themselves disposed to tremble, he will be cowardly and pusillanimous, he will think of this God only with trembling; in a moment when these same fibres shall be more firm, he will contemplate this same God with more coolness. The theologian, or the priest, will call his pusillanimity, *inward feeling, warning from Heaven, secret inspiration;* but he who knows man, will say that this is nothing but a mechanical motion, produced by a physical or natural cause. Indeed, it is by a pure physical mechanism that we can explain all the revolutions which take place frequently from one moment to another in the systems, in all the opinions, and in all the judgments of men: in consequence, we see them sometimes reasoning justly, and sometimes irrationally.

Here is the mode by which, without recurring to grace, to inspirations, to visions, and to supernatural movements, we can render ourselves an account of that uncertain and wavering state into which we sometimes see persons fall, otherwise extremely enlightened, when there is a question of religion. Frequently, in despite of all reasoning, momentary dispositions reconduct them to the prejudices of their infancy, from which on other occasions they appear to be completely undeceived. These changes are very marked, especially in infirmities and sickness, and at the approach of death. The barometer of the understanding is then frequently obliged to fall. Those chimeras which they despised, or which, in a state of health, they set down at their true value, are then realized. They tremble, because the machine is enfeebled; they are irrational, because the brain is incapable of exactly fulfilling its functions. It is evident that these are the true chances which the priests have the knavery to make use of against incredulity, and from which they draw proofs of the reality of their sublime opinions. Those *conversions*, or those changes, which take place in the ideas of men, have always their origin in some physical derangement of their machine, brought on by chagrin, or by some natural and known cause.

Subjected to the continual influence of physical causes, our systems, then, always follow the variations of our body; we reason well when our body is healthy and well-constituted; we reason badly when this body is deranged; from thence our ideas disconnect themselves, we are no longer capable of associating them with precision, of finding our principles, to draw from them just inferences; the brain is shaken, and we no longer see any thing under its true point of view. Such a man does not see his God, in frosty weather, under the same traits as in cloudy and rainy weather: he

does not contemplate him in the same manner in sorrow as in gayety, when in company as when alone. Good sense suggests to us, that it is when the body is sound, and the mind undisturbed by any mist, that we can reason with precision; this state can furnish us with a general standard suitable to regulate our judgments, and even rectify our ideas, when unexpected causes shall make them waver.

If the opinions of the same individual upon his God are wavering and subject to vary, how many changes must they experience in the various beings who compose the human race? If there do not, perhaps, exist two men who see a physical object exactly under the same point of view, what much greater variety must they not have in their modes of contemplating those things which have existence only in their imagination? What an infinity of combinations of ideas must not minds, essentially different, make to themselves, to compose an ideal being, which each moment of life must present under a different form? It would then be an irrational enterprise to attempt to prescribe to men what they ought to think of religion and of God, which are entirely under the cognizance of the imagination, and for which, as we have very frequently repeated, mortals will never have any common standard. To combat the religious opinions of men, is to combat with their imagination, with their organization, and with their habits, which suffice to identify with their existence the most absurd and the least founded ideas. The more imagination men have, the greater enthusiasts will they be in matters of religion, and reason will be less capable of undeceiving them of their chimeras: these chimeras will become a food necessary for their ardent imagination. In fine, to combat the religious notions of men, is to combat the passion which they have for the marvellous. In despite of reason, those persons who have a lively imagination, are perpetually reconducted to those chimeras which habit render dear to them, even when they are troublesome and fatal. Thus, a tender soul has occasion for a God that loves him; the happy enthusiast needs a God who rewards him: the unfortunate enthusiast wants a God, who takes part in his sorrows; the melancholy devotee has occasion for a God who chagrins him, and who maintains him in that trouble which has become necessary to his diseased organization; the frantic penitent needs a cruel God, who imposes on him an obligation to be inhuman towards himself; whilst the furious fanatic would believe himself unhappy if he were deprived of a God who orders him to make others experience the effects of his inflamed humours and of his unruly passions.

He is, without question, a less dangerous enthusiast who feeds himself with agreeable illusions, than he whose soul is tormented by odious spectres. If a virtuous and tender mind does not commit ravages in society, a mind agitated by incommodious passions, cannot fail to become, sooner or later, troublesome to his fellow-creatures. The God of a Socrates, or of a Fenelon, may be suitable to minds as gentle as theirs; but he cannot be the God of a whole nation, in which it will always be extremely rare to find men of their temper. The Divinity, as we have frequently said, will always be for the greater portion of mortals a frightful chimera, calculated to disturb their brain, to set their passions afloat, and to render them injurious to their associates. If honest men only see their God as filled with goodness, vicious, restless, inflexible, and wicked men, will give to their God their own character, and will authorize themselves, from this example, to give a free course to their own passions. Each man can see his chimera only with his own eyes; and the number of those who will paint the Divinity as hideous, afflicting, and cruel, will be always greater and more to be feared, than those who describe him under seducing colours; for one mortal whom this chimera can render happy, there will be thousands which it will make miserable; it will be, sooner or later, an inexhaustible source of divisions, of extravagancies, and of madness; it will disturb the mind of the ignorant, over whom impostors and fanatics will always have an influence; it will frighten the cowardly and the pusillanimous, whom their weakness will incline to perfidy and cruelty; it will

make the most honest tremble, who, even while practising virtue, will fear the displeasure of a fantastical and capricious God; it will not stop the progress of the wicked, who will put it aside, in order to deliver themselves up to crime; or who will even avail themselves of this divine chimera to justify their transgressions. In short, in the hands of tyrants, this God, who is himself a tyrant, will only serve to crush the liberty of the people, and violate, with impunity, the rights of equity. In the hands of priests, this God will be a talisman, suitable to intoxicate, blind, and subjugate equally the sovereign and the subject; in fine, in the hands of the people, this idol will always be a two-edged weapon, with which they will give themselves the most mortal wounds.

On the other hand, the theological God, being, as we have seen, only a heap of contradictions; being represented, in despite of his immutability, sometimes as goodness itself, sometimes as the most cruel and the most unjust of beings; being besides contemplated by men, whose machines experience continual variations; this God, I say, cannot at all times appear the same to those who occupy themselves with him. Those who form the most favourable ideas of him are frequently obliged to acknowledge that the portrait, which they paint to themselves, is not always conformable to the original. The most fervent devotees, the most prepossessed enthusiasts cannot prevent themselves from seeing the traits of their Divinity change; and if they were capable of reasoning, they would feel the want of just inference in the conduct which they unceasingly hold with respect to him. Indeed, would they not see, that his conduct appeared to contradict, every moment, the marvellous perfections which they assign to their God? To pray to the Divinity, is it not doubting of his wisdom, of his benevolence, of his providence, of his omniscience, and of his immutability? Is it not to accuse him of neglecting his creatures, and to ask him to alter the eternal decree of his justice, to change those invariable laws which he has himself determined? To pray to God, is it not to say to him: "O, my

God, I acknowledge your wisdom, your omniscience, and your infinite goodness; nevertheless, you forget me; you lose sight of your creature; you are ignorant, or you feign ignorance of that which he wants; do you not see that I suffer from the marvellous arrangement which your wise laws have made in the universe? Nature, against your commands, actually renders mine existence painful; change, then, I pray you, the essence which your will has given to all beings. See that the elements, at this moment, lose in my favour their distinguishing properties; order it so, that heavy bodies shall not fall, that fire shall not burn, that the brittle frame which I have received from you shall not suffer those shocks which it experiences every instant. Rectify, for my happiness, the plan which your infinite prudence has marked out from all eternity." Such are very nearly the prayers which men form; such are the ridiculous demands which they every moment make to the Divinity, of whom they extol the wisdom, the intelligence, the providence, and the equity, whilst they are hardly ever contented with the effects of his divine perfections.

Men are not more consistent in the thanksgivings which they believe themselves obliged to offer him. Is it not just, say they, to thank the Divinity for his kindness? Would it not be the height of ingratitude to refuse our homage to the author of our existence, and of every thing that contributes to render it agreeable? But I shall say to them, then your God acts from interest; similar to men, who, even when they are the most disinterested, expect at least that we should give them proofs of the impression which their kindness makes upon us. Your God, so powerful, and so great, has he occasion that you should prove to him the sentiments of your acknowledgments? Besides, upon what do you found this gratitude? Does he distribute his benefits equally to all men? Are the greater number among them contented with their condition? you yourself, are you always satisfied with your existence? It will be answered me, without doubt, that this existence alone is the greatest of all benefits. But how can we look upon it as a signal advantage? This

existence, is it not in the necessary order of things? Has it not necessarily entered into the unknown plan of your God? Does the stone owe any thing to the architect for having judged it necessary to his building? Do you know better than this stone the concealed views of your God? If you are a thinking and sensible being, do you not find that this marvellous plan incommodes you every instant; do not even your prayers to the architect of the world prove that you are discontented? You were born without your consent; your existence is precarious; you suffer against your will; your pleasures and your sorrows do not depend upon you; you are not master of any thing; you have not the smallest conception of the plan formed by the architect of the universe whom you never cease to admire, and in which, without your consent, you find yourself placed; you are the continual sport of the necessity which you deify: after having called you into life, your God obliges you to quit it. Where, then, are those great obligations which you believe you owe to Providence? This same God, who gives you the breath of life, who furnishes you your wants, who conserves you, does he not in a moment ravish from you these pretended advantages? If you consider existence as the greatest of all benefits, is not the loss of this existence, according to yourself, the greatest of all evils? If death and sorrow are formidable evils, do not this grief and death efface the benefit of existence, and the pleasure that may sometimes accompany it? If your birth and your funeral, your enjoyments and your sorrows, have equally entered into the views of his providence, I see nothing that can authorize you to thank him. What can be the obligations which you have to a master who, in despite of you, obliges you to enter into this world, there to play a dangerous and unequal game, by which you may gain or lose an eternal happiness?

They speak to us, indeed, of another life, where we are assured that man will be completely happy. But in supposing for a moment the existence of this other life, which has as little foundation as that of the being from whom it is expected, it were necessary, at least, for man to suspend his acknowledgment until he enter into this other life: in the life of which we have a knowledge, men are much more frequently discontented than fortunate; if God, in the world which we occupy, has not been able or willing to permit that his beloved creatures might be perfectly happy, how shall we assure ourselves that he will have the power or the disposition to render them in the end more happy than they are now? They will then cite to us the revelations, the formal promises of the Divinity, who engages to compensate his favourites for the sorrows of the present life. Let us, for an instant, admit the authenticity of these promises; do not these revelations themselves teach us that the divine goodness reserves eternal punishments for the greater number of men? If these menaces be true, do mortals, then, owe acknowledgments to a God who, without consulting them, only gives them their existence, that they may, with the assistance of their pretended liberty, run the risk of rendering themselves eternally miserable? Would it not have been more beneficial for them not to have existed, or at least to have existed only like stones or brutes, from whom it is supposed God exacts nothing, than to enjoy those extolled faculties—the privilege of having merits or demerits—which may conduct intelligent beings to the most frightful misfortunes? In paying attention to the small number of the elect, and to the great number of the condemned, where is the man of feeling who, if he had been the master, had consented to run the risk of eternal damnation?

Thus, under whatever point of view we contemplate the theological phantom, men, if they were consistent, even in their errours, neither owe him prayers, nor homage, nor worship, nor thanksgivings. But in matters of religion, mortals never reason; they only follow the impulse of their fears, of their imagination, of their temperament, of their peculiar passions, or those guides who have acquired the right of controuling their understandings. Fear has made Gods; terrour unceasingly accompanies them; it is impossible to reason when we tremble. Thus men will never reason when

there shall be a question of those objects of which the vague idea will ever be associated to that of terrour. If a mild and honest enthusiast sees his God only as a beneficent father, the greater portion of mortals will only view him as a formidable sultan, a disagreeable tyrant, and a cruel and perverse genius. Thus, this God will always be for the human race a dangerous leaven, suitable to imbitter it, and put it into a fatal fermentation. If, to the peaceable, humane, and moderate devotee, could be left the good God which he has formed to himself after his own heart, the interest of the human race demands that an idol should be overthrown to which fear has given birth, which is nourished by melancholy, of whom the idea and the name are only calculated to fill the universe with carnage and with follies.

We do not, however, flatter ourselves that reason will be at all once capable of delivering the human race from those errours with which so many causes united have conspired to poison it. The vainest of all projects would be the expectation of curing in an instant those epidemical and hereditary errours, rooted during so many ages, and continually fed and corroborated by the ignorance, the passions, the customs, the interests, the fears, and the calamities of nations, always regenerating. The ancient revolutions of the earth have brought forth its first Gods, new revolutions would produce new ones, if the old ones should chance to be forgotten. Ignorant, miserable, and trembling beings, will always form to themselves Gods, or else their credulity will make them receive those which imposture or fanaticism shall announce to them.

Then do not let us propose more to ourselves than to hold reason to those who may be able to understand it; to present truth to those who can sustain its lustre; to undeceive those who shall not be inclined to oppose obstacles to demonstration, and who will not obstinately persist in errour. Let us infuse courage into those who have not the power to break with their illusions. Let us cheer the honest man who is much more alarmed by his fears than the wicked, who, in despite of his opinions, always follows his passions; let us console the unfortunate, who groans under a load of prejudices, which he has not examined; let us dissipate the incertitude of him who doubts, and who, ingenuously seeking after truth, finds in philosophy itself only wavering opinions, little calculated to fix his mind. Let us banish from the man of genius the chimera which makes him waste his time: let us wrest his gloomy phantom from the intimidated mortal, who, duped by his own fears, becomes useless to society: let us remove from the atrabilarious being a God who afflicts him, who exasperates him, and who does nothing more than kindle his anger: let us tear from the fanatic the God who arms him with poniards; let us pluck from impostors and from tyrants a God who serves them to terrify, enslave, and despoil, the human species. In removing from honest men their formidable notions, let us not encourage the wicked, the enemies of society; let us deprive them of those resources upon which they reckon to expiate their transgressions; to uncertain and distant terrours, which cannot stop their excesses, let us substitute those which are real and present; let them blush at seeing themselves what they are; let them tremble at finding their conspiracies discovered; let them have the fear of one day seeing those mortals whom they abuse, cured of the errours of which they avail themselves to enslave them.

If we cannot cure nations of their inveterate prejudices, let us endeavour, at least, to prevent them from again falling into those excesses into which religion has so frequently hurried them: let men form to themselves chimeras; let them think of them as they will, provided their reverles do not make them forget they are men, and that a sociable being is not made to resemble ferocious animals. Let us balance the fictitious interests of heaven, by the sensible interests of the earth. Let sovereigns, and the people, at length acknowledge that the advantages resulting from truth, from justice, from good laws, from a rational education, and from a human and peaceable morality, are much more solid than those which they so vainly expect from their Divinities: let them feel that benefits so real and so precious ought not to be

sacrificed to uncertain hopes, so frequently contradicted by experience. In order to convince themselves, let every rational man consider the numberless crimes which the name of God has caused upon the earth; let them study his frightful history, and that of his odious ministers, who have every where fanned the spirit of madness, discord, and fury. Let princes, and subjects at least, sometimes learn to resist the passions of these pretended interpreters of the Divinity, especially when they shall command them in his name to be inhuman, intolerant, and barbarous; to stifle the cries of nature, the voice of equity, the remonstrances of reason, and to shut their eyes to the interests of society.

Feeble mortals! how long will your imagination, so active and so prompt to seize on the marvellous, continue to seek out of the universe, pretexts to make you injurious to yourselves, and to the beings with whom ye live in society? Wherefore do ye not follow in peace the simple and easy route which your nature has marked out for ye? Wherefore strew with thorns the road of life? Wherefore multiply those sorrows to which your destiny exposes ye? What advantages can ye expect from a Divinity which the united efforts of the whole human species have not been able to make you acquainted with? Be ignorant, then, of that which the human mind is not formed to comprehend; abandon your chimeras; occupy yourselves with truth; learn the art of living happy; perfection your morals, your governments, and your laws; look to education, to agriculture, and to the sciences that are truly useful; labour with ardour; oblige nature by your industry to become propitious to ye, and the Gods will not be able to oppose any thing to your felicity. Leave to idle thinkers, and to useless enthusiasts, the unfruitful labour of fathoming depths from which ye ought to divert your attention: enjoy the benefits attached to your present existence; augment the number of them; never throw yourselves forward beyond your sphere. If you must have chimeras, permit your fellow-creatures to have theirs also; and do not cut the throats of your brethren, when they cannot rave in your own manner. If ye will have Gods, let your imagina-- tion give birth to them; but do not suffer these imaginary beings so far to intoxicate ye as to make ye mistake that which ye owe to those real beings with whom ye live. If ye will have unintelligible systems, if ye cannot be contented without marvellous doctrines, if the infirmities of your nature require an invisible crutch, adopt such as may suit with your humour; select those which you may think most calculated to support your tottering frame, do not insist on your neighbours making the same choice with yourself: but do not suffer these imaginary theories to infuriate your mind: always remember that, among the duties you owe to the *real* beings with whom ye are associated, the foremost, the most consequential, the most immediate, stands a reasonable indulgence for the foibles of others.   •

CHAPTER IX.

Defence of the Sentiments contained in this Work. Of Impiety. Do there exist Atheists?

WHAT has been said, in the course of this work, ought to be sufficient to undeceive those men who are capable of reasoning on the prejudices to which they attach so much importance. But the most evident truths must prove abortive against enthusiasm, habit, and fear; nothing is more difficult than to destroy errour, when long prescription has given it possession of the human mind. It is unassailable when it is supported by general consent, propagated by education, when it has grown inveterate by custom, when it is fortified by example, maintained by authority, and unceasingly nourished by the hopes and the fears of the people, who look upon their errours as a remedy for their sorrows. Such are the united forces which sustain the empire of the Gods in this world, and which appear to render their throne firm and immoveable.

We need not, then, be surprised, to see the greater number of men cherish their own blindness, and fear the truth. Every where we find mortals obstinately attached to phantoms, from

which they expect their happiness, notwithstanding these phantoms are evidently the source of all their sorrows. Smitten with the marvellous, disdaining that which is simple and easy to be comprehended, but little instructed in the ways of nature, accustomed to neglect the use of their reason, the uninformed, from age to age, prostrate themselves before those invisible powers which they have been taught to adore. They address their most fervent prayers to them, they implore them in their misfortunes, they despoil themselves for them of the fruits of their labour, they are unceasingly occupied with thanking these vain idols for benefits which they have not received, or in demanding of them favours which they cannot obtain. Neither experience nor reflection can undeceive them; they do not perceive that their Gods have always been deaf; they ascribe to it their own conduct; they believe them to be irritated; they tremble, they groan, and they sigh at their feet; they strew their altars with presents; they do not see that these beings, so powerful, are subjected to nature, and are never propitious but when this nature is favourable. It is thus that nations are the accomplices of those who deceive them, and are as much opposed to truth, as those who lead them astray.

In matters of religion, there are very few persons who do not partake, more or less, of the opinions of the uninformed. Every man who throws aside the received ideas, is generally looked upon as a madman, a presumptuous being, who insolently believes himself much wiser than others. At the magical names of religion and the Divinity, a sudden and panic terrour takes possession of men's minds; and as soon as they see them attacked, society is alarmed, each imagines that he already sees the celestial monarch lift his avenging arm against the country where rebellious nature has produced a monster, with sufficient temerity to brave his wrath. Even the most moderate persons tax the man with folly and sedition who dares to contest, with this imaginary sovereign, those rights which good sense has never examined. In consequence, whoever undertakes to tear the veil of prejudice, appears an irrational being, and a dangerous citizen; his sentence is pronounced with a voice almost unanimous; the public indignation, stirred up by fanaticism and imposture, renders it impossible for him to be heard; every one believes himself culpable if he does not display his fury against him, and his zeal in favour of a terrible God, whose anger is supposed to be provoked. Thus, the man who consults his reason, the disciple of nature, is looked upon as a public pest; the enemy of an injurious phantom is regarded as the enemy of the human species; he who would establish a lasting peace amongst men, is treated as the disturber of society; they unanimously proscribe him who should be disposed to cheer affrighted mortals by breaking those idols under which prejudice has obliged them to tremble. At the bare name of an atheist, the superstitious man quakes, and the deist himself is alarmed; the priest enters the judgment-seat with fury, tyranny prepares his funeral pile; the uninformed applaud those punishments which irrational laws decree against the true friend of the human species.

Such are the sentiments which every man must expect to excite who shall dare to present to his fellow-creatures that truth which all appear to be in search of, but which all fear to find, or else mistake when we are disposed to show it to them. Indeed, what is an atheist? He is a man, who destroys chimeras prejudicial to the human species, in order to reconduct men back to nature, to experience, and to reason. He is a thinker, who, having meditated upon matter, its energy, its properties, and its modes of acting, has no occasion, in order to explain the phenomena of the universe, and the operations of nature, to invent ideal powers, imaginary intelligences, beings of the imagination, who, far from making him understand this nature better, do no more than render it capricious, inexplicable, unintelligible, and useless to the happiness of mankind.

Thus, the only men who can have simple and true ideas of nature, are considered as absurd or knavish speculators. Those who form to themselves intelligible notions of the mo-

tive-power of the universe, are accused of denying the existence of this power: those who found every thing that is operated in this world, upon constant and certain laws, are accused of *attributing every thing to chance*; they are taxed with blindness and delirium by those enthusiasts whose imagination, always wandering in a vacuum attributes the effects of nature to fictitious causes, which have no existence but in their own brain; to beings of the imagination, to chimerical powers, which they obstinately persist in preferring to real and known causes. No man, in his proper senses, can deny the energy of nature, or the existence of a power, by virtue of which matter acts and puts itself in motion; but no man can, without renouncing his reason, attribute this power to a being placed out of nature, distinguished from matter, having nothing in common with it. Is it not saying that this power does not exist, to pretend that it resides in an unknown being, formed by a heap of unintelligible qualities, of incompatible attributes, from whence necessarily results a whole impossible to have existence? The indestructible elements, the atoms of Epicurus, of which the motion, the meeting, and the combination, have produced all beings, are, without doubt, causes much more real than the theological God. Thus, to speak precisely, they are the partisans of an imaginary and contradictory being, impossible to be conceived, which the human mind cannot compass on any side, who offer us nothing but a vague name, of which nothing can be affirmed; they are those, I say, who make of such a being the creator, the author, the preserver of the universe, who are irrational. Are not those dreamers, who are incapable of attaching any one positive idea to the cause of which they are unceasingly speaking, *true atheists?* Are not those thinkers, who make a pure nothing the source of all the beings, truly blind men? Is it not the height of folly to personify abstractions, or negative ideas, and then to prostrate ourselves before the fiction of our own brain?

Nevertheless, they are men of this temper who regulate the opinions of the world, and who hold out to public scorn and vengeance, those who are more rational than themselves. If you will believe but these profound dreamers, there is nothing short of madness and phrensy that can reject in nature a motive-power, totally incomprehensible. Is it, then, delirium to prefer the known to the unknown? Is it a crime to consult experience, to call in the evidence of our senses, in the examination of the thing the most important to be known? Is it a horrid outrage, to address ourselves to reason: to prefer its oracles to the sublime decisions of some sophists, who themselves acknowledge that they do not comprehend any thing of the God 'whom they announce to us? Nevertheless, according to them, there is no crime more worthy of punishment, there is no enterprise more dangerous against society, than to despoil the phantom, which they know nothing about, of those inconceivable qualities, with which imagination, ignorance, fear, and imposture, have emulated each other in surrounding him; there is nothing more impious and more criminal than to cheer up mortals against a spectre, of which the idea alone has been the source of all their sorrows; there is nothing more necessary, than to exterminate those audacious beings, who have sufficient temerity to attempt to break an invisible charm, which keeps the human species benumbed in errour; —to be disposed to break man's chains, was to rend asunder his most sacred bonds.

In consequence of these clamours, perpetually renovated by imposture, and repeated by ignorance, those nations, which reason, in all ages, has sought to undeceive, have never dared to listen to her benevolent lessons. The friends of mankind were never listened to, because they were the enemies of their chimeras. Thus, the people continue to tremble; very few philosophers have the courage to cheer them; scarcely any person dares brave public opinion, infected by superstition; they dread the power of imposture, and the menaces of tyranny, which always seek to support themselves by illusions. The yell of triumphant ignorance, and haughty fanaticism, at all times stifled the feeble voice of nature;

she was obliged to keep silence, her lessons were quickly forgotten, and when she dared to speak, it was frequently only in an enigmatical language, unintelligible to the greater number of men. How, should the uninformed, who with difficulty compass truths the most evident and the most distinctly announced, have been able to comprehend the mysteries of nature, presented under half words and emblems?

In contemplating the outrageous language which is excited among the theologians, by the opinions of the atheists, and the punishments which at their instigation were frequently decreed against them; should we not be authorized to conclude, that these doctors either are not so certain as they say they are of the existence of their God, or else that they do not consider the opinions of their adversaries to be quite so absurd as they pretend? It is always distrust, weakness, and fear, that render men cruel; they have no anger against those whom they despise: they do not look upon folly as a punishable crime: we should be content with laughing at an irrational mortal, who should deny the existence of the sun; we should not punish him, if we were not irrational ourselves. This theological fury never proves more than the weakness of its cause; the inhumanity of these interested men, whose profession it is to announce chimeras to nations, proves to us, that they alone have an interest in these invisible powers, of whom they successfully avail themselves to terrify mortals.* They are, however, tyrants of the mind, who, but little consistent with their own principles, undo with one hand, that which they rear with the other: they are those, who after having made a Divinity, filled with goodness, wisdom, and equity, traduce, disgrace, and completely annihilate him, by saying, that he is cruel, that he is capricious, unjust, and despotic, that he thirsts after the blood of the unhappy. This granted, these men are truly impious.

* Lucian describes Jupiter, who, disputing with Menippus, is disposed to strike him down with thunder; upon which the philosopher says to him: " Ah! thou waxeth wroth, thou uscst thy thunder! then thou art in the wrong."

He who knows not the Divinity, cannot do him an injury, nor consequently, be called impious. " *To be impious,*" says Epicurus, "*is not to take away from the uninformed the Gods which they have, it is to attribute to these Gods the opinions of the uninformed.*" To be impious, is to insult a God in whom we believe; it is to knowingly outrage him. To be impious, is to admit a good God, whilst at the same time we preach persecution and carnage. To be impious, is to deceive men, in the name of a God, whom we make use of as a pretext for our unworthy passions. To be impious, is to say, that a God, who is supremely happy and omnipotent, can be offended by his feeble creatures. To be impious is to speak falsely on the part of a God whom we suppose to be the enemy of falsehood. In fine, to be impious, is to make use of the Divinity, to disturb society, to enslave them to tyrants; it is to persuade them, that the cause of imposture is the cause of God; it is to impute to God those crimes which would annihilate his divine perfections. To be impious and irrational at the same time, is to make a mere chimera of the God whom we adore.

On the other hand, to be pious, is to serve our country; it is to be useful to our fellow-creatures; to labour to their wellbeing: every one can put in his claim to it, according to his faculties; he who meditates, can render himself useful, when he has the courage to announce truth, to combat errour, to attack those prejudices which every where oppose themselves to the happiness of mankind; it is to be truly useful, and it is even a duty, to wrest from the hands of mortals, those weapons which fanaticism distributes to them, to deprive imposture and tyranny of that fatal empire of opinion, of which they successfully avail themselves at all times and in all places, to elevate themselves upon the ruins of liberty, security, and public felicity. To be truly pious, is to religiously observe the wholesome laws of nature, and to follow faithfully those duties which she prescribes to us; to be pious, is to be humane, equitable, and benevolent; is to respect the *rights of men.* To be pious and rational, is to reject those

reveries, which would lead us to mistake the sober councils of reason.

Thus, whatever fanaticism and imposture may say, he who denies the existence of a God, seeing that it has no other foundation than an alarmed imagination; he who rejects a God perpetually in contradiction with himself; he who banishes from his mind, and his heart, a God continually wrestling with nature, reason, and the happiness of men; he, I say, who undeceives himself on so dangerous a chimera, may be reputed pious, honest, and virtuous, when his conduct shall not deviate from those invariable rules, which nature and reason prescribe to him. Because a man refuses to admit a contradictory God, as well as the obscure oracles which are given out in his name, does it then follow, that such a man, refuses to acknowledge the evident and demonstrable laws of a nature upon which he depends, of which he experiences the power, of which he is obliged to fulfil the necessary duties, under pain of being punished in this world? It is true, that if virtue, by chance, consisted in an ignominious renunciation of reason, in a destructive fanaticism, in useless customs, the atheist could not pass for a virtuous being; but if virtue consist in doing to society all the good of which we are capable, the atheist may lay claim to it; his courageous and tender heart will not be guilty for hurling his legitimate indignation against prejudices, fatal to the happiness of the human species.

Let us listen, however, to the imputations which the theologians lay upon the atheists; let us coolly and without peevishness examine the calumnies which they vomit forth against them: it appears to them that atheism is the highest degree of delirium that can assail the mind, the greatest stretch of perversity that can inflict the human heart; interested in blackening their adversaries, they make absolute incredulity appear to be the effect of crime or folly. We do not, say they to us, see those men fall into the horrours of atheism, who have reason to hope that the future state will be for them a state of happiness. In short, according to our theologians, it is the interest of their passions which makes them seek to doubt the existence of a being, to whom they are accountable for the abuses of this life; it is the fear of punishment alone which is known to atheists; they are unceasingly repeating the words of a Hebrew prophet, who pretends that nothing but folly makes men deny the existence of the Divinity.* If you believe some others, "nothing is blacker than the heart of an atheist, nothing is more false than his mind." "Atheism," according to them, "can only be the offspring of a tortured conscience, that seeks to disengage itself from the cause of its trouble." "We have a right," says Derham, "to look upon an atheist as a monster amongst rational beings, as one of those extraordinary productions which we hardly ever meet with in the whole human species, and who opposing himself to all other men, revolts not only against reason and human nature, but against the Divinity himself."

We shall simply reply to all these calumnies, by saying, that is for the reader to judge if the system of atheism be as absurd as these profound speculators, perpetually in dispute on the uninformed, contradictory, and fantastical productions of their own brain, would have it believed to be?† It is true, perhaps, that hitherto the system of naturalism has not been developed in all its extent; unprejudiced persons will, at least, be enabled to know whether the author has reasoned well or ill, whether he has disguised the most important difficulties, whether he has been disingenuous, whether, like unto the enemies of human reason, he has had recourse to subterfuges, to sophisms, and to subtile distinctions, which ought

---

* Dexit insipiens in corde suo non est Deus. In taking away the negation, the proposition would be nearer truth. Those who shall be disposed to see the abuse which theological spleen knows how to scatter upon atheists, have only to read a work of Doctor Bentley, entitled, *The Folly of Atheism*: it is translated into Latin, in octavo.

† In seeing the theologians so frequently accuse the atheists with being absurd, we should be tempted to believe that they have no idea of that which the atheists have to oppose to them; it is true, they have established an excellent method; the priests say and publish what they please, whilst their adversaries can never defend themselves.

always to make it be suspected of those who use them, either that they do not know, or that they fear the truth. It belongs, then, to candour, to disinterestedness, and to reason, to judge whether the natural principles, which have been here brought forward, be destitute of foundation; it is to these upright judges, that a disciple of nature submits his opinions; he has a right to except against the judgment of enthusiasm, of presumptuous ignorance, and interested knavery. Those persons who are accustomed to think, will, at least, find reasons to doubt many of those marvellous notions, which appear as incontestable truths, only to those who have never examined them by the standard of good sense.

We agree with Derham that atheists are rare; superstition has so disfigured nature, and its rights; enthusiasm has so dazzled the human mind; terrour has so disturbed the hearts of men; imposture and tyranny have so enslaved thought; in fine, errour, ignorance, and delirium, have so perplexed and entangled the clearest ideas, that nothing is more uncommon, than to find men who have sufficient courage to undeceive themselves of notions, which every thing conspires to identify with their existence. Indeed, many theologians, in despite of those invectives with which they attempt to overwhelm atheists, appear frequently to have doubted whether any existed in the world, or if there were persons who could honestly deny the existence of a God.* Their uncertainty was, with-

---

* Those same persons, who at the present day discover atheism to be such a strange system, admit there could have been atheists formerly. Is it, then, that nature has endued us with a less portion of reason than she did men of other times? Or should it be that the God of the present day would be less absurd than the Gods of antiquity? Has the human species then acquired information, with respect to this concealed motive-power of nature? Is the God of modern mythology, rejected by Vanini, Hobbes, Spinosa, and some others, more to be credited than the Gods of the pagan mythology, rejected by Epicurus, Strato, Theodorus, Diagoras, &c. &c.? Tertullian pretended that Christianity had dissipated that ignorance in which the pagans were immersed, respecting the divine essence, and that there was not an artisan among the Christians who did not see God, and who did not know him. Nevertheless, Tertullian himself admitted a corporeal God, and was there-

out doubt, founded upon the absurd ideas which they ascribe to their adversaries, whom they have unceasingly accused of attributing every thing to *chance*, to *blind* causes, to *dead* and *inert* matter, incapable of acting by itself. We have, I think, sufficiently justified the partisans of nature, from these ridiculous accusations; we have, throughout the whole, proved, and we repeat it, that *chance* is a word devoid of sense, which, as well as the word *God*, announces nothing but an ignorance of true causes. We have demonstrated that matter is not dead; that nature, essentially active, and selfexistent, had sufficient energy to produce all the beings which it contains, and all the phenomena which we behold. We have, throughout, proved, that this cause was much more real, and more easy to be conceived than the fictitious, contradictory, inconceivable, and impossible cause, to which theology ascribes the honour of those great effects which it admires. We have made it evident, that the incomprehensibility of natural effects was not a sufficient reason for assigning them a cause, still more incomprehensible than all those of which we can have a knowledge. In fine, if the incomprehensibility of God does not authorize us to deny his existence, it is at least certain that the incompatibility of the attributes which they accord to him, authorizes us to deny that the being who unites them can be any thing more than a chimera, of which the existence is impossible.

This granted, we shall be able to fix the sense that ought to be attached to the name of *atheist*, which, notwithstanding, the theologians, lavish indiscriminately upon all those who deviate in any thing from their revered opinion. If by *atheist*, be designated a man who denies the existence of a power inherent in matter, and without which we cannot conceive nature, and if it be to this power that the name of *God* is given, there do not exist any atheists, and the word under which they are designated would only announce fools: but, if by *atheists*, be understood men without enthusiasm,

fore an atheist, according to the notions of modern theology.—*See the note to chap. iv. of this volume, page 237.*

guided by experience, and the evidence of their senses, who see nothing in nature but that which they find really have existence, or that which they are capacitated to know; who do not perceive, and cannot perceive, any thing but matter, essentially active and moveable, diversely combined, enjoying from itself various properties, and capable of producing all the beings which display themselves to our visual faculties if by *atheists*, be understood, natural philosophers, who are convinced that, without recurring to a chimerical cause, they can explain every thing simply by the laws of motion, by the relations subsisting between beings, by their affinities, their analogies, their attraction, and their repulsion; by their proportions, their composition, and their decomposition:* if by *atheists* be understood those persons who do not know what a *spirit* is, and who do not see the necessity of *spiritualizing*, or of rendering incomprehensible those corporeal, sensible, and natural causes, which they see act uniformly; who do not find that to separate the motivepower from the universe, to give it to a being placed out of the great whole, to a being of an essence totally inconceivable, and whose abode cannot be shown, is a means of becoming better acquainted with it: if, by *atheists*, be understood those men who ingenuously allow that their mind cannot conceive nor reconcile the negative attributes, and the theological abstractions, with the human and moral qualities,

* Dr. Cudworth, in his *Systema Intellectuale*, chap. ii. reckons four species of atheists among the ancients: 1st, The disciples of Anaximander, called Hylopathians, who attributed the formation of every thing to matter, destitute of feeling. 2d, The atomists, or the disciples of Democritus, who attributed every thing to the concurrence of atoms. 3d, The stoical atheists, who admitted a blind nature, but acting under certain laws. 4th, The Hylozoists, or the disciples of Strato, who attributed life to matter. It is well to observe, that the most learned natural philosophers of antiquity have been atheists, either openly or secretly; but their doctrine was always opposed by the superstition of the uninformed, and almost totally eclipsed by the fanatical and marvellous philosophy of Pythagoras, and above all by that of Plato. So true it is, that enthusiasm, and that which is vague and obscure, commonly prevail over that which is simple, natural, and intelligible. --See *Le Clerc's Select Pieces*, vol. ii.

which are attributed to the Divinity; or those men, who pretend that from this incompatible alliance, there can only result an imaginary being, seeing that a pure spirit is destitute of the organs necessary to exercise the qualities and faculties of human nature: if by *atheists* be designated those men who reject a phantom, of whom the odious and discordant qualities are calculated only to disturb the human species, and plunge it into very prejudicial follies: if, I say, thinkers of this sort, are those who are called *atheists*, it is not possible to doubt of their existence; and there would be found a considerable number of them, if the lights of sound natural philosophy, and of just reason, were more generally diffused; from thence they would neither be considered as irrational, nor as furious beings, but as men devoid of prejudice, of whom the opinions, or, if they will, the ignorance, would be much more useful to the human species, than those sciences, and those vain hypotheses, which have so long been the true causes of all man's sorrows.

On the other hand, if by *atheists*, it is wished to designate those men who are themselves obliged to avow that they have no one idea of the chimera whom they adore, or which they announce to others; who cannot render themselves an account, either of the nature, or of the essence of their deified phantom; who can never agree amongst themselves, upon the proofs of the existence of their God, of his qualities, or of his mode of action; who, by dint of negations, have made him a pure *nothing;* who prostrate themselves, or cause others to fall prostrate, before the absurd fictions of their own delirium; if, I say, by *atheists*, be designated men of this kind, we shall be obliged to allow that the world is filled with atheists; and we shall even be obliged to place in this number the most active theologians who are unceasingly reasoning upon that which they do not understand; who are disputing upon a being of whom they cannot demonstrate the existence; who by their contradictions very efficaciously undermine his existence; who annihilate their perfect good being, by the numberless imperfections which they ascribe to him; who rebel

No. X.—39

against this God, by the atrocious character under which they depict him. In short, we shall be able to consider, as true atheists, those credulous people, who, upon hearsay, and from tradition, fall upon their knees before a being of whom they have no other ideas, than those which are furnished them by their spiritual guides, who themselves acknowledge that they comprehend nothing about the matter. An atheist is a man who does not believe the existence of a God; now, no one can be certain of the existence of a being whom he does not conceive, and who is said to unite incompatible qualities.

What has been said, proves that the theologians themselves, have not always known the sense which they would attach to the word *atheist ;* they have vaguely calumniated and combated them as persons, whose sentiments and principles were opposed to their own. Indeed, we find that these sublime doctors, always infatuated with their own particular opinions, have frequently been lavish in their accusations of atheism, against all those whom they were disposed to injure and to blacken, and whose systems they sought to render odious : they were certain of alarming the uninformed and the silly, by vague imputation, or by a word to which ignorance attaches an idea of terrour, because they have no knowledge of its true sense. In consequence of this policy, we have frequently seen the partisans of the same religious sect, the adorers of the same God, reciprocally treat each other as atheist, in the heat of their theological quarrels : to be an atheist, in this sense, is not to have, in every point, exactly the same opinions as those with whom we dispute upon religion. In all times, the uninformed have considered those as atheists, who did not think of the Divinity, precisely in the same manner as the guides whom they were accustomed to follow. Socrates, the adorer of a unique God, was no more than an atheist in the eyes of the Athenian people.

Still more, as we have already observed, those persons have frequently been accused of atheism, who have taken the greatest pains to establish the existence of a God, but who have not produced satisfactory proofs of it

When on a similar subject the proofs were frail and perishable, it was easy for their enemies to make them pass for atheists, who have wickedly betrayed the cause of the Divinity by defending him too feebly. I shall here stop, to show what little foundation there is, for that which is said to be an evident truth, whilst it is so frequently attempted to be proved, and yet can never be verified, even to the satisfaction of those who boast so much of being intimately convinced of it; at least, it is certain, that in examining the principles of those who have essayed to prove the existence of God, they have been generally found weak or false, because they could not be either solid or true; the theologians themselves, have been obliged to discover, that their adversaries could draw from them inductions quite contrary to those notions, which they have an interest in maintaining; in consequence, they have been frequently very highly incensed, against those who believed they had discovered the most forcible proofs of the existence of their God; they did not perceive, that it was impossible not to lay themselves open to attack in establishing principles, or systems, visibly founded upon an imaginary and contradictory being, which each man sees variously.*

In a word, all those who have taken the cause of the theological God in

* What can we think of the sentiments of a man who expresses himself like Paschal, in the eighth article of his thoughts, wherein he discovers a most complete incertitude upon the existence of God? "I have examined," says he, "if this God, of whom all the world speak, might not have left some marks of himself. I look every where, and every where I see nothing but obscurity. Nature offers me nothing, that may not be a matter of doubt and inquietude. If I saw nothing in nature which indicated a Divinity, I should determine with myself to believe nothing about it. If I every where saw the sign of a creator, I should repose myself in peace, in the belief of one. But seeing too much to deny, and too little to assure me of his existence, I am in a situation that I lament, and in which I have a hundred times wished, that if a God does sustain nature, he would give unequivocal marks of it, and that if the signs which he has given be deceitful, that he would suppress them entirely: that he said all or nothing, to the end that I might see which side I ought to follow." Here is the state of a good mind, wrestling with the prejudices that enslave it.

hand, with the most vigour, have been taxed with atheism and irreligion; his most zealous partisans have been looked upon as deserters and traitors; the most religious theologians have not been able to guaranty themselves from this reproach; they have mutually lavished it on each other, and all have, without doubt, merited it, if by atheists be designated those men who have not any idea of their God which does not destroy itself, as soon as they are willing to submit it to the touchstone of reason.†

CHAPTER X.

Is Atheism compatible with Morality?

AFTER having proved the existence of atheists, let us return to the calumnies which are lavished upon them, by the deicolists. "An atheist," according to Abbadie, "cannot be virtuous; to him virtue is only a chimera, probity no more than a vain scruple, honesty nothing but foolishness. He knows no other law than his interest; where this sentiment prevails, conscience is only a prejudice, the law of nature only an illusion, right no more than errour; benevolence has no longer any foundation; the bonds of society are loosened; fidelity is removed; the friend is ready to betray his friend; the citizen to deliver up his country; the son to assassinate his father in order to enjoy his inheritance, whenever he shall find an occasion, and that au-

† Whence we may conclude that errour will not stand the test of investigation—that it will not pass the ordeal of comparison—that it is in its hues a perfect chameleon, that consequently it can never do more than lead to the most absurd deductions. Indeed, the most ingenious systems, when they have their foundations in hallucination, crumble like dust under the rude hand of the essayer: the most sublimated doctrines, when they lack the substantive quality of rectitude, evaporate under the scrutiny of the sturdy examiner who tries them in the crucible. It is not, therefore, by levelling abusive language against those who investigate sophisticated theories, that they will either be purged of their absurdities, acquire solidity, or find an establishment to give them perpetuity. In short, moral obliquities can never be made rectilinear by the mere application of unintelligible terms, or by the inconsiderate jumble of discrepant properties, however gaudy the assemblage.

thority or silence, will shield him from the arm of the secular power, which alone is to be feared. The most inviolable rights, and the most sacred laws, must no longer be considered, but as dreams and visions."[*]

Such, perhaps, would be the conduct, not of a thinking, feeling, and reflecting being, susceptible of reason, but of a ferocious brute, of an irrational creature, who should not have any idea of the natural relations which subsist between beings necessary to their reciprocal happiness. Can it be supposed, that a man, capable of experience, furnished with the faintest glimmerings of good sense, would lend himself to the conduct which is here ascribed to the atheist, that is to say, to a man, who is sufficiently susceptible of reflection to undeceive himself by reasoning upon those prejudices, which every thing strives to show him as important and sacred? Can it, I say, be supposed, that there is, in any polished society, a citizen sufficiently blind not to acknowledge his most natural duties, his dearest interests, the danger which he runs in disturbing his fellow-creatures, or in following no other rule than his momentary appetites? A being, who reasons the least in the world, is he not obliged to feel that society is advantageous to him, that he has need of assistance, that the esteem of his fellow-creatures is necessary to his happiness, that he has every thing to fear from the wrath of his associates, that the laws menace whoever dare infringe them? Every man, who has received a virtuous education, who has in his infancy experienced the tender cares of a father, who has in consequence tasted the sweetness of friendship, who has received kindness, who knows the value of benevolence and equity, who feels the pleasure which the affection of our fellow-creatures procures for us, and the inconveniences which result from their aversion and their contempt, is he not obliged to tremble at losing such manifest advantages, and at incurring by his conduct such visible dangers? Will not the hatred, the fear, the contempt of himself, disturb his repose, every time

that, turning inwardly upon his own conduct, he shall contemplate himself with the same eyes as others? Is there, then, no remorse, but for those who believe in a God? The idea of being seen by a being of whom we have at best very vague notions, is it more forcible, than the idea of being seen by men, of being seen by ourselves, of being obliged to fear, of being in the cruel necessity of hating ourselves, and to blush in thinking of our conduct, and of the sentiments which it must infallibly inspire?

This granted, we shall reply, deliberately, to this Abbadie, that an atheist is a man who knows nature and its laws, who knows his own nature, and who knows what it imposes upon him. An atheist has experience, and this experience proves to him, every moment, that vice can injure him, that his most concealed faults, that his most secret dispositions may be detected and display him in open day; this experience proves to him that society is useful to his happiness; that his interest demands he should attach himself to the country which protects him, and which enables him to enjoy in security the benefits of nature; every thing shows him, that in order to be happy, he must make himself beloved; that his father is for him the most certain of friends; that ingratitude would remove from him his benefactor; that justice is necessary to the maintenance of every association; and that no man, whatever may be his power, can be content with himself, when he knows he is an object of public hatred.

He who has maturely reflected upon himself, upon his own nature, and upon that of his associates, upon his own wants, and upon the means of satisfying them, cannot be prevented from knowing his duties, from discovering that which he owes to himself, and that which he owes to others; then he has morality, he has real motives to conform himself to its dictates; he is obliged to feel that these duties are necessary; and if his reason be not disturbed by blind passions, or by vicious habits, he will feel that virtue is for all men the surest road to felicity. The atheists, or the fatalists, found all their systems upon necessity; thus, their moral speculations, founded upon

---

* See Abbadie on the *Truth of the Christian Religion*, vol. i. chap. xvii.

the necessity of things, are at least, much more permanent and more invariable than those which only rest upon a God who changes his aspect according to the dispositions and the passions of all those who contemplate him. The nature of things, and its immutable laws, are not subject to vary ; the atheist is always obliged to call that which injures him, vice and folly ; to call that which is advantageous to society, or which contributes to its permanent happiness, virtue.

We see, then, that the principles of the atheist are much less liable to be shaken than those of the enthusiast, who founds his morality upon an imaginary being, of whom the idea so frequently varies, even in his own brain. If the atheist deny the existence of a God, he cannot deny his own existence, nor that of beings similar to himself with whom he sees himself surrounded ; he cannot doubt the relations which subsist between them and him, he cannot question the necessity of the duties which flow from these relations ; he cannot, then, be dubious on the principles of morality, which is nothing more than the science of the relations subsisting between beings living together in society.

If, satisfied with a barren speculative knowledge of his duties, the atheist do not apply them to his conduct ; if hurried away by his passions, or by criminal habits, if given up to shameful vices, if possessing a vicious temperament, he appear to forget his moral principles, it does not follow that he has no principles, or that his principles are false ; it can only be concluded from such conduct, that, in the intoxication of his passions, in the confusion of his reason, he does not put in practice speculations extremely true ; that he forgets principles ascertained, to follow those propensities which lead him astray.

Nothing is more common amongst men than a very marked discrepance between the mind and the heart ; that is to say, between the temperament, the passions, the habits, the whims, the imagination, and the mind, or the judgment, assisted by reflection. Nothing is more rare, than to find these things in harmony ; it is then that we see speculation influence practice. The

most certain virtues, are those which are founded upon the temperament of men. Indeed, do we not every day see mortals in contradiction with themselves ? Does not their judgment unceasingly condemn the extravagances to which their passions deliver them up ? In short, does not every thing prove to us, that men, with the best theory, have sometimes the worst practice ; and with the most vicious theory have frequently the most estimable conduct ? In the blindest, the most atrocious superstitions, and those which are the most contrary to reason, we meet with virtuous men ; the mildness of their character, the sensibility of their heart, the excellence of their temperament, reconduct them to humanity, and to the laws of nature, in despite of their furious theories. Amongst the adorers of a cruel, vindictive, and jealous God, we find peaceable minds, who are enemies to persecution, to violence, and to cruelty ; and amongst the disciples of a God filled with mercy and clemency, we see monsters of barbarity and inhumanity. Nevertheless, the one and the other acknowledge that their God ought to serve them for a model : wherefore do they not conform themselves to him ? It is because the temperament of man is always more powerful than his God ; it is because the most wicked Gods cannot always corrupt a virtuous mind, and that the most gentle Gods cannot always restrain hearts driven along by crime. The organization will always be more puissant than religion : present objects, momentary interests, rooted habits, public opinion, have much more power than imaginary beings, or than theories which themselves depend upon the organization of man.

The point in question, then, is to examine if the principles of the atheist are true, and not if his conduct is commendable. An atheist, who, having an excellent theory, founded upon nature, experience, and reason, delivers himself up to excesses, dangerous to himself, and injurious to society, is, without doubt, an inconsistent man. But he is not more to be feared than a religious and zealous man, who, believing in a good, equitable, and perfect God, does not scruple to commit the most frightful excesses in his name.

An atheistical tyrant would not be more to be dreaded than a fanatical tyrant. An incredulous philosopher is not so dreadful as an enthusiastic priest, who fans the flame of discord among his fellow-citizens. Would then an atheist, clothed with power, be equally dangerous as a persecuting king, a savage inquisitor, a whimsical devotee, or a morose bigot? These are assuredly more numerous than atheists, of whom the opinions and the vices are far from being in a condition to have an influence upon society, which is too much blinded by prejudice to be disposed to give them a hearing.

An intemperate and voluptuous atheist, is not a man more to be feared than he who is superstitious, who knows how to connect licentiousness, libertinism, and corruption of morals, with his religious notions. Can it be imagined, with sincerity, that a man, because he is an atheist, or because he does not fear the vengeance of Gods, will be continually intoxicated, will corrupt the wife of his friend, will break open his neighbour's dwelling, and permit himself to commit all those excesses, which are the most prejudicial to himself, or the most deserving of punishment? The blemishes of an atheist, have not, then, any thing more extraordinary in them, than those of the religious man, they have nothing to reproach his doctrine with. A tyrant, who should be incredulous, would not be a more incommodious scourge to his subjects than a religious tyrant; would the people of the latter be more happy from the circumstance that the tiger who governed them believed in a God, heaped presents upon his priests, and humiliated himself at their feet? At least, under the dominion of an atheist, they would not have to apprehend religious vexations, persecutions for opinions, proscriptions, or those strange outrages, for which the interests of Heaven are frequently the pretext, under the mildest princes. If a nation be the victim of the passions and the folly of a sovereign who is an infidel, it will not, at least, suffer from his blind infatuation for theological systems, which he does not understand, nor from his fanatical zeal, which of all the passions that infest kings, is always the most destructive and the most dangerous. An atheistical tyrant, who should persecute for opinions, would be a man not consistent with his principles; he would only furnish one more example, that mortals much more frequently follow their passions, their interests, their temperaments, than their speculations. It is, at least, evident, that an atheist has one pretext less than a credulous prince, for exercising his natural wickedness.

Indeed, if men condescended to examine things coolly, they would find that the name of God is never made use of on earth, but for a pretext to indulge their passions. Ambition, imposture, and tyranny, have formed a league, to avail themselves of its influence, to the end that they may blind the people, and bend them beneath their yoke. The monarch makes use of it, to give a divine lustre to his person, the sanction of Heaven to his rights, and the confidence of its oracles to his most unjust and most extravagant whims. The priest uses it, to give currency to his pretensions, to the end that he may, with impunity, gratify his avarice, pride, and independence. The vindictive and enraged superstitious being introduces the cause of his God, that he may give free scope to his fury, which he qualifies with zeal. In short, religion becomes dangerous, because it justifies, and renders legitimate or commendable those passions and crimes, of which it gathers the fruit: according to its ministers, every thing is permitted to revenge the Most High; thus the Divinity appears to be made only to authorize and palliate the most injurious transgressions. The atheist, when he commits crimes, cannot, at least, pretend that it is his God who commands and approves them; this is the excuse which the superstitious being offers up for his wickedness; the tyrant for his persecutions; the priest for his cruelty and sedition; the fanatic for his excesses; the penitent for his inutility.

" They are not," says Bayle, " the general opinions of the mind, which determine us to act, but the passions." Atheism is a system, which will not make a good man wicked, neither will it make a wicked man good. " Those," says the same author, " who embraced the sect of Epicurus, did not become debauchees because they had embraced the doctrine of. Epicurus; they only embraced the doctrine of Epicurus, then badly understood, because they were debauchees."* In the same manner, a perverse man may embrace atheism, because he will flatter himself, that this system will give full scope to his passions? he will nevertheless be deceived; atheism, if well understood, is founded upon nature and reason, which never will, like religion, either justify or expiate the crimes of the wicked.

From the doctrine which makes morality depend upon the existence and the will of a God who is proposed to men for a model, there unquestionably results a very great inconvenience. Corrupt minds, in discovering how much each of these suppositions are erroneous or doubtful, let loose the rein of all their vices, and concluded that there were no real motives to do good; they imagined that virtue, like the Gods, was only a chimera, and that there was not any reason for practising it in this world. Nevertheless, it is evident, that it is not as creatures of God that we are bound to fulfil the duties of morality ; it is as men, as sensible beings, living together in society, and seeking to secure ourselves a happy existence, that we feel the moral obligation. Whether there exists a God, or whether he exists not, our duties will be the same ; and our nature, if consulted, will prove, that *vice is an evil, and that virtue is a real and substantial good.*†

---

* See Bayle's *Thoughts on Various Subjects*, sec. 177. Seneca has said before him : Ita non ab Epicuro impulsi luxuriantur, sed vitiis dediti, luxuriam suam in philosophiæ sinu absçondunt.— See *Seneca, de vita beata,* chap. xll.

† We are assured, that there have been found philosophers and atheists, who deny the distinction of *vice* and *virtue*, and who have preached up debauchery and licentiousness of manners: in this number, may be reckoned Aristippus, and Theodorus, surnamed the *Atheist*, Bion, the Boristhenite, Pyrrho, &c. amongst the ancients, (see *Diogenes Laertius*,) and amonst the moderns, the author of the *Fable of the Bees*, which, however, could only be intended to show, that in the present constitution of things, vices have identified themselves with nations, and have become necessary to them, in the same manner as strong liquors to those who have habituated themselves to their use. The author who published the *Man Automaton*, has

If, then, there be found atheists, who have denied the distinction of good and evil, or who have dared to strike at the foundation of all morality, we ought to conclude, that upon this point they have reasoned badly ; that they have neither been acquainted with the nature of man, nor known the true source of his duties ; that they have falsely imagined that morality as well as theology, was only an ideal science, and that the Gods once destroyed, there remained no longer any bonds to connect mortals. Nevertheless, the slightest reflection would have proved to them that morality is founded upon the immutable relations subsisting between sensi‘le, intelligent, and sociable beings ; that without virtue no society can maintain itself; that without putting a curb on his desires, no man can conserve himself. Men are constrained from their nature to love virtue, and to dread crime, by the same necessity that obliges them to seek happiness, and fly from sorrow ; thus nature obliges them to place a difference between those objects which please them, and those which injure them. Ask a man who is sufficiently irrational to deny the difference between virtue and vice, if it would be indifferent to him, to be beaten, robbed,

reasoned upon morality like a madman. If all these authors had consulted nature upon morality, as well as upon religion, they would have found that, far from being conducive to vice and depravity, it is conducive to virtue.

Nunquam aliud natura, aliud sapientia dicit.
Juvenal, sat. 14, v. 321.

Notwithstanding the pretended dangers which so many people believe they see in atheism, antiquity did not judge of it so unfavourably. Diogenes Laertius informs us, that Epicurus was in great favour, that his country raised statues to be erected to him, that he had a prodigious number of friends, and that his school subsisted for a very long period. See *Diogenes Laertius*, x. 9. Cicero, although an enemy to the opinions of the Epicureans, gives a brilliant testimony to the probity of Epicurus and his disciples, who were remarkable for the friendship they bore each other. See *Cicero de Finibus*, ii. 25. The philosophy of Epicurus was publicly taught in Athens during many centuries, and Lactantius says, that it was the most followed. Epicuri disciplina multo celebrior semper fuit quam cæterorum. *V. Institut. Divin.* iii. 17. In the time of Marcus Aurelius, there was at Athens a public professor of the philosophy of Epicurus, paid by that emperor, who was him self a stoic.

calumniated, repaid with ingratitude, dishonoured by his wife, insulted by his children, and betrayed by his friend? His answer will prove to you, that, whatever he may say, he makes a difference in the actions of men ; and that the distinction of good and evil does not depend either upon the conventions of men, or upon the ideas which they can have upon the Divinity ; upon the punishments or upon the recompenses which he prepares them in the other life.

On the contrary, an atheist, who should reason with justness, would feel himself much more interested than another, in practising those virtues to which he finds his happiness attached in this world. If his views do not extend themselves beyond the limits of his present existence, he must at least desire to see his days roll on in happiness and in peace. Every man, who, during the calm of his passions, falls back upon himself, will feel that his interest invites him to conserve himself; that his felicity demands that he should take the necessary means to enjoy life peaceably, and exempt from alarm and remorse. Man owes something to man, not because he would offend a God if he were to injure his fellow-creature, but because, in doing him an injury, he would offend a man, and would violate the laws of equity, in the maintenance of which, every being of the human species finds himself interested.

We every day see persons who are possessed of great talents, knowledge, and penetration, join to them the most hideous vices, and have a very corrupt heart: their opinions may be true in some respects, and false in a great many others ; and their principles may be just, but the inductions which they draw from them are frequently defective and precipitate. A man may have at the same time sufficient knowledge to undeceive himself of some of his errours. and too little energy to divest himself of his vicious propensities. Men are only that which their organization, modified by habit, by education, by example, by the government, by transitory or permanent circumstances, makes them. Their religious ideas and their imaginary systems are obliged to yield or accommodate themselves to

their temperaments, their propensities, and their interests. If the system, which makes man an atheist, does not remove from him the vices which he had before, neither does it give him any new ones : whereas, superstition furnishes its disciples with a thousand pretexts for committing evil without remorse, and even to applaud themselves for the commission of crime. Atheism, at least, leaves men such as they are ; it will not render a man more intemperate, more debauched, more cruel, than his temperament before invited him to be ; whereas superstition gives loose to the most terrible passions, or else procures easy expiations for the most dishonourable vices. "Atheism," says Chancellor Bacon, "leaves to man reason, philosophy, natural piety, laws, reputation, and every thing that can serve to conduct him to virtue ; but superstition destroys all these things, and erects itself into a tyranny over the understandings of men : this is the reason why atheism never disturbs the government, but renders man more clear-sighted, as seeing nothing beyond the bounds of this life." The same author adds, that "the times in which men have turned towards atheism have been the most tranquil: whereas superstition has always inflamed their minds and carried them on to the greatest disorders, because it infatuates the people with novelties, which wrest from, and carry with them all the authority of government."*

Men habituated to meditate, and to make study a pleasure, are not commonly dangerous citizens ; whatever may be their speculations, they never produce sudden revolutions upon the earth. The minds of the people, at all times susceptible of being inflamed by the marvellous and by enthusiasm, obstinately resist the most simple truths, and never heat themselves for systems which demand a long train of reflection and reasoning. The system of atheism can only be the result of long and connected study ; of an imagination cooled by experience and reasoning. The peaceable Epicurus never disturbed Greece ; the poem of Lucretius caused no civil wars in Rome;

* See the Moral Essays of Bacon.

Bodin was not the author of the league; the writings of Spinosa have not excited the same troubles in Holland, as the disputes of Gomar and d'Arminius. Hobbes did not cause blood to flow in England, although, in his time, religious fanaticism made a king perish on the scaffold.

In short, we can defy the enemies to human reason to cite a single example which proves, in a decisive manner, that opinions purely philosophical, or directly contrary to religion, have ever excited disturbances in the state. Tumults have always arisen from theological opinions, because both princes and people have always foolishly believed they ought to take a part in them. There is nothing so dangerous as that empty philosophy which the theologians have combined with their systems. It is to philosophy corrupted by priests, to which it peculiarly belongs to fan the flames of discord, invite the people to rebellion, and cause rivers of blood to flow. There is no theological question which has not occasioned immense mischief to man; whilst all the writings of the atheists, whether ancient or modern, have never caused any evil but to their authors, whom omnipotent imposture has frequently immolated at his shrine.

The principles of atheism are not formed for the mass of the people, who are commonly under the tutelage of their priests; they are not calculated for those frivolous and dissipated minds who fill society with their vices and their inutility; they are not suited to the ambitious, to those intriguers, and restless minds, who find their interest in disturbing the harmony of the social compact; much less are they made for a great number of persons enlightened in other respects, who have but very rarely the courage to completely divorce themselves from the received prejudices.

So many causes unite themselves to confirm men in those errours, which they have been made to suck in with their mother's milk, that every step that removes them from these fallacies, costs them infinite pains. Those persons who are most enlightened, frequently cling on some side to the general prejudice. We feel ourselves as it were isolated; we do not speak the language of socie-

ty, when we are alone in our opinions, it requires courage to adopt a mode of thinking that has but few approvers. In those countries where human knowledge has made some progress, and where, besides, a certain freedom of thinking is enjoyed, we can easily find a great number of deists or of incredulous beings, who, contented with having trampled under the foot the grosser prejudices of the uninformed, have not dared to go back to the source, and cite the Divinity himself before the tribunal of reason. If these thinkers did not stop on the road, reflection would quickly prove to them, that the God whom they have not the courage to examine, is a being as injurious, and as revolting to good sense, as any of those doctrines, mysteries, fables, or superstitious customs, of which they have already acknowledged the futility; they would feel, as we have already proved, that all these things are no more than the necessary consequences of those primitive notions which men have indulged respecting their divine phantom, and that, in admitting this phantom, they have no longer any rational cause to reject those inductions which the imagination must draw from it. A little attention would show them that it is precisely this phantom who is the true cause of all the evils of society; that those endless quarrels, and those bloody disputes to which religion and the spirit of party every instant give birth, are the inevitable effects of the importance which they attach to a chimera, ever calculated to kindle the minds of men into combustion. In short, it is easy to convince ourselves that an imaginary being, who is always painted under a terrific aspect, must act in a lively manner upon the imagination, and must produce, sooner or later, disputes, enthusiasm, fanaticism, and delirium.

Many persons acknowledge that the extravagances to which religion gives birth, are real evils; many persons complain of the abuse of religion, but there are very few who feel that this abuse and these evils are the necessary consequences of the fundamental principles of all religion, which can itself be founded only upon those grievous notions which men are obliged to form of the Divinity. We daily see persons undeceived upon religion, who pretend, nevertheless, that this religion *is necessary for the people*, who could not be kept within bounds without it. But, to reason thus, is it not to say, that poison is useful to the people, that it is proper to poison them, to prevent them from making a bad use of their power? Is it not to pretend that it is advantageous to render them absurd, irrational, and extravagant; that they have need of phantoms, calculated to make them giddy, to blind them, and to submit them to fanatics or to impostors, who will avail themselves of their follies to disturb the repose of the world? Besides, is it quite true that religion has a useful influence over the morals of the people? It is very easy to see that it enslaves them without rendering them better; it makes a herd of ignorant slaves, whom their panic terrours keep under the yoke of tyrants and priests; it forms stupid beings, who know no other virtue than a blind submission to futile customs, to which they attach a much greater value than to real virtues, or to the duties of morality, which have never been made known to them. If, by chance, this religion restrains some few timid individuals, it does not restrain the greatest number, who suffer themselves to be hurried along by the epidemical vices with which they are infected. It is in those countries where superstition has the greatest power, wherein we shall always find the least morality. Virtue is incompatible with ignorance, superstition, and slavery; slaves are only kept in subordination by the fear of punishments; ignorant children are intimidated only for an instant by imaginary terrours. To form men, to have virtuous citizens, it is necessary to instruct them, to show them truth, to speak reason to them, to make them feel their interests, to learn them to respect themselves, and to fear shame; to excite in them the ideas of true *honour*, to make them know the value of *virtue*, and the motives for following it. How can these happy effects be expected from religion, which degrades men, or from tyranny which only proposes to itself to vanquish them, to divide them, and to keep them in an abject condition? The false ideas which so many persons have of the utility of religion, which they at least judge to be calculated to restrain the people, arise from the fatal prejudice that there are *useful errours*, and that truth may be dangerous. This principle is completely calculated to eternise the sorrows of the earth; whoever shall have the courage to examine these things, will acknowledge, without hesitation, that all the miseries of the human species are to be ascribed to their errours, and that of these, religious errours must be the most prejudicial from the haughtiness with which they inspire sovereigns, from the importance which is attached to them, from the abject condition which they prescribe to subjects, from the phrensy which they excite among the people: we shall therefore be obliged to conclude, that the sacred errours of men are those of which the interest of mankind demands the most complete destruction, and that it is principally to the annihilation of them, that sound philosophy ought to be employed. It is not to be feared, that this attempt will produce either disorders or revolutions; the more freedom with which *truth* shall be spoken, the more convincing it will appear; the more simple it shall be, the less it will seduce men who are smitten with the marvellous; even those men who seek after truth with the most ardour, have an irresistible inclination, that urges them on, and incessantly disposes them to reconcile errour with its opposite.*

Here is, unquestionably, the reason why atheism, of which, hitherto, the principles have not been sufficiently developed, appears to alarm even those persons who are the most destitute of prejudice. They find the interval too great between the vulgar superstition, and absolute irreligion; they believe they take a wise medium, in com-

The false ideas which so many per-

No. X.—40

* The illustrious Bayle, who teaches us so ably to think, says, with abundant reason, that " there is nothing but a good and solid philosophy, which can like another Hercules, exterminate those monsters called popular errours: it is that alone which can set the mind at liberty." See Thoughts on Various Subjects, § 21. Lucretius had said before him:
Hunc igitur terrorem animi, tenebrasque
    necesse est
Non radii solis, neque lucida tela dici
Discutiant, sed *naturæ* species, ratioque.
                                lib. i. v. 147.

pounding with errour ; they reject the consequence while admitting the principle ; they preserve the phantom without foreseeing that, sooner or later, it must produce the same effects, and give birth, one after another, to the same follies in the heads of human beings. The major part of the incredulous and of the reformers, do no more than prune a cankered tree, to whose roots they have not dared to apply the axe ; they do not see that this tree will, in the end, reproduce the same fruits. Theology, or religion, will always be a heap of combustible matter ; generated in the imagination of mankind, it will always finish by causing conflagrations. As long as the sacerdotal order shall have the privilege of infecting youth, of habituating it to tremble before words, of alarming nations with the name of a terrible God, fanaticism will be master of the mind, imposture will, at its pleasure, sow discord in the state. The most simple phantom, perpetually fed, modified, and exaggerated by the imagination of men, will by degrees become a colossus sufficiently powerful to upset every mind and overthrow empires. Deism is a system at which the human mind cannot stop long ; founded upon a chimera, sooner or later, it will be seen to degenerate into an absurd and dangerous superstition.

Many incredulous beings, and many deists are met with in those countries where liberty of thought reigns ; that is to say, where the civil power has known how to counterbalance superstition. But above all, atheists will be found in those nations, where superstition, backed by the sovereign authority, makes the weight of its yoke felt, and imprudently abuses its unlimited power.* Indeed, when, in this kind

of countries, science, talents, the seeds of reflection are not entirely stifled ; the greater part of the men who think, revolt at the crying abuses of religion, at its multifarious follies, at the corruption and the tyranny of its priests, at those chains which it imposes, believing with reason, that they can never remove themselves too far from its principles ; the God who serves for the basis of such a religion, becomes as odious to them as the religion itself ; if this oppresses them they ascribe it to God. they feel that a terrible, jealous, and vindictive God, must be served by cruel ministers ; consequently, this God becomes a detestable object to every enlightened and honest mind amongst whom are always found the love of equity, liberty, humanity, and indignation against tyranny. Oppression gives a spring to the soul, it obliges man to examine closely the cause of his sorrows ; misfortune is a powerful incentive, that turns the mind to the side of truth. How formidable must not irritated reason be to falsehood ? It tears away its mask, it follows it even into its last entrenchment ; it at least inwardly enjoys its confusion.

_____

CHAPTER XI.

Of the Motives which lead to Atheism? Can this System be Dangerous? Can it be Embraced by the Uninformed?

THE preceding reflections will furnish us wherewith to reply to those who ask what interest men have in not admitting a God? The tyrannies, the persecutions, the numberless outrages committed in the name of this God, the stupidity and the slavery into which

* Atheists are, it is said, more rare in England and in Protestant countries, where toleration is established, than in Roman Catholic countries, where the princes are commonly intolerant and enemies to the liberty of thought. In Japan, in Turkey, in Italy, and above all in Rome, many atheists are found. The more power superstition has, the more those minds which it has not been able to subdue will revolt against it. It is Italy that produced Jordano Bruno, Campanella, Vanini, &c. There is every reason to believe, that had it not been for the persecutions and ill treatment of the synagogue, Spinosa would never have per-

haps promulgated his system. It may also be presumed, that the horrours produced in England by fanaticism, which cost Charles I. his head, pushed Hobbes on to atheism : the indignation which he also conceived at the power of the priests, suggested, perhaps, his principles so favourable to the absolute power of kings. He believed that it was more expedient for a state to, have a single civil despot, a sovereign over religion itself, than to have a multitude of spiritual tyrants, always ready to disturb it. Spinosa seduced by the ideas of Hobbes, fell into the same errour in his *Tractatus Theologico-Politicus*, as well as in his *Treatise de Jure Ecclesiasticorum* the ministers of this God every where plunge the people; the bloody disputes to which this God gives birth; the number of unhappy beings with which his fatal idea fills the world, are they then not motives sufficiently powerful, sufficiently interesting to determine all sensible men who are capable of thinking, to examine the titles of a being who causes so many evils to the inhabitants of the earth?

A theist, very estimable for his talents, asks, *if there can be any other cause than an evil disposition which can make men atheists?** I reply to him, yes, there are other causes; there is the desire of having a knowledge of interesting truths; there is the powerful interests of knowing what opinion to hold upon the object which is announced to us as the most important; there is the fear of deceiving ourselves upon the being who occupies himself with the opinions of men, and who does not permit that they should deceive themselves respecting him with impunity. But when these motives or these causes should not subsist, are not indignation, or, if they will, an *evil disposition*, legitimate causes, good and powerful motives, for closely examining the pretensions and the rights of an invisible tyrant, in whose name so many crimes are committed on the earth? Can any man, who thinks, who feels, who has any elasticity in his soul, prevent himself from being incensed against an austere despot, who is visibly the pretext and the source of all those evils with which the human species is assailed on every side? Is it not this fatal God who is at once the cause and the pretext of that iron yoke which oppresses men, of that

* See Lord Shaftesbury in his Letter on Enthusiasm. Spencer says, that "it is by the cunning of the devil who strives to render the Divinity hateful, that he is represented to us under that revolting character which renders him like unto the head of Medusa, insomuch that men are sometimes obliged to plunge into atheism, in order to disengage themselves from this hideous demon." But it might be said to Spencer, that the *demon who strives to render the Divinity hateful* is the interest of the clergy, which was in all times and in every country, to terrify men, in order to make them the slaves and the instruments of their passions. A God who should not make men tremble would be of no use whatever to the priests.

slavery in which they live, of that blindness which covers them, of that superstition which disgraces them, of those irrational customs which torment them, of those quarrels which divide them, of those outrages which they experience? Must not every mind in which humanity is not extinguished, irritate itself against a phantom, who, in every country, is made to speak only like a capricious, inhuman, and irrational tyrant?

To motives so natural, we shall join those which are still more urgent and personal to every man who reflects: namely, that troublesome fear, which must have birth, and be unceasingly nourished by the idea of a capricious God, so touchy, that he irritates himself against man, even for his most secret thoughts, who can be offended without our knowing it, and whom we are never certain of pleasing; who, moreover, is not restrained by any of the ordinary rules of justice, who owes nothing to the feeble work of his hands, who permits his creatures to have unhappy propensities, who gives them liberty to follow them, to the end that he may have the odious satisfaction of punishing them for faults, which he suffers them to commit? What can be more reasonable, and more just, than to verify the existence, the qualities, and the rights of a judge, who is so severe that he will everlastingly avenge the crimes of a moment? Would it not be the height of folly, to wear without inquietude, like the greater number of mortals, the overwhelming yoke of a God, always ready to crush us in his fury. The frightful qualities with which the Divinity is disfigured by those impostors who announce his decrees, oblige every rational being to drive him from his heart, to shake off his detestable yoke, and to deny the existence of a God, who is rendered hateful by the conduct which is ascribed to him; to scorn a God who is rendered ridiculous by those fables, which in every country are detailed of him. If there existed a God who was jealous of his glory, the crime the most calculated to irritate him would unquestionably be the blasphemy of those knaves who unceasingly paint him under the most revolting character; this God ought to be much more offended against his hideous ministers than against those who deny his existence. The phantom which superstition adores, while cursing him at the bottom of his heart, is an object so terrible that every wise man who meditates upon it, is obliged to refuse him his homage, to hate him, to prefer annihilation to the fear of falling into his cruel hands. *It is frightful*, the fanatic cries out to us, *to fall into the hands of the living God;* and in order that he may escape falling into them, the man who thinks maturely, will throw himself into the arms of nature ; and it is there alone that he will find a safe asylum against those continual storms, which supernatural ideas produce in the mind.

The deist will not fail to tell the atheist that God is not such as superstition paints him. But the atheist will reply to him, that superstition itself, and all the absurd and prejudicial notions to which it gives birth, are only corollaries of those false and obscure principles which are held respecting the Divinity. That his incomprehensibility suffices to authorize the incomprehensible absurdities and mysteries which are told of him, that these mysterious absurdities flow necessarily from an absurd chimera which can only give birth to other chimeras, which the bewildered imagination of mortals will incessantly multiply. This fundamental chimera must be annihilated to assure the repose of man, that he may know his true relations and his duties, and obtain that serenity of soul without which there is no happiness on the earth. If the God of the superstitious be revolting and mournful, the God of the theist will always be a contradictory being, who will become fatal, when he shall meditate on him, or with which, sooner or later, imposture will not fail to abuse him. Nature alone, and the truths which she discovers to us, are capable of giving to the mind and to the heart, a firmness, which falsehood will not be able to shake.

Let us again reply to those who unceasingly repeat, that the interest of the passions alone conduct us to atheism, and that it is the fear of punishments to come, that determine corrupt men to make efforts to annihilate this judge whom they have reason to dread.

We shall, without hesitation, agree that the interests and the passions of men excite them to make inquiries ; without interest no man is tempted to seek ; without passion no man will seek vigorously. The question, then, to be examined here, is, if the passions and interests, which determine some thinkers to examine the rights of God, are legitimate or not ? We have exposed these interests, and we have found that every rational man finds in his inquietudes and his fears, reasonable motives, to ascertain whether or not it be necessary to pass his life in perpetual fears and agonies ? Will it be said, that an unhappy being, unjustly condemned to groan in chains, has not the right of desiring to break them, or to take some means of liberating himself from his prison, and from those punishments which menace him at each instant ? Will it be pretended that his passion for liberty has no legitimate foundation, and that he does an injury to the companions of his misery, in withdrawing himself from the strokes of tyranny, and in furnishing them with assistance to escape from these strokes also ? Is, then, an incredulous man any thing more than one who has escaped from the general prison in which tyrannical imposture detains all mankind ? Is not an atheist who writes, one that has escaped, and furnishes to those of his associates, who have sufficient courage to follow him, the means of setting themselves free from the terrours which menace them ?*

---

* The priests unceasingly repeat that it is pride, vanity, and the desire of distinguishing himself from the generality of mankind, that determines man to incredulity. In this they act like the great, who treat all those as insolent, who refuse to cringe before them. Would not every rational man have a right to ask a priest, where is thy superiority in matters of reasoning ? What motives can I have to submit my reason to thy delirium? On the other hand, may it not be said to the clergy that it is interest which makes them priests : that it is interest which renders them theologians ; that it is the interest of their passions, of their pride, of their avarice, of their ambition, &c., which attaches them to their systems, of which they alone reap the benefits ? Whatever it may be, the priests, contented with exercising their empire over the uninformed, ought to permit those men who think, not to bend their knee before their

We also agree, that frequently the co.ruption of morals, debauchery, licentiousness, and even levity of mind, can conduct men to irreligion or to incredulity; but it is possible to be a libertine, irreligious, and to make a parade of incredulity, without being an atheist on that account: there is unquestionably a difference between thôse who are led to irreligion by dint of reasoning, and those who reject or despise religion, only because they look upon it as a melancholy object, or an incommodious restraint. Many people renounce received prejudices through vanity, or upon hearsay; these pretended strong minds have examined nothing for themselves, they act· on the authority of others, whom they suppose to have weighed things more maturely. This kind of incredulous beings have not, then, any certain ideas, and are but little capacitated to reason for themselves; they are hardly in a state to follow the reasoning of others. They are irreligious in the same manner as the majority of men are religious, that is to say, by credulity, like the people, or through interest, like the priests. A voluptary, a debauchee, buried in drunkenness; an ambitious mortal, an intriguer, a frivolous and dissipated man, a loose woman, a choice spirit of the day, are they personages really capable of judging of a religion which they have not deeply examined and maturely weighed, of feeling the force of an argument, of comparing the whole of a system? If they sometimes discover some faint glimmerings of truth amidst the tempest of their passions, which blind them, these leave on them only some evanescent traces, no sooner received than obliterated. Corrupt men attack the Gods only when they conceive them to be the enemies of their passions.*

The honest man attacks them because he finds they are inimical to virtue, injurious to his happiness, contradictory to his repose, and fatal to the human species.

Whenever our will is moved by concealed and complicated motives, it is extremely difficult to decide what determines it; a wicked man may be led to irreligion or to atheism by those motives which he dare not avow even to himself: he may form to himself an illusion and only follow the interest of his passions in believing he seeks after truth; the fear of an avenging God will perhaps determine him to deny his existence without much examination, uniformly because he is incommodious to him. Nevertheless, the passions happen to be sometimes just; a great interest carries us on to examine things more closely: it may frequently make a discovery of the truth, even to him who seeks after it the least, or who is only desirous of being lulled asleep, and of deceiving himself. It is the same with a perverse man who stumbles upon the truth, as it is with him who, flying from an imaginary danger, should find in his road a dangerous serpent, which in his haste he should kill; he does that by accident, and without design, which a man less troubled in his mind would have done with premeditated deliberation. A wicked man who fears his God, and who would escape from him, may certainly discover the absurdity of those notions which are entertained of him, without discovering for that reason that those same notions in no wise change or alter the evidence and the necessity of his duties.

To judge properly of things, it is necessary to be disinterested; it is necessary to have an enlightened and connected mind to compass a great system. It belongs only to the honest man, to examine the proofs of the existence of a God, and the principles of religion; it belongs only to the man acquainted with nature and its ways, to embrace with intelligence the cause of the System of Nature. The wicked and the ignorant are incapable of judg-

vain idols. Tertullian has said, qui̇s enim philosophum sacrificare compellit! See *Tertull. Apolog. Chap.* 614.

* Arian says, that when men imagine the Gods are in opposition to their passions, they abuse them and overturn their altars.

• The bolder the sentiments of an atheist, and the more strange and suspicious they appear to other men, the more strictly and scrupulously he ought to observe and to perform his duties, especially if he be not desirous that his morals should *calumniate* his

system, which, duly weighed, will make the necessity and the certitude of morality felt, whilst every species of religion tends to render it problematical, or even to corrupt it.

ing with candour; the honest and virtuous are alone competent judges in so weighty an affair. What do I say? is not the virtuous man from thence in a situation to desire the existence of a God who remunerates the goodness of men? If he renounce these advantages which his virtue gives him the right to hope for, it is because he finds them imaginary, as well as the remunerator who is announced to him; and that in reflecting on the character of this God, he is obliged to acknowledge that it is not possible to rely upon a capricious despot, and that the enormities and follies to which he serves as a pretext, infinitely surpass the pitiful advantages that can result from his idea. Indeed, every man who reflects, quickly perceives that for one timid mortal, of whom this God restrains the feeble passions, there are millions whom he cannot curb, and of whom, on the contrary, he excites the fury; for one that he consoles, there are millions whom he affrights, whom he afflicts, whom he obliges to groan; in short, he finds that against one inconsistent enthusiast, which this God, whom he believes good, renders happy, he carries discord, carnage, and affliction, into vast countries, and plunges whole people in grief and tears.

However this may be, do not let us inquire into the motives which may determine a man to embrace a system: let us examine the system, let us convince ourselves if it be true, and if we shall find that it is founded upon truth, we shall never be able to esteem it dangerous. It is always falsehood which injures men; if errour be visibly the source of their sorrows, reason is the truu remedy for them. Do not let us farther examine the conduct of a man who presents us with a system; his ideas, as we have already said, may be extremely sound, when even his actions are highly deserving censure. If the system of atheism cannot render him perverse who is not so by his temperament, it cannot render him good who does not otherwise know the motives which should conduct him to virtue. At least, we have proved that the superstitious man, when he has strong passions and a depraved heart, finds even in his religion a thousand pretexts more than the atheist for injuring the human species. The atheist has not, at least, the mantle of zeal to cover his vengeance, his transports, and his fury; the atheist has not the faculty of expiating, at the expense of money or by the aid of certain ceremonies, the outrages which he commits against society; he has not the advantage of being able to reconcile himself with his God, and by some easy custom, to quiet the remorse of his disturbed conscience; if crime has not deadened every feeling of his heart, he is obliged continually to carry within himself an inexorable judge, who unceasingly reproaches him for his odious conduct, who forces him to blush, to hate himself, and to fear the looks and the resentment of others. The superstitious man, if he be wicked, gives himself up to crime, which is followed by remorse; but his religion quickly furnishes him with the means of getting rid of it; his life is generally no more than a long series of errour and grief, of sin and expiation: still more, he frequently commits, as we have already seen, crimes of greater magnitude, in order to expiate the first: destitute of any permanent ideas of morality, he accustoms himself to look upon nothing as a crime, but that which the ministers and the interpreters of his God forbid him to commit: he considers as virtues, or as the means of effacing his transgressions, actions of the blackest die, which are frequently held out to him as agreeable to this God. It is thus we have seen fanatics expiate, by the most atrocious persecutions, their adulteries, their infamy, their unjust wars, and their usurpations; and to wash away their iniquities, bathe themselves in the blood of those superstitious beings, whose infatuation made them victims and martyrs.

An atheist, if he has reasoned justly, if he has consulted nature, has principles more certain, and always more humane than the superstitious: his religion, whether gloomy or enthusiastic, always conducts the latter either to folly or to cruelty. The imagination of an atheist will never be intoxicated to that degree, to make him believe that violence, injustice, persecution, or assassination, are virtuous or legitimate actions. We every day see that religion, or the cause of Heaven, hoodwinks those persons who are humane, equitable, and rational on every other occasion, so much that they make it a duty to treat with the utmost barbarity those men who step aside from their mode of thinking. A heretic, an incredulous being, ceases to be a man in the eyes of the superstitious. Every society, infected with the venom of religion, offers us innumerable examples of juridicial assassinations, which the tribunals commit without scruple, and without remorse; judges, who are equitable on every other occasion, are no longer so as soon as there is a question of theological chimeras; in bathing themselves in blood, they believe they conform to the views of the Divinity. Almost every where, the laws are subordinate to superstition, and make themselves accomplices in its fury; they legitimate or transform into duties those cruelties which are the most contrary to the rights of humanity.* Are not all these avengers of religion blind men, who, with gayety of heart, and through piety and duty, immolate to it those victims which it appoints? Are they not tyrants, who have the injustice to violate thought, and who have the folly to believe they can enslave it? Are they not fanatics on whom the law, dictated by inhuman prejudices, impose the necessity of becoming ferocious brutes? Are not all those sovereigns, who, to avenge Heaven, torment and persecute their subjects, and sacrifice human victims to the wickedness of their anthrophpoagite Gods, men whom religious zeal has converted into tigers? Are not those priests, so careful of the soul's health, who insolently break into the sanctuary of the thoughts, to the end that they may find in the opinions of man motives for injuring him, odious knaves and disturbers of the mind's repose, whom religion honours, and whom reason detests? What villains

are more odious in the eyes of humanity than those infamous *inquisitors*, who, by the blindness of princes, enjoy the advantage of judging their own enemies, and of committing them to the flames? Nevertheless the superstition of the people makes them respected, and the favour of kings overwhelms them with kindness! Do not a thousand examples prove that religion has every where produced and justified the most unaccountable horrours? Has it not a thousand times armed men with the poniards of homicides, let loose passions much more terrible than those which it pretended to restrain, and broken the most sacred bonds of mortals? Has it not, under the pretexts of duty, of faith, of piety, of zeal, favoured cruelty, stupidity, ambition, and tyranny? Has not the cause of God made murder, perfidy, perjury, rebellion, and regicide legitimate? Have not those princes, who frequently have made themselves the avengers of Heaven, the lictors of religion, hundreds of times been its victims? In short, has not the name of God been the signal for the most dismal follies, and the most frightful and wicked outrages? Have not the altars of the Gods every where swam in blood; and under whatever form they may have shown the Divinity, was he not always the cause or the pretext of the most insolent violation of the rights of humanity?†

---

* The president Grammont relates, with a satisfaction truly worthy a cannibal, the particulars of the punishment of Vanini, who was burnt at Toulouse, although he had disavowed the opinions with which he was accused. This president even goes so far as to find wicked the cries and howlings which torment wrested from this unhappy victim of religious cruelty.

† It is right to remark that the religion of the Christians which boasts of giving to men the most just ideas of the Divinity; which every time that it is accused of being turbulent and sanguinary, only shows its God as on the side of goodness and mercy; which prides itself on having taught the purest system of morality; which pretends to have established for ever concord and peace amongst those who profess it: It is well, I say, to remark that it has caused more divisions and disputes, more political and civil wars, more crimes of every species, than all the other religions of the world united. We will perhaps be told, that the progress of learning will prevent this superstition from producing in future such dismal effects as those which it has formerly done; but we shall reply, that fanaticism will ever be equally dangerous, or that the cause not being removed, the effects will always be the same. Thus so long as superstition shall be held in consideration, and shall have power, there will be disputes, persecutions, regicides, disorders, &c., &c. So long as mankind shall be sufficiently irra-

Never will an atheist, as long as he enjoys his right senses, persuade himself that similar actions can be justifiable; never will he believe that he who commits them can be an estimable man; there is no one but a superstitious being, whose blindness makes him forget the most evident principles of morality, of nature, and of reason, who can possibly imagine that the most destructive crimes are virtues. If the atheist be perverse, he, at least, knows that he does wrong; neither God nor his priests will be able to persuade him that he does right, and whatever crimes he may allow himself to commit, he will never be capable of exceeding those which superstition causes to be committed without scruple, by those whom it intoxicates with its fury, or to whom it holds forth crimes themselves as expiations and meritorious actions.

Thus the atheist, however wicked he may be supposed to be, will at most be only on a level with the devotee, whose religion frequently encourages him to commit crime which it transforms into virtue. As to conduct, if he be debauched, voluptuous, intemperate, and adulterous, the atheist differs in nothing from the most credulous superstitious being, who frequently knows how to connect with his credulity those vices and crimes which his priests will always pardon, provided he render homage to their power. If he be in Hindostan, his bramins will wash him in the Ganges while reciting a prayer. If he be a Jew, upon making an offering, his sins will be effaced; if he be in Japan, he will be cleansed by performing a pilgrimage; if he be a Mahometan, he will be reputed a saint for having visited the tomb of his prophet: if he be a Christian, he will pray, he will fast, he will throw

himself at the feet of his priests and confess his faults to them; these will give him absolution in the name of the Most High, will sell him the indulgences from Heaven, but never will they censure him for those crimes which he shall have committed in support of their several faiths.

We are constantly told, that the indecent or criminal conduct of the priests and of their sectaries proves nothing against the goodness of their religious systems; but wherefore do they not say the same thing of the conduct of the atheist, who, as we have already proved, may have a very good and very true system of morality, even while leading a dissolute life? If it be necessary to judge the opinions of mankind according to their conduct, which is the religion that would bear this scrutiny? Let us, then, examine the opinions of the atheist without approving of his conduct; let us adopt his mode of thinking, if we judge it to be true, useful, and rational; let us reject his mode of acting, if we find it blameable. At the sight of a work filled with truth, we do not embarrass ourselves with the morals of the workman. Of what importance is it to the universe whether Newton were a sober or an intemperate, a chaste or a debauched man? It only remains for us to examine whether he reasoned well, if his principles be certain, if the parts of his system are connected, if his work contains more demonstrable truths than bold ideas. Let us judge in the same manner of the principles of an atheist; if they are strange and unusual, that is a reason for examining them more strictly; if he has spoken truth, if he has demonstrated his positions, let us yield to the evidence; if he be deceived in some parts let us distinguish the true from the false, but do not let us fall into the hackneyed prejudice, which on account of one errour in the detail, rejects a multitude of incontestable truths.* The atheist, when he is deceived, has unquestionably as much

tional to look upon religion as a thing of the first importance to them, the ministers of religion will have the opportunity of confounding every thing on earth under the pretext of serving the interest of the Divinity, which will never be other than their own peculiar interests. The Christian church would only have one mode of wiping away the accusation which is brought against it of being intolerant or cruel, and that would be solemnly to declare *that it is not allowable to persecute or injure any one for his opinions ;* but this is what its ministers will never do.

* Dr. Johnson (the Christian *bear* or *hog*) says in his preface to his dictionary, that "where a man shall have executed his task with all the accuracy possible, he will only be allowed to have done his duty; but if he commit the slightest errour, a thousand snarlers are ready to point it out."

right to throw his faults on the fragility of his nature as the superstitious man. An atheist may have vices and defects, he may reason badly ; but at least his errours will never have the consequences of religious novelties ; they will not, like these, kindle up the fire of discord in the bosom of nations ; the atheist will not justify his vices and his wanderings by religion ; he will not pretend to infallibility, like those self-conceited theologians who attach the divine sanction to their follies, and who suppose that Heaven authorizes those sophisms, those falsehoods, and those errours, which they believe themselves obliged to distribute over the face of the earth.

It will perhaps be said that the refusal to believe in the Divinity, will rend asunder one of the most powerful bonds of society, in making the sacredness of an oath vanish. I reply, that perjury is by no means rare in the most religious nations, nor even amongst those persons who make a boast of being the most thoroughly convinced of the existence of the Gods. Diagoras, superstitious as he was, became, it is said, an atheist on seeing that the Gods had not thundered their vengeance on a man who had taken them as evidence to a falsity. Upon this principle, how many atheists ought to be made among us ? From the principle which has made an invisible and an unknown being the depositary of man's engagements, we do not see it result that their engagements and their most solemn contracts are more solid for this vain formality. Conductors of nations, it is you above all, that I call upon to witness my assertions ! This God, of whom ye say ye are the images, from whom ye pretend to hold the right of governing ; this God, whom ye so often make the witness of your oaths, the guarantee of your treaties ; this God, of whom ye declare ye fear the judgment, has he much weight with ye, whenever there is a question of the most futile interest ? Do ye religiously observe those sacred engagements which ye have made with your allies, and with your subjects ? Princes ! who to so much religion frequently join so little probity, I see the power of truth overwhelms ye ; without doubt, you blush at this question ; and you

are constrained to allow that you equally mock Gods and men. What do I say ? Does not religion itself frequently absolve you from your oaths ? Does it not prescribe that you should be perfidious, and violate plighted faith, above all, when there is a question of its sacred interests, does it not order you to dispense with the engagements you have made with those whom it condemns ? And after having rendered you perfidious and perjured, has it not sometimes arrogated the right of absolving your subjects from those oaths which bound them to you !* If we consider things attentively, we shall see, that under such chiefs, religion and politics are schools of perjury. Therefore, knaves of every condition never recoil when it is necessary to attest the name of God to the most manifest frauds, and for the vilest interests. What end then do oaths answer ? They are snares in which simplicity alone can suffer itself to be caught ; oaths are every where vain formalities, they impose nothing on villains, nor do they add any thing to the engagements of honest men, who, without oaths, would not have had the temerity to violate them. A perfidious and perjured superstitious being, unquestionably has not any advantage over an atheist who should fail in his promises ; neither the one nor the other any longer deserves the confidence of their fellow-citizens nor the esteem of good men : if one does not respect his God in whom he believes, the other neither respects his reason, his reputation, nor public opinion, in which all rational men cannot refuse to believe.†

---

* It is a maxim constantly received in the Roman Catholic religion, that is to say, in that sect of Christianity, the most superstitious and the most numerous, *that no faith is to be held with heretics.* The general council of Constance decided thus, when, notwithstanding the emperor's passport, it decreed John Hus, and Jerome Prague to be burnt. The Roman Pontiff has, it is well known, the right of relieving his secretaries from their oaths, and annulling their vows; the same Pontiff has frequently arrogated to himself the right of deposing kings, and of absolving their subjects from their oaths of fidelity. It is very extraordinary that oaths should be prescribed by the laws of those nations who profess the Christian religion, whilst Christ has expressly prohibited the use of them.

† " An oath," says Hobbes, " adds nothing

It has been frequently asked, if there ever was a nation that had no idea of the Divinity, and if a people uniformly composed of atheists would be able to subsist? Whatever some speculators may say, it does not appear likely that there ever has been upon our globe a numerous people, who have not had an idea of some invisible power, to whom they have shown marks of respect and submission.* Man, inasmuch as he is a fearful and ignorant animal, necessarily becomes superstitious in his misfortunes: either he forms a God for himself, or he admits the God which others are disposed to give him. It does not then appear that we can rationally suppose there may have been, or that there actually is, a people upon the earth a total stranger to the notion of some Divinity. One will show us the sun, or the moon and stars; the other will show us the sea, the lakes, the rivers, which furnish him his subsistence; the trees which afford him an asylum against the inclemency of the air; another will show us a rock of an odd form, a high mountain or volcano that frequently astonishes him; another will present you with his crocodile, whose malignity he fears; his dangerous serpent, the reptile to which he attributes his good or his bad fortune. In short, each man will make you see his phantasm, his domestic or tutelary God with respect.

But from the existence of his Gods, the savage does not draw the same inductions as the civilized and polished

to an obligation, it only augments, in the imagination of him who swears, the fear of violating an engagement, which he would have been obliged to keep ever without any oath."

* It has been sometimes believed that the Chinese were atheists; but this error is due to the Christian Missionaries, who are accustomed to treat all those as *atheists* who do not hold opinions similar to their own upon the Divinity. It always appears that the Chinese are a people extremely superstitious, but that they are governed by chiefs who are not so, without, however, their being atheists for that reason. If the empire of China be as flourishing as it is said to be, it at least furnishes a very forcible proof that those who govern, have no occasion to be superstitious in order to govern with propriety, a people who is so.

It is pretended that the Greenlanders have no idea of the Divinity. Nevertheless, it is difficult to believe it of a nation so savage and so ill-treated by Nature.

man; the savage does not believe it a duty to reason much upon his Divinities; he does not imagine that they ought to influence his morals, nor entirely occupy his thoughts: content with a gross, simple, and exterior worship, he does not believe that these invisible powers trouble themselves with his conduct towards his fellow-creatures; in short, he does not connect his morality with his religion. This morality is coarse, as must be that of all ignorant people; it is proportioned to his wants, which are few; it is frequently irrational, because it is the fruit of ignorance, of inexperience, and of the passions of men, but slightly restrained in their infancy. It is only in numerous, stationary, and civilized societies, where man's wants multiply themselves, and his interests clash, that he is obliged to have recourse to governments, to laws, and to public worship, in order to maintain concord: it is then that men approximating, reason and combine their ideas, refine and subtilize their notions; it is then that those who govern them, avail themselves of the fear of invisible powers to keep them within bounds, to render them docile, and oblige them to obey and live peaceably. It is thus that by degrees, morals and politics find themselves connected with religious systems. The chiefs of nations, frequently superstitious themselves, but little enlightened upon their own interests, but little versed in sound morality, and but little instructed in the true motive-powers of the human heart, believe that they have done every thing for their own authority as well as for the happiness and repose of society, in rendering their subjects superstitious, in menacing them with the wrath of their invisible phantoms, in treating them like children, who are appeased with fables and chimeras. By the assistance of these marvellous inventions, to which even the chiefs and the conductors of nations are themselves frequently the dupes, and which are transmitted from race to race in their duties; sovereigns are dispensed from the trouble of instructing themselves; they neglect the laws, they enervate themselves in ease and sloth, they follow nothing but the caprice, they repose in their deities the care of restraining their subjects; they confide the instruction of the people to priests, who are commissioned to render them good, submissive, and devout, and to teach them, in an early age, to tremble under the yoke of the visible and invisible Gods.

It is thus that nations are kept by their tutors in a perpetual state of infancy, and are only restrained by vain chimeras. It is thus that politics, jurisprudence, education, and morality, are every where infected with superstition. It is thus that men no longer know any duties but those of religion; it is thus that the idea of virtue is falsely associated with that of those imaginary powers to which imposture gave that language which is most conducive to its own immediate interests. It is thus that men are persuaded that without a God there no longer exists any morality for them. It is thus that princes and subjects, equally blind to their true interests, to the duties of nature, and to their reciprocal rights, have habituated themselves to consider religion as necessary to morals, as indispensably requisite to govern men, and as the most certain means of arriving at power and happiness.

It is from these dispositions, of which we have so frequently demonstrated the falsity, that so many persons, otherwise extremely enlightened, look upon it as an impossibility, that a society of atheists could subsist for any length of time. It does not admit a question, that a numerous society who should neither have religion, morality, government, laws, education, nor principles, could not maintain itself, and that it would simply draw together beings disposed to injure each other, or children who would only blindly follow the most fatal impulsions; but then, with all the religion in the world, are not human societies very nearly in this state? Are not the sovereigns of almost every country in a continual state of warfare with their subjects? Are not these subjects, in despite of religion and the terrible notions which it gives them of the Divinity, unceasingly occupied in reciprocally injuring each other, and rendering themselves mutually unhappy? Does not religion itself, with its supernatural notions, unremittingly flatter the vanity and the passions of sovereigns, and throw oil into the fire of dis-

cord between those citizens who are divided in opinion? Could those infernal powers, who are supposed to be ever upon the watch to injure the human species, be capable of producing greater evils upon the earth than spring from fanaticism, and the fury to which theology gives birth? In short, could atheists, assembled together in society, however irrational they may be supposed to be, conduct themselves towards each other in a more criminal manner, than do these superstitious beings, filled with real vices and extravagant chimeras, who have, during so many ages, done nothing more than destroy themselves and cut each other's throats, without reason, and without pity? It cannot be pretended they would; on the contrary, we boldly assert, that a society of atheists, destitute of all religion, governed by wholesome laws, formed by a good education, invited to virtue by recompenses, deterred from crime by equitable punishments, and disentangled from illusions, falsehood, and chimeras, would be infinitely more honest and more virtuous than those religious societies, in which every thing conspires to intoxicate the mind and to corrupt the heart.

When we shall be disposed usefully to occupy ourselves with the happiness of men, it is with the Gods in heaven that the reform must commence; it is by abstracting these imaginary beings, destined to affright people who are ignorant and in a state of infancy, that we shall be able to promise ourselves to conduct man to a state of maturity. It cannot be too often repeated, there is no morality without consulting the nature of man and his true relations with the beings of his species; no fixed principles for man's conduct in regulating it upon unjust, capricious, and wicked Gods; no sound politics, without consulting the nature of man, living in society to satisfy his wants, and to assure his happiness and its enjoyment. No wise government can found itself upon a despotic God, he will always make tyrants of his representatives. No laws will be good without consulting the nature and the end of society. No jurisprudence can be advantageous for nations, if it is regulated upon the caprice and passions of deified tyrants. No education will be rational, unless it be founded upon reason, and not upon chimeras and prejudices. In short there is no virtue, no probity, no talents, under corrupt masters, and under the conduct of those priests who render men the enemies of themselves and of others, and who seek to stifle in them the germs of reason, science, and courage.

It will, perhaps, be asked, if we can reasonably flatter ourselves with ever arriving at the point of making a people entirely forget their religious opinions, or the ideas which they have of the Divinity? I reply, that the thing appears utterly impossible, and that this is not the end which we can propose to ourselves. The idea of a God, inculcated from the most tender infancy, does not appear of a nature to admit eradication from the mind of the majority of mankind: it would, perhaps, be as difficult to give it to those persons, who, arrived at a certain age, should never have heard it spoken of, as to banish it from the minds of those who have been imbued with it from their earliest infancy. Thus it cannot be supposed that it is possible to make a whole nation pass from the abyss of superstition, that is to say, from the bosom of ignorance and of delirium, into absolute atheism, which supposes reflection, study, knowledge, a long series of experience, the habit of contemplating nature, the science of the causes of its various phenomena, of its combinations, of its laws, of the beings who compose it, and of their different properties. In order to be an atheist, or to be assured of the powers of nature, it is necessary to have meditated profoundly ; a superficial glance of the eye will not make us acquainted with her powers; eyes but little exercised, will unceasingly be deceived ; the ignorance of actual causes will make us suppose those which are imaginary ; and ignorance will thus reconduct the natural philosopher himself to the feet of a phantom, in which his limited vision, or his idleness will make him believe he shall find the solution of every difficulty.

Atheism, as well as philosophy and all profound and abstract sciences, then, is not calculated for the uninformed, neither is it suitable for the majority of mankind. There are in all populous and civilized nations, persons whose circumstances enable them to meditate, to make researches, and useful discoveries, which, sooner or later, finish by extending themselves, and becoming beneficial when they have been judged advantageous and true. The geometrician, the mechanic, the chymist, the physician, the civilian, the artisan himself, labour in their closets or in their workshops, seeking the means to serve society each in his sphere ; nevertheless, not one of these sciences or professions are known to the uninitiated, who, however, do not fail in the long run to profit by, and reap the advantages of those labours, of which they themselves have no idea. It is for the mariner that the astronomer labours : it is for him that the geometrician and the mechanic calculate: it is for the mason and the labourer that the skilful architect draws learned designs. Whatever may be the pretended utility of religious opinions, the profound and subtile theologian cannot boast of labouring, of writing, or of disputing for the advantage of the people, whom, however, they contrive to tax exorbitantly for those systems and those mysteries which they will never understand, and which never can at any time be of any utility whatever to them.

It is not, then, for the multitude that a philosopher ought to propose to himself to write or to meditate. The principles of atheism, or the *System of Nature*, are not even calculated, as we have shown, for a great number of persons, extremely enlightened on other points, but frequently too much prepossessed in favour of received prejudices. It is extremely rare to find men who, to an enlarged mind, extensive knowledge, and great talents, join either a well-regulated imagination, or the courage necessary to combat successfully those habitual chimeras with which the brain has been long inoculated. A secret and invincible inclination frequently reconducts, in despite of all reasoning, the most solid and the best-fortified minds to those prejudices which they see generally established, and of which they have themselves drank copiously from their most tender infancy. Nevertheless, by degrees, those principles which then appear strange or revolting, when they have truth on their side, insinuate themselves into the mind, become familiar, extend themselves far and wide, and produce the most advantageous effects over every society : in time, men familiarize themselves with those ideas which originally they had looked upon as absurd and irrational ; at least, they cease to consider those as odious who profess opinions upon things of which experience makes it evident, they may be permitted to have doubts without danger to the public.

The diffusion of ideas, then, amongst men, is not to be dreaded. Are they useful ? By degrees they will fructify. The man who writes, must neither fix his eyes upon the time in which he lives, nor upon his actual fellow-citizens, nor upon the country which he inhabits. He must speak to the human species, he must foresee future generations ; in vain will he expect the applauses of his contemporaries ; in vain shall he flatter himself with seeing his precocious principles received kindly by prejudiced minds ; if he has told truth, the ages that shall follow will render justice to his efforts ; meantime, let him content himself with the idea of having done well, or with the secret suffrages of those few friends to truth who inhabit the earth. It is after his death that the writer of truth triumphs ; it is then that the 'stings of hatred and the shafts of envy, either exhausted or blunted, give place to truth, which being eternal, must survive all the errours of the earth.*

Besides, we shall say with Hobbes, that " We cannot do men any harm by proposing our ideas to them ; the worst mode is to leave them in doubt and dispute ; indeed, are they not so already ?" If an author who writes be deceived, it is because he may have reasoned badly. Has he laid down false principles ? It remains to examine them. Is his system false and ridiculous ? It will serve to make truth appear in its greatest splendour ; his work will fall into contempt ; and the writer, if he be witness to its fall, will be sufficiently punished for his temerity ; if he be dead, the living cannot disturb his ashes. No man writes with a design to injure his fellow-creatures ; he always proposes to himself to merit their suffrages, either by amusing them, by exciting their curiosity, or by communicating to them discoveries which he believes useful. No work can be dangerous ; above all, if it contains *truth*. It would not be so, even if it contained principles evidently contrary to experience and good sense. Indeed, what would result from a work that should now tell us the sun is not luminous ; that parricide is legitimate ; that robbery is allowable ; that adultery is not a crime ? The smallest reflection would make us feel the falsity of these principles, and the whole human race would protest against them. Men would laugh at the folly of the author, and presently his book and his name would be known only by their ridiculous extravagancies. There is nothing but religious follies that are pernicious to mortals ; and wherefore ? It is because authority always pretends to establish them by violence, to make them pass for virtues, and rigorously punishes those who should be disposed to laugh at or to examine them. If men were more rational, they would

---

* It is a problem with a great many people, if *truth* may not be injurious. The best intentioned persons are themselves frequently in great doubt upon this important point. *Truth* never injures any but those who deceive men : these have the greatest interest in being undeceived. *Truth* may be injurious to him who announces it, but no *truth* can possibly injure the human species, and never can it be too clearly announced to beings always little disposed to listen to, or comprehend it. If all those who write to announce important truths, *which are always considered as the most dangerous*, were sufficiently warmed with the public welfare to speak freely, even at the risk of displeasing their renders, the human race would be much more enlightened and much happier than it is. To write in ambiguous words, is frequently to write to nobody. The human mind is idle, we must spare it as much as possible the trouble and embarrassment of reflecting. What time and study does it not require at the present day to unravel the ambiguous oracles of the ancient philosophers, whose true sentiments are almost entirely lost to us ! If *truth* be useful to men, it is an injustice to deprive them of it ; if *truth* ought to be admitted, we must admit its consequences, which also are *truths*. Men, for the most part, are fond of *truth*, but its consequences inspire them with so much fear, that frequently they prefer remaining in *errour*, of which habit prevents them from feeling the deplorable effects.

consider religious opinions and theological systems with the same eyes as systems of natural philosophy, or problems in geometry : these latter never disturb the repose of society, although they sometimes excite very warm disputes amongst some 'of the learned. Theological quarrels would never be attended with any evil consequences, if men could arrive at the desirable point of making those who have power in their hands, feel that they ought not to have any other sensations than those of indifference and contempt, for the disputes of persons who do not themselves understand the marvellous questions upon which they never cease disputing.

It is at least, this indifference, so just, so rational, so advantageous for states, that sound philosophy proposes to introduce by degrees upon the earth. Would not the human species be much happier, if the sovereigns of the world, occupied with the welfare of their subjects, and leaving to superstition its futile contests, submitted religion to politics ; obliged its haughty ministers to become citizens ; and carefully prevented their quarrels from interrupting the public tranquillity ? What advantages would there not result to science, to the progress of the human mind, to the perfectionating of morality, of jurisprudence, of legislation, of education, from the liberty of thought ? At present genius every where finds shackles ; religion continually opposes itself to its course : man, enveloped with bandages, does not enjoy any one of his faculties ; his mind itself is tortured, and appears continually wrapped up in the swaddling clothes of infancy. The civil power, leagued with the spiritual power, appears only disposed to rule over brutalized slaves, confined in an obscure prison, where they make each other reciprocally feel the effects of their mutual ill-humour. Sovereigns detest liberty of thought, because they fear truth ; this truth appears formidable to them, because it would condemn their excesses ; these excesses are dear to them, because they know no more than their subjects their true interests, which ought to blend themselves into one.

Let not the courage of the philosopher, however, be abated by so many

united obstacles, which would appear to exclude for ever truth from its dominion ; reason from the mind of man, and nature from its rights. The thousandth part of those cares which are bestowed to infect the human mind, would be sufficient to make it whole. Do not then let us despair, do not let us do man the injury to believe that truth is not made for him ; his mind seeks after it incessantly ; his heart desires it ; his happiness demands it loudly ; he fears it or mistakes it, only because religion, which has overthrown all his ideas, perpetually keeps the bandeau of delusion over his eyes, and strives to render him a total stranger to virtue.

Maugre the prodigious exertions that are made to drive truth, reason, and science, from the earth ; time, assisted by the progressive knowledge of ages, may be able one day to enlighten even those princes who are so outrageous against truth, such enemies to justice, and to the liberty of mankind. Destiny will, perhaps, one day conduct them to the throne of some enlightened, equitable, courageous, and benevolent sovereign, who acknowledging the true source of human miseries, shall apply to them the remedies with which wisdom has furnished him : perhaps he will feel that those Gods, from whom he pretends he derives his power, are the true scourges of his people ; that the ministers of these Gods are his own enemies and rivals ; that the religion which he looks upon as the support of his power, does, in fact, only weaken and shake it ; that superstitious morality is false, and serves only to pervert his subjects, and to give them the vices of slaves, in lieu of the virtues of the citizen ; in short, he will see in religious errours, the fruitful source of the sorrows of the human species ; he will feel that they are incompatible with every equitable administration.

Until this desirable epoch for humanity, the principles of *naturalism* will be adopted only by a small number of thinkers, they cannot flatter themselves with having a great many approves or proselytes ; on the contrary, they will find ardent adversaries, or contemners, even in those persons who, upon every other subject, dis cover the most acute minds and display the greatest knowledge. Those men who have the greatest share of talents, as we have already observed, cannot always resolve to divorce themselves completely from their religious ideas; imagination, so necessary to splendid talents, frequently forms in them an insurmountable obstacle to the total extinction of prejudice; this depends much more on the judgment than on the mind. To this disposition, already so prompt to form illusions for them, is also joined the power of habit; to a great many men it would be wresting from them a portion of themselves to take away their ideas of God; it would be depriving them of an accustomed aliment; it would be plunging them into a vacuum, and obliging their distempered minds to perish for want of exercise.*

Let us not, then, be surprised if very great and learned men obstinately shut their eyes, or run counter to their ordinary sagacity, every time there is a question respecting an object which they have not the courage to examine with that attention which they have lent to many others. Lord Chancellor Bacon pretends that " *a little philosophy disposes men to atheism, but that great depth reconducts them to religion.*" If we analyze this proposition, we shall find it to signify, that very moderate and indifferent thinkers are quickly enabled to perceive the gross absurdities of religion, but that little accustomed to meditate, or destitute of those certain principles which could serve to guide them, their imagination presently replaces them in the theological labyrinth, from whence reason, too weak, appeared disposed to withdraw them. Timid souls fear even to take courage again; minds accustomed to be satisfied with theological

solutions, no longer see in nature any thing but an inexplicable enigma, an abyss which it is impossible to fathom. Habituated to fix their eyes upon an ideal and mathematical point, which they have made the centre of every thing, the universe becomes a jumble to them, whenever they lose sight of it; and in the confusion in which they find themselves involved, they rather prefer returning to the prejudices of their infancy, which appear to explain every thing, than to float in the vacuum, or quit that foundation which they judge to be immoveable. Thus, the proposition of Bacon, appears to indicate nothing, except it be, that the most experienced persons cannot defend themselves against the illusions of their imagination, the impetuosity of which resists the strongest reasoning.

Nevertheless, a deliberate study of nature is sufficient to undeceive every man who will calmly consider things: he will see that every thing in the world is connected by links invisible to the superficial and to the too impetuous observer, but extremely intelligible to him who views things with coolness. He will find that the most unusual, and the most marvellous, as well as the most trifling and ordinary effects are equally inexplicable; but that they must flow from natural causes, and that supernatural causes, under whatever name they may be designated, with whatever qualities they may be decorated, will do no more than increase difficulties, and make chimeras multiply. The simplest observation will incontestably prove to him that every thing is necessary, that all the effects which he perceives are material, and can only originate in causes of the same nature, when even he should not be able, by the assistance of the senses, to recur to these causes. Thus his mind will every where show him nothing but matter acting sometimes in a manner which his organs permit him to follow, and sometimes in a mode imperceptible to him: he will see that all beings follow constant and invariable laws, by which all combinations form and destroy themselves, all forms change, whilst the great whole ever remains the same. Then cured of the notions with which he was imbued, undeceived in those erroneous

* Menage has remarked, that history speaks of very few incredulous women, or female atheists. This is not surprising, their organization renders them fearful, the nervous system undergoes periodical variations in them, and the education which they receive, disposes them to credulity. Those amongst them who have a sound constitution, and imagination, have occasion for chimeras suitable to occupy their idleness; above all, when the world abandons them, devotion and its ceremonies then become a business or an amusement for them.

ideas, which, from habit, he attached to imaginary beings, he will cheerfully consent to be ignorant of that which his organs cannot compass; ne will know that obscure terms, devoid of sense, are not calculated to explain difficulties; and guided by reason, he will throw aside all hypotheses of the imagination, to attach himself to those realities which are confirmed by experience.

The greater number of those who study nature, frequently do not consider, that with the eyes of prejudice they will never discover more than that which they have resolved beforehand to find; as soon as they perceive facts contrary to their own ideas, they quickly turn aside, and believe their eyes have deceived them; or else, if they turn back, it is in hopes to be able to reconcile them with those notions with which their mind is imbued. It is thus we find enthusiastic philosophers, whose prepossessions show them, even in those things which most openly contradict their opinions, incontestable proofs of those systems with which they are preoccupied. Hence those pretended demonstrations of the existence of a good God, which are drawn from final causes, from the order of nature, from his kindness to man, &c., &c. Do the same enthusiasts perceive disorder, calamities, and revolutions? They induct new proofs from the wisdom, the intelligence, the bounty of their God, whilst all these things as visibly contradict these qualities, as the first appear to confirm or to establish them. These prejudiced observers are in an ecstasy at the sight of the periodical motion and order of the stars, at the productions of the earth, at the astonishing harmony of the parts of animals; they forget, however, the laws of motion, the powers of attraction and repulsion, and of gravitation, and assign all these great phenomena to an unknown cause of which they have no idea! In short, in the heat of their imagination, they place man in the centre of nature; they believe him to be the object and the end of all that exists; that it is for him that every thing is made; that it is to rejoice and please him that every thing has been created; whilst they do not perceive that very frequently the whole of na-

ture appears to be loosed against him, and that destiny obstinately persists in rendering him the most miserable of beings.*

Atheism is only so rare because every thing conspires to intoxicate man, from his most tender age, with a dazzling enthusiasm, or to puff him up with a systematic and organized ignorance, which is of all ignorance the most difficult to vanquish and to root out. Theology is nothing more than a science of words, which, by dint of repetition, we accustom ourselves to substitute for things; as soon as we feel disposed to analyze them, we find that they do not present us with any actual sense. There are very few men in the world who think deeply, who render to themselves an account of their ideas, and who have penetrating minds; justness of intellect is one of the rarest gifts which nature bestows on the human species.† Too lively an imagination, an over-eager curiosity, are as powerful obstacles to the discovery of truth, as too much phlegm, a slow conception, indolence of mind, the want of a thinking habit. All men have more or less imagination, curiosity, phlegm, bile, indolence, activity: it is from the just equilibrium, which nature

---

* The progress of sound philosophy will always be fatal to superstition, which nature will continually contradict. Astronomy has caused judiciary astrology to vanish; experimental philosophy, the study of natural history and chymistry, render it impossible for jugglers, priests, and sorcerers, to perform miracles. Nature, deeply studied, must necessarily cause that phantom, which ignorance has substituted in its place, to disappear.

† It is not to be understood here that nature has any choice in the formation of her beings, it is merely to be considered that the circumstances which enable the junction of a certain quantity of those atoms or parts necessary to form a human machine in such due proportions that one disposition shall not overbalance the other, and thus render the judgment erroneous by giving it a particular bias, very rarely occur. We know the process of making gunpowder; nevertheless, it will sometimes happen, that the ingredients have been so happily blended, that this destructive article is of a superior quality to the general produce of the manufactory, without, however, the chymist being on that account entitled to any particular commendation; circumstances have been favourable, and these seldom occur.

has observed in their organization, that justness of mind depends. Nevertheless, as we have heretofore said, the organization of man is subject to change, and the judgment of his mind varies with the changes which his machine is obliged to undergo: hence those almost perpetual revolutions which take place in the ideas of mortals, above all when there is a question concerning those of objects upon which experience does not furnish them with any fixed basis whereon to support them.

To seek and discover truth, which every thing strives to conceal from us, and which (the accomplices of those who lead us astray) we are frequently disposed to dissimulate to ourselves, or which our habitual terrours make us fear to find, there needs a just mind, an upright heart, in good faith with itself, and an imagination tempered with reason. With these dispositions, we shall discover truth, which never shows itself either to the enthusiast, smitten with his reveries; to the superstitious being, nourished with melancholy; to the vain man, puffed up with his presumptuous ignorance; to the man devoted to dissipation and to his pleasures; or to the reasoner, disingenuous with himself, who is only disposed to form illusions to his mind. With these dispositions the attentive philosopher, the geometrician, the moralist, the politician, the theologian himself, when he shall sincerely seek truth, will find that the angular stone, which serves for the foundation of all religious systems, evidently supports falsehood. The philosopher will find in matter a sufficient cause of his existence, of his motion, of his combination, of his modes of acting, always regulated by general laws incapable of varying. The geometrician will calculate the active force of matter, and without quitting nature, he will find that, to explain her phenomena, it is not necessary to have recourse to a being or to a power incommensurable with all known powers. The politician, instructed in the true motive-powers, which can act on the mind of nations, will feel that it is not necessary to recur to imaginary motive-powers, whilst there are real ones to act upon the will of the citizens, and to determine them to labour to the maintenance of their

association; he will acknowledge that a fictitious motive-power is only calculated to slacken or disturb the motion of a machine so complicated as that of society. He who shall more honour truth than the subtilties of theology, will quickly perceive that this vain science is nothing more than an unintelligible heap of false hypotheses, begging of principles, of sophisms, of vitiated circles, of futile distinctions, of captious subtilties, of disingenuous arguments, from which it is not possible there should result any thing but puerilities, or endless disputes. In short, all men who have sound ideas of morality, of virtue, of that which is useful to man in society, whether to conserve himself, or to conserve the body of which he is a member, will acknowledge that men, in order to discover their relations and their duties, have only to consult their own nature, and ought to be particularly careful not to found them upon a contradictory being, or to borrow them from a model which will do more than disturb their minds, and render them uncertain of their proper mode of acting.

Thus every rational thinker, in renouncing his prejudices, may feel the inutility and the falsity of so many abstract systems, which hitherto have only served to confound all our notions and render doubtful the clearest truths. In re-entering his proper sphere, and quitting the regions of the empyreum, where his mind can only bewilder itself; in consulting reason, man will discover that of which he needs a knowledge, and undeceive himself of those chimerical causes which enthusiasm, ignorance, and falsehood, have every where substituted to true causes and to real motive-powers, that act in a nature out of which the human mind can never ramble without going astray, and without rendering itself miserable.

The Deicolists, and all theologians, unceasingly reproach their adversaries, with their taste for *paradoxes*, or for *systems*, whilst they themselves found all their reasoning upon imaginary hypotheses, and make a principle of renouncing experience, of despising nature, of setting down as of no account the evidence of their senses, and of submitting their understanding to the yoke of authority. Would not,

then the *disciples of nature* be justified in saying to these men:—"We
only assure ourselves of that which
we see; we yield to nothing but evidence; if we have a system, it is
founded only upon facts. We perceive in ourselves and every where
else nothing but matter, and we conclude from it, that matter can both feel
and think. We see that every thing
operates in the world after mechanical
laws, by the properties, by the combination, by the modification of matter,
and we seek no other explication of
the phenomena which nature presents.
We conceive only a single and unique
world, in which every thing is linked
together, where each effect is due to
a natural cause, either known or unknown, which it produces according
to necessary laws. We affirm nothing
that is not demonstrable, and which
you are not obliged to admit as well
as us; the principles which we lay
down are clear and evident: they are
facts; if some things be obscure and
unintelligible to us, we ingenuously
agree to their obscurity; that is to say,
to the limits of our own knowledge.*
But we do not imagine an hypothesis
in order to explain these effects; we
either consent to be for ever ignorant
of them, or else we wait until time,
experience, and the progress of the
human mind shall throw a light upon
them. Is not our manner of philoso-
phizing the true one? Indeed, in every
thing which we advance on the subject
of nature, we proceed precisely in the
same manner as our adversaries themselves proceed in all the other sciences,
such as *natural history, natural
philosophy, mathematics, chymistry,
morality, and politics.* We scrupulously confine ourselves to that which
is known to us through the medium of
our senses, the only instruments which
nature has given us to discover truth.
What is the conduct of our adversaries? In order to explain things
which are unknown to them, they
imagine beings still more unknown
than those things which they are desirous of explaining; beings of whom
they themselves acknowledge they have
no one notion! They invert, then, the

true principles of logic, which consist
in proceeding from that which is most
known to that with which we are least
acquainted. But upon what do they
found the existence of these beings by
whose aid they pretend to resolve all
difficulties? It is upon the universal
ignorance of men, upon their inexperience, upon their terrours, upon their
disordered imaginations, upon a pretended *intimate sense,* which is in
reality only the effect of ignorance,
fear, the want of a reflecting habit,
and the suffering themselves to be
guided by authority. Such, O theologians, are the ruinous foundations upon
which ye build the edifice of your doctrine! After this, ye find it impossible
to form to yourselves any precise idea
of those Gods who serve for the basis
of your systems; ye are unable to
comprehend either their attributes,
their existence, the nature of their residence, or their manner of acting.
Thus, even by your own confession,
ye are in a state of profound ignorance
on the primary elements (of which it
is indispensably requisite to have a
knowledge) of a thing which ye constitute the cause of all that exists.
Thus, under whatever point of view
ye are contemplated, it is ye that build
systems in the air, and of all systema-
tizers ye are the most absurd; because,
in relying on your imagination to create a cause, this cause ought at least
to diffuse light over the whole; it is
upon this condition alone that its incomprehensibility could be pardoned:
but can this cause serve to explain any
thing? Does it make us conceive more
clearly the origin of the world, the nature of man, the faculties of the soul,
the source of good and of evil? No,
unquestionably, this imaginary cause
either explains nothing, multiplies of
itself the difficulties to infinity, or
throws embarrassments and obscurity
on all those matters in which they have
made it interpose. Whatever may be
the question agitated, it becomes complicated as soon as they introduce the
name of God: this name envelops the
clearest sciences in clouds, and renders
the most evident notions complicated
and enigmatical. What idea of morality does your Divinity present to
man, upon whose will and example
you found all the virtues? Do not all

---

* Nescire quædam magna pars est sa-
pientiæ.

your revelations show him to us under the character of a tyrant who sports with the human species; who commits evil for the pleasure of doing it, who only governs the world according to the rules of his unjust caprices? All your ingenious systems, all your mysteries, all the subtleties which ye have invented, are they capable of clearing your God, whom ye say is so perfect, from that blackness and atrocity with which good sense cannot fail to accuse him? In short, is it not in his name that ye disturb the universe, that ye persecute, that ye exterminate all who refuse to subscribe to those systematical reveries which ye have decorated with the pompous name of religion. *Acknowledge, then, O theologians! that ye are, not only systematically absurd, but also that ye finish by being atrocious and cruel from the importance which your pride and your interest attach to those ruinous systems, under which ye equally overwhelm human reason and the felicity of nations.*"

CHAPTER XII.

A Summary of the Code of Nature.

Truth is the only object worthy the research of every wise man; since that which is false cannot be useful to him: whatever constantly injures him cannot be founded upon truth; consequently, ought to be for ever proscribed. It is, then, to assist the human mind, truly to labour for his happiness, to point out to him the clew by which he may extricate himself from those frightful labyrinths in which his imagination wanders; from those sinuosities whose devious course makes him err, without ever finding a termination to his incertitude. Nature alone, known through experience, can furnish him with this desirable thread; her eternal energies can alone supply the means of attacking the Minotaur; of exterminating the figments of hypocrisy; of destroying those monsters, who during so many ages, have devoured the unhappy victims, which the tyranny of the ministers of a pretended God have exacted as a cruel tribute from affrighted mortals. By steadily grasping this ines-

timable clew, man can never be led astray—will never ramble out of his course; but if, careless of its invaluable properties, for a single instant he suffers it to drop from his hand; if, like another Theseus, ungrateful for the favour, he abandons the fair bestower, he will infallibly fall again into his ancient wanderings; most assuredly become the prey to the cannibal offspring of the White Bull. In vain shall he carry his views towards heaven, to find resources which are at his feet; so long as man, infatuated with his religious notions, shall seek in an imaginary world the rule of his earthly conduct, he will be without principles; while he shall pertinaciously contemplate the regions of a fanciful heaven, so long he will grope in those where he actually finds himself; his uncertain steps will never encounter the welfare he desires; never lead him to that repose after which he so ardently sighs, nor conduct him to that surety which is so decidedly requisite to consolidate his happiness.

But man, blinded by his prejudices, rendered obstinate in injuring his fellow, by his enthusiasm, ranges himself in hostility even against those who are sincerely desirous of procuring for him the most substantive benefits. Accustomed to be deceived, he is in a state of continual suspicion; habituated to mistrust himself, to view his reason with diffidence, to look upon truth as dangerous, he treats as enemies even those who most eagerly strive to encourage him; forewarned in early life against delusion, by the subtlety of imposture, he believes himself imperatively called upon to guard, with the most sedulous activity, the bandeau with which they have hoodwinked him; he thinks his future welfare involved in keeping it for ever over his eyes; he therefore wrestles with all those who attempt to tear it from his obscured optics. If his visual organs, accustomed to darkness, are for a moment opened, the light offends them; he is distressed by its effulgence; he thinks it criminal to be enlightened; he darts with fury upon those who hold the flambeau by which he is dazzled. In consequence, the atheist is looked upon as a malignant pest, as a public poison, which like another Upas, destroys every thing withiu the vortex of its influence; he who dares to arouse mortals from the lethargic habit which the narcotic doses administered by the theologians have induced, passes for a perturbator; he who attempts to calm their frantic transports, to moderate the fury of their maniacal paroxysms, is himself viewed as a madman, who ought to be closely chained down in the dungeons appropriated to lunatics; he who invites his associates to rend their chains asunder, to break their galling fetters, appears only like an irrational, inconsiderate being, even to the wretched captives themselves: who have been taught to believe, that nature formed them for no other purpose than to tremble: only called them into existence that they might be loaded with shackles. In consequence of these fatal prepossessions, the *Disciple of Nature* is generally treated as an assassin; is commonly received by his fellow-citizens in the same manner as the feathered race receive the doleful bird of night, which, as soon as it quits its retreat, all the other birds follow with a common hatred, uttering a variety of doleful cries.

No, mortals blended by terrour! The friend of nature, is not your enemy; its interpreter is not the minister of falsehood; the destroyer of your vain phantoms is not the devastator of those truths necessary to your happiness; the disciple of reason is not an irrational being, who either seeks to poison you, or to infect you with a dangerous delirium. If he wrests the thunder from the hands of those terrible Gods that affright ye, it is that ye may discontinue your march, in the midst of storms, over roads that ye can only distinguish by the sudden, but evanescent glimmerings of the electric fluid. If he breaks those idols, which fear has served with myrrh and frankincense—which superstition has surrounded by gloomy despondency—which fanaticism has imbrued with blood; it is to substitute in their place those consoling truths that are calculated to heal the desperate wounds ye have received; that are suitable to inspire you with courage, sturdily to oppose yourselves to such dangerous errours; that have power to enable you to resist such

formidable enemies. If he throws down the temples, overturns the altars, so frequently bathed with the bitter tears of the unfortunate, blackened by the most cruel sacrifices, smoked with servile incense, it is that he may erect a fane sacred to peace; a hall dedicated to reason; a durable monument to virtue, in which ye may at all times find an asylum against your own phrensy; a refuge from your own ungovernable passions; a sanctuary against those powerful men, by whom ye are oppressed. If he attacks the haughty pretensions of deified tyrants, who crush ye with an iron sceptre, it is that ye may enjoy the rights of your nature; it is to the end that ye may be substantively freemen, in mind as well as in body; that ye may not be slaves, eternally chained to the oar of misery; it is that ye may at length be governed by men who are citizens, who may cherish their own semblances, who may protect mortals like themselves, who may actually consult the interests of those from whom they hold their power. If he battles with imposture, it is to re-establish truth in those rights which have been so long usurped by fiction. If he undermines the base of that unsteady, fanatical morality, which has hitherto done nothing more than perplex your minds, without correcting your hearts; it is to give to ethics an immoveable basis, a solid foundation, secured upon your own nature; upon the reciprocity of those wants which are continually regenerating in sensible beings: dare, then, to listen to his voice; you will find it much more intelligible than those ambiguous oracles, which are announced to you as the offspring of a capricious Divinity; as imperious decrees that are unceasingly at variance with themselves. Listen, then, to nature, she never contradicts her own eternal laws.

"O thou!" cries this nature to man, "who, following the impulse I have given you, during your whole existence, incessantly tend towards happiness, do not strive to resist my sovereign law. Labour to your own felicity; partake without fear of the banquet which is spread before you, and be happy; you will find the means legibly written on your own heart. Vainly dost thou, O superstitious being! seek after thine happiness beyond the limits of the universe, in which my hand hath placed thee: vainly shalt thou ask it of those inexorable phantoms, which thine imagination, ever prone to wander, would establish upon my eternal throne: vainly dost thou expect it in those celestial regions, to which thine own delirium hath given a locality and a name: vainly dost thou reckon upon capricious deities with whose benevolence thou art in such ecstasies, whilst they only fill thine abode with calamity—thine heart with dread—thy mind with illusions—thy bosom with groans. Dare, then, to affranchise thyself from the trammels of religion, my self-conceited, pragmatic rival, who mistakes my rights; renounce those Gods, who are usurpers of my privileges, and return under the dominion of my laws. It is in my empire alone that true liberty reigns. Tyranny is unknown to its soil; equity unceasingly watches over the rights of all my subjects, maintains them in the possession of their just claims; benevolence, grafted upon humanity, connects them by amicable bonds; truth enlightens them, and never can imposture blind them with his obscuring mists. Return, then, my child, to thy fostering mother's arms! Deserter, trace back thy wandering steps to nature! She will console thee for thine evils; she will drive from thine heart those appalling fears which overwhelm thee; those inquietudes that distract thee; those transports which agitate thee; those hatreds that separate thee from thy fellow-man, whom thou shouldst love as thyself. Return to nature, to humanity, to thyself! Strew flowers over the road of life: cease to contemplate the future; live to thine own happiness; exist for thy fellow-creatures; retire into thyself, examine thine own heart, then consider the sensitive beings by whom thou art surrounded, and leave those Gods who can effect nothing towards thy felicity. Enjoy thyself, and cause others also to enjoy those comforts which I have placed with a liberal hand, for all the children of the earth, who all equally emanate from my bosom: assist them to support the sorrows to which destiny has submitted them in common with thyself. Know, that I approve thy pleasures, when without injuring thyself, they are not fatal to thy brethren, whom I have rendered indispensably necessary to thine own individual happiness. These pleasures are freely permitted thee, if thou indulgest them with moderation; with that discretion which I myself have fixed. Be happy, then, O man! Nature invites thee to participate in it; but always remember, thou canst not be so alone; because I invite all mortals to happiness as well as thyself; thou will find it is only in securing their felicity that thou canst consolidate thine own. Such is the decree of thy destiny: if thou shalt attempt to withdraw thyself from its operation, recollect that hatred will pursue thee; vengeance overtake thy steps; and remorse be ever ready at hand to punish the infractions of its irrevocable decrees.

"Follow, then, O man! in whatever station thou findest thyself, the routine I have described for thee, to obtain that happiness to which thou hast an indispensable right to challenge pretension. Let the sensations of humanity interest thee for the condition of other men, who are thy fellow-creatures; let thine heart have commiseration for their misfortunes; let thy generous hand spontaneously stretch forth to lend succour to the unhappy mortal who is overwhelmed by his destiny; always bearing in thy recollection, that it may fall heavy upon thyself, as it now does upon him. Acknowledge, then, without guile, that every unfortunate has an inalienable right to thy kindness. Above all, wipe from the eyes of oppressed innocence the trickling crystals of agonizing feeling; let the tears of virtue in distress fall upon thy sympathizing bosom; let the genial glow of sincere friendship animate thine honest heart; let the fond attachment of a mate, cherished by thy warmest affection, make thee forget the sorrows of life: be faithful to her love, responsible to her tenderness, that she may reward thee by a reciprocity of feeling; that under the eyes of parents, united in virtuous esteem, thy offspring may learn to set a proper value on practical virtue; that after having occupied thy riper years, they may comfort thy declining age, gild with content thy setting sun, cheer the evening of thine existence, by a dutiful return of that care which thou shalt have bestowed on their imbecile infancy.

"Be just, because equity is the support of human society! Be good, because goodness connects all hearts in adamantine bonds! Be indulgent, because feeble thyself, thou livest with beings who partake of thy weakness! Be gentle, because mildness attracts attention! Be thankful, because gratitude feeds benevolence, nourishes generosity! Be modest, because haughtiness,is disgusting to beings at all times well with themselves. Forgive injuries, because revenge. perpetuates hatred! Do good to him who injureth thee, in order to show thyself more noble than he is; to make a friend of him, who was once thine enemy! Be reserved in thy demeanour, temperate in thine enjoyment, chaste in thy pleasures, because voluptuousness begets weariness, intemperance engenders disease; forward manners are revolting: excess at all times relaxes the springs of thy machine, will ultimately destroy thy being, and render thee hateful to thyself, contemptible to others.

"Be a faithful citizen; because the community is necessary to thine own security; to the enjoyment of thine own existence; to the furtherance of thine own happiness. Be loyal, and submit to legitimate authority; because it is requisite to the maintenance of that society which is necessary to thyself. Be obedient to the laws; because they *are*, or *ought to be*, the expression of the public will, to which thine own particular will ought ever to be subordinate. Defend thy country with zeal; because it is that which renders thee happy, which contains thy property, as well as those beings dearest to thine heart: do not permit this common parent of thyself, as well as of thy fellow-citizens, to fall under the shackles of tyranny; because from thence it will be no more than thy common prison. If thy country, deaf to the equity of thy claims, refuses thee happiness—if, submitted to an unjust power, it suffers thee to be oppressed, withdraw thyself from its bosom in silence, but never disturb its peace.

"In short, be a man; be a sensible, rational being; be a faithful husband; a tender father; an equitable master; a zealous citizen; labour to serve thy country by thy prowess, by thy talents, by thine industry; above all, by thy virtues. Participate with thine associates those gifts which nature has bestowed upon thee; diffuse happiness among thy fellow-mortals; inspire thy fellow-citizens with content; spread joy over all those who approach thee, that the sphere of thine actions, enlivened by thy kindness, illumined by thy benevolence, may react upon thyself; be assured that the man who makes others happy, cannot himself be miserable. In thus conducting thyself, whatever may be the injustice of others, whatever may be the blindness of those beings with whom it is thy destiny to live, thou wilt never be totally bereft of the recompense which is thy due: no power on earth will be able to ravish from thee that never-failing source of the purest felicity, inward content; at each moment thou wilt fall back with pleasure upon thyself; thou wilt neither feel the rankling of shame, the terrour of internal alarm, nor find thy heart corroded by remorse. Thou wilt esteem thyself; thou wilt be cherished by the virtuous, applauded and loved by all good men, whose suffrages are much more valuable than those of the bewildered multitude. Nevertheless, if externals occupy thy contemplation, smiling countenances will greet thy presence; happy faces will express the interest they have in thy welfare; jocund beings will make thee participate in their placid feelings. A life so spent, will each moment be marked by the serenity of thine own mind, by the affection of the beings who environ thee; will be made cheerful by the friendship of thy fellows; will enable thee to rise a contented, satisfied guest from the general feast, conduct thee gently down the declivity of life, lead thee peaceably to the period of thy days, for die thou must: but already thou wilt survive thyself in thought; thou wilt always live in the remembrance of thy friends; in the grateful recollection of those beings whose comforts have been augmented by thy friendly attentions; thy virtues will, beforehand, have erected to thy fame an imperishable monument. If Heaven occupied itself with thee, it would 'feel satisfied with thy conduct, when it shall thus have contented the earth.

"Beware, then, how thou complainest of thy condition ; be just, be kind, be virtuous, and thou canst never be wholly destitute of felicity. Take ♦heed how thou enviest the transient pleasure of seductive crime ; the deceitful power of victorious tyranny ; the specious tranquillity of interested imposture ; the plausible manners of venal justice ; the showy, ostentatious parade of hardened opulence. Never be tempted to increase the number of sycophants to an ambitious despot ; to swell the catalogue of slaves to an unjust tyrant ; never suffer thyself to be allured to infamy, to the practice of extortion, to the commission of outrage, by the fatal privilege of oppressing thy fellows ; always recollect it will be at the expense of the most bitter remorse thou wilt acquire this baneful advantage. Never be the mercenary accomplice of the spoilers of thy country ; they are obliged to blush secretly whenever they meet the public eye.

"For, do not deceive thyself, it is I who punish, more surely than the Gods, all the crimes of the earth ; the wicked may escape the laws of man, but they never escape mine. It is I who have formed the hearts, as well as the bodies of mortals ; it is I who have fixed the laws which govern them. If thou deliverest thyself up to voluptuous enjoyment, the companions of thy debaucheries may applaud thee ; but I shall punish thee with the most cruel infirmities ; these will terminate a life of shame with deserved contempt. If thou givest thyself up to intemperate indulgences, human laws may not correct thee, but I shall castigate thee severely by abridging thy days. If thou art vicious, thy fatal habits will recoil on thine own head. Princes, those terrestrial Divinities, whose power places them above the laws of mankind, are nevertheless obliged to tremble under the silent operation of my decrees. It is I who chastise them ; it is I who fill their breasts with suspicion ; it is I who inspire them with terrour ; it is I who make them writhe under inquietude ; it is I who make them shudder with horrour, at the very name of august truth ; It is I who, amidst the crowd of nobles who surround them, make them feel the inward workings of shame ; the keen anguish of guilt ; the poisoned arrows of regret ; the cruel stings of remorse ; it is I who, when they abuse my bounty, diffuse weariness over their benumbed souls ; it is I who follow uncreated, eternal justice ; it is I who, without distinction of persons, know how to make the balance even ; to adjust the chastisement to the fault ; to make the misery bear its due proportion to the depravity ; to inflict punishment commensurate with the crime. The laws of man are just, only when they are in conformity with mine ; his judgments are rational, only when I have dictated them : my laws alone are immutable, universal, irrefragable ; formed to regulate the condition of the human race, in all ages, in all places, under all circumstances.

"If thou doubtest mine authority, if thou questionest the irresistible power I possess over mortals, contemplate the vengeance I wreak on all those who resist my decrees. Dive into the recesses of the hearts of those various criminals, whose countenances, assuming a forced smile, cover minds torn with anguish. Dost thou not behold ambition tormented day and night, with an ardour which nothing can extinguish ? Dost thou not see the mighty conqueror become the lord of devastated solitudes ; his victorious career, marked by a blasted cultivation, reign sorrowfully over smoking ruins ; govern unhappy wretches who curse him in their hearts ; while his mind, gnawed by remorse, sickens at the gloomy aspect of his own triumphs ? Dost thou believe that the tyrant, encircled with his flatterers, who stun him with their praise, is unconscious of the hatred which his oppression excites ; of the contempt which his vices draw upon him ; of the sneers which his inutility call forth ; of the scorn which his debaucheries entail upon his name? Dost thou think that the haughty courtier does not inwardly blush at the galling insults he brooks, and despise, from the bottom of his heart, those meannesses by which he is compelled to purchase favours ? ' Contemplate the indolent child of wealth, behold him a prey to the lassitude of unmeasured enjoyment, corroded by the satiety which always follows his exhausted pleasures. View the miser with an emaciated countenance, the consequence of his own penurious disposition, whose callous heart is inaccessible to the calls of misery, groaning over the accumulating load of useless treasure, which at the expense of himself, he has laboured to amass. Behold the gay voluptuary, the smiling debauchee, secretly lament the health they have so inconsiderately damaged, so prodigally thrown away: see disunion, joined to hatred, reign between those adulterous married couples. See the liar deprived of all confidence; the knave stript of all trust; the hypocrite fearfully avoiding the penetrating looks of his inquisitive neighbour; the impostor trembling at the very name of formidable truth. Bring under your review the heart of the envious, uselessly dishonoured; that withers at the sight of his neighbour's prosperity. Cast your eyes on the frozen heart of the ungrateful wretch, whom no kindness can warm, no benevolence thaw, no beneficence convert into a genial fluid. Survey the iron feelings of that monster whom the sighs of the unfortunate cannot mollify. Behold the revengeful being nourished with venomous gall, whose very thoughts are serpents; who in his rage consumes himself. Envy, if thou canst, the waking slumbers of the homicide; the startings of the iniquitous judge; the restlessness of the oppressor of innocence; the fearful visions of the extortioner; whose couches are infested with the torches of the furies. Thou tremblest without doubt at the sight of that distraction which, amidst their splendid luxuries, agitates those farmers and receivers of taxes, who fatten upon public calamity—who devour the substance of the orphan—who consume the means of the widow—who grind the hard earnings of the poor: thou shudderest at witnessing the remorse which rends the minds of those reverend criminals, whom the uninformed believe to be happy, whilst the contempt which they have for themselves, the unerring shafts of secret upbraidings, are incessantly revenging an outraged nation. Thou seest, that content is for ever banished the heart —quiet for ever driven from the habitations of those miserable wretches on whose minds I have indelibly engraved the scorn, the infamy, the chastisement which they deserve. But, no! thine eyes cannot sustain the tragic spectacle of my vengeance. Humanity obliges thee to partake of their merited sufferings; thou art moved to pity for these unhappy people, to whom consecrated errours renders vice necessary; whose fatal habits make them familiar with crime. Yes; thou shunnest them without hating them; thou wouldst succour them, if their contumacious perversity had left thee the means. When thou comparest thine own condition, when thou examinest thine own mind, thou wilt have just cause to felicitate thyself, if thou shalt find that peace has taken up her abode with thee; that contentment dwells at the bottom of thine own heart. In short, thou seest accomplished upon them, as well as upon thyself, the unalterable decrees of destiny, which imperiously demand, that crime shall punish itself, that virtue never shall be destitute of remuneration."

Such is the sum of those truths which are contained in the *Code of Nature;* such are the doctrines, which its disciples can announce. They are unquestionably preferable to that supernatural religion which never does any thing but mischief to the human species. Such is the worship that is taught by that sacred reason, which is the object of contempt with the theologian—which meets the insult of the fanatic, who only estimates that which man can neither conceive nor practise; who makes his morality consist in fictitious duties; his virtue in actions generally useless, frequently pernicious to the welfare of society; who, for want of being acquainted with nature, which is before their eyes, believe themselves obliged to seek in ideal worlds imaginary motives, of which every thing proves the inefficacy. The motive which the morality of nature employs, is the self-evident interest of each individual, of each community, of the whole human species, in all times, in every country, under all circumstances. Its worship is the sacrifice of vice, the practice of real virtues; its object is the conservation of the human race, the happiness of the individual, the peace of mankind; its recompenses are affection, esteem, and glory; or in their default, contentment of mind, with merited self-esteem, of which no power will ever be able to deprive virtuous mortals; its punishments, are hatred, contempt, and indignation; which society always reserves for those who outrage its interests; from which even the most powerful can never effectually shield themselves.

Those nations who shall be disposed to practise a morality so wise, who shall inculcate it in infancy, whose laws shall unceasingly confirm it, will neither have occasion for superstition, nor for chimeras. Those who shall obstinately prefer figments to their dearest interests, will certainly march forward to ruin. If they maintain themselves for a season, it is because the power of nature sometimes drives them back to reason, in despite of those prejudices which appear to lead them on to certain destruction. Superstition, leagued with tyranny for the waste of the human species, are themselves frequently obliged to implore the assistance of a reason which they contemn; of a nature which they disdain: which they debase; which they endeavour to crush under the ponderous bulk of their false Divinities. Religion, in all times so fatal to mortals, when attacked by reason, assumes the sacred mantle of public utility; it rests its importance on false grounds, founds its rights upon the indissoluble alliance which it pretends subsists between morality and itself, although it never ceases for a single instant to wage against it the most cruel hostility. It is, unquestionably, by this artifice, that it has seduced so many sages. In the honesty of their hearts, they believe it useful to politics; necessary to restrain the ungovernable fury of the passions; thus hypocritical superstition, in order to mask to superficial observers its own hideous character, always knows how to cover itself with the sacred armour of utility; to buckle on the invulnerable shield of virtue; it has, therefore, been believed imperative to respect it, and favour imposture, because it has artfully entrenched itself behind the altars of truth. It is from this intrenchment we ought to drive it; it should

No. XI.—43

be dragged forth to public view; stripped of its surreptitious panoply; exposed in its native deformity; in order that the human race may become acquainted with its dissimulation; that mankind may have a knowledge of its crimes; that the universe may behold its sacrilegious hands, armed with homicidal poniards, stained with the blood of nations, whom it either intoxicates with its fury, or immolates without pity to the violence of its passions.

The Morality of Nature is the only religion which her interpreter offers to his fellow-citizens, to nations, to the human species, to future races, weaned from those prejudices which have so frequently disturbed the felicity of their ancestors. The friend of mankind cannot be the friend of God, who at all times has been a real scourge to the earth. The Apostle of Nature will no be the instrument of deceitful chimeras, by which this world is made only an abode of illusions; the adorer of truth will not compromise with falsehood; • he will make no covenant with errour, conscious it must always be fatal to mortals. He knows that the happiness of the human race imperiously exacts that the dark unsteady edifice of superstition should be razed to its foundations, in order to elevate on its ruins a temple to nature suitable to peace—a fane sacred to virtue. He feels it is only by extirpating, even to the most slender fibres, the poisonous tree, that during so many ages has overshadowed the universe, that the inhabitants of this world will be able to use their own eyes—to bear with steadiness that light which is competent to illumine their understanding—to guide their wayward steps—to give the necessary ardency to their minds. If his efforts should be vain; if he cannot inspire with courage beings too much accustomed to tremble; he will, at least, applaud himself for having dared the attempt. Nevertheless, he will not judge his exertions fruitless, if he has only been enabled to make a single mortal happy: if his principles have calmed the conflicting transports of one honest mind; if his reasonings have cheered up some few virtuous hearts. At least he will have the advantage of having banished from his own mind the importunate terrour of superstition;

of having expelled from his own heart the gall which exasperates zeal : of having trodden under foot those chimeras with which the uninformed are tormented. Thus, escaped from the peril of the storm, he will calmly contemplate from the summit of his rock, those tremendous hurricanes which superstition excites; he will hold forth a succouring hand to those who shall be willing to accept it; he will encourage them with his voice; he will second them with his unwearied exertions, and in the warmth of his own compassionate heart, he will exclaim :—

" O Nature, sovereign of all beings ! and ye, her adorable daughters, Virtue, Reason, and Truth: it is to you that our only Divinities: it is to you that belong the praises of the human race : to you appertains the homage of the earth. Show us, then, O Nature! that which man ought to do, in order to obtain the happiness which thou makest him desire. Virtue! Animate him with thy beneficent fire. Reason! Conduct his uncertain steps through the paths of life. Truth! Let thy torch illumine his intellect, dissipate the darkness of his road. Unite, O assisting Deities! your powers, in order to submit the hearts of mankind to your dominion. Banish errour from our mind; wickedness from our hearts; confusion from our footsteps; cause knowledge to extend its salubrious reign; goodness to occupy our minds; serenity to dwell in our bosoms. Let imposture, confounded, never again dare to show its head. Let our eyes, so long either dazzled or blindfolded, be at length fixed upon those objects we ought to seek. Dispel for ever those mists of ignorance, those hideous phantoms, together with those seducing chimeras,

which only serve to lead us astray. Extricate us from that dark abyss into which we are plunged by superstition; overthrow the fatal empire of delusion; crumble the throne of falsehood ; wrest from their polluted hands the power they have usurped. Command men, without sharing your authority with mortals : break the chains that bind them down in slavery : tear away the bandeau by which they are hoodwinked ; allay the fury that intoxicates them; break in the hands of sanguinary, lawless tyrants, that iron sceptre with which they are crushed ; exile to the imaginary regions, from whence fear has imported them, those Gods by whom they are afflicted. Inspire the intelligent being with courage; infuse energy into his system, that, at length, he may feel his own dignity ; that he may dare to love himself; to esteem his own actions when they are worthy ; that a slave only to your eternal laws, he may no longer fear to enfranchise himself from all other trammels ; that blest with freedom, he may have the wisdom to cherish his fellow-creature ; and become happy by learning to perfection his own condition ; instruct him in the great lesson, that the high road to felicity, is prudently to partake himself, and also to cause others to enjoy, the rich banquet which thou, O Nature ! hast so bountifully set before him. Console thy children for those sorrows to which their destiny submits them, by those pleasures which wisdom allows them to partake; teach them to yield silently to necessity. Conduct them without alarm to that period which all beings must find ; *let them learn that time changes all things, that consequently they are made neither to avoid death nor to fear its arrival.*"

APPENDIX.

TRUE MEANING

OF THE

SYSTEM OF NATURE.

INTRODUCTION.

MAN, unfortunately for himself, wishes to exceed the limits of his sphere, and to transport himself beyond the visible world. He neglects experience, and feeds himself with conjectures. Early prepossessed by artful men against reason, he neglects its cultivation. Pretending to know his fate in another world, he is inattentive to his happiness in the present. The author's object is, to recal man to reason by rendering it dear to him,—to dissipate the clouds which obscure the way to this happiness,—to offer reflections useful to his peace and comfort, and favourable to mental improvement.

So far from wishing to destroy the duties of morality, it is the author's object to give them double force, and establish them on the altar of virtue, which alone merits the homage of mankind.

CHAPTER I.
On Nature.

MAN is the work of nature, and subject to her laws, from which he cannot free himself, nor even exceed in thought. A being formed by nature, he is nothing beyond the great whole of which he forms a part. Beings supposed to be superior to, or distinguished from, nature, are mere chimeras, of which no real idea can he formed.

Man is a being purely physical. The moral man is only the physical man, considered in a certain point of view. His organization is the work of nature; his visible actions and invisible movements are equally the natural effects and consequences of his mechanism. His inventions are the effect of his essence. His ideas proceed from the same cause. Art is only nature, acting by instruments which she has herself made—all is the impulse of nature.

It is to physics and experience, that man in all his researches ought to have recourse. Nature acts by simple laws. When we quit experience, imagination leads us astray. 'Tis from want of experience that men have formed wrong ideas of matter.*

* Men have fallen into a thousand errours, by ascribing an existence to the objects of our interior perceptions, distinct from ourselves, in the same manner as we conceive them separately. It becomes of importance, therefore, to examine the nature of the distinctions which subsist among those objects. Some of these are so distinct from others, that they cannot exist together. The surface of a body cannot at the same time be both white and black in all its parts: nor can one body be more or less extended than another of the same dimensions. Two ideas, thus distinguished, necessarily exclude one another: since the existence of one of them necessarily infers the non-existence of the other, and, consequently, its own separate and independent existence. This class I call real or exclusive existence.

But there is another class, which, in opposition to the former, I call fictitious, or imaginary existence. While a body is passing

Indolence is gratified in following example: habit, and authority, rather than experience, which demands activity, or reason, which requires reflection. Hence an aversion to every thing that deviates from ordinary rules, and

from one colour or shape to another, we successively experience different sensations: yet it is evident that we remain the same, it only being that body which changes colour or shape. But the body is neither its colour nor shape, since it could exist without them, and still be the same body. Neither is the shape or figure of a body, its colour, motion, extent, nor hardness; because those qualities are distinct from each other, and any of them can exist separate from and independent of the rest. But as they can exist together, they are not distinguished like those which cannot exist together at the same time. They cannot have a separate and distinct existence from bodies whose properties they are. The same power by which a white body exists, is that by which its whiteness also exists. What we call whiteness cannot exist of itself, separate from a body. This is the distinction between things capable of being separated, though found joined together, and which, though exciting in us different impressions, may yet be separately considered, and become so many distant objects of perception. This class of imaginary or fictitious objects, existing only in our mind, must not be confounded with the first class of objects, which have a real, exclusive, and independent existence of their own.

Innumerable errours have arisen by confounding those distinctions. In mathematics, for example, we hear every moment of points and lines, or extensions without length, and surfaces having length and breadth without depth,—though geometers themselves confess, that such bodies neither do nor can exist, but in the mind, while every body in nature is truly extended in every sense. Unskilful materialists have fallen into gross absurdities, by mistaking, for real and distinct existences, the different properties of extension, separately considered by mathematicians. Hence, they formed the world of atoms, or small bodies, without either bulk or extension, yet possessing infinite hardness, and a great variety of forms. Bodies such as those can only exist in the minds of atomists.

If even able men can be so clumsily deceived, by not distinguishing between the real existence of external bodies, and the fictitious existence of perceptions, existing only in the mind, it is not to be wondered at, that a multitude of errours should have arisen, in comparing, not only those perceptions themselves, but even their mutual relations with one another.

I do not say, that sensations can exist separate from ourselves. The sentiments of pleasure and pain, though not distinct from him who feels them, certainly are so from my mind, which perceives, reflects upon, and

an implicit respect for ancient institutions. Credulity proceeds from in experience. By consulting experience and contemplating the universe, we shall only find in it matter and motion.

compares them with other sensations. As the sentiment of real existence is clearer than that of imaginary or fictitious, we imagine that a similar distinction exists between all the objects that the mind conceives. Hence the operations of mind, and its different properties, have been considered, like real beings, as so many entities having a real existence of their own, and have thus acquired a physical existence, which they do not possess of themselves. Hence our mind has been distinguished from ourselves, as the part is from its whole. The mind itself has been separated from the soul, or that which animates, from that which makes us live. In the mind, a distinction has been made between the understanding and the will; in other words, between that which perceives and that which wills, that which wills and that which wills not. Our perceptions have been distinguished from ourselves, and from one another; hence thoughts, ideas, &c., which are nothing but the faculty of perception itself, viewed in relation to some of its functions. All these, however, are only modifications of our essence, and no more distinguished from themselves, nor from us, than extension, solidity, shape, colour, motion, or rest, from the same body. Yet absolute distinctions have been made between them, and they have been considered as so many small entities, of which we form the assemblage. According, therefore, to those philosophers, we are composed of thousands of little bodies, as distinct from one another as the different trees in a forest, each of which exists by a particular and independent power.

With regard to things really distinct from us, not only their properties, but even the relations of those properties, have been distinguished from themselves, and from one another; and to these a real existence has been given. It was observed, that bodies act upon, strike and repel one another, and, in consequence of their action and reaction, changes were produced in them. When, for example, I put my hand to the fire, I feel what is called heat; in this case, fire is the cause, and heat the effect. To abridge language, general terms, applying to particular ideas of a similar nature, were invented. The body that produces the change in another, was called the cause,— and the body suffering the change, the effect. As those terms produce in the mind some idea of existence, action, reaction, and change, the habit of using them makes men believe that they have a clear and distinct perception of them. By the continual use of these words, me nhave at length believed, that there can exist a cause, neither a substance, nor a body; a cause, though distinct from all matter, without either action or reaction, yet capable of producing every supposable effect.

CHAPTER II.

Of Motion and its Origin.

IT is motion which alone forms the connexions between our organs and external and internal objects. A cause is a being that puts another in motion, or which produces the change that one body effects upon another by means of motion. We only know the manner in which a body acts upon us by the change it produces. It is from actions only that we can judge of interior motions, as thoughts, and other sentiments—when we see a man flying we conclude him to be afraid.

The motion of bodies is a necessary consequence of their essence. Every being has laws of motion peculiar to itself.

Every body in the universe is in motion. Action is essential to matter. All beings but come into existence, increase, diminish, and ultimately perish; metals, minerals, &c., are all in action. The stones which lie upon the ground act upon it by pressure. Our sense of smell is acted upon by emanations from the most compact bodies.

Motion is inherent in nature, which is the great whole, out of which nothing can exist, and is essential to it. Matter moves by its own energy, and possesses properties, according to which it acts.

In attributing the motion of matter to a cause, we must suppose, that matter itself has come into existence—a thing impossible; for since it cannot be annihilated, how can we imagine it to have had a beginning? Whence has matter come? It has always existed. What is the original cause of its motion? Matter has always been in motion, as motion is a consequence of its existence, and existence always supposes properties in the existing body. Since matter possesses properties, its manner of action necessarily flows from its form of existence. Hence a heavy body must fall.

CHAPTER III.

Of Matter and its Motion.

THE changes, forms, and modifications of matter alone proceed from

motion. By motion, every body in nature is formed, changed, enlarged, diminished, and destroyed.

Motion produces a perpetual transmigration, exchange, and circulation of the particles of matter. These particles separate themselves to form new bodies. One body nourishes other bodies; and those afterwards restore to the general mass the elements which they had borrowed from it. Suns are produced by the combinations of matter; and these wonderful bodies, which man in his transitory existence only sees for a moment, will one day, perhaps, be dispersed by motion.

CHAPTER IV.

Laws of Motion common to all Beings—Attraction and Repulsion—Necessity.

WE consider effects as natural, when we see their acting cause. When we see an extraordinary effect, whose cause is unknown to us, we have recourse to imagination, which creates chimeras.

The visible end of all the motions of bodies, is the preservation of their actual form of existence, attracting what is favourable, and repelling what is prejudicial to it. From the moment of existence, we experience motions peculiar to a determined essence.

Every cause produces an effect, and there cannot be an effect without a cause. If every motion, therefore, be ascribable to a cause; and these causes being determined by their nature, essence and properties; we must conclude, that they are all necessary, and that every being in nature, in its given properties and circumstances, can only act as it does. Necessity is the infallible and constant tie of causes to their effects: and this irresistible power, universal necessity, is only a consequence of the nature of things, in virtue of which the whole acts by immutable laws.*

* Changes are produced in bodies by their action and reaction upon one another. The same body, at present a cause, was previously an effect; or, in other words, the body which produces a change in another, by acting upon it, has itself undergone a change by the action of another body. One body may, in relation to others, be, at the same time,

CHAPTER V.
Of Order and Disorder—Intelligence and Chance.

THE view of the regular motions of nature produces, in the human mind, the idea of order. This word only expresses a thing relative to ourselves. The idea of order or disorder is no proof that they really exist in nature, since there every thing is necessary. Disorder in relation to a being is nothing but its passage into a new order or form of existence. Thus, in our eyes, death is the greatest of all disorders; but death only changes our essence. We are not less subject afterwards to the laws of motion.

Intelligence is called the power of acting according to an end, which we know the being possesses to whom we ascribe it. We deny its existence in beings whose forms of action are different from ours.

When we do not perceive the connexion of certain effects with their causes, we attribute them to chance. When we see, or think we see, what is called order, we ascribe it to an intelligence, a quality borrowed from ourselves, and from the particular form in which we are affected.

An intelligent being thinks, wills, and acts, to arrive at an end. For this purpose, organs, and an end similar to our own, are necessary. They would above all be necessary to an intelligence supposed to govern nature, as without organs, there can neither be ideas, intuition, thought, will, plan nor action. Matter, when combined in a certain manner, assumes action, intelligence, and life.*

both cause and effect. While I push forward a body with the stick in my hand, the motion of the stick, which is the effect of my impulse, is the cause of the progression of the body that is pushed. The word *cause*, only denotes the perception of the change which one body produces in another, considered in relation to the body that produces it; and the word *effect*, signifies nothing more than the perception of the same change, considered relatively to the body that suffers it. The absurdity of supposing the existence of independent and absolute causes, which neither are nor can be effects, must appear obvious to every unbiassed understanding.

The infinite progression of bodies which have been in succession, cause and effect, soon fatigued men desirous of discovering a general cause for every particular effect. They all at once, therefore, ascended to a first cause, supposed to be universal, in relation to which every particular cause is an effect, though not itself the effect of any cause. The only idea they can give of it is, that it produced all things; not only the form of their existence, but even their existence itself. It is not, according to them, either a body, or a being like particular beings; in a word, it is the universal cause. And this is all they can say about it.

From what has been said, it must appear, that this universal cause is but a chimera, a mere phantom, at most an imaginary or fictitious being, only existing in the minds of those who consider it. It is, however, the Destiny of the Greeks,—the God of philosophers, Jews, and Christians,—the Benevolent Spirit of the new Parisian sect of Saint-Simonians; the only sect which has ever yet attempted to found a worship upon principles bearing any resemblance to morality, reason, or common sense.

Those who, without acknowleaging this universal cause, content themselves with particular causes, have generally distinguished them from material substances. Seeing the same change often produced by different actions or causes, they conceived the existence of particular causes, distinct from sensible bodies. Some have ascribed to them intelligence and will,—hence gods, demons, genii, good and bad spirits. Others, who cannot conceive the existence of a mode of action different from their own, have imagined certain virtues to proceed from the influence of the stars, chance, and a thousand other dark, unintelligible terms, which signify nothing more than blind and necessary causes.

* Among the innumerable errours into which men are continually falling, by confounding fictitious with real objects, is that of supposing an infinite power, cause, wisdom, or intelligence, to exist, from only considering the properties of wisdom, power, and intelligence, in the beings whom they see. The term *infinite* is totally incompatible with the existence of any thing finite, positive or real: in other words, it carries with it the impossibility of real existence. Those who call a power, quantity, or number infinite, speak of something undetermined, of which no just idea can be formed; because, however extended the idea may be, it must fall short of the thing represented. An infinite number, for example, can neither be conceived nor expressed. Admitting for a moment, the existence of such a number, it may be asked, whether a certain part, the half for example, may not be taken from it? This half is finite, and may be counted and expressed; but by doubling it, we make a sum equal to an infinite number, which will then be determined, and to which a unit may at least be added. This sum will then be greater than it was before, though infinite, or that to which nothing could be added, yet we can make no addition to it! It is, therefore, at the same time, both

CHAPTER VI.

Of Man, his Physical and Moral Distinctions—His Origin.

MAN is always subject to necessity. His temperament is independent of him, yet it influences his passions. His blood more or less abundant or warm, his nerves more or less relaxed, the aliments upon which he feeds, all act upon him and influence him.

Man is an organized whole, composed of different matters, which act according to their respective properties. The difficulty of discovering the causes of his motions and ideas, produced the division of his essence into two natures. He invented words, because ignorant of things.

Man, like every thing else, is a production of nature. What is his origin? We want experience to answer the question.

Has he always existed, or is he an instantaneous production of nature? Either of the cases is possible. Matter is eternal, but its forms and combinations are transitory. It is probable, that he was produced at a particular period of our globe, upon which he, like its other productions, varies according to the difference of climate. He was doubtless produced male and female, and will exist so long as the globe remains in its present state. When that is changed, the human species must give way to new beings, capable of incorporating themselves with the new qualities which the globe will then possess.

When we are unable to account for the production of man, to talk of God and of creation, is but confessing our ignorance of the energy of nature.

infinite and finite, and consequently possesses properties exclusive of one another. We might, with equal propriety, conceive the existence of a white body which is not white, or, in other words, a mere chimera; all we can say of which is, that it neither does nor can exist. What has been said of an infinite number, equally applies either to an infinite cause, intelligence, or power. As there are different degrees of causation, intelligence, and power, those degrees must be considered as units, the sum of which will express the quantity of the power, and intelligence, of such causes. An infinity of power, action, or intelligence, to which nothing can be added, nor conceived, is impossible, never has existed, and never can exist.

Man has no right to believe himself a privileged being in nature. He is subject to the same vicissitudes as its other productions. The idea of human excellence is merely founded on the partiality which man feels for himself.

CHAPTER VII.

Of the Soul and its Spirituality.

WHAT is called the soul moves with us. Now, motion is a property of matter. The soul also shows itself material in the invincible obstacles which it encounters on the part of the body. If the soul causes me to move my arm when there is no obstacle in the way, it ceases doing so when the arm is pressed down by a heavy weight. Here then is a mass of matter which annihilates an impulse given by a spiritual cause, which, being unconnected with matter, ought to meet with no resistance from it.

Motion supposes extent and solidity in the body that is moved. When we ascribe action to a cause, we must therefore consider that cause to be material.

While I walk forward, I do not leave my soul behind me. Soul, therefore, possesses one quality in common with the body and peculiar to matter. The soul makes a part of the body, and experiences all its vicissitudes, in passing through a state of infancy and of debility, in partaking of its pleasures and pains; and with the body exhibiting marks of dulness, debility, and death. In short, it is only the body viewed in relation to some of its functions.

What sort of substance is it which can neither be seen nor felt? An immaterial being, yet acting upon matter! How can the body inclose a fugitive being, which eludes all the senses.

CHAPTER VIII.

Of the Intellectual Faculties—All derived from Sensation.

SENSATION is a manner of being affected, peculiar to certain organs of animated bodies, occasioned by the presence of a material object. Sensibility is the result of an arrangement peculiar to animals. The organs reciprocally communicate impressions to one another.

Every sensation is a shock given to organs; a perception, that shock communicated to the brain; an idea the image of the object which occasioned the sensation and perception. If our organs, therefore, be not moved, we can neither have perceptions nor ideas.

Memory produces imagination. We form a picture of the things we have seen, and, by imagination, transport ourselves to what we do not see.

Passions are movements of the will, determined by the objects which act upon it, according to our actual form of existence.

The intellectual faculties attributed to the soul, are modifications ascribable to the objects which strike the senses. Hence a trembling in the members, when the brain is affected by the movement called fear.*

---

* Man is born with a disposition to know, or to feel and receive impressions from the action of other bodies upon him. Those impressions are called sensations, perceptions, or ideas. These impressions leave a trace or vestige of themselves, which are sometimes excited in the absence of the objects which occasioned them. This is the faculty of memory, or the sentiment by which a man has a knowledge of former impressions, accompanied by a perception of the distinction between the time he received, and that in which he remembers them.

Every impression produces an agreeable or disagreeable sensation. When lively, we call it pleasure, or pain; when feeble, satisfaction, ease, inconvenience, or uneasiness. The first of these sentiments impels us towards objects, and makes us use efforts to join and attach them to ourselves, to augment and prolong the force of the sensation, to renew and recall it when it ceases. We love objects which produce such sensations, and are happy in possessing them: we seek and desire their possession, and are miserable upon losing them. The sentiment of pain induces us to fly and shun objects which produce it, to fear, hate, and detest their presence.

We are so constituted, as to love pleasure and hate pain; and this law, engraven by nature on the heart of every human being, is so powerful, that in every action of life it forces our obedience. Pleasure is attached to every action necessary to the preservation of life, and pain to those of an opposite nature. Love of pleasure, and hatred of pain, induce us, without either examination or reflection, to act so as to obtain possession of the former and the absence of the latter.

The impressions once received, it is not in man's power either to prolong or to render them durable. There are certain limits beyond which human efforts cannot exceed. Some impressions are more poignant than others, and render us either happy or miserable. An impression, pleasant at its commencement, frequently produces pain in its progress. Pleasure and pain are so much blended together, that it is seldom that the one is felt without some part of the other.

Man, like every other animal, upon coming into the world, abandons himself to present impressions, without foreseeing their conse-

quences or issue. Foresight can only be acquired by experience, and reflection upon the impressions communicated to us by objects. Some men, in this respect, continue infants all their lives, never acquiring the faculty of foresight; and even among the most wise, few are to be found, upon whom, at some periods of life, certain violent impressions, those of love, for example, the most violent of all, have not reduced into a state of childhood, foreseeing nothing, and permitting themselves to be guided by momentary impulses.

As we advance in years, we acquire more experience in comparing new and unknown objects with the idea or image of those whose impression memory has preserved. We judge of the unknown from the known, and consequently, know whether those ought to be sought for or avoided.

The faculty of comparing present with absent objects, which exist only in the memory, constitutes reason. It is the balance with which we weigh things; and by recalling those that are absent, we can judge of the present, by their relations to one another. This is the boasted reason which man, upon I know not what pretext, arrogates to himself to the exclusion of all other animals. We see all animals possessing evident marks of judgment and comparison. Fishes resort to the same spot at the precise hour in which they have been accustomed to receive food. The weaker animals form themselves into societies for mutual defence. The sagacity of the dog is generally known, and the foresight of the bee has long been proverbial. The bears of Siberia, and the elephants of India, seem to possess a decided superiority in understanding over the human savages and slaves, who inhabit those countries.

Some philosophers suppose the existence of the sense of touch in man, in a superior degree than in other animals, sufficient to account for his superiority over them. If to that we add, the advantage of a greater longevity, and a capacity of supporting existence all over the globe, an advantage peculiar to the human species, perhaps we have enumerated all the causes of superiority which man ever received from nature, whatever may be his pretensions. Speech, or the power of communicating ideas, is common to almost all animals. Some of them even possess it

CHAPTER IX.

Diversity of the Intellectual Faculties—They depend, like the Moral Qualities, on Physical Causes.—Natural Principles of Society.

TEMPERAMENT decides the moral qualities. This we have from nature, and from our parents. Its different kinds are determined by the quality of the air we breathe, by the climate we inhabit, by education, and the ideas it inspires. By making mind spiritual, we administer to it improper remedies. Constitution, which can be changed, corrected and modified, should alone be the object of our attention.

Genius is an effect of physical sensibility. It is the faculty possessed by some human beings, of seizing, at one glance, a whole and its different parts. By experience, we foresee effects not yet felt—hence prudence and foresight. Reason is nature modified by experience.

The final end of man is self-preservation, and rendering his existence happy. Experience shows him the need he stands in of others to attain that object, and points out the means of rendering them subservient to his views. He sees what is agreeable or disagreeable to them, and these experiences give him the idea of justice, &c. Neither virtue nor vice are founded on conventions, but only rest upon relations subsisting among all human beings.

Men's duties to one another arise from the necessity of employing those means which tend to the end proposed by nature. It is by promoting the happiness of other men, that we engage them to promote our own.

Politics should be the art of directing the passions of men to the good of society. Laws ought to have no other object than the direction of their actions also to the same object.

Happiness is the uniform object of all the passions. These are legitimate and natural, and can neither be called good or bad, but in so far as they affect

in a higher degree than man in certain states of society. Damp erre describes a nation, whose speech consisted in the howling of a few guttural sounds, and whose vocabulary did not contain more than thirty words.

No. XI.—44

other men. To direct the passions to virtue, it is necessary to show mankind advantages resulting from its practice.

CHAPTER X.

The Mind draws no Ideas from itself—We have no Innate Ideas.

IF we can only form ideas of material objects, how can their cause be supposed immaterial?.

To this, dreams are opposed as an objection; but in sleep the brain is filled with a crowd of ideas which it received when awake. Memory always produces imagination. The cause of dreams must be physical, as they most frequently proceed from food, humours, and fermentations, unanalogous to the healthy state of man.

The ideas supposed to be innate, are those which are familiar to, and, as it were, incorporated with us; but it is always through the medium of the senses that we acquire them. They are the effect of education, example, and habit. Such are the ideas formed of God, which evidently proceed from the descriptions given of him.

Our moral ideas are the fruits of experience alone. The sentiments of paternal and filial affection are the result of reflection and habit.

Man acquires all his notions and ideas. The words beauty, intelligence, order, virtue, grief, pain and pleasure, are, to me, void of meaning, unless I compare them with other objects. Judgment presupposes sensibility; and judgment itself is the fruit of comparison.

CHAPTER XI.

Of the System of Man's Liberty.

MAN is a physical being, subject to nature, and consequently to necessity. Born without our consent, our organization is independent of us, and our ideas come to us involuntarily. Action is the sequel of an impulse communicated by a sensible object.

I am thirsty, and see a well; can I hinder myself from wishing to drink of it? But I am told, the water is poisoned, and I abstain from drinking

Will it be said, that in this case I am free? Thirst necessarily determined me to drink; the discovery of poison necessarily determines me not to drink. The second motive is stronger than the first, and I abstain from drinking. But an imprudent man, it may be said, will drink. In this case his first impulse will be strongest. In either case, the action is necessary. He who drinks is a madman; but the actions of madmen are not less necessary than those of other men.

A debauchee may be persuaded to change his conduct. This circumstance does not prove that he is free; but only, that motives can be found, sufficient to counteract the effect of those which formerly acted upon him.

Choice by no means proves liberty; since hesitation only finishes when the will is determined by sufficient motives; and man cannot hinder motives from acting upon his will. Can he prevent himself from wishing to possess what he thinks desirable? No; but we are told he can resist the desire, by reflecting upon its consequences. But has he the power of reflecting? Human actions are never free; they necessarily proceed from constitution, and from received ideas, strengthened by example, education, and experience. The motive which determines man is always beyond his power.

Notwithstanding the system of human liberty, men have universally founded their systems upon necessity alone. If motives were thought incapable of influencing the will, why make use of morality, education, legislation, and even religion? We establish institutions to influence the will; a clear proof of our conviction, that they must act upon it. These institutions are necessity demonstrated to man.

The necessity that governs the physical, governs also the moral world, where every thing is also subject to the same law.

————

CHAPTER XII.

Examination of the Opinions which maintain the System of Necessity to be Dangerous.

IF men's actions are necessary, by what right, it is asked, are crimes punished, since involuntary actions are never the objects of punishment?

Society is an assemblage of sensible beings, susceptible of reason, who love pleasure, and hate pain. Nothing more is necessary to engage their concurrence to the general welfare. Necessity is calculated to impress all men. The wicked are madmen against whom others have a right to defend themselves. Madness is an involuntary and necessary state, yet madmen are confined. But society should never excite desires, and afterwards punish them. Robbers are often those whom society has deprived of the means of subsistence.

By ascribing all to necessity, we are told the ideas of just and unjust, of good and evil, are destroyed. No; though no man acts from necessity, his actions are just and good relative to the society whose welfare he promotes. Every man is sensible that he is compelled to love a certain mode of conduct in his neighbour. The ideas of pleasure and pain, vice and virtue, are founded upon our own essence.

Fatalism neither emboldens crime, nor stifles remorse, always felt by the wicked. They have long escaped blame or punishment, they are not on that account better satisfied with themselves. Amidst perpetual pangs, struggles, and agitations, they can neither find repose nor happiness. Every crime costs them bitter torments and sleepless nights. The system of fatality establishes morality, by demonstrating its necessity.

Fatality, it is said, discourages man, paralyzes his mind, and breaks the ties that connect him with society. But does the possession of sensibility depend upon myself? My sentiments are necessary, and founded upon nature. Though I know that all men must die, am I on that account, the less affected by the death of a wife, a child, a father, or a friend?

Fatalism ought to inspire man with a useful submission and resignation to his fate. The opinion, that all is necessary, will render him tolerant. He will lament and pardon his fellowmen. He will be humble and modest, from knowing that he has received every thing which he possesses.

Fatalism, it is said, degrades man

into a mere machine. Such language is the invention of ignorance, respecting what constitutes his true dignity. Every machine is valuable, when it performs well the functions to which it is destined. Nature is but a machine, of which the human species makes a part. Whether the soul be mortal or immortal, we do not the less admire its grandeur and sublimity in a Socrates.

The opinion of fatalism is advantageous to man. It prevents useless remorse from disturbing his mind. It teaches him the propriety of enjoying with moderation, as pain ever accompanies excess. He will follow the paths of virtue, since every thing shows its necessity for rendering him estimable to others and contented with himself.

CHAPTER XIII.

Of the Soul's Immortality—The Dogma of a Future State—Fear of Death.

THE soul, step by step, follows the different states of the body. With the body, it comes into existence, is feeble in infancy, partakes of its pleasures and pains, its states of health and disease, activity or depression; with the body, is asleep or awake, and yet it has been supposed immortal!

Nature inspires man with the love of existence, and the desire of its continuation produced the belief of the soul's immortality. Granting the desire of immortality to be natural, is that any proof of its reality? We desire the immortality of the body, and this desire is frustrated. Why should not the desire of the soul's immortality be frustrated also?

The soul is only the principle of sensibility. To think, to suffer, to enjoy, is to feel. When the body, therefore, ceases to live, it cannot exercise sensibility. Where there are no senses, there can be no ideas. The soul only perceives by means of the organs: how then is it possible for it to feel, after the dissolution?

We are told of divine power—but divine power cannot make a thing exist and not exist at the same time. It

cannot make the soul think without the means necessary to acquire thoughts.

The destruction of his body always alarms man, notwithstanding the opinion of the soul's immortality; a sure proof that he is more affected by the present reality, than by the hope of a distant futurity.

The very idea of death is revolting to man, yet he does every thing in his power to render it more frightful. It is a period which delivers us up defenceless to the undescribable rigours of a pitiless despot. This, it is said, is the strongest rampart against human irregularities. But what effect have those ideas produced upon those who are, or at least pretend to be, persuaded of their truth? The great bulk of mankind seldom think of them; never, when hurried along by passion, prejudice, or example. If they produce any effect, it is only upon those to whom they are unnecessary in urging to do good, and restraining from evil. They fill the hearts of good men with terrour, but have not the smallest influence over the wicked.

Bad men may be found among infidels, but infidelity by no means implies wickedness. On the contrary, the man who thinks and meditates, better knows motives for being good, than he who permits himself to be blindly conducted by the motives of others. The man who does not expect another state of existence, is the more interested in prolonging his life, and rendering himself dear to his fellow-men, in the only state of existence with which he is acquainted. The dogma of a future state destroys our happiness in this life; we sink under calamity, and remain in errour, in expectation of being happy hereafter.

The present state has served as the model of the future. We feel pleasure and pain—hence a heaven and a hell. A body is necessary for enjoying heavenly pleasures—hence the dogma of a resurrection.

But whence has the idea of hell arisen? Because, like a sick person who clings even to a miserable existence, man prefers a life of pain to annihilation, which he considers as the greatest of calamities. That notion was besides counterbalanced by the idea of divine mercy.

Did not men, by a happy inconsistency, deviate in their conduct from those insolent ideas, the terrours ascribed to a future state are so strong, that they would sink into brutality, and the world become a desert.

Although this dogma may operate upon the passions, do we see fewer wicked men among those who are the most firmly persuaded of its truth? Men who think themselves restrained by those terrours, impute to them effects ascribable only to present motives, such as timidity, and apprehension of the consequences of doing a bad action. Can the fears of a distant futurity restrain the man upon whom those of immediate punishment produce no effect?

Religion itself destroys the effect of those terrours. The remission of sin emboldens the wicked man to his last moment. This dogma is consequently opposed to the former.

The inspirers of those terrours admit them to be ineffectual; priests are continually lamenting that man is still hurried on by his vicious inclinations. In fine, for one timid man who is restrained by those terrours, there are millions whom they render ferocious, useless, and wicked, and turn aside from their duties to society, which they are continually tormenting.

CHAPTER XIV.

Of Education—Morality and Laws sufficient to Restrain Man—Desire of Immortality—Suicide.

LET us not seek motives to action in this world, in a distant futurity. It is to experience and truth that we ought to have recourse, in providing remedies to those evils which are incident to our species. There, too, must be sought those motives which give the heart inclinations useful to society.

Education, above all, gives the mind habits, useful to the individual and to society. Men have no need either of celestial rewards or supernatural punishments.

Government stands in no need of fables for its support. Present rewards and punishments are more efficacious than those of futurity, and they only

ought to be employed. Man is every where a slave, and consequently void of honour; base, interested, and dissimulating. These are the vices of governments. Man is every where deceived, and prevented from cultivating his reason; he is consequently stupid and unreasonable: every where he sees vice and crime honoured; and therefore concludes the practice of vice to lead to happiness, and, that of virtue, a sacrifice of himself. Every where he is miserable, and compelled to wrong his neighbours, that he may be happy. Heaven is held up to his view, but the earth arrests his attention. Here he will, at all events, be happy. Were mankind happier and better governed, there would be no need of resorting to fraud for governing them.

Cause man to view this state as alone capable of rendering him happy; bound his hopes to this life, instead of amusing him with tales of a futurity; show him what effect his actions have over his neighbours; excite his industry; reward his talents; make him active, laborious, benevolent and virtuous; teach him to value the affection of his contemporaries, and let him know the consequences of their hatred.

However great the fear of death may be, chagrin, mental affliction, and misfortunes, cause us sometimes to regard it as a refuge from human injustice.

Suicide has been variously considered. Some have imagined that man has no right to break the contract which he has entered into with society. But upon examining the connexions which subsist between man and nature, they will be found neither to be voluntary on the one part, nor reciprocal on the other. Man's will had no share in bringing him into the world, and he goes out of it against his inclination. All his actions are compulsatory. He can only love existence upon condition that it renders him happy.

By examining man's contract with society, we shall find that it is only conditional and reciprocal, and supposes mutual advantages to the contracting parties. Convenience is the bond of connexion. Is it broken? Man from that moment becomes free. Would we blame the man who, finding himself destitute of the means of subsistence in the city retires into the country? He who dies, only retires into solitude.

The difference of opinion upon this, as well as other subjects, is necessary. The suicide will tell you, that in his situation, your conduct would be precisely similar: but to be in the situation of another, we must possess his organization, constitution, and passions; be, in short, himself, placed in the same circumstances, and actuated by the same motives. These maxims may be thought dangerous: but maxims alone do not lead men to the adoption of such violent resolutions. It is a constitution whetted by chagrin, a vicious organization, a derangement of the machine—in a word, necessity. Death is a resource of which oppressed virtue should never be deprived.

CHAPTER XV.

Of Man's Interest, or the Ideas he forms of Happiness—Without Virtue he cannot be Happy.

INTEREST is the object to which every man, according to his constitution, attaches happiness. The same happiness does not suit all men, as that of every man depends upon his peculiar organization. It may, therefore, be easily conceived, that in beings of such different natures, what constitutes the pleasure of one man, may be indifferent, or even disgusting to another. No man can determine what will constitute the happiness of his neighbour.

Compelled, however, to judge of actions from their effects upon ourselves, we approve of the interest which animates them, according to the advantage which they produce to the human species. Thus, we admire valour, generosity, talents, and virtue.

It is the nature of man to love himself, to preserve his existence, and to render it happy. Experience and reason soon convince him, that he cannot alone command the means of procuring happiness. He sees other human beings engaged in the same pursuit, yet capable of assisting him to attain his desired object. He perceives, that they will favour his views in so far only as they coincide with their own

interest. He will then conclude, that to secure his own happiness, he must conciliate their attachment, approbation, and assistance; and that it is necessary to make them find advantages in promoting his views. The procuring of those advantages to mankind constitutes virtue. The wise man finds it his interest to be virtuous. Virtue is nothing more than the art of rendering a man happy, by contributing to the happiness of others. Merit and virtue are founded upon the nature and wants of man.

The virtuous man is always happy. In every face he reads the right which he has acquired over the heart. Vice is compelled to yield to virtue, whose superiority she blushingly acknowledges. Should the man of virtue sometimes languish in contempt or obscurity, the justice of his cause forms his consolation for the injustice of mankind. This consolation is denied to the wicked, whose hearts are the abode of anxiety, shame, and remorse.

CHAPTER XVI.

The Erroneous Opinions entertained by Man of Happiness are the True Causes of his Misery.

NOTHING can be more frivolous than the declamations of a gloomy philosopher against the love of power, grandeur, riches, or pleasure. Every thing which promises advantages is a natural object of desire.

Paternal authority, those of rank, riches, genius, and talents are founded upon those advantages. It is only on account of the advantages they produce, that the sciences are estimable. Kings, rich and great men, may impose upon us by show and splendour, but it is from their benefits alone that they have legitimate power over us.

Experience teaches us, that the calamities of mankind have sprung from religious opinions. The ignorance of natural causes created Gods, and imposture made them terrible. Man lived unhappy, because he was told that God had condemned him to misery. He never entertained a wish of breaking his chains, as he was taught, that stupidity, the renouncing of reason, mental debility, and spiritual debasement, were the means of obtaining eternal felicity. Kings, transformed by men into Gods, seemed to inherit the right of government: and politics became the fatal art of sacrificing the happiness of all to the caprice of an individual.

The same blindness pervaded the science of morality. Instead of founding it upon the nature of man, and the relations which subsist between him and his fellows, or upon the duties resulting from those relations, religion established an imaginary connexion between man and invisible beings. The Gods, always painted as tyrants, became the model of human conduct. When man injured his neighbour, he thought he had offended God, and believed that he could pacify him by presents and humility. Religion corrupted morality, and the expiations of piety completed its destruction. Religious remedies were disgusting to human passions, because unsuited to the nature of man: and they were called divine. Virtue appeared hateful to man, because it was represented to him as inimical to pleasure. In the observance of his duties, he saw nothing but a sacrifice of every thing dear; and real motives to induce such a sacrifice were never shown him. The present prevailed over the future, the visible over the invisible. Man became wicked, as every thing told him, that to enjoy happiness it was necessary to be so.

Melancholy devotees, finding the objects of human desire incapable of satisfying the heart, decried them as pernicious and abominable. Blind physicians! who take the natural state of man for that of disease! Forbid man to love and to desire, and you wrest from him his being! Bid him hate and despise himself, and you take away his strongest motives to virtue.

In spite of our complaints against fortune, there are many happy men in this world. There are also to be found sovereigns, ambitious of making nations happy; elevated souls who encourage genius, succour indigence, and possess the desire of engaging admiration.

Poverty itself is not excluded from happiness. The poor man, habituated to labour, knows the sweets of repose.

With limited knowledge, and few ideas, he has still fewer desires.

The sum total of good exceeds that of evil. There is no happiness in the gross, though much of it in the detail. In the whole course of a man's life few days are altogether unhappy.— Habit lightens our sorrows, and suspended grief is enjoyment. Every want, at the moment of its gratification, becomes a pleasure. Absence of pain and of sickness is a happy state, which we enjoy without being sensible of it. Hope assists us to support calamity. In short, the man who thinks himself the most unhappy, sees not the approach of death without terrour, unless despair has, to his eyes, disfigured the whole of nature. When nature denies us any pleasure, she leaves open a door for our departure; and should we not make use of it, it is because we still find a pleasure in existence.

———

CHAPTER XVII.

Origin of our Ideas concerning the Divinity.

Evil is necessary to man, since without it he would be ignorant of what is good. Without evil, he could neither have choice, will, passions, nor inclinations; he could neither have motives for loving nor hating. He would then be an automaton, and no longer man. The evil which he saw in the universe, suggested to man the idea of a Divinity. A crowd of evils, such as plagues, famines, earthquakes, inundations, and conflagrations, terrified him. But what ideas did he form of the cause which produced such effects? Man never imagined nature the cause of the calamities which afflicted herself. Finding no agent on earth, capable of producing such effects, he directed his attention to heaven, the imagined residence of beings, whose enmity destroyed his felicity in this world.

Terrour was always associated with the idea of those powerful beings.

From known objects, men judge of unknown. Man gave, from himself, a will, intelligence, and passions similar to his own, to every unknown cause which acted upon him. Influenced himself by submission and presents, he employed these to gain the favour of the Divinity.

The business relative to those offerings was confided to old men, and much ceremony was used in making them. The ceremonies were continued, and became custom. Thus religion and priestcraft were introduced into the world.

The mind of man (whose essence it is to labour incessantly upon unknown objects, to which it originally attached consequence, and dares not afterwards coolly examine) soon modified those systems.

By a necessary consequence of those opinions, nature was soon stripped of all power. Man could not conceive the possibility of nature's permitting him to suffer, were she not herself subject to a power, inimical to his happiness, and having an interest in punishing and afflicting him.

This is the origin of the rebellious angels. Notwithstanding his power, he could not subdue them. He is understood to be in the same situation with regard to those men who offend him.

Having thus, in their own opinion, satisfactorily accounted for human misery, another difficulty occurred. It could not be denied, that just men were sometimes included in the punishments of God.

It was then said, that because man had sinned, God might avenge himself upon the innocent—like those wicked princes, who proportion punishment more to the grandeur and power of the party offended, than to the magnitude and reality of the offence. The most wicked men, and the most tyrannical governments, have been the models of a Divinity, and his divine administration.

CHAPTER XVIII.

Of Mythology and Theology.

MAN originally worshipped nature. All things were spoken of allegorically, and every part of nature was personified. Hence a Saturn, Jupiter, Apollo, &c. The vulgar did not perceive that it was nature and her parts which were thus allegorized. The source from which Gods were taken was soon forgotten. An incomprehensible being was formed from the power of nature, and called its mover. Thus nature was separated from herself, and became considered as an inanimate mass incapable of action.

It became necessary to ascribe qualities to this moving power. This being, or, latterly, spirit, intelligence, incorporeal being; that is to say, a substance different from any that we know, was seen by nobody. Men could only ascribe it to qualities from themselves. What they called human perfection was the model in miniature of the perfection of the Divinity.

But, on the other hand, in viewing the calamities and disorders to which the world was so subject, why not attribute to him malice, imprudence, and caprice? This difficulty was thought removed in creating enemies to him.

CHAPTER XIX.

Absurd and Extraordinary Theological Opinions.

GOD, we are told, is good,—but God is the author of all things. All the calamities which afflict mankind, must, of course, be imputed to him. Good and evil suppose two principles : if there be only one, he must alternately be good and wicked.

God, say theologians, is just, and evil is a chastisement for the injuries which men have done him. To offend any one, supposes the existence of connexions between the offending and offended parties. To offend is to cause pain ; but how can a feeble creature like man, who has received his very existence from God, act against an infinite power, which never consents to sin or disorder ?

Justice supposes the disposition of rendering to every one his due ; and we are told, that God owes us nothing: that, without prejudice to his equity, he may plunge the work of his hand into an abyss of misery. Evils are said only to be temporary—surely, then, they are unjust, during a certain period. God chastises his friends for their good : but if God be good, can he permit them to suffer, even for a moment?

If God be omniscient, why try his friends, from whom he knows he has nothing to fear? If omnipotent, why be disturbed by the petty plots raised against him?

What good man does not wish to render his fellow-creatures happy? Why does not God make man happy? No man has reason to be contented with his lot. What can be said to all this? God's judgments are impenetrable. In this case, how can men pretend to reason about him? Since unsearchable, upon what foundation can a single virtue be attributed to him? What idea can we form of a justice which bears no resemblance to that of man?

His justice is said to be balanced by his mercy, but his mercy derogates from his justice. If unchangeable, can he for a moment alter his designs?

God, say the priests, created the world for his own glory. But already superior to every thing, was any addition wanting to his glory? The love of glory is the desire of being distinguished among our equals. If God be susceptible of it, why does he permit any one to abuse his favours? or why are they insufficient to make us act according to his wishes? Because he has made me a free agent. But why grant me a liberty which he knows I will abuse?

In consequence of this freedom, men will be eternally punished in the other world, for the faults they have committed in this life. But why punish eternally the faults of a moment? what would we think of the king, that eternally punished one of his subjects, who, in the moment of intoxication, had offended his pride, without however doing him any real injury, especially had he himself previously intoxicated him? Would we consider the monarch as all-powerful, who is forced to permit all his subjects, with the exception of a few faithful friends, to insult his laws, and even his own person, and thwart him in every measure?

It is said, that the qualities of God are so unlike to those of man, and so eminent, that no resemblance whatever subsists between them. But in this case, how can we form an idea of them? Why does theology presume to announce them?

But God has spoken, and made himself known to man. When and to whom? where are those divine oracles? in absurd and contradictory collections, where the God of wisdom speaks an obscure, insidious, and foolish language; where the God of benevolence is cruel and sanguinary; where the God of justice is unjust, partial, and ordains iniquity; where the God of mercy decrees the most horrid punishment to the victims of his wrath.

The relations subsisting between God and man, can be only founded upon moral qualities. But if man be ignorant of these, how can they serve as the model for his conduct? how can he possibly imitate them?

There is no proportion between God and man; and where that is wanting, there can be no relations. If God be incorporeal, how can he act upon bodies? how can they act upon him, so as to give him offence, disturb his repose, and excite his anger? If the potter be displeased with the bad shape of the vessel he has made, whom has he but himself to blame for it?

If God owes man nothing, man owes him as little. Relations must be reciprocal, and duties are founded upon mutual wants. If these are useless to God, he cannot owe any thing for them, and man cannot him. God's authority can only be founded upon the good which he bestows upon men; and their duties must solely rest upon the favours which they expect from him. If God do not owe man happiness, every relation between them is annihilated.

How can we reconcile the qualities ascribed to God with his metaphysical attributes? How can a pure spirit act like man, a corporeal being? A pure spirit can neither hear our prayers, nor be softened by our miseries. If immutable, he cannot change. If all nature, without being God, can exist in conjunction with him, he cannot be infinite. If he either suffers, or cannot prevent, the evils and disorders of the world, he cannot be omnipotent. He cannot be every where, if he is not in man while he commits sin, or goes out of him at the moment of its commission.

A revelation would prove malice in the Deity. It supposes, that he has for a long time denied man a knowledge necessary to his happiness. If it be made to a small number only, it is a partiality inconsistent with his justice. Revelation would destroy God's immutability, as it supposes him to have done at one period what he wished not to do at another. What kind of revelation is it, which cannot be understood? If one man only were incapable of understanding it, that circumstance would be alone sufficient to convict God of injustice.

CHAPTER XX.

Examination of Dr. Clarke's Proofs of the Existence of a Deity.

ALL men, it is said, believe in the existence of a deity, and the voice of nature is alone sufficient to establish it. It is an innate idea.

But what proves that idea to be acquired is, the nature of the opinion, which varies from age to age, and from nation to nation. That it is unfounded, is evident from this, that men have perfected every science, which has a real object, while that of God has been always in nearly the same state. There is no subject upon which men have entertained such a variety of opinions.

Admitting every nation to have a form of worship, that circumstance by no means proves the existence of a God. The universality of an opinion does not prove its truth. Have not all nations believed in the existence of witchcraft and of apparitions? Previous to Copernicus, did not all men believe that the earth was immoveable, and that the sun turned round it?

The ideas of God and his qualities are only founded upon the opinions of our fathers, infused into us by education; by habits contracted in infancy, and strengthened by example and authority. Hence the opinion, that all men are born with an idea of the Divinity. We retain those ideas, without ever having reflected upon them.

Dr. Clarke has adduced the strongest arguments which have ever yet been advanced in support of the existence of a Deity. His propositions may be reduced into the following:—

1. "Something has existed from all

eternity." Yes; but what is it? Why not matter, rather than spirit? When a thing exists, existence must be essential to it. That which cannot be annihilated, necessarily exists: such is matter. Matter, therefore, has always existed.

2. "An independent and unchangeable being has existed from all eternity."

First of all, what is this being? Is it independent of its own essence? No; for it cannot make the beings whom it produces act otherwise than according to their given properties. One body only depends upon another, in so far as it owes existence and form of action to it. By this trifle alone can matter be dependent. But if matter be eternal, it cannot be indebted for its existence to another being; and if eternal and self-existing, it is evident that, in virtue of those qualities, it contains within itself every thing requisite for action. Matter being eternal, has no need of a maker. Is this being unchangeable? No; as such a being could neither will nor produce successive actions. If this being created matter, there was a time in which it had resolved that matter should not exist, and another that it should. This being, therefore, cannot be unchangeable.

3. "This eternal, immutable, and independent being is self-existence." But since matter is eternal, why should it not be self-existent?

4. "The essence of a self-existent being is incomprehensible." True, and such is the essence of matter.

5. "A necessarily self-existing being is necessarily eternal." But it would have that property in common with matter? Why, then, separate this being from the universe?

6. "The self-existing being must be infinite, and every where present." Infinite! be it so; but we have no reason to think that matter is finite. Every where present! No; matter certainly occupies a part of space, and from that part, at least, the Divinity must be excluded.

7. "The necessarily self-existent being must be one." Yes, if nothing can exist out of it. But can any one deny the existence of the universe?

8. "The self-existent being is necessarily intelligent." But intelligence is

a human quality. To have intelligence, thoughts and senses are necessary. A being that has senses is material, and cannot be a pure spirit. But does this being, this great whole, possess a particular intelligence which puts it in motion; Since nature contains intelligent beings, why strip her of intelligence?

9. "The self-existent being is a free agent." But does God find no difficulty in executing his plans? Does he wish the continuance of evil, or can he not prevent it? In that case, he either permits sin or is not free. He can only act according to the laws of his essence. His will is determined by the wisdom and qualities which are attributed to him: He is not free.

10. "The supreme cause of all things possesses infinite power." But if man be free to commit sin, what becomes of God's infinite power?

11. "The author of all things is necessarily wise." If he be the author of all things, he is author of many things which we think very foolish.

12. "The supreme cause necessarily possesses every moral perfection." The idea of perfection is abstract. It is relative to our mode of perception that a thing appears perfect to us. When injured by his works, and forced to lament the evils we suffer, do we think God perfect? Is he so in respect to his works, where we universally see confusion blended with order?

If it be pretended that we cannot know God, and that nothing positive can be said about him, we may well be allowed to doubt of his existence. If incomprehensible, can we be blamed for not understanding him?

We are told that common sense and reason are sufficient to demonstrate his existence; but we are also told that, in these matters, reason is an unfaithful guide. Conviction, besides, is always the effect of evidence and demonstration.

———

CHAPTER XXI.

Examinations of the Proofs of the Existence of a Divinity.

No variety, it is said, can arise from a blind physical necessity, which must always be uniform; that the variety we

see around us can only proceed from the will and ideas of a necessarily existing being.

Why should not this variety arise from natural causes—from a self-acting matter, whose motion joins and combines various and analogous elements? Is not a loaf of bread produced from the combination of meal, yeast, and water? Blind necessity is a name which we give to a power with whose energy we are unacquainted.

But it is said that the regular movements and admirable order of the universe, and the benefits daily bestowed upon man, announce wisdom and intelligence. Those movements are the necessary effects of the laws of nature, which we call either good or bad, as they effect ourselves.

Animals, it is asserted, are a proof of the powerful cause which created them. The power of nature cannot be doubted. Are animals, on account of the harmony of their parts, the work of an invisible being? They are continually changing, and finally perish. If God cannot form them otherwise, he is neither free nor powerful; if he change his mind, he is not immutable; if he allow machines, whom he has created sensible, to experience sorrow, he is destitute of bounty; if he cannot make his works more durable, he is deficient in skill.

Man, who thinks himself the chief work in nature, proves either the malice or incapacity of his pretended author. His machine is more subject to derangement than that of other beings. Who, upon the loss of a loved object, would not rather be a beast or a stone than a human being? Better be an inanimated rock than a devotee, trembling under the yoke of his God, and foreseeing still greater torments in a future state of existence!

Is it possible, say theologians, to conceive the universe to be without a maker, who watches over his workmanship? Show a statue or a watch to a savage, which he has not before seen, and he will at once conclude it to be the work of a skilful artist.

1. Nature is very powerful and industrious; but we are as little acquainted with the manner in which she forms a stone or a mineral as a brain organized like that of Newton. Nature can

do all things, and the existence of any thing proves itself to be one of her productions. Let us not conclude that the works which most astonish us are not of her production.

2. The savage to whom a watch is shown will either have ideas of human industry or he will not. If he has, he will at once consider it to be the production of a being of his own species; if not, he will never think it the work of a being like himself. He will consequently attribute it to a genius or spirit, i. e. to an unknown power, whom he will suppose capable of producing effects beyond those of human beings. By this, the savage will only prove his ignorance of what man is capable of performing.

3. Upon opening and examining the watch, the savage will perceive that it must be a work of man. He will at once perceive its difference from the immediate works of nature, whom he never saw produce wheels of polished metal. But he will never suppose a material work the production of an immaterial being. In viewing the world, we see a material cause of its phenomena, and this cause is nature, whose energy is known to those who study her.

Let us not be told, that we thus attribute every thing to blind causes, and to a fortuitous concourse of atoms: we call those causes blind of which we are ignorant: we attribute effects to chance, when we do not perceive the tie which connects them with their causes. · Nature is neither a blind cause, nor does she act by chance: all her productions are necessary, and always the effect of fixed laws. There may be ignorance on our part, but the words Spirit, God, and Intelligence, will not remedy, but only increase that ignorance.

This is a sufficient answer to the eternal objection made to the partisans of nature, of attributing every thing to chance. Chance is a word void of meaning, and only exposes the ignorance of those who use it. We are told that a regular work cannot be formed by the combinations of chance; that an epic poem, like the Iliad, can never be produced by letters thrown together at random. Certainly not. It is nature that combines, according to fixed laws, an organized head capable of producing

such a work. Nature bestows such a temperament and organization upon a brain, that a head, constituted like that of Homer, placed in the same circumstances, must necessarily produce a poem like the Iliad, unless it be denied that the same causes produce the same effects.

Every thing is the effect of the combinations of matter. The most admirable of her productions which we behold, are only the natural effects of her parts, differently arranged.*

CHAPTER XXII.

Of Deism, Optimism, and Final Causes.

ADMITTING the existence of a God, and even supposing him possessed of views and of intelligence, what is the result to mankind? What connexion can subsist between us and such a being? Will the good or bad effects proceeding from his omnipotence and

* Whatever may be their pretensions, the partisans of religion can only prove, that every thing is the effect of a cause; that we are often ignorant of the immediate causes of the effects we see; that even when we discover them, we find that they are the effects of other causes, and so on, *ad infinitum*. But they neither have proved, nor can they prove, the necessity of ascending to a first eternal cause, the universal cause of all particular ones, producing not only the properties, but even the existence of things, and which is independent of every other cause. It is true, we do not always know the tie, chain, and progress of every cause; but what can be inferred from that? Ignorance can never be a reasonable motive either of belief or of determination.

I am ignorant of the cause that produces a certain effect, and cannot assign one to my own satisfaction. But must I be contented with that assigned by another more presumptuous, though no better informed than I, who says he is convinced; especially when I know the existence of such a cause to be impossible? The watch of a shipwrecked European having fallen into the hands of an Indian tribe, they held a consultation to discover the cause of its extraordinary movements. For a long time, they could resolve upon nothing. At length, one of the group, bolder than the rest, declared it to be an animal of a species different from any with which they were acquainted; and as none of them could convince him that the movements of the watch could proceed from any other principle than that which produces animal life and action, he thought himself entitled to oblige the assembly to accept of his explication.

providence be other than those of his wisdom, justice, and eternal decrees? Can we suppose that he will change his plans on our account? Overcome by our prayers, will he cause the fire to cease from burning, or prevent a falling building from crushing those who are passing beneath it? What can we ask of him, if he be compelled to give a free course to the events which he has ordained? Opposition, on our part, would be phrensy.

Why deprive me of my God, says the happy enthusiast, who favours me, whom I view as a benevolent sovereign continually watching over me? Why, says the unfortunate man, deprive me of my God, whose consoling idea dries up my tears?

I answer by asking them, on what do they found the goodness which they attribute to God? For one happy human being, how many do we not see miserable? Is he good to all men? How many calamities do we not daily see, while he is deaf to our prayers? Every man, therefore, must judge of the Divinity according as he is affected by circumstances.

In finding every thing good in the world, where good is necessarily attended with evil, the optimists seem to have renounced the evidence of their senses. Good is, according to them, the end of the whole. But the whole can have no end: if it had, it would cease being the whole.

God, say some men, knows how to benefit us by the evils which he permits us to suffer in this life. But how do they know this? Since he has treated us ill in this life, what assurance have we of a better treatment in a future state? What good can possibly result from the plagues and famines which desolate the earth? It is necessary to create another world to exculpate the Divinity from blame for the calamities he makes us suffer in the present.

Some men suppose that God, after creating matter out of nothing, abandoned it for ever to its primary impulse. These men only want a God to produce matter, and suppose him to live in complete indifference as to the fate of his workmanship. Such a God is a being quite useless to man.

Others have imagined certain duties to be due by man to his Creator. Oth-

ers suppose that, in consequence of his justice, he will reward and punish. They make a man of their God. But these attributes contradict each other; for, by supposing him the author of all things, he must, consequently, be the author of both good and evil. We might as well believe all things.

It is asked of us, would you rather depend upon blind nature than on a good, wise, and intelligent being? But, 1. Our interest does not determine the reality of things. 2. This being, so supereminently wise and good, is presented to us as a foolish tyrant, and it would be better for man to depend upon blind nature than upon such a being. 3. Nature, when well studied, teaches us the means of becoming happy, so far, at least, as our essence will permit. She informs us of the proper means of acquiring happiness.

CHAPTER XXIII.

Examination of the Supposed Advantages which result to Man from the Notions of a Divinity, or their Influence upon Morals, Politics, Science, the Welfare of Nations, and of Individuals.

MORALITY, originally having only for its object the self-preservation of man, and his welfare in society, had nothing to do with religious systems. Man, from his own mind, found motives for moderating his passions and resisting his vicious inclinations, and for rendering himself useful and estimable to those of whom he constantly stood in need.

Those systems which describe God as a tyrant cannot render him an object of imitation to man.. They paint him jealous, vindictive, and interested. Thus religion divides men. They dispute with and persecute one another, and never reproach themselves with crimes committed in the name of God.

The same spirit pervades religion. There we hear of nothing but victims; and even the pure Spirit of the Christians must have his own son murdered to appease his fury.

Man requires a morality, founded upon nature and experience.

Do we find real virtue among priests? Are these men, so firmly persuaded of God's existence, the less addicted to debauchery and intemperance? Upon seeing their conduct, we are apt to think that they are entirely undeceived in their opinions of the Divinity.

Does the idea of a rewarding and avenging God impose upon those princes who derive their power, as they pretend, from the Divinity himself? Are those wicked and remorseless monarchs who spread destruction around them atheists? They call the Divinity to witness, at the very moment when they are about to violate their oaths.

Have religious systems bettered the morals of the people? Religion, in their opinion, supersedes every thing. Its ministers, content with supporting dogmas and rites, useful to their own power, multiply troublesome ceremonies, with a view of drawing profit by their slaves transgressing them. Behold the work of religion and priestcraft in a sale of the favours of Heaven! The unmeaning words, impiety, blasphemy, sacrilege, and heresy, were invented by priests; and those pretended crimes have been punished with the greatest severities.

What must be the fate of youth under such preceptors? From infancy the human mind is poisoned with unintelligible notions and disturbed by phantoms, genius is cramped by a mechanical devotion, and man wholly prejudiced against reason and truth.

Does religion form citizens, fathers, or husbands? It is placed above every thing. The fanatic is told that he must obey God, and not man; consequently, when he thinks himself acting in the cause of Heaven, he will rebel against his country, and abandon his family.

Were education directed to useful objects, incalculable benefits would arise therefrom to mankind. Notwithstanding their religious education, how many men are subject to criminal habits. In spite of a hell, so horrid even in description, what crowds of abandoned criminals fill our cities! Those men would recoil with horrour from him who expressed any doubts of God's existence. From the temple, where sacrifices have been made, divine oracles uttered, and vice denounced in the name of Heaven, every man returns to his former criminal courses.

Are condemned thieves and murderers either atheists or unbelievers?— those wretches believe in a God. They have continually heard him spoken of; neither are they strangers to the punishment which he has destined to crimes. But a hidden God and distant punishments are ill calculated to restrain crimes, which present and certain chastisements do not always prevent.

The man who would tremble at the commission of the smallest crime in the face of the world, does not hesitate for a moment when he thinks himself only seen by God. So feeble is the idea of divinity when opposed to human passions.

Does the most religious father, in advising his son, speak to him of a vindictive God? His constitution destroyed by debauchery, his fortune ruined by gaming, the contempt of society—these are the motives he employs.

The idea of a God is both useless and contrary to sound morality:—it neither procures happiness to society nor to individuals. Men always occupied with phantoms, live in perpetual terrour. They neglect their most important concerns, and pass a miserable existence in groans, prayers, and expiations. They imagine that they appease God by subjecting themselves to every evil. What fruit does society derive from the lugubrious notions of those pious madmen? They are either misanthropes, useless to themselves and to the world, or fanatics who disturb the peace of nations. If religious ideas console a few timid and peaceable enthusiasts, they render miserable during life millions of others, infinitely more consistent with their principles. The man who can be tranquil under a terrible God must be a being destitute of reason.

CHAPTER XXIV.

Religious Opinions cannot be the Foundation of Morality—Parallel between Religious and Natural Morality—Religion impedes the Progress of the Mind.

ARBITRARY and inconsistent opinions, contradictory notions, abstract and unintelligible speculations, can never serve as a foundation to morality; which must rest upon clear and evident principles, deduced from the nature of man, and founded upon experience and reason. Morality is always uniform, and never follows the imagination, passions, or interests of man. It must be stable and equal for all men, never varying with time or place.• Morality, being the science of the duties of man living in society, must be founded on sentiments inherent in our nature. In a word, its basis must be necessity.

Theology is wrong in supposing that mutual wants, the desire of happiness, and the evident interests of societies and of individuals are insufficient motives to influence man. • The ministers of religion subject morality to human passions by making it flow from God. They found morality upon nothing by founding it upon a chimera.

The ideas entertained of God, owing to the different views which are taken of him, vary with the fancy of every man, from age to age, from one country to another.

Compare the morality of religion with that of nature, and they will be found essentially different. Nature invites men to love one another, to preserve their existence, and to augment their happiness. Religion commands him to love a terrible God, to hate himself, and sacrifice his soul's most precious joys to his frightful idol. Nature bids man consult his reason; religion tells him that reason is a fallible guide. Nature bids him search for truth; religion prohibits all investigation. Nature bids man be sociable, and love his neighbours; religion commands him to shun society, and sequester himself from the world. Nature enjoins tenderness and affection to the husband; religion considers matrimony as a state of impurity and corruption. Nature bids the wicked man resist his shameful propensities, as destructive to his happiness; religion, while she forbids pleasure, think no little as the vulgar; crime, promises pardon to the criminal, by humbling himself before its ministers, by sacrifices, offerings, ceremonies, and prayers.

The human mind, perverted by religion, has hardly advanced a single step in improvement. Logic has been uniformly employed to prove the most palpable absurdities. Theology has inspired kings with false ideas of their rights, by telling them that they hold their power from God. The laws became subject to the caprices of religion. Physics, anatomy, and natural history

were only permitted to see with the eyes of superstition. The most clear facts were refuted, when inconsistent with religious hypothesis.

Is a question in natural philosophy solved by saying, that phenomena, such as volcanoes or deluges, are proofs of Divine wrath? Instead of ascribing wars and famines to the anger of God, would it not have been more useful to show men that they proceeded from their own folly, and from the tyranny of their princes? Men would then have sought a remedy to their evils in a better government. Experience would have convinced man of the inefficacy of fasts, prayers, sacrifices, and processions, which never produced any good.

CHAPTER XXV.

Man, from the ideas which are given of the Deity, can conclude nothing—Their Absurdity and Uselessness.

SUPPOSING the existence of an intelligence, like that held out by theology, it must be owned that no man has hitherto corresponded to the wishes of providence. God wishes himself to be known by men, and even the theologians can form no idea of him. Admitting that they did so, that his being and attributes are evident to them, do the rest of mankind enjoy the same advantages?

Few men are capable of profound and constant meditation. The common people of both sexes, condemned to toil for subsistence, never reflect. People of fashion, all females, and young people of both sexes, only occupied about their passions and their pleasures, think as little as the vulgar. There are not, perhaps, ten men of a million of people, who have seriously asked themselves what they understand by God; and even fewer can be found who have made a problem of the existence of a Divinity: yet conviction supposes evidence, which can alone produce certainty. Who are the men that are convinced of God's existence? Entire nations worship God upon the authority of their fathers and their priests. Confidence, authority, and habit, stand in the stead of conviction and proof. All rests upon authority; reason and investigation are universally prohibited.

Is the conviction of the existence of a God, so important to all men, reserved only to priests and the inspired? Do we find the same unanimity among them as with those occupied with studying the knowledge of useful arts? If God wishes to be known to all men, why does he not show himself to the whole world, in a less equivocal and more convincing manner than he has hitherto done in those relations which seem to charge him with partiality? Are fables and metamorphoses the only means which he can make use of? Why have not his name, attributes, and will, been written in characters legible by all men?

By ascribing to him contradictory qualities, theology has put its God in a situation where he cannot act. Admitting that he existed with such extraordinary and contradictory qualities, we can neither reconcile to common sense nor to reason the conduct and worship prescribed towards him.

If infinitely good, why fear him? if infinitely wise, why interest ourselves about our fate? if omniscient, why tell him of our wants, or fatigue him with our prayers? if every where, why erect to him temples? if master of all, why make him sacrifices and offerings? if just, whence has arisen the belief that he will punish man, whom he has created weak and feeble? if reasonable, why be angry with a blind creature like man? if immutable, why do we pretend to change his decrees? and if inconceivable, why presume to form any idea of him?

But if, on the other hand, he be irascible, vindictive, and wicked, we are not bound to offer up to him our prayers. If a tyrant, how can we love him? How can a master be loved by his slaves, whom he has permitted to offend him that he might have the pleasure of punishing them? If all-powerful, how can man fly from his wrath? If unchangeable, how can man escape his fate?

Thus, in whatever point of view we consider God, we can neither render him prayers nor worship.

Even admitting the existence of a Deity, full of equity, reason, and benevolence, what would a virtuous atheist have to fear, who should unexpectedly find himself in the presence of a being whom, during life, he had misconceived and neglected? "O, God!" he might say, "inconceivable being, whom I could not discover, pardon, that the limited understanding thou hast given me has been inadequate to thy discovery! How could I discover thy spiritual essence by the aid of sense alone? I could not submit my mind to the yoke of men, who, confessedly not more enlightened than I, agreed only among themselves in bidding me renounce the reason which thou hast given! But, O God! if thou lovest thy creatures, I have also loved them! If virtue pleaseth thee, my heart ever honoured it. I have consoled the afflicted; never did I devour the substance of the poor. I have ever been just, bountiful, and compassionate."

In spite of reason, men are often, by disease, brought back to the prejudices of infancy. This is most frequently the case with sick people: upon the approach of death, they tremble, because the machine is enfeebled; the brain being unable to perform its functions, they of course fall into deliriums. Our systems experience the changes of our body.

CHAPTER XXVI.

Apology for the Sentiments contained in this Work.

MEN tremble at the very name of an atheist. But who is an atheist? The man who brings mankind back to reason and experience, by destroying prejudices inimical to their happiness; who has no need of resorting to supernatural powers in explaining the phenomena of nature.

It is madness, say theologians, to suppose incomprehensible motions in nature. Is it madness to prefer the known to the unknown?—to consult experience and the evidence of our senses?—to address ourselves to reason, and prefer her oracles to the decision of sophists, who even confess themselves ignorant of the God they announce?

When we see priests so angry with atheistical opinions, should we not sus-

pect the justice of their cause? Spiritual tyrants! 'tis ye who have defamed
the Divinity, by besmearing him with
the blood of the wretched! You are
the truly impious. Impiety consists in
insulting the God in whom it believes.
He who does not believe in a God cannot injure him, and cannot of course be
impious.

On the other hand, if piety consists
in serving our country, in being useful
to our fellow-creatures, and in observ-
ing the laws of nature, an atheist is
pious, honest, and virtuous, when his
conduct is regulated by the laws which
reason and virtue prescribe to him.

. Men, we are told, who have reason
to expect future happiness, never fall
into atheism. The interest of the passions and the fear of punishment alone
make atheists. But men who endeavour to enlighten that reason which im-
prints every idea of virtue, are not calculated to reject the existence of a
future state, from an apprehension of
its chastisements.

It is true, the number of atheists is
inconsiderable, because enthusiasm has
dazzled the human mind, and the progress of errour has been so very great,
that few men have courage to search
for truth. If by atheists are meant
those who, guided by experience and
the evidence of their senses, see nothing
in nature but what really exists; if by
atheists are meant natural philosophers,
who think every thing may be accounted for by the laws of motion, without
having recourse to a chimerical power;
if by atheists are meant those who know
not what a spirit is, and who reject a
phantom whose opposite qualities only
disturb mankind; doubtless, there are
many atheists: and their number would
be greater, were the knowledge of
physics and sound reason more generally disseminated.

An atheist does not believe in the
existence of a God. No man can be
certain of the existence of an inconceivable being, in whom inconsistent
qualities are said to be united. In this
sense, many theologians would be atheists, as well as those credulous beings
who prostrate themselves before a being of whom they have no other idea
than that given them by men avowedly
comprehending nothing of him themselves.

CHAPTER XXVII.

Is Atheism Compatible with Sound Morality?

THOUGH the atheist denies the existence of a God, he neither denies his
own existence nor that of other men;
he cannot deny the existence of relations which subsist between men, nor
the duties which necessarily result from
those relations. He cannot doubt the
existence of morality, or the science of
the relations which subsist between
men living in society. Though he
may sometimes seem to forget the
moral principles, it does not follow
that they do not exist. He may act
inconsistently with his principles, but
a philosophical infidel is not so much
an object of dread as an enthusiastic
priest. Though the atheist disbelieves
in the existence of a God, can it be
thought that he will indulge to excesses
dangerous to himself and subject to
punishments?

Whether would men be happier under an atheistical prince, or a believing
tyrant, continually bestowing presents
upon priests? Would we not have to
fear religious quarrels from the latter?
Would not the name of God, of which
the monarch avails himself, sometimes
serve as an excuse for the persecutions
of the tyrant? Would he not at least
hope to find in religion a pardon for his
crimes? .

Much inconveniency may arise from
making morality depend upon the existence of a God. When corrupt minds
discover the falsehood of those suppositions, they will think virtue itself,
like the Deity, a mere chimera, and see
no reason to practise it in life. It is,
however, as beings living in society,
that we are bound by morality. Our
duties must always be the same, whether a God exist or not.

If some atheists deny the existence
of good and of evil, it only proves their
own ignorance. A natural sentiment
causes man to love pleasure and hate
pain. Ask the man who denies the
existence of virtue and vice, would he
be indifferent at being robbed, calumniated, betrayed, and insulted? His
answer will prove that he makes a distinction between men's actions; that
the distinctions of good and evil depend
neither upon human conventions nor
the idea of a Deity; neither do they

depend upon the rewards or punishments of a future state of existence.

The atheist, believing only in the present life, at least wishes to live happy. Atheism, says Bacon, renders man prudent, as it limits his views to this life. Men accustomed to study and meditation never are bad citizens.

Some men, undeceived themselves in religious matters, pretend that religion is useful to the people, since, without it, they could not be governed. But has religion had a useful influence upon popular manners? It enslaves, without making obedient; it makes idiots, whose sole virtue consists in a blind submission to paltry and silly ceremonies, to which more consequence is attached than to real virtue or pure morality. Children are only frightened for a moment by imaginary terrours. It is only by showing men the truth that they can appreciate its value, and find motives for cultivating it.

It is chiefly among nations where superstition, aided by authority, makes its heavy yoke be felt, and imprudently abuses its power, that the number of atheists is considerable. Oppression infuses energy into the mind, and occasions a strict investigation into the causes of its evils. Calamity is a powerful goad, stimulating the mind to the side of truth.

CHAPTER XXVIII.

Motives which lead to Atheism—Can this be Dangerous?

WHAT interest, we are asked, can men have to deny the existence of God? But are the tyrannies exercised in his name, and the slavery in which men groan under priests, sufficient motives for determining us to examine into the pretensions of a class that occasions so much mischief in the world? Can there be a stronger motive than the incessant dread excited by the belief in a being who is angry with our most secret thoughts, whom we may unknowingly offend, who is never pleased with us, who gives man evil inclinations that he may punish him for them, who eternally punishes the crime of a moment? The deist will tell us that we only paint superstition; but such a supposition will never prove the existence of a Deity. If the God of superstition be a disgusting being, that of deism must always be inconsistent and impossible.

The depraved devotee finds in religion a thousand pretexts for being wicked. The atheist has no cloak of zeal to cover his vengeance and fury.

No sensible atheist thinks that the cruel actions caused by religion are capable of being justified. If the atheist be a bad man, he knows when he is committing wrong. Neither God nor his priests can then persuade him that he has been acting properly.

The indecent and criminal conduct of his ministers, say some men, proves nothing against religion. May not the same thing be said of an atheist of good principles and a bad practice? Atheism, it is said, destroys the force of oaths; but perjury is common enough with those nations who boast the most of their piety. Are the most holy kings faithful to their oaths? Does not religion itself sometimes grant a dispensation from them, especially when the perjury is beneficial to the holy cause? Do criminals refrain from swearing, when necessary to their justification? Oaths are a foolish formality, which neither impose upon villains nor add any thing to the engagements of good men.

It has been asked, whether a people ever existed that had not some idea of a Deity; and could a nation of atheists exist?

A timid and ignorant animal, like man, necessarily becomes superstitious under calamity. He either creates a God himself or takes that which is offered him by another. But the savage does not draw the same conclusion from the existence of his Gods as the polished citizen. A nation of savages content themselves with a rude worship, and never reason about the Divinity. It is only in civilized states that men subtilize those ideas.

A numerous society, without either religion, morality, government, laws, or principles, doubtless cannot exist, since it would only be an assemblage of men mutually disposed to injure one another. But, in spite of all religions in the world, are not all human societies nearly in that state? A society of atheists, governed by good laws, whom

No. XII.—46

rewards excite to virtue, and punishments deter from crime, would be infinitely more virtuous than those religious societies in which every thing tends to disturb the mind and to deprave the heart.

We cannot expect to take away from a whole nation its religious ideas, because they have been inculcated from the tenderest infancy. But the vulgar, in the long run, may reap advantages from labours, of which they at present have no idea. Atheism, having truth on its side, will gradually insinuate itself into the mind, and become familiar to man.

CHAPTER XXIX.

Abridgement of the System of Nature.

O YE, says Nature, who, according to the impulse which I have given you, tend every instant towards happiness, do not resist my sovereign law! labour at your felicity ; enjoy without fear ; be happy.

Return, O devotee, to Nature! She will banish from thy heart the terrours which are overwhelming thee. Cease to contemplate futurity. Live for thyself and thy fellow-creatures. I approve of thy pleasures, while they neither injure thee nor others, whom I have rendered necessary to thy happiness.

Let humanity interest thee in the fate of thy fellow-creature. Consider that, like him, thou mayest one day be miserable. Dry up the tears of distressed virtue and injured innocence. Let the mild fervour of friendship, and the esteem of a loved companion, make thee forget the pains of life.

Be just, since equity supports the human race. Be good, as bounty attaches every heart. Be indulgent, since thou livest among beings weak like thyself. Be modest, as pride hurts the self-love of every human being. Par-

don injuries, as vengeance eternizes hatred. Do good to him who injures thee, that thou mayest show thyself greater than he, and also gain his friendship. Be moderate, temperate, and chaste, since voluptuousness, intemperance, and excess, destroy thy being, and render thee contemptible.

It is I who punish the crimes of this world. The wicked man may escape human laws, but mine he can never fly from. Abandon thyself to intemperance, and man will not punish thee, but I will punish thee, by shortening thy existence. If addicted to vice, thou wilt perish under thy fatal habits. Princes, whose power surpasseth human laws, tremble under mine. I punish them by infusing suspicion and terrour into their minds. Look into the hearts of those criminals, whose smiling countenances conceal an anguished soul. See the covetous miser, haggard and emaciated, groaning under wealth, acquired by the sacrifice of himself. View the gay voluptuary, secretly writhing under a broken constitution ; see the mutual hatred and contempt which subsist between the adulterous pair! The liar, deprived of all confidence ; the icy heart of ingratitude, which no act of kindness can dissolve ; the iron soul of the monster whom the sight of misfortune could never soften ; the vindictive being, nourishing in his bosom the gnawing vipers which are consuming him! Envy, if thou darest, the sleep of the murderer, the iniquitous judge, or the oppressor, whose couches are surrounded by the torches of the furies! But no! humanity obliges thee to partake of their merited torments. Comparing thyself with them, and finding thy bosom the constant abode of peace, thou wilt find a subject of self-congratulation. Finally, behold the decree of destiny fulfilled on all! She wills that virtue shall never go unrewarded, but crime be ever its own punishment.

CONTENTS

OF THE

FIRST VOLUME.

CONTENTS

OF THE ·

SECOND VOLUME.

APPENDIX.